Model Predictive Vibration Control

Gergely Takács · Boris Rohal'-Ilkiv

Model Predictive Vibration Control

Efficient Constrained MPC Vibration Control for Lightly Damped Mechanical Structures

Springer

Gergely Takács
Faculty of Mechanical Engineering
Institute of Automation
Measurement and
Applied Informatics
Slovak University of Technology in
Bratislava
Námestie slobody 17
812 31 Bratislava 1
Slovakia

Boris Rohal'-Ilkiv
Faculty of Mechanical Engineering
Institute of Automation
Measurement and
Applied Informatics
Slovak University of Technology in
Bratislava
Námestie slobody 17
812 31 Bratislava 1
Slovakia

ISBN 978-1-4471-2332-3
e-ISBN 978-1-4471-2333-0
DOI 10.1007/978-1-4471-2333-0
Springer London Heidelberg New York Dordrecht

British Library Cataloguing in Publication Data
A catalogue record for this book is available from the British Library

Library of Congress Control Number: 2011941708

Printed on acid-free paper

Springer is part of Springer Science+Business Media (www.springer.com)

"Prediction is very difficult,
especially about the future."

Niels Bohr

This book is dedicated to our dearest better halves Janka Sipos and Marta Rohal'ová-Ilkivová whose passion, love and support fuels our lives; to our parents Erika and András Takács, Ol'ga and the memory of Ivan Rohal'-Ilkiv who have raised us with respect for knowledge, wisdom and science; and to our students who unlike our family and loved ones will have to read the rest of the book too…

Preface

This book provides insight into the model predictive control of lightly damped vibrating structures. Conclusions of ongoing research in the field, up-to-date experimental results and the doctoral dissertation thesis titled "Efficient Model Predictive Control Applied on Active Vibration Attenuation" by Gergely Takács have been summarized into a clearly presented and accessible form. The book is intended for use in undergraduate or graduate level university curricula or for industrial practitioners interested in computationally efficient predictive control utilized in active vibration attenuation. It is assumed the reader has a very basic understanding of linear control theory and vibration mechanics.

The control strategy discussed in this book is based on the idea of using a mathematical model to predict the future behavior of a vibrating system and selecting the best control moves based on an optimization procedure using this predicted information. This method is known as model predictive control (MPC) and due to its intense computational requirements has been so far used mainly to control processes with very slow dynamics. The control moves computed by MPC will not only be ideal in a sense of damping performance, but they will also respect process constraints arising from physical actuator limitations, safety or economic reasons. This title will introduce the current state and the theoretical particulars behind this advanced control strategy and show how it can be implemented using piezoelectric actuators to lightly damped vibrating structures, in order to eliminate or attenuate undesired vibrations.

Using more than 170 illustrations, photographs, diagrams and several tables, the book will take the reader through the necessary steps in understanding the fundamentals of active vibration control (AVC), give a thorough review of the current state of model predictive control and finally will also introduce the implementation of computationally efficient MPC algorithms and compare different predictive control strategies in simulation and experiment.

Both active vibration attenuation and model predictive control have been treated in numerous excellent books already. So why would we need another publication on these topics? Works discussing the field of (active) vibration control are generally limited to presenting traditional control methods ranging from positive position feedback (PPF) to linear quadratic control (LQ). The progress of control theory has not stopped with providing us the tools to synthesize and implement simple controllers such as proportional integrating derivative (PID). Modern optimization-based control methods, such as model predictive control are generally not considered for active vibration control applications. This can be partly attributed to the fact that the results of control theory tend to be transferred to real-life applications very slowly. The other major reason is due to the obvious implementation limitations: the sampling speeds usually encountered in AVC are too fast for real-time deployment. Advantages of predictive controllers over traditional controllers are not limited to an increased performance, but these methods also handle process, actuator and safety-related constraints on an algorithmic level.

On the other hand, either the books published on the topic of model predictive control are focused exclusively on the theory in deep mathematical detail or, even if practical implementation examples are given, they are limited to processes with slow dynamics. The reason for this is that implementation of predictive controllers on petrochemical plants, heaters and other slow processes do not invoke the computational time issue. Applying predictive control in active vibration attenuation is therefore not a topic of these publications. This book is distinct from general works on AVC or MPC because it presents the multi-disciplinary area of predictive control applied in vibration control, treating the subject as one compound problem. We offer a specific cross-section of these two actual and attractive engineering fields and suggest solutions for the research and industrial community.

Gergely Takács is currently a research engineer at the Institute of Automation, Measurement and Applied Informatics of the Faculty of Mechanical Engineering of the Slovak University of Technology in Bratislava. He has received his PhD degree in Mechatronics from the Slovak University of Technology in 2009. His recent doctoral studies and commencing academic career have been fully devoted to the application of computationally efficient model predictive controllers in the active vibration control of lightly damped structures. His research interests include active vibration control, smart materials, advanced actuators, and computationally efficient model predictive control. Gergely Takács is a member of IEEE.

Boris Rohal'-Ilkiv has received his degrees from the Slovak University of Technology in Bratislava, the Faculty of Mechanical Engineering, in control engineering. Currently, he is a tenured professor at the Institute of Automation, Measurement and Applied Informatics at the Faculty of Mechanical Engineering, where he is an active lecturer and researcher in the area of dynamical systems modeling and control. He has devoted the majority of his academic career to model predictive control, with a special attention focused at practical real-time controller implementation issues. Boris Rohal'-Ilkiv is a member of IEEE.

Bratislava, August 2011

Gergely Takács
Boris Rohal'-Ilkiv

Contact

Feel free to contact us with any matter related to this book or its general topic. Let us know what you think, if you have found mistakes, missing references or anything else. Your feedback is highly appreciated and we will do our best to rectify any insufficiencies for further editions. Unsolicited grant proposals and agencies are welcome as well:

Gergely Takács: gergely.takacs@stuba.sk
Boris Rohal'-Ilkiv: boris.rohal-ilkiv@stuba.sk

Acknowledgments

First we would like to express our deepest gratitude towards Basil Kouvaritakis and Mark Cannon of the University of Oxford. We both had the possibility to spend time with these two great scientists: Gergely was staying a term at the Department of Engineering Science as a visiting scholar in three separate occasions, while Boris stayed with the control group as a part of an earlier British–Slovak joint research project. When Gergely arrived at Oxford as a first-year doctoral student (with a lot to desire even in basic knowledge on control theory), Basil and Mark answered even the silliest questions with immense patience and sacrificed one of their most scarce assets: time. Later on, during numerous interesting discussions we could share their insight into model predictive control, learning how to look at issues in control theory from a new perspective. We cannot emphasize the importance of these rewarding conversations and would like to thank their hospitality this way.

We appreciate every piece of advice and assistance, which came from our colleagues at the Institute of Automation, Measurement and Applied Informatics at the Faculty of Mechanical Engineering of the Slovak University of Technology in Bratislava. Together they create an environment that almost feels like home. It is wonderful to do research at such a friendly place. In particular, we would like to express our gratitude towards the head of this institute, Gabriel Hulkó. He has always followed our research activity with serious interest, providing important advice and useful hints. We would like to thank not only for the professional encouragement but also for the good personal relationship to: Tomáš Polóni, Ján Vachálek, Tomáš Volenský, Cyril Belavý, the army of procrastinating PhD students, and many others.

We would like to thank for the financial support granted by the Slovak Research and Development Agency (SRDA) within the framework of the projects "Predictive control of mechatronic systems with fast dynamics and constraints" (APVV-0280-06), "Advanced methods for modeling, control and design of mechatronic systems as lumped-input and distributed-output systems" (APVV-0160-07) and "Model predictive control and joint state and parameter estimation methods for fast nonlinear mechatronic systems" (APVV-0090-10). Such grant agencies as the SRDA are the modern day patrons of science; their funding makes possible the progress of research. At the same time we cannot forget to thank the panel of anonymous reviewers, who—as we imagine—sitting in their secret lairs with dark hoods on their heads, make the favorable decisions on our project proposals.

We are both very grateful for the hard work of Claire Protherough, our former editor at Springer who took us through the initial steps of the publication process. She was always available when we needed answers, and was very understanding about our requests on deadline extensions. In the final stage of the manuscript preparation and in the following production phase Grace Quinn and Christine Velarde took over this project at Springer, thus we would like to thank for their outstanding guidance as well. All of this exceptional help and support from Claire, Grace and Christine unfortunately means that we cannot blame them for any mistakes left in the book. Oh well.

We would like to thank for the hard work of Adrian Wills of the School of Electrical Engineering and Computer Science at The University of Newcastle, Australia. A journal article by Adrian and his colleagues inspired us to pursue our research in the application of model predictive control in vibration attenuation in the first place. His opinion is very much valued and his reflections are included in the book. We would also like to express our gratitude towards Lars Imsland of the Department of Engineering Cybernetics at the Norwegian University of Science and Technology (NTNU) for going through our book manuscript and providing us with pointers on how we could improve the quality of the material even further. Lars has been closely involved with efficient model predictive control algorithms, therefore his opinion on the contents of the book was important to us. His valuable comments are reflected in the final version of the manuscript. Finally, we would like to thank the assistance of Hans Joachim Ferreau, member of the Optimization in Engineering Center (OPTEC), Department of Engineering (ESAT) at the Katholieke Universiteit Leuven, Belgium. Joachim developed the qpOASES efficient quadratic programming solver aimed at model predictive control that has been extensively used throughout this book, therefore his observations based on reviewing the manuscript have been extremely helpful.

Neither of us is a native speaker of the English language, so we have passed on the manuscript to a third person who has learned English as a third language. But definitely better than we did. So our thanks goes to Janka Sipos, who despite being an English Literature major has chewed through the highly technical manuscript and corrected our spelling and grammatical mistakes. Any typos, spelling and grammatical errors left here are intentional and congratulations to those who have found them. No, seriously, please let us know.

During the preparation of the book manuscript, we have been continually seeking the rightful owners of those photographs and illustrations we have found in other sources and decided to include here. We would like to express our gratitude to the individuals and institutions who took the time to reply to our requests and have explicitly given permission to use these materials; including the United States National Aeronautics and Space Administration (NASA), the European Space Agency (ESA), The Boeing Company, Liebherr-International Deutschland

GmbH, the CEDRAT Group; Thomas Hugin of the Noise & Vibration Control Ltd,; Kazuhiko Yamashita of the Japan Society of Maintenology, Thomas Huber of the Gustavus Adolphus College and Bishakh Bhattacharya of the Indian Institute of Technology.

Acknowledgements: G. Takács

I would also like to thank my former thesis supervisor and current co-author and colleague, Boris Rohal'-Ilkiv, who provided me with invaluable support and help throughout the progress of research, which has been eventually summarized in this book. The vital advice, suggestions and constructive criticism offered by him made it possible to proceed with this work so far. Apart from the outstanding scientific and professional work of Prof. Rohal', he is an understanding and patient person. It has been a pleasure to work with him.

On a more personal level, thanks goes to everyone who supported me throughout my studies and beginning my academic career by encouraging me, and making sure I did not take life too seriously. Thanks for the good times together to my circle of friends and close ones including but not limited to: Anna and Tamás Kiss, Tibor Renczés, Szilvia Tóth, Szilárd Izsák, Erik Putz, Júlia Németh, Dóri and Árpád Zirig, Eszter Fekete, Eszter Varga, Béla Edmár, Dávid Lovas, Dávid Derzsi, Marianna and Róbert Renczés, Ferenc Varga and Ákos Kovács of "PlusPlus", Krisztina Sebő, László Lubinszky, Viktória Izsák, Gábor Horváth, Erzsébet and László Slíž of "Golf", Daniel Nyéky, Miroslav Mates, Mária Sedláková, Nhumai Hothi, Ashley Napier, Ankor Raithatha, and many others...

I am grateful to my parents Erika and András Takács for supporting me through my ongoing academic endeavors and helping me through life. Their unconditional care and love is highly appreciated. Finally yet importantly, I would like to thank my girlfriend Janka Sipos for the perpetual love, understanding and support she has given me. Thanks for putting up with all the odd working habits and hours, time lost because of my trips around the globe and of course with me being a wired squirrel occasionally. Janka is a wonderful person, and a key to my happiness.

Dunajská Streda, Slovakia, July 2011 Gergely Takács

Acknowledgements: B. Rohal'-Ilkiv

I highly appreciate the efforts and perseverance of my close colleague and first author of this monograph, Gergely Takács, with which he embarked on a broad theme of active vibration damping. I am grateful to him for numerous discussions, many constructive suggestions, developing of all computer programs, design and performing all the experiments presented on these pages.

Last but not least, I would like to thank my family, wife Marta and daughters Martina, Tatjana and Hanka, who have always supported and encouraged me in every way.

Bratislava, Slovakia, June 2011 Boris Rohal'-Ilkiv

The authors gratefully acknowledge the financial support granted by the *Slovak Research and Development Agency* under the contracts **APVV-0280-06**, **APVV-0160-07** and **APVV-0090-10**.

SLOVAK RESEARCH
AND DEVELOPMENT
AGENCY

Contents

Part III Model Predictive Vibration Control

Abbreviations

AAMPC	Autonomous augmented MPC
A/D	Analog to digital
ADP	Ammonium dihydrogen phosphate
AFM	Atomic force microscope
AMB	Active magnetic bearing
ANC	Active noise control
ANN	Artificial neural network
ANSYS	Analysis system (software)
APB-IPM	Approximate primal-barrier interior-point method
APDL	ANSYS parametric design language
ARE	Algebraic Ricatti equation
AS	Active set
ATLAS	Automatically tuned linear algebra software (software)
ATW	Aerostructures test wing
AVA	Active vibration attenuation
AVC	Active vibration control
AVD	Active vibration damping
AVI	Active vibration insulation
BLAS	Basic linear algebra subprograms (software)
BNC	Bayonet Neil-Concelman connector
BNT	Bismuth sodium titanate
BST	Binary search tree
CAVF	Constant amplitude velocity feedback
CCD	Charge coupled device
CGVF	Constant gain velocity feedback
CPU	Central processing unit
CRHPC	Constrained receding horizon predictive control
D/A	Digital to analog
DAQ	Data acquisition
DARE	Discrete time algebraic Ricatti equation
DMC	Dynamic matrix control

DNA	Deoxyribonucleic acid
DOF	Degrees of freedom
ENRMPC	Extended Newton–Raphson MPC
EAP	Electroactive polymer
EGR	Exhaust gas recirculation
eMPC	Explicit MPC, see MPMPC
EP	Equality problem
ER	Electrorheological
ES	Electrostrictive
ESSP	Electro-statically stricted polymer
EVE	Electro-viscoelastic elastomer
FE	Finite element
FEA	Finite element analysis
FEM	Finite element method
FFT	Fast Fourier transformation
FIR	Finite impulse response
FPE	Final prediction error
FPGA	Field-programmable gate array
FSR	Finite step response
GA	Genetic algorithm
GPC	Generalized predictive control
HDA	High displacement actuator
HEB	European standard wide flange beam designation
HVAC	Heating, ventilation and air conditioning
HW	Hardware
I/O	Input–output
IP	Interior point
IPMC	Ionic polymer metallic composite
ISS	International space station
KKT	Karush–Kuhn–Tucker
LAPACK	Linear algebra package (software)
Laser	Light amplification by the stimulated emission of radiation
LDV	Laser Doppler vibrometer
LMI	Linear matrix inequality
LP	Linear programming
LQ	Linear quadratic
LQG	Linear quadratic Gaussian
LQR	Linear quadratic regulator
LTI	Linear time-invariant
MAC	Model algorithmic control
MAGLEV	Magnetic levitation
MDOF	Multiple degrees of freedom
MEMS	Micro-electromechanical systems
MFC	Macrofiber composite
MFM	Magnetic force microscope

MHO	Moving horizon observer
MIMO	Multiple-input, multiple-output
MP	Multi-parametric
MPC	Model predictive control
MPMPC	Multi-parametric model predictive control, also known as explicit MPC
MPT	Multi-parametric toolbox (software)
MR	Magnetorheological
MRI	Magnetic resonance imaging
MS	Magnetostrictive
NiTi	Nickel–titanium
NiTiNOL	Nickel–titanium NOL (alloy), nitinol
NMPC	Nonlinear model predictive control
NOL	Naval Ordnance Laboratory, see nitinol and terfenol
NP	Non-deterministic polynomial-time (hard)
NPF	Negative position feedback
NR	Newton–Raphson (procedure)
NRMPC	Newton–Raphson MPC
$NRMPC_{k+1}$	Original extended Newton–Raphson MPC (1 step)
$NRMPC_{k+2}$	Extended Newton–Raphson MPC (2 steps)
$NRMPC_{k+3}$	Extended Newton–Raphson MPC (3 steps)
OKID	Observer Kalman filter identification
ORM	Optimal region merging
PAF	Payload adapter fitting
PCI	Peripheral component interconnect
PF	Position feedback
PID	Proportional-integral-derivative
PPF	Positive position feedback
PLC	Programmable logic controller
PLZT	Lead lanthanum zirconate titanate
PMN	Lead magnesium niobate
PMN-PT	Lead magnesium niobate–lead titanate
PVF	Positive position feedback
PSD	Proportional-summing-derivative
PT	Lead titanate
PVF	Positive velocity feedback
PWA	Piecewise-affine
PWL	Piecewise-linear
PZT	Lead zirconate-titanate
PZT5A	A certain type of PZT material
QP	Quadratic programming
qpOASES	A type of online active set quadratic programming solver (software)
QPMPC	Quadratic programming MPC
RAM	Random access memory

RMS	Root mean square
RTW	Real-time workshop (software)
SDOF	Single degree of freedom
SDP	Semidefinite programming
SGPC	Stable generalized predictive control
SeDuMi	Self-dual-minimization (software)
SI	International system of units (Système international d'unités)
SIORHC	Stabilizing input/output receding horizon control
SISO	Single-input, single output
SMA	Shape memory alloy
SOCP	Second order cone programming
STOL	Short takeoff and landing
TCP/IP	Transmission control protocol / Internet protocol
TB	Tungsten bronze
TEFLON	Tetrafluorethylene
Terfenol-D	Terbium ferrum NOL dysprosium
TET	Task execution time
UAV	Unmanned aerial vehicle
UNF	Unified fine thread
USB	Universal serial bus
VTOL	Vertical takeoff and landing
xPC	xPC Target—rapid prototyping platform (software)
YALMIP	Yet another LMI parser (software)

Symbols

Greek Lower Case Letters

α	Rayleigh mass damping factor
α_l	Line search coefficient
α_s	Eigenvector matrix scaler
β	Rayleigh stiffness damping factor
γ	Prediction performance bound
γ_s	Shear strain
δ	Eigenvector
δ_d	Damping ratio (with physical dimension)
ε	Mechanical strain
ε_σ	Permittivity under constant or no stress
ε_S	Permittivity under constant strain
ζ	Dimensionless damping ratio
ζ_{fil}	Filter damping
η	Available degrees of freedom in the terminal state
η_v	Viscosity coefficient
θ	Angle in radians
θ_p	Active set coefficient vector
ι	Lagrange multiplier
ι_{ms}	Magnetostrictive coefficient
κ	Eigenvalue
λ	Lagrange multiplier
μ	Numerical vector scaler
μ_σ	Magnetic permeability at constant stress
ν	Lagrange multipliers for the terminal condition
ν_{xz}	Poisson's ratio in a given direction
ξ	Modal participation factor
ξ_{fil}	Diagonal matrix containing filter damping terms
π	Ratio of a circle's area to the square of its radius
ρ	Density

σ	Stress
σ_s	Singular value
τ	Shear stress
τ_d	Delay
τ_Y	Yield strength
υ	Simplification constant in distributed parameter systems
ϕ	Phase shift
χ	Root function defined by the NRMPC problem
ψ	Boundary function in interior point QP algorithms
ω_f	Angular frequency of a forced input
ω_{fil}	Frequency of a filter
ω_n	Angular natural frequency
ω_i	i-th angular natural frequency
ω_d	Damped angular natural frequency

Greek Capital Letters

Γ	Definition matrix of ellipsoids
Γ_f	f partition of $\Gamma_{\mathbf{z}}^{-1}$
Γf_x	Mixed f, x partition of $\Gamma_{\mathbf{z}}^{-1}$
Γ_x	x partition of $\Gamma_{\mathbf{z}}^{-1}$
Γ_{xf}	Mixed x, f partition of $\Gamma_{\mathbf{z}}^{-1}$
$\Gamma_{\mathbf{x}}$	Definition matrix of the invariant ellipsoid E_x in state-space, intersection of E_z with the state-space
$\Gamma_{\mathbf{xz}}$	Definition matrix of the invariant ellipsoid E_{xz} created by the projection of the augmented invariant ellipsoid E_z
$\Gamma_{\mathbf{z}}^{-1}$	Definition matrix of the augmented invariant ellipsoid E_z
$\bar{\Gamma}_z^{-1}$	One step ahead iteration of $\Gamma_{\mathbf{z}}^{-1}$
$\tilde{\Gamma}_z^{-1}$	Two steps ahead iteration of $\Gamma_{\mathbf{z}}^{-1}$
Δ	Eigenvector matrix
Θ	Simplification matrix in NRMPC
Θp	Active set coefficient matrix
Λ	Eigenvalue matrix
Λ_{fil}	Diagonal matrix of the squares of filter frequencies
Ξ	Boundary condition
Π	Target set
Π_{na}	Maximal invariant target set created by a finite constraint checking horizon
Π_s	Simplified invariant target set defined as a low complexity polytope
Π_∞	Maximal invariant target set
Σ	Transformation matrix in optimized dynamics NRMPC
Φ	Closed loop system matrix
Φ_{eq}	Equality constraint in general

Φ_{in}	Inequality constraint in general
Ψ	Closed loop matrix of the augmented system
Ω	Separation function

Latin Lower Case Letters

a_i	Elements of the active set
b	Viscous damping coefficient
b_c^i	i-th row of the constraint vector $\mathbf{b}_0$
c	Scalar perturbation, perturbation vector for $n_u > 1$
d	Element of the piezoelectric matrix
d_0	Initial deflection of the beam tip at $t = 0$ s
d_t	Deflection of the beam tip at time t
d_{ts}	Deflection boundary after which the state is the part of the target set
e	Scalar error, difference between reference and measured output
f	Generic function
f_0	First or dominant mechanical resonance frequency
g	Generic function
h	Scaler
i	i-th element in a series, index, generic iteration
k	Linear stiffness coefficient/discrete time step
l	Length
m	Point mass
n	Discrete dimension
n_b	Block adaptive update speed
n_c	Control horizon
n_{min}	Number of samples necessary to enter the target set from the given initial condition
n_p	Prediction horizon
n_u	Input dimension
n_x	State dimension
n_y	Output dimension
o_i	Perpendicular distance of the i-th hyperplane from the origin
p	Iteration of the quadratic programming algorithm
q	Displacement coordinate
r	Scalar input penalty/setpoint
s	Mechanical compliance in a given direction
t	Continuous time
t_{avg}	Average task execution time
t_{max}	Maximal task execution time
t_{min}	Minimal task execution time
t_{std}	Standard deviation of the average task execution time
u	Input

u_{max}	Maximal (peak) input
$\bar{u}$	Upper input constraint
$\underline{u}$	Lower input constraint
Δu	First difference of input
$\Delta\bar{u}$	Upper input rate constraint
$\Delta\underline{u}$	Lower input rate constraint
$u(s)$	Input in Laplace domain
$v(s)$	Disturbance in Laplace domain
$w(s)$	Reference in Laplace domain
x	State
$\bar{x}$	Upper state constraint
$\underline{x}$	Lower state constraint
$\tilde{x}$	Augmented state (integral action)
$\check{x}$	Previous value of x, where $x = (\mathbf{A} + \mathbf{BK})\check{x}$
Δx	First difference of state
$\Delta\bar{x}$	Upper state rate constraint
$\Delta\underline{x}$	Lower state rate constraint
x_0	Initial state, state x_k at time $k = 0$
y	Output
$\bar{y}$	Upper output constraint
$\underline{y}$	Lower output constraint
Δy	First difference of output
$\Delta\bar{y}$	Upper output rate constraint
$\Delta\underline{y}$	Lower output rate constraint
y_{avg}	Arithmetic mean of beam tip vibrations
y_{max}	Peak value of mean compensated beam tip vibrations
$y(s)$	Output in Laplace domain

Latin Capital Letters

A	Area
$\bar{A}$	Amplitude gain in constant amplitude velocity feedback
A_n	Integration constant in distributed parameter systems
A_c^i	i-th row of the constraint matrix $\mathbf{A}_c$
B_n	Integration constant in distributed parameter systems
E	Modulus of elasticity (Young's modulus)
E_x	Intersection of the augmented ellipsoid E_z with the x space
E_{xz}	Projection of the augmented ellipsoid E_z to the x space
E_z	Augmented invariant ellipsoid
$F(s)$	Transfer function of the closed-loop system
F_b	Damping force
F_k	Spring force
F_m	Inertial force
$G(s)$	Transfer function of the controller

$H(s)$	Transfer function of the plant
J	Cost and cost function
J_{nc}	Finite horizon cost, implying n_c optimization variables
J_{NRMPC}	Infinite horizon cost in the NRMPC formulation given as $J_{\mathrm{NRMPC}} = \mathbf{f}_k^T \mathbf{f}_k$
J_T	Terminal cost $J_T = x_{n_c}^T \mathbf{P}_f x_{n_c}$
J_∞	Infinite horizon cost $J_\infty = J_{n_c} + J_T$
K	Control gains in constant gain velocity feedback
K_p	Proportional gain in a PID controller
K_i	Integral gain in a PID controller
K_d	Derivative gain in a PID controller
M	Bending moment
M_w	Moment magnitude scale
N	Number of elements in a series, final element
T	Shear force
T_c	Time constant of a first order transfer
T_d	Derivative time constant
T_i	Integral time constant
T_s	Sampling period
V	Lyapunov function

Latin Bold Lower Case Letters

$\mathbf{a}$	Active set
$\mathbf{a}^*$	Active set at the solution
$\mathbf{b_c}$	Inequality constraint vector
$\mathbf{b_e}$	Equality constraint vector
$\mathbf{c_E}$	Stiffness coefficient in a constant electric field
$\mathbf{d}$	Piezoelectric coupling matrix
$\mathbf{e}$	Vector of control errors in tracking
$\mathbf{e_i}$	i-th column of the identity matrix
$\mathbf{e_p}$	Piezoelectric coupling matrix in the stress-charge form
$\mathbf{f}$	Vector of future perturbations
$\mathbf{g}$	Magnetostrictive coupling matrix
$\mathbf{h}$	Vertex representation of regions $\mathscr{X}_i$ in MPMPC
$\mathbf{j}$	Search directions
$\mathbf{k}$	Product of $\mathbf{V_s}\Phi\mathbf{V_s^{-1}}$ in simplified polyhedral target sets
$\mathbf{l}$	Product of $\mathbf{V_s}x$ in simplified polyhedral target sets
$\mathbf{m^i}$	Vector aiding the simplification of the NRMPC online process
$\mathbf{m_v}$	Vector aiding the simplification of the NRMPC online process
$\mathbf{o}$	Definition vector of the piecewise-affine MPC law in MPMPC
$\mathbf{p}$	Electric degrees of freedom
$\tilde{\mathbf{p}}$	Definition vector of the half space bounding region $\mathscr{X}_i$ in MPMPC

$\mathbf{q}$	Vector of displacements
$\dot{\mathbf{q}}$	Vector of velocities
$\ddot{\mathbf{q}}$	Vector of accelerations
$\mathbf{r}$	Vector of reference values
$\mathbf{s}$	Vector of Laplace operators
$\mathbf{s_H}$	Mechanical compliance at constant magnetic field
$\mathbf{s_E}$	Mechanical compliance at constant electric field
$\mathbf{u}$	Sequence of inputs
$\tilde{\mathbf{u}}$	Tail of the sequence of inputs $\mathbf{u}$
$\mathbf{u_d}$	Vector of desired inputs in tracking
$\mathbf{u_e}$	Vector of input errors in tracking
$\mathbf{u}(s)$	Vector of inputs in Laplace domain
$\Delta\mathbf{u}$	Incremental improvement in the solution of the QP
$\mathbf{u}^*$	Optimal sequence of inputs/optimal solution of a constrained optimization problem
$\mathbf{u}^\Delta$	Optimal solution of an unconstrained optimization problem
$\mathbf{v}$	Vector aiding the simplification of the NRMPC online process
$\mathbf{v}_{\mathrm{fil}}$	Vector of individual filter outputs
$\mathbf{v}(s)$	Vector of disturbances in Laplace domain
$\mathbf{w}$	Generic optimization variable
$\mathbf{w}(s)$	Vector of references in Laplace domain
$\mathbf{x}$	State vector
$\mathbf{y}$	Output vector
$\mathbf{y}(s)$	Vector of outputs in Laplace domain
$\mathbf{z}$	Augmented state vector

Latin Bold Capital Letters

$\mathbf{A}$	State transition matrix
$\mathbf{A_c}$	Inequality constraint matrix
$\mathbf{A_c^a}$	Matrix of active inequality constraints at the current active set $\mathbf{a}_p$
$\mathbf{A_e}$	Equality constraint matrix
$\mathbf{B}$	Input matrix
$\mathbf{B_0}$	Inequality constraint matrix
$\mathbf{B_d}$	Structural damping matrix
$\mathbf{B_m}$	Magnetic flux
$\mathbf{B_p}$	Dielectric loss
$\mathbf{C}$	Output matrix
$\mathbf{D}$	Direct input–output feed-through matrix
$\mathbf{D}_{\mathrm{e}}$	Electric displacement
$\mathbf{E}$	Vector selecting the first (block) element/augmented dynamics vector in NRMPC
$\mathbf{E_e}$	Electric field strength
$\mathbf{E}_{\mathrm{fil}}$	Rectangular matrix to use more filters than actuators

$\mathbf{F}$	State cost component matrix
$\mathbf{G}$	Constant in the QP problem/mixed input and state cost component matrix
$\mathbf{H}$	Hessian in the QP problem/input cost component matrix
$\mathbf{H_m}$	Magnetic field strength
$\mathbf{I}$	Matrix with ones on the main diagonal, *eye* matrix
$\mathbf{K}$	Linear quadratic gain
$\mathbf{K_p}$	Anisotropic permittivity
$\mathbf{K_s}$	Stiffness matrix
$\mathbf{K_z}$	Piezoelectric coupling matrix
$\mathbf{L}$	Cost transformation matrix
$\mathbf{L_c}$	Perturbation partition of the cost transformation matrix
$\mathbf{L_{cx}}$	Mixed perturbation-state partition of the cost transformation matrix
$\mathbf{L_e}$	Electrical input to the piezoelectric dynamics equation
$\mathbf{L_x}$	State partition of the cost transformation matrix
$\mathbf{L_{xc}}$	Mixed state-perturbation partition of the cost transformation matrix
$\mathbf{M}$	Free state prediction matrix, except in Chap. 2 where it refers to the concept of mass matrix
$\tilde{\mathbf{M}}$	Matrix aiding the mathematical transformation of the optimized augmented dynamics in NRMPC into a convex optimization problem
$\mathbf{N}$	Forced state prediction matrix
$\tilde{\mathbf{N}}$	Matrix aiding the mathematical transformation of the optimized augmented dynamics in NRMPC into a convex optimization problem
$\mathbf{O}$	Definition matrix of the piecewise-affine MPC law in MPMPC
$\mathbf{P}$	Solution of the (discrete-time algebraic) Ricatti equation
$\mathbf{P}_f$	Terminal weighting matrix
$\tilde{\mathbf{P}}$	Definition matrix of the half space bounding region $\mathscr{X}_i$ in MPMPC
$\mathbf{Q}$	State penalty weighting matrix
$\mathbf{R}$	Input penalty weighting matrix
$\mathbf{S}$	Strain
$\mathbf{S_v}$	Eigenvalues in a vector form
$\mathbf{T}$	Perturbation shift matrix in the augmented state representation
$\mathbf{U}$	Matrix aiding the mathematical transformation of the optimized augmented dynamics in NRMPC into a convex optimization problem
$\mathbf{V}$	Matrix aiding the mathematical transformation of the optimized augmented dynamics in NRMPC into a convex optimization problem
$\mathbf{W_i}$	Concatenating, truncating and simplifying matrices in the online NRMPC

$\mathbf{V}_S$	Definition matrix of the simplified polyhedral target set
$\mathbf{X}$	Matrix aiding the mathematical transformation of the optimized augmented dynamics in NRMPC into a convex optimization problem
$\mathbf{Y}$	Matrix aiding the mathematical transformation of the optimized augmented dynamics in NRMPC into a convex optimization problem
$\mathbf{Z}$	Generic optimization variable

Latin Calligraphic Letters

$\mathscr{D}$	Block diagonal matrix with I and R on its main diagonal
$\mathscr{H}_\infty$	$\mathscr{H}_\infty$ optimal control gain
$\mathscr{H}_2$	$\mathscr{H}_2$ optimal control gain
$\mathscr{K}$	Unconstrained MPC gain
$\mathscr{L}$	Laplace transform
$\mathscr{Q}$	Multiple DOF displacements in Laplace domain
$\mathscr{P}$	Transfer function of the plant
$\mathscr{R}$	Block diagonal matrix with input penalty $\mathbf{R}$ on the main diagonal
$\mathscr{S}$	Sensitivity function
$\mathscr{T}$	Complementary sensitivity function
$\mathscr{X}$	Polyhedral region
$\tilde{\mathscr{X}}$	Polyhedral region after clipping
$\mathscr{X}_u$	Unsaturated polyhedral region
$\mathscr{X}_s$	Saturated polyhedral region
$\mathscr{Y}$	Convex polytope
$\mathscr{Z}$	Z-transform

Blackboard Bold Letters

$\mathbb{N}$	Set of natural numbers
$\mathbb{Z}$	Set of integers
$\mathbb{Q}$	Set of rational numbers
$\mathbb{R}$	Set of real numbers

Others

Designation and meaning of other variables, marks, indices, sub- and superscripts are defined locally. Select variables may have more than one meaning, please refer to the context for explanation. Select concepts may be marked with different variables, depending on whether it is a theoretical consideration, implementation example or actual code segment.

Relation signs $>$, $\geq$ and $<$, $\leq$ are used in the scalar binary relation sense (larger, larger or equal, smaller, smaller or equal) for scalar values and element-wise relation sense for vectors. In case the equation uses matrix notation and can be considered a linear matrix inequality, relation signs $>$, $\geq$ are used to define positive definiteness and positive semidefiniteness, while signs $<$, $\leq$ denote the concepts of negative definiteness and negative semidefiniteness. Although linear matrix inequalities in mathematics generally use the notation $\succ$, $\succeq$ and $\prec$, $\preceq$ the use of these symbols will be avoided in this book.

The notation $\kappa\{\mathbf{A}\}$ denotes the eigenvalues of matrix $\mathbf{A}$. Although most scientific literature uses the notation λ for eigenvalues, due to the interdisciplinary nature of this book the Greek letter of is λ is reserved for a different concept.

Depending on the context the number 1 may denote the mathematical concept of a vector of ones, which is an n elements long vector containing only the numbers 1: $1 = \mathbf{1} = [1\ 1\ 1 \ldots 1]^T$. This book does not utilize the bold notation $\mathbf{1}$ for a vector of ones.

Part I
Vibration Control

Chapter 2
Basics of Vibration Dynamics

Vibrations are mechanical oscillations about an equilibrium position. There are cases when vibrations are desirable, such as in certain types of machine tools or production lines. Most of the time, however, the vibration of mechanical systems is undesirable as it wastes energy, reduces efficiency and may be harmful or even dangerous. For example, passenger ride comfort in aircraft or automobiles is greatly affected by the vibrations caused by outside disturbances, such as aeroelastic effects or rough road conditions. In other cases, eliminating vibrations may save human lives, a good example is the vibration control of civil engineering structures in an earthquake scenario.

All types of vibration control approaches—passive, semi-active and active—require analyzing the dynamics of vibrating systems. Moreover, all active approaches, such as the model predictive control (MPC) of vibrations considered in this book require simplified mathematical models to function properly. We may acquire such mathematical models based on a first principle analysis, from FEM models and from experimental identification. To introduce the reader into the theoretical basics of vibration dynamics, this chapter gives a basic account of engineering vibration analysis.

There are numerous excellent books available on the topic of analyzing and solving problems of vibration dynamics. This chapter gives only an outline of the usual problems encountered in vibration engineering and sets the ground for the upcoming discussion. For those seeking a more solid ground in vibration mechanics, works concentrating rather on the mechanical view can be very valuable such as the work of de Silva [10] and others [4, 22]. On the other hand, the reader may get a very good grip of engineering vibrations from the books discussing active vibration control such as the work of Inman [21] and others [15, 18, 37, 38].

The vibration of a point mass may be a simple phenomenon from the physical viewpoint. Still, it is important to review the dynamic analysis beyond this phenomenon, as the vibration of a mass-spring-damper system acts as a basis to understand more complex systems. A system consisting of one vibrating mass has one natural frequency, but in many cases, in a controller it is sufficient to replace a continuous

G. Takács and B. Rohal'-Ilkiv, *Model Predictive Vibration Control*,
DOI: 10.1007/978-1-4471-2333-0_2,

structure with complex geometry. The vibration dynamics of point mass and other comparably simple models may represent a surprisingly large portion of real-life mechanical systems [8, 21]. We will begin our analysis in the first section with a case in which damping is not considered, then gradually build a more detailed representation of the physics of vibrations. Section 2.2 will introduce damping to the simple vibrating point mass, and following this, Sect. 2.3 considers the forced vibration of this essential mechanical system.

Multiple degree of freedom systems will be introduced in Sect. 2.4 including a concise treatment of the eigenvalue problem and modal decomposition. Since vibration dynamics of the continuum is a complex and broad topic, Sect. 2.5 will only make a brief excursion to distributed parameter systems. The transversal vibration of cantilever beams will be covered, as in upcoming chapters such a demonstration system will be utilized to test the implementation of model predictive controllers. Finally, this chapter ends with a discussion on the models used in vibration control in Sect. 2.6. This section covers transfer function models, state-space models, identification from FEM models and experimental identification. The aim of this chapter is to briefly introduce the mathematical description of vibration phenomena, in order to characterize the nature of the mechanical systems to be controlled by the model predictive control strategy presented in the upcoming chapters of this book.

Figure 2.1 illustrates the Venus Express spacecraft[1] under preparation for experimental vibration dynamics analysis [12]. The body of the spacecraft is equipped with accelerometers while outside disturbance is supplied to the structure via a shake table. Gaining knowledge on the vibration properties of mechanical systems is essential for both active and passive vibration engineering, as unexpected vibratory response may jeopardize mission critical performance or structural integrity.

2.1 Free Vibration Without Damping

The simplest possible example that may help understand the dynamics of vibrations is the oscillating point mass, which has one degree of freedom. Such a system is schematically illustrated in Fig. 2.2. Let us assume for now that damping is negligible and there is no external force acting on the system. The vibrating mass, often referred to as the simple harmonic oscillator, has a mass of m and is sliding on a frictionless surface. The position of the mass is denoted by one time-dependent coordinate, $q(t)$. The mass is connected to a surface with a linear spring, having the spring constant k.

According to Newton's second law of motion, there is an inertial force generated by the mass, which is proportional to its acceleration [10]:

$$F_m = m\frac{d^2q}{dt^2} = m\ddot{q}(t) \tag{2.1}$$

[1] Courtesy of the European Space Agency (ESA) and European Aeronautic Defence and Space Company (EADS)-Astrium.

Fig. 2.1 Venus Express spacecraft is under preparation for experimental vibration dynamics analysis on a shake table [12]

where $\ddot{q}(t)$ is the acceleration and F_m is the inertial force of the mass. There is another force acting against this, which is proportional to the spring constant k [4, 22, 41]:

$$F_k = kq(t) \tag{2.2}$$

Because there is no other energy source or sink, the sum of these two forces will be zero. We can now assemble our equation of motion, which is an ordinary differential equation (ODE) given by [4, 10, 21, 49, 51]:

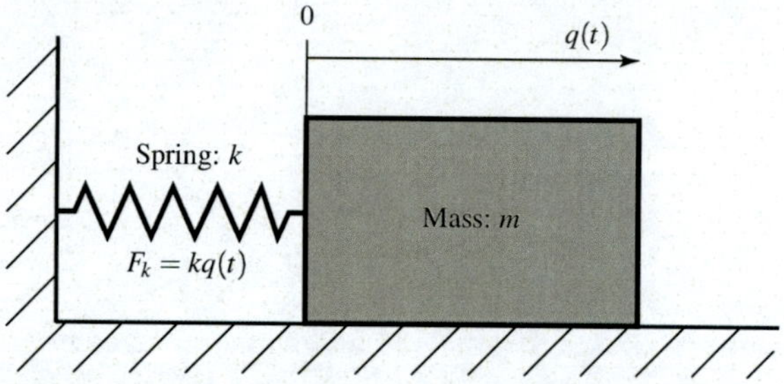

Fig. 2.2 Free vibration of a point mass without damping

$$m\ddot{q}(t) + kq(t) = 0 \tag{2.3}$$

One may classify mechanical vibrations according to whether an outside force is present:

- free vibration
- forced vibration

In free vibration, a mechanical system is excited by an initial condition, such as a displacement, velocity or acceleration and then is allowed to vibrate freely without further force interaction. A mechanical system in free vibration will oscillate with its natural frequency and eventually settle down to zero due to damping effects. In forced vibration, an external force is supplied to the system.

We may inspect how the moving mass will physically behave by imagining that we deflect our spring and move the mass to an initial position of $q(0)$, then let it go without inducing an initial velocity or acceleration. The mass will start to vibrate back and forth, and since there is no energy dissipation, its position will oscillate between $q(0) = \pm\tilde{q}$. If we plot its position in relation with time, we will get harmonic motion, which can be described by a trigonometric function. If we multiply a sine function shifted from zero by ϕ radians by our amplitude, we have an oscillating harmonic motion between $\pm\tilde{q}$. In addition to the amplitude, this function has a period of oscillations as well. Let us denote the angular frequency by ω_n, which in fact expresses the frequency of oscillations. Now we have a full mathematical description of the assumed motion of the mass [15, 21]:

$$q(t) = \tilde{q}\sin(\omega_n t + \phi) \tag{2.4}$$

where t is the progressing time, $\tilde{q}$ is the amplitude and ω_n is the angular velocity expressing the period of the oscillations. The constant amplitude $\tilde{q}$ and phase shift ϕ are constants that can be uniquely determined based on the initial conditions. We can substitute this trial solution in (2.4) back into the equation of motion in (2.3) and get:

$$m\frac{d^2\left(\tilde{q}\sin(\omega_n t + \phi)\right)}{dt^2} + k\left(\tilde{q}\sin(\omega_n t + \phi)\right) = 0 \tag{2.5}$$

after double differentiating the first term we will get

$$-m\omega_n^2 \left(\tilde{q}\sin(\omega_n t + \phi)\right) + k\left(\tilde{q}\sin(\omega_n t + \phi)\right) = 0 \tag{2.6}$$

we can further simplify this to calculate ω_n and get [41, 49]:

$$\omega_n = \sqrt{\frac{k}{m}} \tag{2.7}$$

Substituting this back to the original trial solution (2.4) and using the initial conditions, we get a solution of our ODE. We can convert the angular or circular period of vibration ω_n expressed in rad/sec into more familiar units [15, 51, 49]:

$$f_n = \frac{1}{2\pi}\omega_n = \frac{1}{2\pi}\sqrt{\frac{k}{m}} \tag{2.8}$$

where f_n gives oscillations per second or Hz (Hertz), or [49]

$$T_n = \frac{1}{f_n} = \frac{2\pi}{\omega_n} = 2\pi\sqrt{\frac{m}{k}} \tag{2.9}$$

which gives us the period of one oscillation in seconds.

If we divide the equation of motion in (2.3) by the mass, we can express it in terms of the angular natural frequency ω_n:

$$\ddot{q}(t) + \omega_n^2 q(t) = 0 \tag{2.10}$$

The solution in (2.4) is in fact a *trial solution*, which is a type of educated engineering guess; nevertheless if it works then it is the solution itself [21]. In our case, the trial solution in Eq. (2.4) works and it is a valid solution of the vibrating point mass. Although it is a product of a logical deduction, there are other ways to express the expected solution of the ODE describing the equation of motion of the point mass. A common alternative way to express the displacement of the point mass is to use an exponential function of time. This is a more mathematical representation of the same concept [21]:

$$q(t) = \tilde{q}e^{\kappa t} \tag{2.11}$$

where $\tilde{q}$ is the complex vibration amplitude. This representation is called the *phasor* representation, where a phasor can be simply understood as a rotating vector [15]. Note that in the field of vibration mechanics instead of κ the phasor representation uses λ. In order to keep the notation consistent throughout the book, this custom has been changed in order to reserve λ for concepts used in predictive control. The rotation velocity of the vector is included in the complex variable κ, which is essentially an eigenvalue. The real part of the amplitude $\tilde{q}$ is the physical amplitude, while the real component of the phasor describes the harmonic motion of the point mass.

Substituting the trial solution in (2.11) to the original equation of motion and differentiating results the same expressions for ω_n, however κ can assume both negative and positive values. The undamped natural frequency ω_n is the positive of the two κ solutions.

$$\kappa = \pm j\omega_n t \tag{2.12}$$

The reason that $e(x)$ or the natural exponent is often used in vibration analysis instead of simple trigonometric functions comes from the fact that it is mathematically easier to manipulate the exponential function and solve differential equations by expecting solutions in this form. The complex exponential is simply an eigenfunction of differentiation. Although trigonometric functions naturally come to mind when describing oscillatory motion, the natural exponential function $e(x)$ appears commonly in trial solutions of ODE describing vibration phenomena. The equivalence between trigonometric functions and the exponential function is given by Euler's formula. The general solution of the equation of motion after substituting (2.11) and solving for κ will be

$$q(t) = Ae^{-j\omega_n t} + Be^{j\omega_n t} \tag{2.13}$$

where A and B are integration constants determined by the initial conditions. The general solution can be equivalently described by an equation using trigonometric functions [4, 22]:

$$q(t) = A\cos\omega_n t + B\sin\omega_n t \tag{2.14}$$

where A and B are again integration constants to be determined based on the initial conditions.

2.2 Free Vibration with Damping

The previous section discussed free vibration of a point mass without damping. This means that a mass connected to a spring and deflected to the initial position of $q(0) = \tilde{q}$ would oscillate with the same amplitude indefinitely. As we can see, this is a very unrealistic model—we have to add some sort of energy dissipation mechanism, or in other words, damping.

Damping is a complex phenomenon, not very well understood and modeled in science. Different damping models exist; however, these represent reality only under certain conditions and assumptions. Probably the most popular damping model is *viscous damping*, which expresses the damping force that is proportional to velocity by a constant b. This force can be expressed by [21, 50]:

$$F_b = b\frac{d\dot{q}(t)}{dt} = b\dot{q}(t) \tag{2.15}$$

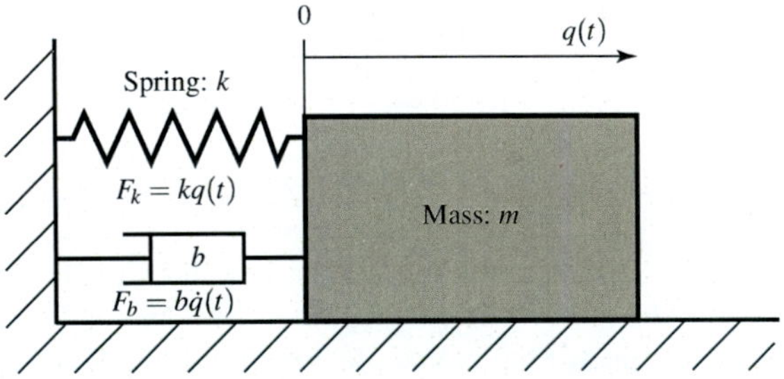

Fig. 2.3 Free vibration of a point mass with viscous damping

We can improve our previous model by adding this damping force to our system. Let us have the same vibrating mass m connected by a spring with the linear spring constant k to a fixed surface. Displacement is measured by $q(t)$ and let us add a viscous damper with the constant b to our representation. This is represented schematically by Fig. 2.3.

Now that we know the viscous damping force is expressed by (2.15), we can add it to the original equation of motion. The spring force F_k and the damping force F_b act against the inertial force F_m. The sum of these forces is zero, if we express this, we obtain an ODE again [4, 10, 21, 22]:

$$m\ddot{q}(t) + b\dot{q}(t) + kq(t) = 0 \tag{2.16}$$

Dividing the whole equation of motion by m results in the following term:

$$\ddot{q}(t) + \frac{b}{m}\dot{q}(t) + \frac{k}{m}q(t) = 0 \tag{2.17}$$

Let us call half of the ratio of the viscous damping constant b and mass m as δ_d or [49]:

$$\delta_d = \frac{1}{2}\frac{b}{m} \tag{2.18}$$

and use (2.7) to substitute for k/m with yields [51, 52]:

$$\ddot{q}(t) + 2\delta_d\dot{q}(t) + \omega_n^2 q(t) = 0 \tag{2.19}$$

Another common representation of the damping both in mechanical vibration analysis and vibration control is proportional damping ζ[2] which is expressed as a percentage of critical damping [18, 51, 52]:

$$\frac{b}{m} = 2\zeta\omega_n = 2\frac{b}{b_c}\sqrt{\frac{k}{m}} \tag{2.20}$$

[2] Note that ζ is not the same as δ_d.

where critical damping is denoted by $b_c = 2\sqrt{km}$. Instead of expressing the simplified differential equation in the terms of the damping coefficient δ_d as in Eq. (2.19) we may express it using proportional damping ζ and get [4, 10, 21]:

$$\ddot{q}(t) + 2\zeta\omega_n\dot{q}(t) + \omega_n^2 q(t) = 0 \tag{2.21}$$

Let us now assume that the trial solution to the ODE will come in the form of (2.11). Remember, this is the more mathematical representation of our solution but the differentiation of an exponential function is the exponential itself which makes the evaluation process a little simpler. After substituting the trial solution into (2.21) we obtain the following equation:

$$\frac{d^2(e^{\kappa t})}{dt^2} + 2\zeta\omega_n\frac{d(e^{\kappa t})}{dt} + \omega_n^2(e^{\kappa t}) = 0 \tag{2.22}$$

and after differentiating this will be reduced to

$$\kappa^2 + 2\zeta\omega_n\kappa + \omega_n^2 = 0 \tag{2.23}$$

Solving this using simple algebra will yield the solution for κ. The roots of this equation will be [49]:

$$\kappa_{1,2} = -\zeta\omega_n \pm \omega_n\sqrt{\zeta^2 - 1} = -\zeta\omega_n \pm j\omega_d \tag{2.24}$$

The damped natural frequency in terms of proportional damping ζ will be then:

$$\omega_d = \omega_n\sqrt{1 - \zeta^2} \tag{2.25}$$

In this interpretation the overdamped, underdamped and critically damped oscillations are defined by the magnitude of ζ. As ζ is a percentage of critical damping, $\zeta < 1$ will result in an underdamped decaying periodic motion, $\zeta > 1$ will result in an overdamped aperiodic motion, while $\zeta = 1$ will result in a periodic and critically damped motion. Similarly, by substituting the same trial solution into Eq. (2.19) will yield:

$$\frac{d^2(e^{\kappa t})}{dt^2} + 2\delta_d\frac{d(e^{\kappa t})}{dt} + \omega_n^2(e^{\kappa t}) = 0 \tag{2.26}$$

which after differentiation will be reduced to

$$\kappa^2 + 2\delta_d\kappa + \omega_n^2 = 0 \tag{2.27}$$

in the terms of δ_d expressing the amount of damping in the system. The roots $\kappa_{1,2}$ of this characteristic equation are expressed by:

$$\kappa_{1,2} = -\delta_d \pm \sqrt{\delta_d^2 - \omega_n^2} \tag{2.28}$$

The second term is the damped natural frequency ω_d or

$$\omega_d = \sqrt{\delta_d^2 - \omega_n^2} \tag{2.29}$$

Depending on the magnitude of the damped natural frequency ω_d we may have overdamping, critical damping or underdamping. Overdamping is the case, when the initial conditions such as initial displacement or velocity result in an aperiodic motion. From now on we will assume that $\zeta < 1$ or equivalently $\omega_d^2 > \delta_d^2$ which results in a periodic vibration with a constantly decaying amplitude.

Let us now interpret this in the physical sense: we cannot express the solution as a simple harmonic function anymore. Because of the energy dissipation, for an underdamped system, the vibration amplitudes will gradually decay and the system will settle at equilibrium. We have to introduce an exponential term, to simulate decay caused by the damping. Our previously assumed solution general solution in (2.4) will be changed to [21]:

$$q(t) = \tilde{q}e^{-\zeta\omega_n t}\sin(\omega_d t + \phi) \tag{2.30}$$

The first exponential term $e^{-\delta_d \omega_n t}$ introduces the exponential decay and emulates the damping effect. Using trigonometric identities, this can also be written as [22]:

$$q(t) = e^{-\zeta\omega_n t}(A\cos(\omega_d t) + B\sin(\omega_d t)) \tag{2.31}$$

where A and B are integration constants which can be determined from the initial conditions. The general solution of the free vibration of the underdamped point mass can also be written in terms of δ_d by stating that [51, 52]:

$$q(t) = \tilde{q}e^{-\delta_d t}\sin(\omega_d t + \phi) \tag{2.32}$$

or equivalently as

$$q(t) = e^{-\delta_d t}(A\sin(\omega_d t) + B\cos(\omega_d t)) \tag{2.33}$$

2.3 Forced Vibration of a Point Mass

The damped vibration of a point mass is a passive representation of dynamic systems. Its motion is only controlled by initial conditions such as deflecting the spring into an initial displacement $q(0)$ or adding an initial velocity $\dot{q}(0)$ to the vibrating mass. We have to introduce an outside force in order to model a controllable active vibrating system. Let us assume that—in addition to the spring, damping and inertial forces—an external force $f_e(t)$ can also supply energy to the system. Combining the equation of motion in (2.16) for the damped vibrating point mass with this external force $f_e(t)$ we will get the following new equation of motion incorporating an outside force effect, for example an actuator or a disturbance [49]:

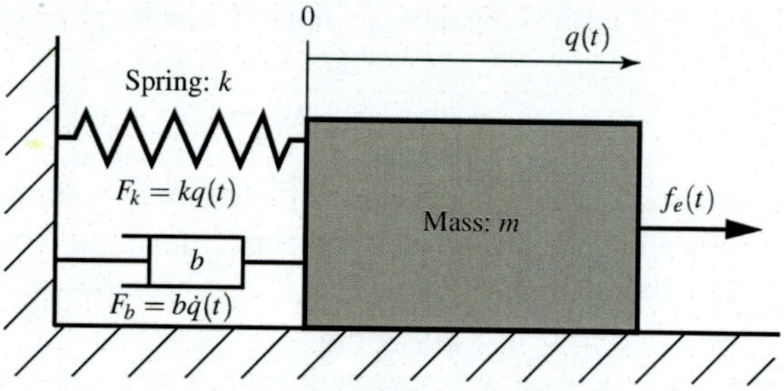

Fig. 2.4 Forced vibration of a point mass with viscous damping

$$m\ddot{q}(t) + b\dot{q}(t) + kq(t) = f_e(t) \tag{2.34}$$

This is a second order ordinary differential equation, just like in the previous cases.

The type of the outside excitation force can be arbitrary and its course may be mathematically described by a step, impulse, harmonic, random or any other functions. To evaluate an analytic solution of the forced equation of motion, we have to have at least some sort of knowledge about the excitation. Let us therefore assume that our excitation force is harmonic, generated for example by a rotating imbalance so we may describe it by [4, 21]:

$$f_e(t) = \tilde{f}_e \sin \omega_f t \tag{2.35}$$

where $\tilde{f}_e$ is the amplitude of the excitation force and ω_f is the angular frequency of the outside disturbance. If we substitute this back into our original equation of motion for the forced response of the point mass in (2.34) we will get

$$m\ddot{q}(t) + b\dot{q}(t) + kq(t) = \tilde{f}_e \sin \omega_f t \tag{2.36}$$

It would be natural to assume that, after an initial transient phase, the vibrating point mass will start to copy the harmonic motion of the outside excitation. This physical assumption can be translated into an expected trial solution of the time response given by:

$$q(t) = \tilde{q} e^{j(\omega_f t + \phi)} \tag{2.37}$$

where $\tilde{q}$ is the amplitude of our vibrations. Substitute this back into the ODE expressing the forced response of the point mass in (2.36), differentiate and simplify to get the amplitude [21]:

$$\tilde{q} = \tilde{f}_e \frac{1}{-m\omega_f^2 + j\omega_f b + k} \tag{2.38}$$

This is of course just a solution for one type of external force. This representation looks much like a transfer function, and in fact, it is easy to apply Laplace transformation to get transfer functions for controller design. For control purposes, it is also possible to transform our ODE into a decoupled form, which is referred to as the state-space representation.

The solution of the equation of motion consists of two parts. The transient response describes the passing effects, while the steady-state response will characterize the response after the initial effects have settled. The total time response of an under-damped system with $\zeta < 1$ will be [21]:

$$q(t) = e^{-\zeta\omega_n t}(A \sin \omega_d t + B \cos \omega_d t) + \tilde{q}(\sin \omega_f t - \phi) \tag{2.39}$$

which is a sum of the steady state and the transient solution. Note that this general solution contains integration constants A and B which in general are not the same as the ones for free vibration. Furthermore, note the three angular frequencies in this equation: the angular natural frequency ω_n, the damped angular natural frequency ω_d and the frequency of the periodic excitation ω_n.

The analytic solution for other types of excitation is analogous to the periodic case. As this work is interested rather in the control aspects of (forced) vibrations, tools known from control engineering such as transfer function and state-space representations will be used to evaluate the response of a system to an outside excitation. To those interested in more details on the analytic representation of forced vibrations the books by de Silva [10] and others [4, 22] may be recommended.

2.4 Multiple Degree of Freedom Systems

The very simple single degree of freedom mass-spring-damper system introduced in the previous sections gives a good foundation for the analysis of vibrating systems. It is possible to simplify the essential dynamic behavior of many real mechanical systems to SDOF and replace it with an analysis procedure similar to the one introduced previously [8, 21].

In the vibration analysis of mechanical systems with multiple degrees of freedom (MDOF), instead of one vibrating mass, we replace our real structure with two or more oscillating masses. If the real system has well-defined separate moving parts, we can consider it as a lumped interconnected parameter system. The degrees of freedom of a lumpedparameter system are equal to the number of vibrating mass points and this is also true for the number of resonant frequencies. A mechanical system or structure which does not have well-defined separately oscillating parts but consists of a continuously spread mass and geometry is a distributed system. Continuum or distributed parameter systems have an infinite amount of resonant frequencies and corresponding vibration shapes. It is, however, possible to discretize the system into large amounts of lumped interconnected parameters and approximate its behavior with methods commonly used for lumped parameter systems. This idea is in fact used in FEM software to extract the vibration dynamics of distributed mechanical systems defined with complex three-dimensional geometry.

Let us choose a very simple example, which has more than one degree of freedom and therefore may be considered as a MDOF system. The system of connected

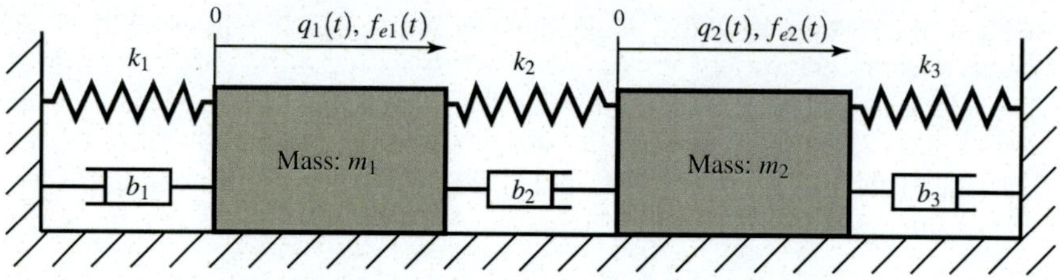

Fig. 2.5 Multiple degrees of freedom system: connected set of two masses

moving masses illustrated in Fig. 2.5 is sliding on a frictionless surface. Now instead of one coordinate defining the movement, we have two for each moving mass: $q_1(t)$ and $q_2(t)$. The two moving masses m_1 and m_2 are connected to each other and the surrounding fixed wall through a series of springs and dampers with the spring and damping coefficients k_1, k_2, k_3 and b_1, b_2, b_3. There are external force inputs associated with individual masses denoted by f_{e1} and f_{e2} .

Using a simple mechanical analysis, we may create a free body diagram for each mass and analyze the forces acting on them. After assembling the equations of motion, we obtain the following set of equations for our two masses [4, 10, 18]:

$$m_1\ddot{q}_1 + (b_1 + b_2)\dot{q}_1 - b_2\dot{q}_2 + (k_1 + k_2)q_1 - k_2q_2 = f_1 \tag{2.40}$$

$$m_2\ddot{q}_2 - b_2\dot{q}_1 + (b_2 + b_3)\dot{q}_2 - k_2q_1 + (k_2 + k_3)q_2 = f_2 \tag{2.41}$$

It is possible to rewrite the one equation for motion per moving mass into a compact set, using matrix notation [10, 37, 52]:

$$\begin{aligned}\begin{bmatrix} m_1 & 0 \\ 0 & m_2 \end{bmatrix}\begin{bmatrix} \ddot{q}_1 \\ \ddot{q}_2 \end{bmatrix} + \begin{bmatrix} b_1 + b_2 & -b_2 \\ -b_2 & b_2 + b_3 \end{bmatrix}\begin{bmatrix} \dot{q}_1 \\ \dot{q}_2 \end{bmatrix} \\ + \begin{bmatrix} k_1 + k_2 & -k_2 \\ -k_2 & k_2 + k_3 \end{bmatrix}\begin{bmatrix} q_1 \\ q_2 \end{bmatrix} = \begin{bmatrix} f_{e1} \\ f_{e2} \end{bmatrix}\end{aligned} \tag{2.42}$$

Note the similarity between this equation of motion and the SDOF forced equation motion in (2.34). We have a matrix containing the masses, which is multiplied by a vector of accelerations. Similarly, we have matrices containing damping elements and spring constants. We can in fact use a matrix notation to create from this [21]:

$$\mathbf{M}\ddot{\mathbf{q}} + \mathbf{B_d}\dot{\mathbf{q}} + \mathbf{K_s}\mathbf{q} = \mathbf{f_e} \tag{2.43}$$

where matrix $\mathbf{M}$ is the mass matrix,[3] $\mathbf{B_d}$ is the structural damping matrix and $\mathbf{K_s}$ is the stiffness matrix. Vector $\mathbf{q}$ contains the displacement coordinates for each degree of freedom. For an N degree of freedom system the constant matrices $\mathbf{M}$, $\mathbf{B_d}$ and $\mathbf{K_s}$ will all have $N \times N$ elements.

---

3 It is customary to denote the mass matrix with **M** however in the upcoming chapter this symbol will be reserved for an entirely different concept.

The solution of such systems is in fact very similar to the solution of SDOF systems. To illustrate this, let us consider a case without damping and with no outside force. Removing these effects from the equation of motion in matrix from (2.43), we get [38]:

$$\mathbf{M}\ddot{\mathbf{q}} + \mathbf{K_s}\mathbf{q} = 0 \tag{2.44}$$

for which we have to find a solution. Similar to the SDOF systems, our solution can be expected in a form of a set of harmonic functions, which mathematically is simply an amplitude multiplied by a complex exponential:

$$\mathbf{q} = \tilde{\mathbf{q}}e^{j\omega_n t} = \tilde{\mathbf{q}}e^{\kappa t} \tag{2.45}$$

As introduced previously, the term $e^{j\omega_n t}$ or analogously $e^{\kappa t}$ is just a mathematical trick to solve differential equations by using the so-called phasor form for the solution. If we take the real part of Euler's formula, we essentially expect a cosine function. Let us substitute this solution to our matrix equation of motion, and differentiate to get:

$$\left(-\omega_n^2\mathbf{M} + \mathbf{K_s}\right)\tilde{\mathbf{q}}e^{j\omega_n t} = 0 \tag{2.46}$$

2.4.1 *The Eigenvalue Problem*

To solve the equation expressed by (2.46) we can assume that the exponential part $e^{j\omega t}$ cannot be zero, therefore we will reduce our expression to [37, 38]:

$$\left(-\omega_n{}^2\mathbf{M} + \mathbf{K_s}\right)\tilde{\mathbf{q}} = 0 \tag{2.47}$$

This is a problem often encountered in mathematics, called the *eigenvalue problem* which in general mathematics assumes the form [22]:

$$(\mathbf{A} - \kappa\mathbf{I})\,\delta = 0 \tag{2.48}$$

where κ contains the eigenvalues of the system while δ is the eigenvector. To get our problem (2.47) into a similar form, we have to multiply it by the inverse of the mass matrix $\mathbf{M}^{-1}$ to get

$$\left(\mathbf{M}^{-1}\mathbf{K_s} - \omega_n^2\mathbf{M}^{-1}\mathbf{M}\right)\tilde{\mathbf{q}} = 0 \tag{2.49}$$

which is essentially just given by [10, 52]

$$\left(\mathbf{M}^{-1}\mathbf{K_s} - \omega_n^2\mathbf{I}\right)\tilde{\mathbf{q}} = 0 \tag{2.50}$$

The solution to the eigenvalue problem expressed by our physical vibrating system is a set of N *eigenvalues* ranging from $\omega_1^2, \omega_2^2 \ldots \omega_N^2$, where N is the number of degrees of freedom. These eigenvalues have a very well-defined physical meaning: they contain the (square of the) angular natural frequencies associated with the individual masses. Substituting these eigenvalues back into the original equation, we get a set of amplitudes $\tilde{\mathbf{q}}$ called the *eigenvectors*. Each eigenvalue or natural frequency has an associated eigenvector. This eigenvector expresses the mode shapes of the system, in other words, the geometrical shape of the vibration within a given resonant frequency. The magnitude of the eigenvectors is not expressed in physical coordinates—instead, modal shapes are scaled by a method of choice. To avoid confusion, we will substitute the notation $\mathbf{q}_i$ with δ_i referring to the fact that these amplitudes have a physically valid magnitude only in relation to each other but not globally.

Eigenvalues and eigenvectors expressing the angular natural vibration frequency of individual masses and the vectors of modal shape associated with those frequencies can be assembled in a compact notation:

$$\Lambda = \mathrm{diag}(\omega_i^2) = \mathrm{diag}(\kappa_i) = \begin{bmatrix} \omega_1^2 & \cdots & 0 \\ \vdots & \ddots & \vdots \\ 0 & \cdots & \omega_N^2 \end{bmatrix} = \begin{bmatrix} \kappa_1 & \cdots & 0 \\ \vdots & \ddots & \vdots \\ 0 & \cdots & \kappa_N \end{bmatrix} \tag{2.51}$$

$$\Delta = \begin{bmatrix} \delta_1 \; \delta_2 \; \delta_3 \; \ldots \; \delta_N \end{bmatrix} = \begin{bmatrix} \tilde{\mathbf{q}}_1 \; \tilde{\mathbf{q}}_2 \; \tilde{\mathbf{q}}_3 \; \ldots \; \tilde{\mathbf{q}}_N \end{bmatrix} \tag{2.52}$$

where Λ is a diagonal matrix with the square of the individual eigenfrequencies ω_i^2 on its main diagonal. Solving the eigenvalue problem, we get the modal shapes which are expressed by the amplitudes δ_i associated with the eigenfrequencies.

2.4.2 Modal Decomposition

It is possible to simplify the solution of a multi-degree freedom system by substituting it with a set of single degree freedom systems. Eigenvectors have a mathematical property called *orthogonality*, which is the basis of this simplification. If Δ is the set of eigenvectors, it can be shown that when we use it to multiply the mass matrix from both sides we obtain [22, 38, 52]:

$$\begin{aligned} \Delta^T \mathbf{M} \Delta &= \begin{bmatrix} \delta_1 \; \delta_2 \; \delta_3 \; \ldots \; \delta_N \end{bmatrix}^T \mathbf{M} \begin{bmatrix} \delta_1 \; \delta_2 \; \delta_3 \; \ldots \; \delta_N \end{bmatrix} \\ &= \begin{bmatrix} \delta_1^T \mathbf{M} \delta_1 & \delta_1^T \mathbf{M} \delta_2 & \ldots & \delta_1^T \mathbf{M} \delta_N \\ \delta_2^T \mathbf{M} \delta_1 & \delta_2^T \mathbf{M} \delta_2 & \ldots & \delta_2^T \mathbf{M} \delta_N \\ \vdots & \ddots & \ldots & \vdots \\ \delta_N^T \mathbf{M} \delta_1 & \delta_N^T \mathbf{M} \delta_2 & \ldots & \delta_N^T \mathbf{M} \delta_N \end{bmatrix} = \mathbf{I} \end{aligned} \tag{2.53}$$

while multiplying the stiffness matrix by Δ from both sides we get

$$\Delta^T \mathbf{K_s} \Delta = \left[\delta_1\ \delta_2\ \delta_3\ \dots\ \delta_N\right]^T\ \mathbf{K_s}\left[\delta_1\delta_2\delta_3\dots\delta_N\right]$$
$$= \begin{bmatrix} \delta_1^T \mathbf{K_s}\delta_1 & \delta_1^T \mathbf{K_s}\delta_2 & \dots & \delta_1^T \mathbf{K_s}\delta_N \\ \delta_2^T \mathbf{K_s}\delta_1 & \delta_2^T \mathbf{K_s}\delta_2 & \dots & \delta_2^T \mathbf{K_s}\delta_N \\ \vdots & \ddots & \dots & \vdots \\ \delta_N^T \mathbf{K_s}\delta_1 & \delta_N^T \mathbf{K_s}\delta_2 & \dots & \delta_N^T \mathbf{K_s}\delta_N \end{bmatrix} = \Lambda \tag{2.54}$$

Keeping in mind the orthogonality properties of the modal matrices, we can introduce a coordinate transformation, which changes the original displacement coordinates into the so-called *modal coordinates* or *modal participation factors*:

$$\mathbf{q} = \Delta\xi \tag{2.55}$$

This transformation may be interpreted in a way that we take the vibration amplitude in physical coordinates as a linear sum of modal shapes. The coordinate ξ is also called the modal participation factor, because it determines how each mode participates in the final vibration. Mathematically this is:

$$\mathbf{q} = \xi_1\delta_1 + \xi_2\delta_2 + \xi_3\delta_3 + \dots + \xi_i\delta_i + \dots + \xi_N\delta_N \tag{2.56}$$

We can use (2.55) substitute for $\mathbf{q}$ in the matrix equation of motion for free, undamped systems to get [22, 38]:

$$\mathbf{M}\Delta\ddot{\xi} + \mathbf{K_s}\Delta\xi = 0 \tag{2.57}$$

Let us now multiply the equation by Δ^T from the left to get

$$\Delta^T\mathbf{M}\Delta\ddot{\xi} + \Delta^T\mathbf{K_s}\Delta\xi = 0 \tag{2.58}$$

Using the orthogonality properties introduced in (2.53) and (2.54) we can simplify this equation to get

$$\ddot{\xi} + \Lambda\xi = 0 \tag{2.59}$$

Instead of having a large coupled multiple degree of freedom, this decomposes the original system into a set of several single degree of freedom systems [22]:

$$\xi_i + \omega_i^2\xi_i = 0 \tag{2.60}$$

where ξ_i are the individual modal participation factors associated with the given mass m_i and ω_i is the angular natural frequency.

Solutions for the free and forced vibration for both damped and undamped systems can be developed using similar methods.

2.5 Distributed Parameter Systems

In practice, the vibration of continuously distributed parameter systems is solved and analyzed through the finite element method. As in the case of other fields of science, in FEM vibration analysis the continuous structure and its geometry are discretized into finite portions, the *elements*. The continuously distributed structure is then considered a large lumped parameter system with hundreds, thousands and even millions of degrees of freedom. If one aims to perform a modal analysis on a continuous system with complex geometry, the FEM software first creates a discretized version of the original structure. The equation of motion is then expressed in the matrix form of (2.43) and then the eigenvalue problem is solved. The solution of the eigenvalue problem with large matrices is not a trivial task, fortunately numerical mathematics have provided us with tools to speed up this process.

2.5.1 *Exact Solution*

Exact analytical solution for the vibration response of distributed parameter systems is complex and it is worked out only for certain shapes with primitive geometry, such as beams, bars, plates or discs. As we will be using a clamped cantilever beam with a fixed and free end in our examples, the foundations and basics for formulating the exact solution of this problem will be introduced here.

Let us therefore consider a clamped cantilever beam according to Fig. 2.6. The beam is vibrating in a transversal fashion, up and down in the y direction. The vibration is predominant in the y direction and we will neglect vibration in the x direction. The vibration in the direction perpendicular to the beam length is commonly referred to as *flexural* or *transverse* vibration. We may separate the beam to create infinitesimal slices dx. The position of this element is denoted by the coordinate x. The beam is made of the homogeneous material with the density ρ, its constant cross section is given by A, Young's modulus by E and its second moment of area by I.

2.5.1.1 Equation of Motion for the Transversal Movement of a Beam

Figure 2.7 denotes a separate element of this beam. The beam element has a centerline known as the neutral axis marked by a dashed line. The horizontal position of the beginning of the center line is marked by the coordinate x, while the vertical is marked by y. As it has been noted before, the width of this slice is dx. There are two types of force effects acting on the right side of this slice: a shear force T and a bending moment M. The left size is subjected to an infinitesimally larger shear force and bending moment.

As the element is vibrating in the transversal direction, its vertical position is expressed by y. There is also rotation involved in the element, which can be described by $\phi(x, t)$ or [10]:

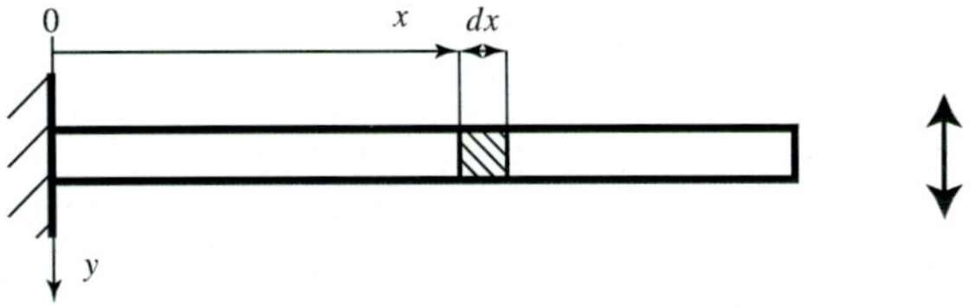

Fig. 2.6 Schematic representation of a clamped beam under transversal vibration

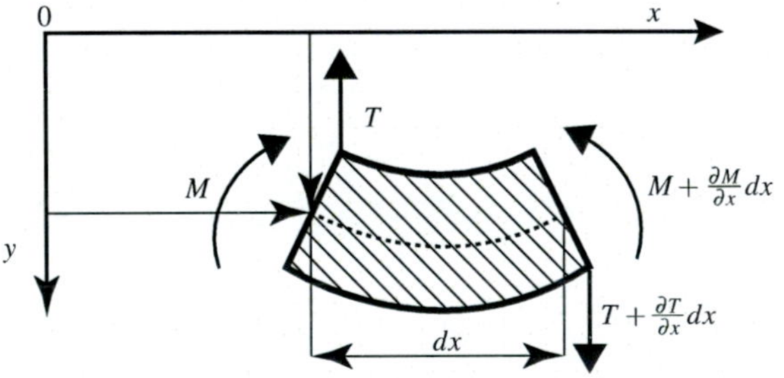

Fig. 2.7 Forces and moments acting on the infinitesimally small portion of the clamped beam under transversal vibration

$$\theta(x,t) = \frac{dy(x,t)}{dx} \tag{2.61}$$

therefore both the position of the element and its rotation is expressed by just one coordinate and its derivative with respect to x. Let us now express the second moment of inertia I of the element dx defined in the plane perpendicular to the direction of the transversal motion using the polar moment of inertia defined at the neutral axis:

$$I = \rho J \tag{2.62}$$

If we look at the forces and moments acting on the element, we can write the equation of motion for the beam as [22, 10, 51, 52]:

$$\rho A \frac{\partial^2 y(x,t)}{\partial t^2} = -T + T + \frac{\partial T}{\partial x}dx \tag{2.63}$$

$$\rho J \frac{\partial^2}{\partial t^2}\left(\frac{\partial y(x,t)}{\partial x}\right) = T\frac{dx}{2} + M + \left(T + \frac{\partial T}{\partial x}dx\right)\frac{dx}{2} - M - \frac{\partial M}{\partial x}dx \tag{2.64}$$

where the first equation describes the effects of the shear force T and the second equation describes the moment effects. If we take a close look at the first equation, the term on the left is nothing more than the mass of the element (ρA) multiplied by the transversal acceleration expressed as the second derivative of the y coordinate, thus creating an inertial force. The right side of the equation is merely a sum of shear forces acting on both sides of this element. The second equation is very similar, with the inertial moment on the left and the sum of all moments acting on the elements on the right.

We may discard some of the second order terms that do not contribute to the solution significantly and after rearranging the equations we finally get [21, 52]:

$$\rho A \frac{\partial^2 y(x,t)}{\partial t^2} = \frac{\partial T}{\partial x} \tag{2.65}$$

$$\rho J \frac{\partial^3 y(x,t)}{\partial t^2 \partial x} = T - \frac{\partial M}{\partial x} \tag{2.66}$$

2.5.1.2 Simplifying the Equation of Motion

It is possible to collect these two equations into one, by expressing T from the second equation (2.66) and substituting it back into the first one after differentiation. We will get one equation of motion given by

$$\rho A \frac{\partial^2 y(x,t)}{\partial t^2} = \frac{\partial^2 M}{\partial x^2} + \rho J \frac{\partial}{\partial x}\left(\frac{\partial^3 y(x,t)}{\partial t^2 \partial x}\right) \tag{2.67}$$

In statics, the curvature r of the deflection curve marked by the dotted line in the middle of Fig. 2.7 can be expressed by:

$$\frac{1}{r} \approx \frac{d^2 y}{dx^2} = -\frac{M}{EJ} \tag{2.68}$$

We can express M from this and substitute it back into our simplified equation of motion in (2.67) so we get [22]:

$$\rho A \frac{\partial^2 y(x,t)}{\partial t^2} = -EJ \frac{\partial^2}{\partial x^2}\left(\frac{\partial^2 y(x,t)}{\partial t^2}\right) + \rho J \frac{\partial}{\partial x}\left(\frac{\partial^3 y(x,t)}{\partial t^2 \partial x}\right) \tag{2.69}$$

This is the equation of motion for a beam vibrating in the transversal direction. To simplify notation, let us mark the time differentiation of $y(x,t)$ with respect to t by dots as in $\dot{y}$ and the position differentiation with respect to x by Roman numerals as in y^{ii}. If we ignore the effects of rotational inertia, we may denote the simplified equation motion for the free transversal vibration of a beam with constant cross section by [4, 52]:

$$\rho A \ddot{y} + EJ y^{iv} = 0 \tag{2.70}$$

We may further simplify this by dividing the whole equation by ρA and introducing

$$c = \frac{EJ}{\rho A} \tag{2.71}$$

where c is a constant[4] encompassing the square of the longitudinal wave and the square of the radius of quadratic moment of inertia. We finally arrive at the following equation of motion [3, 10, 21]:

---

[4] Certain literature divides this constant to a $c = c_0^2 i^2$, where $c_0 = \frac{E}{\rho}$ is the speed of the longitudinal wave and i is the radius of quadratic moment of inertia given by $i = \frac{J}{A}$.

$$\ddot{y} + c^2 y^{iv} = 0 \tag{2.72}$$

This equation expresses the free transversal vibration of a beam, neglecting the dynamic effects of the longitudinal forces and rotational inertia. Clearly, there is a lot of simplification assumed in this representation. If the above equation of motion would also include the effects of the rotational inertia, it would be according to Rayleigh's beam theorem. On the other hand, if it would include both the effects of the rotational inertia and the longitudinal forces, it would be Timoshenko's beam theorem [52]. The equation presented here is thus a special version of the Timoshenko beam theory also called the Euler–Bernoulli beam equation—or the classical beam theory.

2.5.1.3 Solving the Equation of Motion

The equation of motion in (2.72) merely gives a simplified representation of beam dynamics. The solution depends on the problem, as we also have to introduce boundary conditions and initial conditions. As the position differentiation of y is of the fourth degree, we can have four types of boundary conditions at the beginning and at the end [51]:

$$\begin{aligned} y(0,t) &= \Xi_1(t) & y(l,t) &= \Xi_5(t) \\ y^i(0,t) &= \Xi_2(t) & y^i(l,t) &= \Xi_6(t) \\ y^{ii}(0,t) &= \Xi_3(t) & y^{ii}(l,t) &= \Xi_7(t) \\ y^{iii}(0,t) &= \Xi_4(t) & y^{iii}(l,t) &= \Xi_8(t) \end{aligned} \tag{2.73}$$

These boundary conditions express the position of the beam (and its derivatives) at any given time. In more practical terms, the zeroth derivation is a deflection position, while the first is the angle of the tangent line to the neutral axis. Moreover, the second and third derivatives can be expressed using the moment and shear force as:

$$y^{ii}(0,t) = -\frac{M}{EJ} \tag{2.74}$$

$$y^{iii}(0,t) = -\frac{T}{EJ} \tag{2.75}$$

In addition to the boundary condition, we also have initial conditions, expressing geometrical configuration at zero time:

$$y(x,0) = \Psi_1(x) \quad \dot{y}(x,0) = \Psi_2(x) \tag{2.76}$$

As our main interest is vibration dynamics, let us assume that we want to find the resonant frequencies and mode shapes of a beam under free transversal vibration. Furthermore, let us now expect to arrive at a solution in the following form [21, 22]:

$$y(x,t) = a(x)V(t) \tag{2.77}$$

meaning that the position at a given place and time is composed of a combination of function $a(x)$, which is only dependent on the horizontal position and $V(t)$ which is only dependent on time. We may substitute this expected form back to (2.72) and get a new equation of motion:

$$\frac{a^{iv}}{a} = -c\frac{\ddot{V}}{V} \tag{2.78}$$

Since the right-hand side of the equation is only a function of time and the left-hand side of the equation is only the function of position x, each side of the equation must be equal to a constant. Let us call this constant Ω^2, which will help us separate the partial differential equation in (2.78) into two ordinary differential equations [22]:

$$a^{iv} - \frac{\Omega^2}{c}a = 0 \tag{2.79}$$

$$\ddot{V} + \Omega^2 V = 0 \tag{2.80}$$

In order to keep the notation simple, let us introduce the new constant υ^4 as:

$$\upsilon^4 = \frac{\Omega^2}{c} \tag{2.81}$$

Now we can expect the solution of Eq. (2.79) in the following form, with a constant A and an exponential term:

$$a(x) = Ae^{rx} \tag{2.82}$$

Substituting this back into (2.79) will yield a characteristic equation with the following roots:

$$r_{1,2} = \pm\upsilon \quad r_{3,4} = \pm j\upsilon \tag{2.83}$$

The solution of (2.79) will now assume the form [10, 22]:

$$a(x) = A_1e^{\upsilon x} + A_2e^{-\upsilon x} + A_3e^{j\upsilon x} + A_3e^{-j\upsilon x} \tag{2.84}$$

Utilizing the well-known Euler's formula establishing a relationship between trigonometric and complex exponential functions

$$e^{\pm j\upsilon x} = \cos \upsilon x \pm j \sin \upsilon x \tag{2.85}$$

$$e^{\pm \upsilon x} = \cosh \upsilon x \pm \sinh \upsilon x \tag{2.86}$$

and substituting this back into (2.84) we will get the solution in the following form [4, 10]:

$$a(x) = C_1 \cosh \upsilon x + C_2 \sinh \upsilon x + C_3 \cos \upsilon x + C_4 \sin \upsilon x \tag{2.87}$$

Constants C_1, C_2, C_3 and C_4 are integration constants and can be uniquely determined from the boundary conditions.

As different beam setups have different boundary conditions, we will pick a clamped cantilever beam, which is fixed at one end and free to vibrate at the other, as an example. For this clamped cantilever beam, the boundary conditions are given by [10, 21]:

$$\begin{aligned} y(0,t) &= 0 \qquad y^{ii}(l,t) = 0 \\ y^{i}(0,t) &= 0 \qquad y^{iii}(l,t) = 0 \end{aligned} \tag{2.88}$$

or, in other words, the beam cannot move at the clamped end and there are no shear forces or moments at the free end. Substituting these boundary conditions into (2.87), we will get a set of equations for the integration constants [51, 52]:

$$C_1 + C_3 = 0 \tag{2.89}$$

$$\upsilon(C_2 + C_4) = 0 \tag{2.90}$$

$$\upsilon^2(C_1 \cosh \upsilon x + C_2 \sinh \upsilon x - C_3 \cos \upsilon x - C_3 \sin \upsilon x) = 0 \tag{2.91}$$

$$\upsilon^2(C_1 \cosh \upsilon x + C_2 \sinh \upsilon x + C_3 \cos \upsilon x - C_3 \sin \upsilon x) = 0 \tag{2.92}$$

where l is the overall length of the beam implied by the boundary conditions. Moreover, for a nonzero υ the first two equations also imply that

$$C_3 = -C_1 \qquad C_4 = C_2 \tag{2.94}$$

Substituting for C_3 and C_4 in the remaining two equations yields a set of two homogeneous equations with C_1 and C_2 as unknowns:

$$(\cosh \upsilon l + \cos \upsilon l)C_1 + (\sinh \upsilon l + \sin \upsilon l)C_2 = 0 \tag{2.95}$$

$$(\sinh \upsilon l - \sin \upsilon l)C_1 + (\cosh \upsilon l + \cos \upsilon l)C_2 = 0 \tag{2.96}$$

In order for Eq. (2.95) to have nontrivial solutions, its determinant has to equal zero [21, 51]. Computing this will yield the following frequency equation [21]:

$$\cos \upsilon l \cosh \upsilon l + 1 = 0 \tag{2.97}$$

The resonant frequencies of the beam can be calculated based on (2.81) by computing the roots of (2.97) and substituting into [22]:

$$\Omega_n = (\upsilon_n l)^2 \sqrt{\frac{EJ}{\rho A l^4}} \tag{2.98}$$

The equations given by (2.89) will not make it possible to compute the integrating constants, though it is possible to compute their ratios and substitute it back into (2.87) to get $a_n(x)$. The general solution describing the mode shapes of vibration will be finally given by [22, 52]:

$$y(x,t) = \sum_{n=1}^{\infty} (A_n \cos \Omega_n t + B_n \sin \Omega_n t) a_n x \tag{2.99}$$

where the integration constants A_n and B_n are given by

$$A_n = \frac{2}{l} \int_0^l \Psi_1(x) a_n(x) dx \tag{2.100}$$

$$B_n = \frac{2}{l\Omega_n} \int_0^l \Psi_2(x) a_n(x) dx \tag{2.101}$$

The process of obtaining the resonant frequency and mode shapes for other types of distributed parameter systems with simple geometry is analogous to the above introduced process. The interested reader is kindly referred to works discussing this topic in more depth [10, 21]. It is easy to see that working out a solution is a fairly time-consuming and complicated process, even for systems with simple geometry. This is why most practitioners prefer to utilize finite element analysis or experimental procedures to assess the vibration properties of such systems.

2.5.2 Damping in Distributed Systems Simulated by FEA

The damping of distributed mechanical systems is a very complex phenomenon. Unfortunately, energy dissipation in materials is not entirely explored by science at present. One of the simplest methods to approximate damping is to use the so-called Rayleigh damping which is often utilized in FEM simulations. This involves calculating the damping matrix as a sum of the mass and stiffness matrices, multiplied by the damping constants α and β:

$$\mathbf{B_d} = \alpha \mathbf{M} + \beta \mathbf{K_s} \tag{2.102}$$

The damping constants α and β are not known directly, instead they are calculated from the modal damping ratios ζ_i. This is actually the ratio of actual damping to critical damping for a particular mode of vibration. In case ω_i is the natural circular frequency for a given mode of vibration, we can calculate the constants utilizing [51, 52]:

$$\zeta_i = \frac{\alpha}{2\omega_i} + \frac{\beta\omega_i}{2} \tag{2.103}$$

It is often assumed that α and β are constant over a range of frequencies. For a given ζ_i and a frequency range, two simultaneous equations may be solved to obtain the Rayleigh damping constants. In most real-life structural applications mass damping can be neglected, therefore setting constant $\alpha = 0$. In these cases, β may be evaluated from the known values of ζ_i and ω_i:

$$\beta = \frac{2\zeta_i}{\omega_i} \tag{2.104}$$

2.6 Creating Models for Vibration Control

Different control systems call for different models and representations. There are numerous popular methods suitable to create a mathematical representation of the real system such as transfer functions or the state-space representation. We may choose to begin with a completely analytical fashion, and build our model based on the exact underlying physics. A mathematical model created on such first principle models using the underlying physics and equations is known as a white-box model. Models of vibrating structures are derived on a phenomenological basis for example in [26, 36, 48].

Active vibration control often utilizes advanced materials bonded into the structure. Moreover, the materials can have a coupled nature, having an intense electromechanical interaction with the structure—for example piezoceramics. If the underlying model is too complicated or it is too formidable to create a first principles model we may use experimental identification. If one is aware of the structure of the underlying physics, they may choose a given model and fit its parameters on an experimental measurement. This is known as a grey-box model. If nothing is known about the physics or one does not choose a model with a fixed structure, black-box identification is carried out.

In active vibration control there is an outside energy source acting on our vibrating mechanical system, such as a piezoelectric actuator. The model representing the behavior of the system therefore must represent forced vibration, regardless of whether it is created based on first principles or experimental methods.

2.6.1 Transfer Function Models

We will introduce a transfer function representation based on the dynamics analysis presented previously in Sect. 2.3 on the forced vibration of a point mass. Recall that the equation of motion of a point mass system with damping and a generic outside force is

$$m\ddot{q}(t) + b\dot{q}(t) + kq(t) = f_e \tag{2.105}$$

Now in order to create a transfer function from this, we must apply Laplace transformation to our equation. Let us denote the Laplace operator with s. The position coordinate $q(t)$ will be now denoted instead with $Q(s)$ and our external force f_et with $F_e(s)$. Differentiation in the Laplace domain is simply multiplication with the Laplace operator, therefore we will get [22]:

$$ms^2 Q(s) + bsQ(s) + kQ(s) = F_e(s) \tag{2.106}$$

The transfer function $H(s)$ will be simply a ratio of the Laplace transform of the output divided by the Laplace transform of the input. In this case, we may state that our transfer function expressing the effect to a force input is simply

$$H(s) = \frac{\mathscr{L}\{q(t)\}}{\mathscr{L}\{f_e(t)\}} = \frac{Q(s)}{F_e(s)} = \frac{1}{ms^2 + bs + k} \tag{2.107}$$

In the Laplace domain, we can compute the output of the mass-spring-damper system by the linear combination of the transfer function and an arbitrary force signal by evaluating

$$Q(s) = H(s)F(s) \tag{2.108}$$

Another valid representation of the transfer function is given in the term of the natural frequency ω_n and the proportional damping ζ [18]. For this, we divide the transfer function in (2.107) by the mass to get

$$H(s) = \frac{\frac{1}{m}}{s^2 + 2\zeta\omega_n s + \omega_n^2} \tag{2.109}$$

where $c/m = 2\zeta\omega_n$ and ζ is proportional damping given as a percentage of critical damping.

2.6.1.1 Multiple DOF Transfer Functions

Lumped parameter systems with several vibrating masses or equivalent representations of distributed systems can be expressed using a transfer function representation as well. Let us consider a matrix representation of the forced equation of motion for an N degrees of freedom system as in (2.43) and perform a Laplace transform on the equation in the following sense:

$$\mathbf{M}\mathbf{s}^2\mathscr{Q} + \mathbf{B_d}\mathbf{s}^2\mathscr{Q} + \mathbf{K_s}\mathscr{Q} = \mathbf{F_e} \tag{2.110}$$

where $\mathbf{s}\mathscr{Q}$ denotes an n elements long vector of Laplace operators and output displacements, $\mathbf{s}^2\mathscr{Q}$ expresses the elementwise square of the Laplace operators multiplied by the output displacements $\mathscr{Q}$[5] in the Laplace domain. After the transformation and rearranging the terms, the equation of motion will be given in a matrix form in the s-domain:

$$\mathscr{H}(s)\mathscr{Q} = \mathbf{F_e} \tag{2.111}$$

where $\mathscr{H}(s)$ is a matrix of partial dynamics contributions. To get the transfer function for particular input and output points, one must rearrange this term. It is highly recommended to use computer symbolic algebra to arrive at the final transfer functions given large systems with many DOFs [18]. The transfer function representation of systems with several inputs or outputs (MIMO, SIMO, MISO) is not common, most engineering practitioners prefer to use the state-space representation to model the behavior of more complex systems.

2.6.1.2 Poles and Zeros

From a control viewpoint, we may further analyze the transfer function representation in (2.109), to see where the poles and zeros of lightly damped systems are located and what the physical interpretation of the poles and zeros of the transfer functions is. For this, let us assume that the proportional damping is approaching zero $\zeta \to 0$ to get

$$H(s) = \frac{\frac{1}{m}}{s^2 + \omega_n^2} \tag{2.112}$$

As it is usual in control, the denominator of the transfer function is extremely important. By setting the denominator equal to zero, we will get the *characteristic equation*. The roots of this equation are the so-called *poles* of transfer function, which amongst others affect the dynamic behavior and stability of the system. The poles represent the resonant frequencies of the vibrating system, in other words, the frequency values at which the inputs will be amplified. The characteristic equation of the vibrating point mass is

$$s^2 + \omega_n^2 = 0 \tag{2.113}$$

from which we may calculate the poles

$$s = \pm j\omega_n \tag{2.114}$$

It is clear that the poles of an undamped system will be located only on the imaginary axis (no real part) on complex plane. Similarly, the poles for a lightly damped system will be located close to the imaginary axis with very small negative real parts.

---

[5] In order to avoid confusion with **Q** used in later chapters as *input penalty*, the multiple DOF displacements transformed into the Laplace domain are denoted as $\mathscr{Q}$ here.

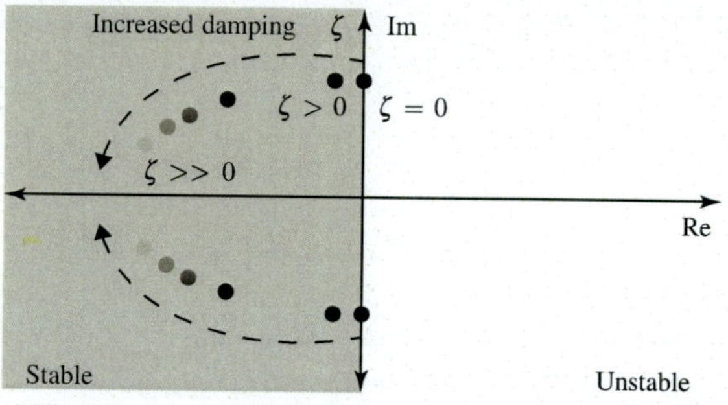

Fig. 2.8 Location of the poles for the vibrating point mass system depending on the amount of damping

Vibrating mechanical systems are inherently open-loop stable, therefore the poles of the transfer function must have negative real parts. Poles, and therefore the resonant frequency of a system, only depend on the distribution of mass, damping and stiffness and are independent on the excitation point or the measurement of the output [18].

Setting the numerator of a transfer function equal to zero and calculating roots will yield the *zeros* of the system. Zeros of the transfer function do depend on the point and type of excitation and the output. Some transfer functions may not even have zeros. In a physical interpretation, at zeros the vibrating system will attenuate the input. Similar to the poles, the zeros would also have no real part in the absence of damping. A pair of poles located at zero indicates a so-called *rigid body mode* [18]. A rigid body mode is the case when in the presence of a static or low frequency outside force the whole mechanical system moves without moving the parts relative to each other. In the case of the two vibrating masses example presented earlier in Fig. 2.5 a rigid body mode would occur if none of the masses were connected to a ground, but just to each other. At low frequencies or quasi-static forces, the two masses would move as one without considering the spring and damper in between them.

The location of the poles in the complex plane for the continuous transfer function of a simple second order vibrating system is demonstrated in Fig. 2.8. The left half of the plane is stable; this is where open-loop physical vibrating system poles are located. Without damping, the poles are on the imaginary axis, on the verge of stability (*right*). Lightly damped systems have poles very close to the imaginary axis (*middle*) while a more considerable amount of proportional damping would place the poles further away (*left*). Generally speaking, this is also true for the zeros of the system. The poles tend to be further away from each other in a lightly damped system, while increasing damping values will bring them closer together in the direction of the imaginary axis.

2.6.1.3 Discrete Transfer Functions

As digital control systems work in a sampled fashion instead of a continuous-time transfer function, we often need to use a discrete transfer function. In a discrete-time system, the transfer function is created using the *Z-transform*, which is an analogy to the Laplace transform. For our case, a discrete time transfer function would be defined by:

$$H(z) = \frac{\mathscr{Z}\{q(kT)\}}{\mathscr{Z}\{f_e(kT)\}} = \frac{Q(z)}{F_e(z)} \tag{2.115}$$

A continuous time transfer function may be converted into discrete time using various algorithms using zero or first order hold to extract the function values in the time domain. The definition of the Z-transform essentially sums the pairs of the function values multiplied by a corresponding delay operator z^{-k}. Numerous numerical algorithms implement this procedure very effectively. Another example of converting a continuous transfer function into discrete is the so-called *bilinear transformation*. If T is the sampling time, to convert from continuous into discrete known as the Tustin transformation we can use: A continuous time transfer function may be converted into discrete time using various algorithms using zero or first order hold to extract the function values in the time domain. The definition of the Z-transform essentially sums the pairs of the function values multiplied by a corresponding delay operator z^{-k}. Numerous numerical algorithms implement this procedure very effectively. Another example of converting a continuous transfer function into discrete is the so-called *bilinear transformation*. If T is the sampling time, to convert from continuous into discrete known as the Tustin transformation we can use:

$$s = \frac{2}{T}\frac{z-1}{z+1} \tag{2.116}$$

while the reverse operation into convert from discrete to continuous is

$$z = \frac{2+sT}{2-sT} \tag{2.117}$$

Software tools such as for example Matlab also aid the conversion between continuous and discrete time models. In Matlab, this can be carried out using the *c2d* command.

A transfer function representation can be used to design or tune a controller for the vibrating system. A common approach used in practice is to utilize a software tool such as Matlab or Simulink to simulate the response of a system and the controller.

2.6.1.4 Experimental Identification

We may create a transfer function following a dynamical analysis and fill in the parameters such as m, b and k from direct measurements and consider them as physical quantities. However, in experimental identification, we may choose a given generic form of a transfer function and not expect to use parameters, which have a physical sense. Instead, we can use system identification algorithms, which compute these parameters so that a test signal input matches a measured output. This makes sense for example when we would like to replace the dynamics of a continuous clamped beam with a single mass-spring-damper model. We cannot directly measure these quantities, rather we should find a generic second order transfer function that properly describes the first resonant mode and is given by

$$H(s) = \frac{\mathscr{L}\{q(t)\}}{\mathscr{L}\{f_e(t)\}} = \frac{Q(s)}{F_e(s)} = \frac{1}{a_1 s^2 + a_2 s + a_3} \tag{2.118}$$

where parameters a_1, a_2 and a_3 are found experimentally using an identification algorithm.

2.6.1.5 State-Space Representation

A *state-space* representation of a dynamic system may be perceived as a way to decouple high order differential equations into simpler low order equivalents, collected in a matrix representation. The set of first order differential equations then relates the inputs and outputs of a system by the aid of an intermediate variable called the *state*. If for example a single DOF vibrating system is described by a second order differential equation, we may use a state-space equivalent consisting of two first order differential equations. Analogously, if an MDOF vibrating system has its modes decomposed according to the guidelines presented previously in Sect. 2.4.2, then the behavior of each mode is described by N second order differential equations, which may be changed into $2N$ first order equations in the state-space representation. Most modern optimization-based control algorithms such as MPC utilize this type of mathematical model, and the numerical simulation of time responses is more convenient as well. To demonstrate the state-space representation of dynamic systems, we may divide the second order differential equation describing the forced response of the spring-mass-damper system in (2.34) into two first order ones.

2.6.1.6 State-Space Representation for a Single DOF System

Let us begin the rewriting of our second order differential equation by leaving the inertial force at the left side of our equation of motion (2.34) and moving the rest to the right which will result in:

$$m\ddot{q}(t) = -b\dot{q}(t) - kq(t) + f_e \tag{2.119}$$

while dividing the whole by m is

$$\ddot{q}(t) = -\frac{b}{m}\dot{q}(t) - \frac{k}{m}q(t) + \frac{1}{m}f_e \tag{2.120}$$

Because our original ODE is second order, let us choose two new variables $x_1(t)$ and $x_2(t)$. These two variables are the so-called state variables and we may combine them to create a column state vector $\mathbf{x}(t)$. We may also choose that $x_1(t)$ is equivalent to our original position variable q, while $x_2(t)$ is simply a differentiation of it.

$$\mathbf{x}(t) = \begin{bmatrix} x_1(t) \\ x_2(t) \end{bmatrix} = \begin{bmatrix} q(t) \\ \dot{q}(t) \end{bmatrix} \tag{2.121}$$

Let us now write a set of two first order ordinary differential equations with the help of our state variables $x_1(t)$ and $x_2(t)$. Substitute these into the transformed equation of motion and simply state that $\dot{x}_1(t) = x_2(t)$ [5]:

$$\dot{x}_1(t) = x_2(t) \tag{2.122}$$

$$\dot{x}_2(t) = -\frac{b}{m}x_2(t) - \frac{k}{m}x_1(t) + \frac{1}{m}f_e \tag{2.123}$$

which can be equivalently written in the following matrix form [5, 10]:

$$\begin{bmatrix} \dot{x}_1(t) \\ \dot{x}_2(t) \end{bmatrix} = \begin{bmatrix} 0 & 1 \\ -\frac{b}{m} & -\frac{k}{m} \end{bmatrix} \begin{bmatrix} x_1(t) \\ x_2(t) \end{bmatrix} + \begin{bmatrix} 0 \\ \frac{1}{m} \end{bmatrix} \begin{bmatrix} 0 & f_e \end{bmatrix} \tag{2.124}$$

We may use a more compact matrix notation for the state-space variables $\mathbf{x}(t)$; moreover, let us define an input vector $\mathbf{u}(t)$ which contains the outside excitation force f_e as its first element: $\mathbf{u} = \begin{bmatrix} f_e & 0 \end{bmatrix}$. According to this, the matrix form of the state-space equation can be written as follows [5, 10]:

$$\dot{\mathbf{x}}(t) = \mathbf{A}\mathbf{x}(t) + \mathbf{B}\mathbf{u}(t) \tag{2.125}$$

$$\mathbf{y}(t) = \mathbf{C}\mathbf{x}(t) \tag{2.126}$$

where $\mathbf{A}$ is the state transition matrix, $\mathbf{B}$ is the input matrix and $\mathbf{C}$ is the output matrix. Figure 2.9 illustrates the matrix block algebra of the continuous state-space dynamic system representation. The output vector $\mathbf{y}$(t) may contain both the position and velocity elements of the state or one of them according to the output matrix $\mathbf{C}$. For example a valid choice of $\mathbf{C}$ is $\mathbf{C} = \begin{bmatrix} 1 & 0 \end{bmatrix}$ which would result in a scalar displacement output.

$$\mathbf{y}(t) = q(t) = \begin{bmatrix} 1 & 0 \end{bmatrix} \mathbf{x}(t) = \begin{bmatrix} 1 & 0 \end{bmatrix} \begin{bmatrix} x_1(t) \\ x_2(t) \end{bmatrix} = \begin{bmatrix} 1 & 0 \end{bmatrix} \begin{bmatrix} q(t) \\ \dot{q}(t) \end{bmatrix} \tag{2.127}$$

2.6.1.7 Discrete State-Space Representation

In the discrete time equivalent of the state-space equation we replace the continuous time variable t by its sampled analogy $t = kT$, where T is the sample time:

$$\mathbf{x}(k+1) = \mathbf{A}\mathbf{x}(k) + \mathbf{B}\mathbf{u}(k) \tag{2.128}$$

$$\mathbf{y}(k) = \mathbf{C}\mathbf{x}(k) \tag{2.129}$$

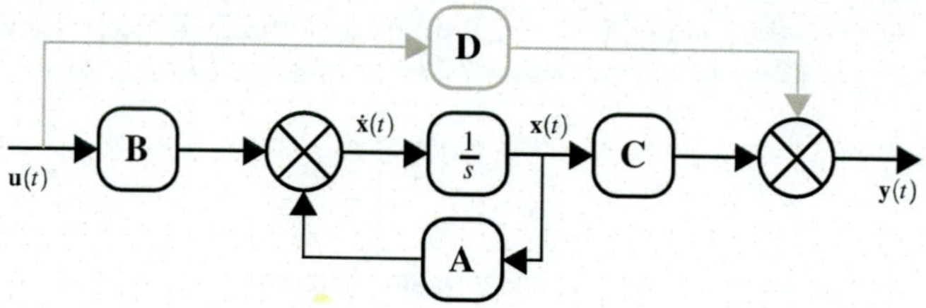

Fig. 2.9 Matrix block algebra of the continuous state-space dynamic system representation

Note that the contents of the matrices **A**, **B** and **C** also change if a system representation is converted from continuous into discrete. The Matlab command *c2d* is useful for the state-space representation as well.

Just as in the case of transfer functions, we may choose to apply an analytic approach and substitute for model parameters in a physical sense. Replacing complex vibrating systems with a mass-spring-damper equivalent, however, calls for parameters, which do not necessarily reflect physical quantities. In that case, we just choose our model, for example a second order state-space representation:

$$\dot{\mathbf{x}}(t) = \begin{bmatrix} a_{11} & a_{12} \\ a_{21} & a_{22} \end{bmatrix} \mathbf{x}(t) + \begin{bmatrix} b_1 \\ b_2 \end{bmatrix} \mathbf{f}_e \tag{2.130}$$

$$\mathbf{y}(t) = \begin{bmatrix} c_1 & c_2 \end{bmatrix} \mathbf{x}(t) \tag{2.131}$$

and fill in the parameters using experimental identification procedures, based on test data.

2.6.1.8 State-Space Representation of Multiple DOF Systems

The previously introduced approach illustrated the process of obtaining a second order state-space model for a single degree of freedom system. As shown earlier, the equation of motion for multiple degree of freedom systems can be compactly written as:

$$\mathbf{M}\ddot{\mathbf{q}} + \mathbf{B}\dot{\mathbf{q}} + \mathbf{K_s}\mathbf{q} = \mathbf{f_e} \tag{2.132}$$

The order of systems expressed by this compact matrix notation can range from lumped parameter systems with model orders under ten, to several thousands and even millions for finite element models after discretization. To obtain a full order state-space model from the equation of motion in (2.132), at first one needs to transfer the terms $\mathbf{B}\dot{\mathbf{q}} + \mathbf{K_s}\mathbf{q}$ to the right side, while leaving $\mathbf{M}\ddot{\mathbf{q}}$:

$$\mathbf{M}\ddot{\mathbf{q}} = -\mathbf{B}\dot{\mathbf{q}} - \mathbf{K_s}\mathbf{q} + \mathbf{f_e} \tag{2.133}$$

The equation of motion is then multiplied by the inverse of the mass matrix $\mathbf{M}^{-1}$ according to [10, 21]:

$$\ddot{\mathbf{q}} = -\mathbf{M}^{-1}\mathbf{B}\dot{\mathbf{q}} - \mathbf{M}^{-1}\mathbf{K_s}\mathbf{q} + \mathbf{M}^{-1}\mathbf{f_e} \tag{2.134}$$

In the next step, we will re-define the position output $\mathbf{q}$ and the outside excitation force $\mathbf{f_e}$ in terms of state variables and input according to:

$$\begin{bmatrix} \mathbf{x} \\ \dot{\mathbf{x}} \end{bmatrix} = \mathbf{q} \quad \mathbf{u} = \mathbf{f_e} \tag{2.135}$$

Substituting our new state variables and input into the modified equation of motion in (2.134) will result in:

$$\ddot{\mathbf{x}} = -\mathbf{M}^{-1}\mathbf{B}\dot{\mathbf{x}} - \mathbf{M}^{-1}\mathbf{K_s}\mathbf{x} + \mathbf{M}^{-1}\mathbf{u} \tag{2.136}$$

which may be equivalently stated using the matrix notation

$$\begin{bmatrix} \dot{\mathbf{x}} \\ \ddot{\mathbf{x}} \end{bmatrix} = \begin{bmatrix} -\mathbf{M}^{-1}\mathbf{B} \\ -\mathbf{M}^{-1}\mathbf{K_s} \end{bmatrix} \begin{bmatrix} \mathbf{x} \\ \dot{\mathbf{x}} \end{bmatrix} + \mathbf{M}^{-1}\mathbf{u} \tag{2.137}$$

further, this is equivalent to stating that

$$\begin{bmatrix} \dot{\mathbf{x}} \\ \ddot{\mathbf{x}} \end{bmatrix} = \begin{bmatrix} \mathbf{0} & \mathbf{I} \\ -\mathbf{M}^{-1}\mathbf{B} & -\mathbf{M}^{-1}\mathbf{K_s} \end{bmatrix} \begin{bmatrix} \mathbf{x} \\ \dot{\mathbf{x}} \end{bmatrix} + \begin{bmatrix} \mathbf{0} \\ \mathbf{M}^{-1} \end{bmatrix} \mathbf{u} \tag{2.138}$$

where $\mathbf{I}$ is an identity matrix and $\mathbf{0}$ is a zero matrix of conforming dimensions. The state-space matrices $\mathbf{A}$ and $\mathbf{B}$ from this are:

$$\mathbf{A} = \begin{bmatrix} \mathbf{0} & \mathbf{I} \\ -\mathbf{M}^{-1}\mathbf{B} & -\mathbf{M}^{-1}\mathbf{K_s} \end{bmatrix} \quad \mathbf{B} = \begin{bmatrix} \mathbf{0} \\ \mathbf{M}^{-1} \end{bmatrix} \tag{2.139}$$

The resulting state-space system will have an order n_x, twice the DOF of the original system. In a small lumped parameter system, this will create a system model of manageable dimensions.

Typical FEM models of complex vibrating systems with ten to hundred thousand nodes can be decoupled and directly transformed into the state-space representation. Nevertheless, this results in extremely large state-space models not suitable for direct controller design. The order of these state-space systems needs to be reduced through the method of *model reduction*. State-space systems commonly used in control engineering have an order of 10–100 states, which may be even lower for the computationally intensive model predictive control approach. Although the order of the new and reduced state-space system will be severely truncated, it may still fully represent the dynamic behavior of the original system described by the much larger FEM model.

2.6.2 Experimental Identification Procedures

In case the modeling procedure of a controlled system is irrelevant or a first principles model would be too complex to develop, experimental identification may be used.

Experimental identification procedures are often utilized in order to create models for control design [7, 9, 54, 56, 63]. Control engineering uses system identification to build mathematical models of dynamical systems using statistical methods to find parameters in a set of experimental data.

2.6.2.1 Experiment Design

The quality of the identified model greatly depends on the quality of the experiment. In this relation, we have to emphasize the need for a properly designed input excitation signal. For example, in control engineering it is entirely acceptable to utilize a step change of temperature levels in a thermal related system—such as in heating. The resulting change of output carries enough information, so that an identification algorithm can extract a simple first or second order transfer function. In this case, the output levels are satisfactory and there are no resonant phenomena.

However, in the case of vibrating systems the previously considered step or impulse change at the actuators would not produce a satisfactory output signal. Due to its nature, one of the most important aspects of vibrational systems are not the precision of a static deflection after a step change in input, rather the dynamic response in and around the resonant frequency. A step or impulse input into a vibration attenuation system with bonded piezoelectrics would only result in vibrations with small amplitudes. In reality, the piezoelectric actuators may seriously affect and amplify vibration amplitudes in resonance. Therefore, it is wiser to use test signals that excite the vibrating system around the resonance frequencies of interest.

Such signals can be generated through the so-called *swept sine* or *chirp* function. The chirp signal is a type of signal with a constant amplitude, while the frequency changes from a low to a high limit in a given time. The frequency content of the signal covers the bandwidth of chirp signal—resulting in an ideally flat response in the frequency domain. An example of a chirp signal in the time domain is given in Fig. 2.10a. The frequency content of a different signal is shown in Fig. 2.10b, which indicates that the bandwidth of the chirp test signal is evenly spread out through the range of interest. A spectrogram featured in Fig. 2.10c relates the progression of time to the frequency content of the same signal. Chirp signals are commonly used to excite vibrating systems in academic works [23, 25, 28, 30, 31, 53, 57].

Another popular signal choice to excite vibrating systems in order to extract dynamic models is a pseudo-random binary signal [6, 25, 32, 39, 44, 55]. A pseudo-random signal has two levels changing in a random but repeatable fashion. The frequency content of the signal can also be influenced to concentrate on the bandwidth of interest.

It is possible to supply time-domain measurement data directly to identification software, but by transforming it using fast Fourier transform (FFT) the identification procedure may be carried out in the frequency domain as well. Therefore, if the system identification software allows it, the mathematical model may be created based on fitting it to either the time or the frequency domain data set. As the FFT transform is a unitary transform, the frequency domain data set will contain exactly the same

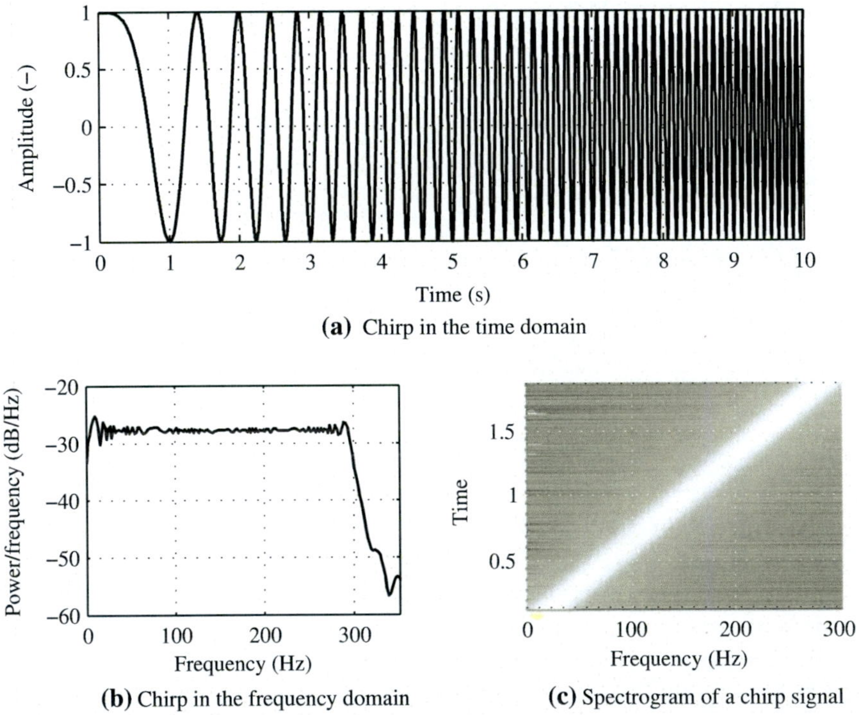

(**a**) Chirp in the time domain

(**b**) Chirp in the frequency domain

(**c**) Spectrogram of a chirp signal

Fig. 2.10 An example chirp signal with a ± 1 (-) amplitude ranging from 0 to 10 Hz in 10 s is shown plotted in the time domain in (**a**). A different signal ranging from DC to 300 Hz in 2 s is shown in the frequency domain on a periodogram in (**b**) while the spectrogram of the same signal is featured in (**c**)

amount of data as the time domain, implying the same computational load for both domains. In case one requires to identify a very large time domain data set with clear resonance peaks in the frequency domain, it is advised to use a frequency domain data with non-equidistant frequency resolution. Leaving more data points in the vicinity of the peaks of the frequency domain data while reducing the number of data points elsewhere may significantly reduce the computational load and still produce high quality models. A common practice is to take the low and high frequency of interest and divide the region in between by logarithmically spaced frequencies.

2.7 Identification via Software Packages

There are off-the-shelf solutions for identification of mathematical models based on experimental test procedures. One of the most convenient and accessible is the System Identification Toolbox [59], a part of the Matlab software suite. In addition

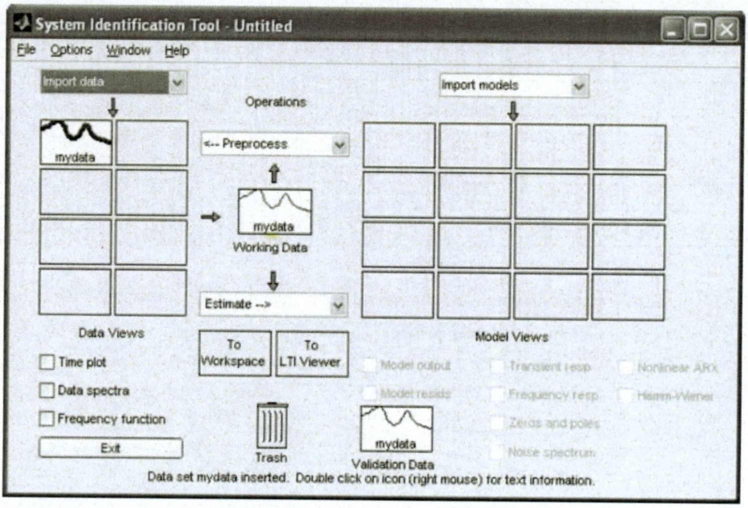

Fig. 2.11 The graphical user interface frontend of the Matlab System Identification Toolbox

to the general use, the System Identification Toolbox is also commonly used for creating models of vibrating mechanical systems [3, 14, 24, 27, 40, 47, 58, 60]. The System Identification Toolbox is largely based on the work of Ljung [29] and implements common techniques used in system identification. The toolbox aids the user to fit both linear and nonlinear models to measured data sets known as black box modeling [20]. Gray box modeling which tunes the parameters of a model with known structure is also offered by the suite. The types of available models are low order process models, transfer functions, state-space models, linear models with static nonlinearities, nonlinear autoregressive models, etc. The toolbox is made use of in the upcoming chapters of this book; see the system identification process in Sect. 5.2.

In addition to the usual Matlab command line interface, identification in the System Identification Toolbox may be carried out via the graphical user interface (GUI). A screenshot of the GUI frontend is featured in Fig. 2.11. The identification tasks are divided into separate parts. After creating an identification and validation data set, the data is pre-processed. Identification is initialized by selecting and setting up the proper model type. Finally the models can be validated using numerous techniques: comparing model response with measurement data, step response, a pole-zero plot, etc.

Transfer functions and low order process models are suitable for many controller types. It is also possible to create an MPC controller based on transfer functions, although state-space models are used more often in advanced control schemes. The MPC controller considered in this work utilizes a state-space representation as well. The aim of the identification process is therefore: given the input and output data set one needs to identify the contents of matrices **A**, **B**, **C**. The Matlab System Identification toolbox offers two estimation methods for state-space models:

- subspace identification
- iterative prediction-error minimization method (PEM).

The order of a state-space system depends on many factors and it has to be determined by the control engineer at the time of the design of the representative mathematical model. It is always favorable to use the simplest possible system, which still describes the identified phenomena on a satisfactory level. MPC can be a computationally intensive operation in real-time, one must keep this in mind when creating a state-space model. In other words, the larger the model dimensionality or order n_x is, the more time it takes to perform the MPC optimization algorithm at each time step k. Surprisingly many practicing control engineers believe that most real phenomena involving single-input single-output (SISO) control can be approximated by simple second order state-space systems. For vibrating systems, this can be true mainly when one vibration mode is dominating over the others. Good examples are lightly damped vibrating systems where the first vibration mode is much more dominant than the others. In case the response of a second order system is unsatisfactory, one needs to increase the system order and inspect whether the response characteristics improve. In controlled mechanical vibrations the order of the identified system n_x should be an even number[6] and will contain an $f_i = n_x/2$ resonant frequencies. Given that one attempts to control a nonlinear vibrating system, there are ways to describe the phenomena at certain working points by several models. This however increases both the level of complexity of the linear MPC controller and the expected computational load. The application of nonlinear MPC (NMPC) in fast sampling application such as vibration control is still under development, as both the theoretical basis of NMPC and the computational hardware requires significant improvement for practical deployment.

One shall always verify whether the model produced by the experimental identification procedure is stable. In discrete systems, the poles of the transfer function must always reside within the unit circle. In other words, the absolute value or length of the vector denoting the pole position must be smaller than one. This condition can be restated for the case of state-space systems by saying that the absolute value of the eigenvalues of the matrix **A** shall be smaller than one. In practice, for vibrating mechanical systems, the absolute value of the eigenvalues will be a number which is smaller than one, albeit not by too much. This is understandable, since if the physical vibrating system is very lightly damped its behavior closely resembles that of a marginally stable system.[7]

Other system identification software is for example the System Identification Toolbox (or ID Toolbox)[8] [35], which is named identically to the official Mathworks

[6] Unless of course one uses an augmented system model with filters, observers etc.

[7] A vibrating system without outside energy cannot be marginally stable, since that would create a system without energy dissipation and without damping.

[8] Also known as The University of Newcastle Identification Toolbox (UNIT).

supported identification tool [33, 34]. This is however a free and entirely different software supporting a wide range of standard identification techniques. The Matlab suite developed by Ninness et al. also contains support for novel system identification algorithms [16, 43, 61, 62]. Similar to the System Identification Toolbox, the SMI Toolbox [19] is based on subspace identification techniques as well. Although this package has proven its worth over the years, its development has been on halt for a very long time and is slightly outdated. SLIDENT, which is incorporated into the Fortran 77 Subroutine Library in Control Theory (SLICOT) is suited for the identification of large-scale problems [45, 46]. ITSIE (Interactive Software Tool for System Identification Education) [17] is rather suited for the education process than research work or practical engineering deployment. SOCIT developed by NASA Langley Research Center [13] uses an eigensystem realization algorithm method based on Markov parameters and singular value decomposition techniques.

2.8 FEM-Based Identification

The study of physical systems described by ordinary or partial differential equations established through mathematical–physical analysis is not always possible. The reason for this is that there may be a lack of exact analytical solution, or boundary conditions can be too complex for a realistic solution. In some cases, analytical solution is possible, but a finite element (FE) model seems to be faster and more practical.

Models based on a finite element method (FEM) analysis may be used to create simplified mathematical representations of physical systems. Transient simulation results from the commercial FEM software package ANSYS were for example identified by Dong et al. in [11]. One may choose to extract and simplify the dynamic matrices assembled by the FEM software directly, or it is also possible to perform simulation analyses to get responses ready for an experimental identification procedure.

Direct output from an ANSYS harmonic analysis contains frequency data and the real and imaginary part of the amplitude response to a sinusoidal excitation in the frequency domain. First this data file is read into Matlab. The real and imaginary parts of the input can be generally described by

$$\begin{aligned} F_r &= F_0 \cos \Psi \\ F_i &= F_0 \sin \Psi \end{aligned} \tag{2.140}$$

where F_0 is the amplitude component of the signal, F_r is the real part of the response and F_i is the imaginary. The angle Ψ includes phase information. This formulation produced by ANSYS is not suitable for direct processing, therefore it has to be converted into amplitude and phase form. This may be done utilizing the following relation [1, 2]:

$$F = \sqrt{F_r + F_i}$$
$$\Psi = \tan^{-1}\left(\frac{F_i}{F_r}\right) \tag{2.141}$$

The conversion also involves manipulation with the direct results to ensure correct identification functionality. The converted raw data is made into a data object suitable for the System Identification package using the frequency function option [59]. Frequencies are input in rad/s, phase angles in degrees. The raw data from ANSYS must be also subjected to pre-processing such as filtering the bandwidth of interest.

The technique described above can be viable for system identification, given a good quality FEM model that has also been verified experimentally. In practice, however, the physical properties of mechanical systems (such as damping) are not always exactly known or cannot be directly measured. Moreover, a FEM model assumes Rayleigh damping, which distorts the amplitude levels especially at higher frequencies. According to the experience of the authors, if a structure with bonded piezoelectric transducers is modeled including the bonding glue or resin layer, the FEM model and measured frequency response may show a wide variation in the harmonic analysis results when properties of the explicitly modeled glue layer are adjusted. Using a control model based on a FEM model certainly cannot substitute a well-designed experiment; it may be however used at early stages of control system design-before the controlled plant is physically realized.

References

1. Ansys Inc (2005) Release 10.0 Documentation for ANSYS. Ansys Inc., Canonsburg
2. Ansys Inc (2009) Release 12.1 Documentation for ANSYS. Ansys Inc. /SAS IP Inc., Canonsburg
3. Bandyopadhyay B, Manjunath T, Umapathy M (2007) Modeling, control and implementation of smart structures: a FEM-state space approach, 1st edn. Springer Verlag, Berlin
4. Beards CF (1996) Structural vibration: analysis and damping, 1st edn. Arnold, London
5. Benaroya H, Nagurka ML (2010) Mechanical vibration: analysis, uncertainities and control, 3rd edn. CRC Press, Taylor&Francis Group, Boca Raton
6. Carra S, Amabili M, Ohayon R, Hutin P (2008) Active vibration control of a thin rectangular plate in air or in contact with water in presence of tonal primary disturbance. Aerosp Sci Technol 12(1):54–61. doi: 10.1016/j.ast.2007.10.001, http://www.sciencedirect.com/science/article/B6VK2-4PXDM8C-1/2/db87a30acd2bfaefa3f97e3896bc9232, Aircraft Noise Reduction
7. Cavallo A, De Maria G, Natale C, Pirozzi S (2007) Gray-box identification of continuous-time models of flexible structures. IEEE Trans Control Syst Technol 15(5):967–981. doi:10.1109/TCST.2006.890284
8. Chiang RY, Safonov MG (1991) Design of $\mathcal{H}_\infty$ controller for a lightly damped system using a bilinear pole shifting transform. In: American control conference, pp 1927–1928
9. Clijmans L, Swevers J, Baerdemaeker JD, Ramon H (2000) Sprayer boom motion, Part 1: Derivation of the mathematical model using experimental system identification theory. J Agric Eng Res 76(1):61–69. doi:10.1006/jaer.2000.0530, http://www.sciencedirect.com/science/article/B6WH1-45F4RPT-2T/2/a326ca97e631682a5f8ae6e2120d30ed

10. de Silva CW (2006) Vibration: fundamentals and practice, 2nd edn. CRC Press, Boca Raton
11. Dong X, Meng G, Peng J (2006) Vibration control of piezoelectric smart structures based on system identification technique: numerical simulation and study. J Sound Vib 297:680–693
12. European space agency (ESA) and European aeronautic defence and space company (EADS)-Astrium (2004) Venus Express being prepared for vibration tests by Intespace facilities, Toulouse. Image ID: SEMW4L808BE
13. Fang B, Kelkar AG, Joshi SM, Pota HR (2004) Modelling, system identification, and control of acoustic-structure dynamics in 3-D enclosures. Control Eng Pract 12(8): 989–1004. doi:10.1016/j.conengprac.2003.10.003, http://www.sciencedirect.com/science/article/B6V2H-4BYR2KY-1/2/8a78e1d6cbd24a286de16096e8cf88e7, Special section on emerging technologies for active noise and vibration control systems
14. Fraanje R, Verhaegen M, Doelman N, Berkhoff A (2004) Optimal and robust feedback controller estimation for a vibrating plate. Control Eng Pract 12(8):1017–1027. doi:10.1016/j.conengprac.2003.09.007, http://www.sciencedirect.com/science/article/B6V2H-4B0PNNN-1/2/2879914555f3304fea841d08ef5898de, Special section on emerging technologies for active noise and vibration control systems
15. Fuller CR, Elliott SJ, Nelson PA (1996) Active control of vibration, 1st edn. Academic Press, San Francisco
16. Gibson S, Wills A, Ninness B (2005) Maximum-likelihood parameter estimation of bilinear systems. IEEE Trans Autom Control 50(10):1581–1596
17. Guzmán J, Rivera D, Dormido S, Berenguel M (2010) ITSIE: an interactive software tool for system identification education. Available http://aer.ual.es/ITSIE/
18. Hatch MR (2000) Vibration simulation using MATLAB and ANSYS, 1st edn. Chapman and Hall / CRC press, Boca Raton
19. Haverkamp B, Verhaegen M (1997) SMI Toolbox, 1st edn. Delft University of Technology
20. Hellerstein JL, Diao Y, Parekh S, Tilbury DM (2004) Feedback control of computing systems. Wiley/IEEE Press, Hoboken
21. Inman DJ (2006) Vibration with control. Wiley, Chichester
22. Inman DJ (2007) Engineering vibrations, 3rd edn. Pearson International Education (Prentice Hall), Upper Saddle River
23. Ji H, Qiu J, Zhu K, Badel A (2010) Two-mode vibration control of a beam using nonlinear synchronized switching damping based on the maximization of converted energy. J Sound Vib 329(14):2751–2767. doi:10.1016/j.jsv.2010.01.012, http://www.sciencedirect.com/science/article/B6WM3-4YG1R6R-2/2/88406f934e48ccfe56a6188409cc989e
24. Kaiser OE, Pietrzko SJ, Morari M (2003) Feedback control of sound transmission through a double glazed window. J Sound Vib 263(4):775–795. doi:10.1016/S0022-460X(02)01259-2, http://www.sciencedirect.com/science/article/B6WM3-48D38FJ-2/2/7f8ee2d801360f2532e76c9c53353872
25. Landau ID, Constantinescu A, Rey D (2005) Adaptive narrow band disturbance rejection applied to an active suspension–an internal model principle approach. Automatica 41(4):563–574. doi:10.1016/j.automatica.2004.08.022, http://www.sciencedirect.com/science/article/B6V21-
26. Lara A, Bruch J, Sloss J, Sadek I, Adali S (2000) Vibration damping in beams via piezo actuation using optimal boundary control. Int J Solids Struct 37:6537–6554
27. Li M, Lim TC, Lee JH (2008) Simulation study on active noise control for a 4-T MRI scanner. Magn Reson Imaging 26(3):393–400. doi:10.1016/j.mri.2007.08.003. http://www.sciencedirect.com/science/article/B6T9D-4R8KT3W-2/2/2797c565f329cf6cd1e567eefb69607e
28. Liu J, Liu K (2006) A tunable electromagnetic vibration absorber: characterization and application. J Sound Vib 295(3–5):708–724. doi:10.1016/j.jsv.2006.01.033. http://www.sciencedirect.com/science/article/B6WM3-4JP9FXN-6/2/0b961839d0b922bbd94dcc5ce5c5f9e4

29. Ljung L (1999) System identification: theory for the user, 2nd edn. PTR Prentice Hall, Upper Saddle River
30. Luo Y, Xie S, Zhang X (2008) The actuated performance of multi-layer piezoelectric actuator in active vibration control of honeycomb sandwich panel. J Sound Vib 317(3–5):496–513. doi:10.1016/j.jsv.2008.03.047, http://www.sciencedirect.com/science/article/B6WM3-4SJR2GN-1/2/04c4aad317afe74e20e6f5810f689674
31. Mahmoodi SN, Jalili N, Khadem SE (2008) An experimental investigation of nonlinear vibration and frequency response analysis of cantilever viscoelastic beams. J Sound Vib 311(3–5):1409–1419. doi:10.1016/j.jsv.2007.09.027, http://www.sciencedirect.com/science/article/B6WM3-4R113SP-1/2/4baf1df12aa15dbfcbdd0e4f13149b17
32. Moshrefi-Torbati M, Keane A, Elliott S, Brennan M, Anthony D, Rogers E (2006) Active vibration control (AVC) of a satellite boom structure using optimally positioned stacked piezoelectric actuators. J Sound Vib 292(1–2):203–220. doi:10.1016/j.jsv.2005.07.040, http://www.sciencedirect.com/science/article/pii/S0022460X05005171
33. Ninness B, Wills A (2006) An identification toolbox for profiling novel techniques. In: 14th IFAC symposium on system identification, pp 301–307
34. Ninness B, Wills A, Gibson S (2005) The University of Newcastle identification toolbox (UNIT). In: IFAC World Congress, pp 1–6
35. Ninness B, Wills A, Mills A (2009) System identification toolbox (ID Toolbox). Available http://sigpromu.org/idtoolbox/
36. Pradhan S (2005) Vibration suppression of FGM shells using embedded magnetostrictive layers. Int J Solids Struct 42(9–10):2465–2488. doi:10.1016/j.ijsolstr.2004.09.049, http://www.sciencedirect.com/science/article/B6VJS-4F6SSGN-1/2/b6f9e2e6ffc65bfc0c4af5083e37df0b
37. Preumont A (2002) Vibration control of active structures, 2nd edn. Kluwer Academic Publishers, Dordrecht
38. Preumont A, Seto K (2008) Active control of structures, 3rd edn. Wiley, Chichester
39. Rashid A, Nicolescu CM (2006) Active vibration control in palletised workholding system for milling. Int J Mach Tools Manuf 46(12–13):1626–1636. doi:10.1016/j.ijmachtools.2005.08.020, http://www.sciencedirect.com/science/article/B6V4B-4HGD76C-2/2/273540b1466f54bf47cc11a241007364
40. Rastgaar M, Ahmadian M, Southward S (2010) Experimental application of orthogonal eigenstructure control for structural vibration cancellation. J Sound Vib 329(19):3873–3887. doi:10.1016/j.jsv.2010.03.015, http://www.sciencedirect.com/science/article/B6WM3-502WG3V-1/2/a4d2874628bb1caebddd51cc6366b58c
41. Rossing TD, Moore RF, Wheeler PA (2001) The Science of Sound, 3rd edn. Addison Wesley, San Francisco
42. Rossiter JA (2003) Model-based predictive control: a practical approach, 1st edn. CRC Press, Boca Raton
43. Schon T, Wills A, Ninness B (2011) System identification of nonlinear state-space models. Automatica 37(1):39–49
44. Seba B, Nedeljkovic N, Paschedag J, Lohmann B (2005) $\mathscr{H}_\infty$ feedback control and Fx-LMS feedforward control for car engine vibration attenuation. Appl Acoust 66(3):277–296. doi:10.1016/j.apacoust.2004.07.015, http://www.sciencedirect.com/science/article/B6V1S-4DTKD2W-1/2/d413b004e2a2e14e9df7fcf75f2df02f
45. Sima V, Simal D, Van Huffel S (2002) SLICOT system identification software and applications. In: 2002 IEEE international symposium computer aided control system design proceedings, pp 45–50. doi:10.1109/CACSD.2002.1036927
46. Sima V, Sima DM, Huffel SV (2004) High-performance numerical algorithms and software for subspace-based linear multivariable system identification. J Comput Appl Math 170(2):371–397. doi:10.1016/j.cam.2003.12.046, http://www.sciencedirect.com/science/article/B6TYH-4BYR6W6-B/2/e0433bca6148658f54c9a683d281c020
47. Skullestad A, Hallingstad O (1998) Vibration parameters identification in a spacecraft subjected to active vibration damping. Mechatronics 8(6):691–705.

doi:10.1016/S0957-4158(97)00051-2, http://www.sciencedirect.com/science/article/B6V43-3W18XD5-4/2/d6e2d2a478a77ff09e9f6f7dfe7fd503
48. Sloss J, Bruch J, Sadek I, Adali S (2003) Piezo patch sensor/actuator control of the vibrations of a cantilever under axial load. Compos Struct 62:423–428
49. Šolek P (2009) Techincká Mechanika II, 1st edn. Slovenská technická univerzita v Bratislave, Nakladateľstvo STU, Bratislava (Technical Mechanics II) in Slovak language.
50. Starek L (1985) Vyššia dynamika, 2nd edn. SVŠT, Bratislava (Advanced Dynamics) in Slovak language.
51. Starek L (2006) Kmitanie mechanických sústav. Slovenská technická univerzita v Bratislave, Nakladateľstvo STU, Bratislava (Vibration of Mechanical Systems) in Slovak language.
52. Starek L (2009) Kmitanie s riadením, 1st edn. Slovenská technická univerzita v Bratislave, Nakladateľstvo STU, Bratislava (Vibration with Control) in Slovak language.
53. Sumali H, Meissner K, Cudney HH (2001) A piezoelectric array for sensing vibration modal coordinates. Sens Actuators A 93(2):123–131. doi: 10.1016/S0924-4247(01)00644-6, http://www.sciencedirect.com/science/article/B6THG-43N2J67-5/2/b7e5396a71e7ee7fae9ef93cb254b7b7
54. Takács G, Rohaľ-Ilkiv B (2008) Experimental identification of a structure with active vibration cancelling. Acta Mech Slovaca 12(3-B):795–803
55. Takács G, Rohaľ-Ilkiv B (2009) MPC with guaranteed stability and constraint feasibility on flexible vibrating active structures: a comparative study. In: Hu H (ed) Proceedings of the eleventh IASTED international conference on control and applications, Cambridge
56. Takács G, Rohaľ-Ilkiv B (2009) Newton-Raphson based efficient model predictive control applied on active vibrating structures. In: Proceedings of the European Control Control Conference, Budapest
57. Takács G, Rohaľ-Ilkiv B (2009) Newton-Raphson MPC controlled active vibration attenuation. In: Hangos KM (ed) Proceedings of the 28. IASTED international conference on modeling, identification and control, Innsbruck
58. Takács G, Rohaľ-Ilkiv B (2010) Capacitive proximity sensor position feedback in active vibration control of lightly damped cantilevers. In: Shokin YI, Bychkov I, Potaturkin O (eds) Proceedings of the 3rd IASTED international multiconference ACIT-CDA, Novosibirsk, pp 692–700
59. The Mathworks (2011) Matlab system identification toolbox v7.4.2 (R2011a), Natick. Software. Available http://www.mathworks.com/help/toolbox/ident/
60. Weber B, Feltrin G (2010) Assessment of long-term behavior of tuned mass dampers by system identification. Eng Struct 32(11):3670–3682. doi: 10.1016/j.engstruct.2010.08.011. http://www.sciencedirect.com/science/article/B6V2Y-5119FWV-2/2/20320a260725a111ed30767718b9b50b
61. Wills A, Ninness B (2008) On gradient-based search for multivariable system estimates. IEEE Trans Autom Control 53(1):298–306
62. Wills A, Ninness B, Gibson S (2009) Maximum likelihood estimation of state space models from frequency domain data. IEEE Trans Autom Control 54(1):19–33
63. Wills AG, Bates D, Fleming AJ, Ninness B, Moheimani SOR (2008) Model predictive control applied to constraint handling in active noise and vibration control. IEEE Trans Control Syst Technol 16(1):3–12

Chapter 3
Smart Materials in Active Vibration Control

Every feedback control system has essential components like the hardware computing control input via the strategy of our choice, sensors to provide feedback to this controller and actuators to carry out the required changes in plant dynamics. This chapter is concerned with the latter two components, that is sensors and actuators. More specifically, here we take a closer look at some of the advanced engineering materials that can be used as actuators and in some cases as sensors in active vibration control applications (AVC).

There are many well-known traditional actuating components such as electromagnetic devices, pneumatic actuators, rotary and linear motors etc., which may be effectively utilized in vibration control as well. Unlike the previously mentioned devices, modern engineering materials which are often referred to as *intelligent* or *smart* have the advantage of being lightweight and more importantly they can be seamlessly structurally integrated. For example, a composite aeroelastic wing equipped with thin piezoelectric wafers cast directly into the structure enables us to suppress undesirable vibration without adding a considerable mass or changing the shape of the wing. On the other hand, advanced materials like the magnetorheological fluid may add unprecedented properties to already existing components, for example creating automotive dampers with automatically adjusted damping properties. Figure 3.1 illustrates[1] an experimental actuator capable of providing displacements exceeding the usual range of simple piezoelectric materials [80]. The robust and low-cost high displacement actuator (HDA) made of pre-stressed polymeric materials and piezoelectric ceramics is an excellent example of advanced engineering smart materials. The aim of this chapter is to introduce the reader to some of these cutting-edge materials and their use in vibration control. Actuators like the aforementioned electromagnetic linear motors, pneumatic devices and others will not be covered here.

Thanks to the reciprocal physical effects experienced in some of these materials, actuating elements can also be used in a sensor configuration. Just as in the case of actuators, many feedback sensing systems exist other than the ones using smart

---

[1] Courtesy of NASA.

G. Takács and B. Rohal'-Ilkiv, *Model Predictive Vibration Control*,
DOI: 10.1007/978-1-4471-2333-0_3,

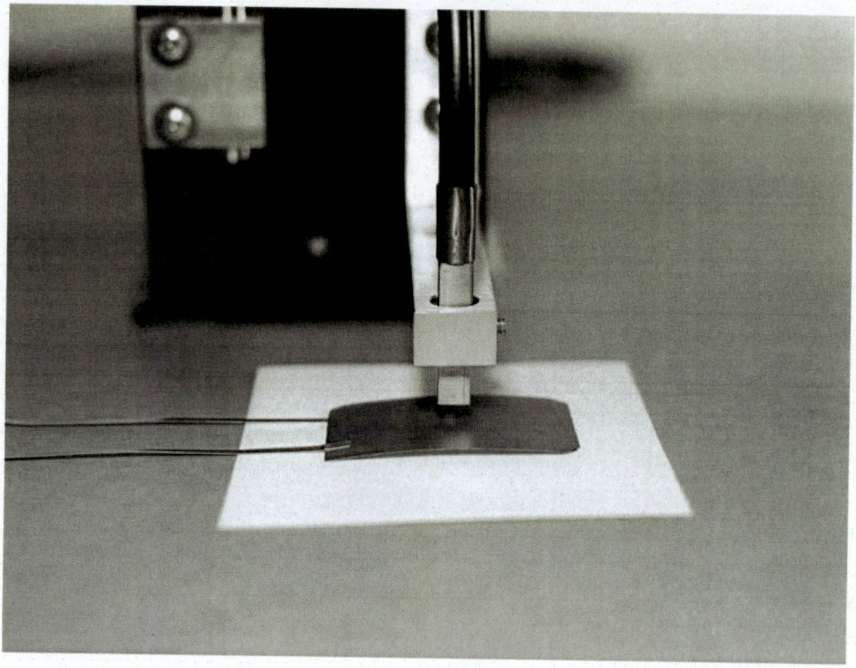

Fig. 3.1 An experimental high displacement actuator (HDA) developed at NASA Langley Research Center [80] is placed under a fiber optic displacement sensor

materials. Some of these are among others accelerometers,[2] strain sensors based on resistance wires, or more advanced devices like industrial laser triangulation heads or laser Doppler vibrometers (LDV). This chapter will concentrate on the materials themselves and the underlying physical aspects of the sensors in vibration attenuation, while introducing other possible feedback methods in a practical example in Sects. 5.3.4 and 5.5.3.

There are several engineering materials available nowadays, which exhibit some very desirable properties for use in AVC. So what is the criterion of classifying a material to be *smart?* The keyword here is coupling. From the structural point of view, the behavior of classical materials can be sufficiently described by their elastic constants: the elastic constant relates stress and strain, the thermal constant relates temperature and strain. In smart materials, coupling also exists between the either two (or even more) of the following fields: electric charge, strain, magnetic, temperature, chemical and light. This coupling is also obvious between the constitutive equations describing the behavior of these materials. The most common smart materials which are used in active structures are shape memory alloys, magneto- and electrostrictive materials, semi-smart magneto- and electro-rheological fluids where the coupling is one directional, electrochemical materials and of course piezoelectrics.

[2] However, these are also based on the piezoelectric effect and use piezoelectric materials [46].

The chapter begins with a discussion on the shape memory effect and shape memory alloy materials. In addition to the shape memory effect, the passive albeit still very interesting superelastic nature of these materials is also introduced. After characterizing the interactions between the applied temperature, stress and strain; the utilization of shape memory alloys in vibration control is reviewed. Section 3.2 addresses the magnetostrictive and electrostrictive effect and their use in vibration attenuation. Following this, the popular magnetorhoeologic fluids and the related electrorheological fluids are introduced, also with a short review of their existing and potential applications in active vibration control. The next section, that is 3.4, deals with the piezoelectric effect and piezoelectric materials. In addition to the physical basis, some of the mathematical modeling concepts and finite element analysis aspects are also covered. Our selection of smart materials ends with a section on the emerging electrochemical polymer based actuators in Sect. 3.5. A couple of short paragraphs on other types of materials and actuators end our discussion on smart materials in AVC.

3.1 Shape Memory Alloys

Shape memory alloys (SMA) demonstrate apparent plastic deformation and recovery to the original shape after heating. SMA can recover as much as 5% strain, which compared to materials like piezoceramics is a considerable shape change. The main advantage of this type of smart material is the ability to perform complex movements with few elements. The shape change and the resulting movement can be achieved by a small temperature change, and causes the SMA to undergo a type of solid state phase transformation. This change is the so-called martensitic deformation in metals. Shape memory alloys may be used to supply energy to systems with very slow dynamics and thus induce vibrations, effectively creating an active vibration control system. It is also common to utilize the SMA as a type of slowly changing adaptive part to form a semi-active vibration suppression system [33]. Figure 3.2 illustrates[3] the use of shape memory alloy materials to create slow-speed morphing wing surfaces on aircraft and SMA-wire based linear motors.

3.1.1 SMA Materials and Properties

The most common SMA material is an alloy of nickel and titanium, which is often referred to as nitinol.[4] In the nickel and titanium (NiTi)-based alloys, the two elements are present in approximately equal atomic percentages. Several other alloys exist of which we list FeMnSi, copper-based alloys such as CuZnAl or CuAlNi and some

[3] Courtesy of NASA.

[4] This nickel and titanium alloy was discovered and developed by Buechler et al. in 1963 at the U.S. Naval Ordnance Laboratory, thus the name NiTiNOL.

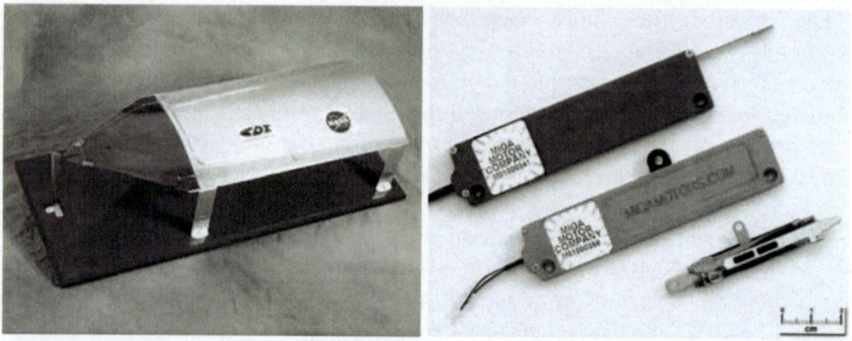

(a) A wing-like SMA demonstration device **(b)** SMA based actuators

Fig. 3.2 A SMA-based demonstration device resembling an aircraft surface is featured in (**a**) [78], while (**b**) shows small actuators in a linear motor mode based on SMA wires [79]

steel formulations [28, 33]. The advantage of NiTi-based SMA is its high electric resistivity, thus allowing the material to be rapidly heated upon the application of electric current.

Shape memory alloys present two interesting macroscopic properties, these are:

- superelasticity
- shape memory effect

The former, superelasticity is the ability of this type of material to return to its original shape after a considerable amount of mechanical stress and deformation. This process needs no temperature change to be completed, and it is called the mechanical memory effect. Elasticity is approximately 20 times higher than other elastic metallic materials [68]. Objects manufactured from the superelastic version of nitinol find their application mainly as medical instruments, there are several laboratory experiments investigating the use of superelastic (austenitic) nitinol as means for passive vibration damping.

The latter property is more interesting for the control community, as nitinol can be effectively used as an actuator. Because of the shape memory effect or the thermal memory effect, the plastically deformed SMA material returns to its original memorized shape after applying a small amount of heat as illustrated in Fig. 3.3. The deformation is not limited to pure bending as in bi-metallic structures, but may include tensional and torsional deformations or their mixtures [68]. A 4 mm diameter nitinol wire may lift even a 1000 kg load; however, it will lose its memory effect because of this large loading. To prevent this, a load limit is usually enforced, for example, in this case a 150 kg load would not induce a loss of the memory effect while still being a very high force output [68]. Enforcing such load limits to prevent the loss of the memory effect call for control systems encompassing constraint handling for which model predictive control (MPC) is an ideal candidate.

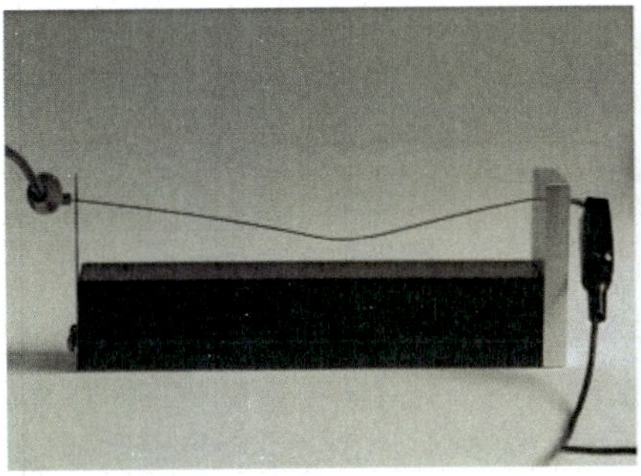

(**a**) SMA wire prior to activation

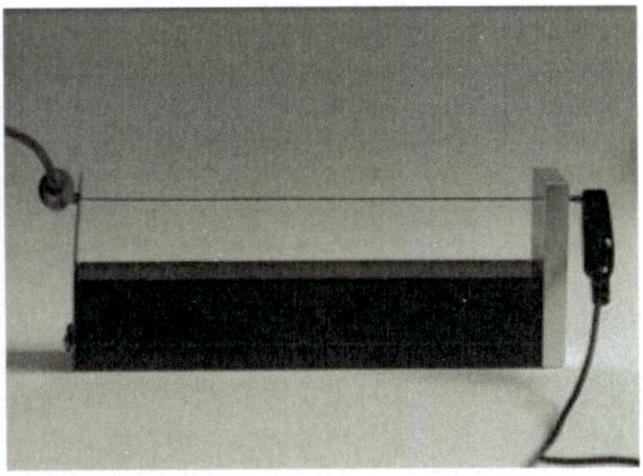

(**b**) SMA wire after activation

Fig. 3.3 An SMA wire is placed in between a spring steel blade and a rigid aluminum clamp. The wire is loose prior to activation as shown in (**a**). Due to the applied current (9 V battery) the wire temperature is raised above the activation temperature. The wire regains its original straight shape in (**b**) and exerts a force, which is enough to deform the spring steel

3.1.2 Stress, Strain and Temperature

Both the superelastic and shape memory effects are due to a phase change from *austenite*, which is the higher temperature and stronger phase, to *martensite* which is the lower temperature and softer phase. Unlike the phase changes that come to mind like the change from solid to liquid and gas, this is a solid phase change. The austenitic solid phase is stable at elevated temperatures and has a strong body centered cubic crystal structure. The martensitic phase has a weaker asymmetric parallelogram structure, having up to 24 crystal structure variations [101]. When martensitic nitinol is subject to external stress it goes through different variations of the possible crystal structures and eventually settles at the one allowing for maximal deformation. This mechanism is called *detwinning*. There are four temperatures characterizing the shape memory effect of SMA:

- M_f: martensite finish—this is the lowest temperature, below all of the material has the soft martensitic structure
- M_s: martensite start—an intermediate temperature, when the martensite phase starts to appear in the prevalently austenitic phase
- A_s: austenite start—an intermediate temperature, when the austenite phase starts to appear in the prevalently martensitic phase
- A_f: austenite finish—this is the highest temperature, above which all of the material has the hard martensitic structure. Superelastic SMA are designed to work over this temperature, while the thermal-induced memory effect finishes at this temperature

These temperature characteristics and limits may be set upon manufacturing the alloy. For example, it is possible to create an alloy with a reshaping temperature close to the normal temperature of the human body.

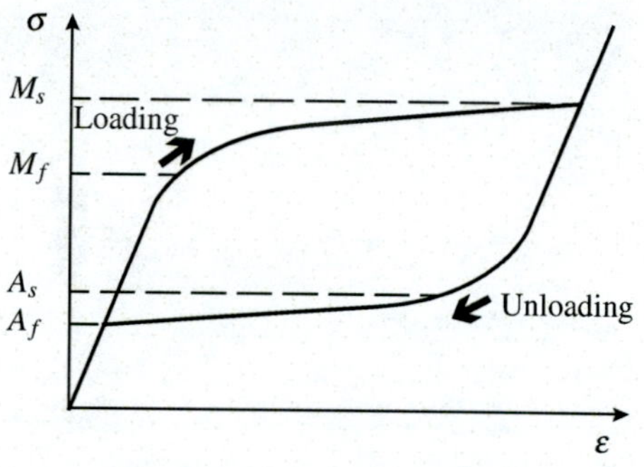

Fig. 3.4 Pseudoelastic behavior of shape memory alloys, illustrated on a stress-strain hysteresis curve

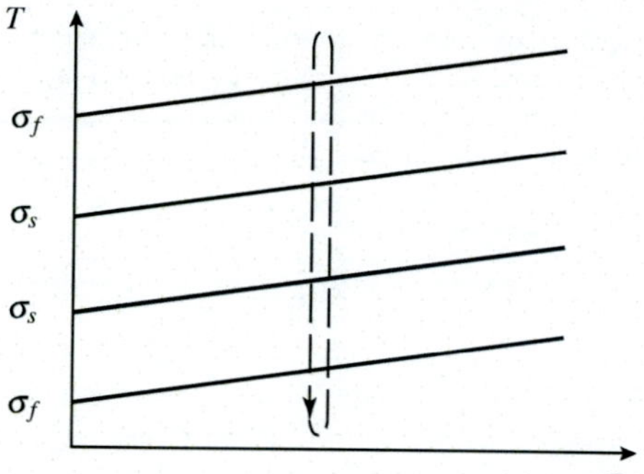

Fig. 3.5 Thermal memory effect of SMA in relation to constant stress and variable temperature

For the case of uniaxial loading, the stress–strain curve for SMA is denoted in Fig. 3.4 [99]. The curve shows a pseudoelastic behavior, where the applied load takes the material from the austenite phase to the martensite phase along the upper curve. This is the stress-induced superelastic behavior of austenitic nitinol, therefore we may state that the temperature here is a constant $T > A_f$. The reverse transformation occurs in unloading the SMA material, when the material transforms from martensite into austenite along the lower curve, thus forming a hysteresis loop [99]. In Fig. 3.4 ε denotes strain, σ denotes stress. Martensite starts to form at M_s and finishes at M_f, while the austenite starts to form at A_s and finishes at A_f.

The dashed line in Fig. 3.5 denotes a scenario, where the SMA is subject to a temperature change in constant stress [99]. Note, however that the phase change start and finishing temperatures are linearly dependent on the loading stress. Temperature is marked by T while stress is σ.

Finally, Fig. 3.6 illustrates the percentual composition of martensite and austenite phases in a temperature-induced martensitic deformation [101]. The curve starts from below the low temperature M_f and takes the right side of the hysteresis path. At a certain A_s temperature the phase change to austenite begins, while the martensite composition decreases. Eventually the material gets to the A_f temperature where 100% of it is converted into the austenite phase.

Shape setting of an SMA actuator can be done in a high temperature oven. The heat treatment is performed in two steps: first the material is constrained into the desired

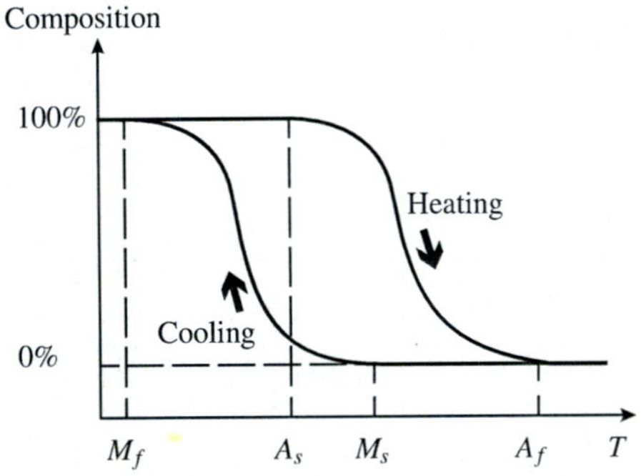

Fig. 3.6 Temperature dependent deformation: austenitic-martensitic phase transformation hysteresis of shape memory alloys

form, and heated at 525°C for 5–10 min depending on the size of the clamp. After this, the material is quenched with air cooling, water or oil. This is then followed by a repeated heating at 475°C for about 30–60 min [69]. Figure 3.7 illustrates SMA materials memorized to different shapes. While the plates on the left and the wire is memorized to a flat (straight) shape, the third and darker plate is memorized into a curved shape.

3.1.3 SMA in Vibration Control

The free and/or forced vibration behavior of plates and other structures with embedded SMA materials is studied using analytic or FEM methods in [35, 54, 99, 85, 130]. The cited works focus on modeling issues for the need of optimal design for classical vibration response manipulation, without actively controlled components. The inclusion of SMA elements in plates, beams and other mechanisms can be understood as a form of semi-active control. SMA has been already considered as passive or semi-active vibration damping devices in civil engineering structures [43, 126, 101].

Although several models have been proposed for SMA, the constitutive description of the complex pseudoelastic and shape memory effect phenomena cannot be developed by classical plasticity theory. Models based on the nonlinear generalized plasticity have been successfully applied for SMA [28].

SMA as an actuator is suitable for low frequency and low precision applications, therefore, their usage in active vibration attenuation applications is questionable. It is interesting enough to note that SMA can also be used as a type of sensor. The work of Fuller et al. pointed out that embedded SMA wires in a Wheatstone configuration may give accurate estimation of strain levels due to oscillations in a beam [33]. The use of SMA as sensors is, however, atypical as piezoelectric or resistance-wire based sensors are also cheap and readily available.

Active vibration control is proposed utilizing an SMA actuator in [22]. Here, the temperature of the SMA is manipulated to change mechanical properties.

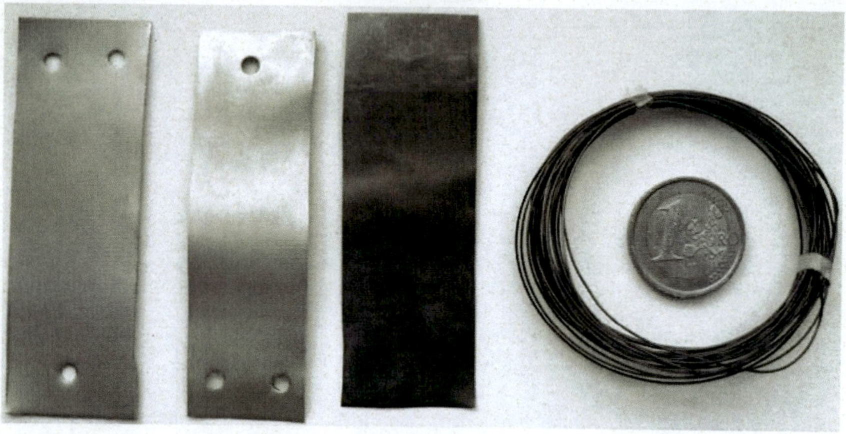

Fig. 3.7 SMA materials can be memorized to different shapes. The wire and the plates on the *left* are set to a straight shape, while the darker plate is memorized to a curved shape

Vibration damping is achieved combining active and passive methods. In a review article Bars et al. lists shape memory alloys as a particularly interesting tool for smart structures and states the need for advanced control algorithms such as MPC to tackle issues such as multi-point inputs and outputs, delays and possibly actuator nonlinearity [13].

Shape memory alloy materials are utilized in [23] for vibration damping purposes. According to the step response of the material, upon the application of a constant current jump the SMA wire exerts force, which can be approximated according to a first order response [15, 23]:

$$T_c \frac{df(t)}{dt} + f(t) = i(t) \tag{3.1}$$

where the force exerted by the SMA wire is denoted by $f(t)$, the actuating current by $i(t)$ while T_c is the time constant of the first order transfer. The temperature in an SMA wire actuator is approximately linearly dependent on the applied current [54]. Unfortunately, the time constant is different in the heating and cooling cycles [23, 24]. The time constant is also highly dependent on the prestrain applied to the wire. Because of these parameter variations it is likely that an MPC control-based SMA system would require the explicit handling of model uncertainties. The above cited work of Choi et al. utilized sliding mode controlled nitinol wires to damp the first modal frequency of a building-like structure in the vicinity of 5 Hz, providing certain basis to use SMA for lightly damped structures with a low first resonant frequency. Here, the time constant was approximated to be 125 ms that would indicate an approximately 8 Hz bandwidth.

A very interesting possibility is utilizing an adaptive passive approach instead of actively controlling the vibration amplitudes, velocities or accelerations. Using a structure or mechanism with integrated SMA parts, one could tune its vibration

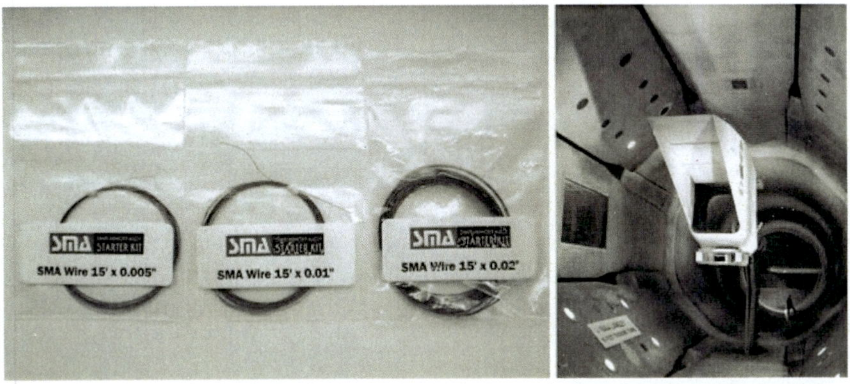

(a) SMA wire **(b)** SMA actuated F-15 inlet

Fig. 3.8 Spools of shape memory alloy wires with different diameters are shown in (**a**), while (**b**) shows a full-scale F-15 inlet (modified flight hardware) with integrated shape memory alloy actuators installed in the NASA Langley Research Center 16-foot Transonic Tunnel [81]

frequency in real-time according to the outside excitation [33]. By this method, the resonant frequency of the structure could actively adapt to the quality and character of the measured outside excitation. Using the idea an actively controlled steel structure has been presented in [86]. The resonant frequency of the structure could be shifted about 32% of its nominal value through the application of heat into the SMA.

An overview of the civil engineering applications of SMA materials is given in [47]. John and Hariri investigate the effect of shape memory alloy actuation on the dynamic response of a composite polymer plate in [49]. The work examines the stiffness change and thus the shift of natural frequencies in a composite plate both in simulation and in experiment, founding a basis for the future application of SMA-enhanced active materials for vibration attenuation. Spools of SMA wire with different diameters are illustrated in Fig. 3.8a, while an SMA actuated F-15 aircraft inlet is shown[5] in Fig. 3.8b.

3.2 Magneto- and Electrostrictive Materials

Both magnetostrictive and electrostrictive materials demonstrate a shape change upon the application of magnetic or electric fields. This small shape change is believed to be caused by the alignment of magnetic/electric domains within the material upon the application of the fields. The advent of specialized engineering materials enables the use of these materials in active vibration control applications thanks to the increased deformation strains.

[5] Courtesy of NASA.

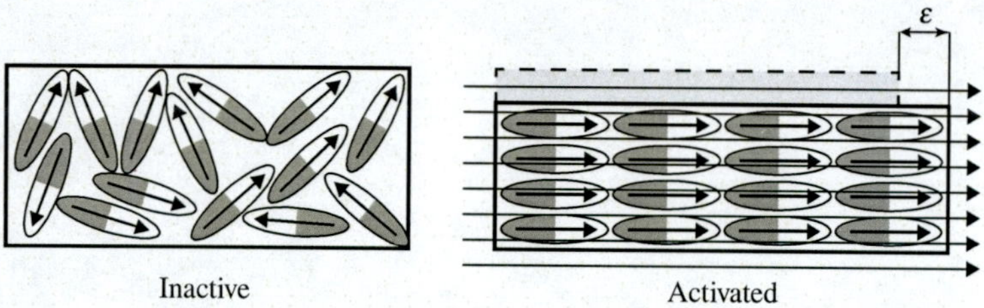

Fig. 3.9 Inactive magnetostrictive material with randomly ordered dipoles is shown on the *left*, while the *right* side shows an activated magnetostrictive material under a magnetic field. The dipoles have been ordered along the magnetic flux lines causing the material to expand

3.2.1 Magnetostrictive Materials

Magnetostrictive (MS) materials change their shape when subjected to a magnetic field. A common example of the magnetostrictive effect in everyday life is the humming noise emitted by electric transformers. This is due to the expansion and contraction of metallic parts in response to magnetostriction, induced by the changing electromagnetic field. Nearly all ferromagnetic materials demonstrate this property, but the shape and volume change is very small. Ferromagnetic materials have a structure divided into magnetic domains exhibiting uniform magnetic polarization. The applied magnetic field causes the rotation of these domains and in return a slight shape change on a macroscopic level. Figure 3.9 illustrates the randomly oriented magnetic domains within the material and the reorientation after the magnetic field is applied. The reciprocal phenomenon to magnetostriction is called the *Villari effect*. This describes the change of magnetic properties under applied load.

The deformation in magnetostrictive materials is characterized with the magnetostrictive coefficient ι_{ms}, which expresses the fractional length change upon applying a magnetic field. The shape change of the material is zero at zero magnetic field, however upon the application of the field it grows linearly according to the magnetostrictive coefficient until the material reaches magnetostrictive saturation.

The application of certain rare earth materials into an alloy allowed using the effect of magnetostriction in real-life engineering applications. Early types of magnetostrictive alloys demonstrated large magnetostriction, but only by applying high magnetic fields or at cryogenic temperatures [51]. These difficulties were eliminated by the introduction of *Terfenol-D*,[6] which continues to be the most common magnetostrictive material [1]. Terfenol-D exhibits about 2000 $\mu\varepsilon$ at room temperature, while Cobalt, which demonstrates the largest magnetostrictive effect of the pure elements, exhibits only 60 $\mu\varepsilon$ strain. Another common material goes by the trade name *Metglas 2605SC*, yet another by *Galfenol*.

---

6 Similar to nitinol, it has been invented at the United States Naval Ordnance Laboratory (NOL). Terfenol-D stands for Terbium Ferrum NOL Dysprosium; with the chemical composition $Tb_xDy_{1-x}Fe_2$.

The typical recoverable strain of magnetostrictive materials like Terfenol-D is in the order of 0.15%. Maximal response is presented under compressive loads. Magnetostrictive actuators have a long life span and may be used in high precision applications. Actuators may be used in compression alone as load carrying elements. Pre-stressing the actuators may increase both efficiency and the coupling effects [18].

3.2.2 *Electrostrictive Materials*

Electrostriction is closely related to magnetostriction. Due to electrostriction, all dielectrics change their shape upon the application of an electric field. The physical effect is similar to magnetostriction as well: non-conducting materials have randomly aligned polarized electrical domains. If the material is subjected to a strong electric field, the opposing sides of these domains become charged with a different polarity. The domains will be attracted to each other, thus reducing material thickness in the direction of the applied field and elongating it in a perpendicular direction.

All dielectrics exhibit some level of electrostriction; however, a class of engineering ceramics does produce higher strains than other materials. Such materials are known as *relaxor ferroelectrics* [56, 116] for example: lead magnesium niobate (PMN), lead magnesium niobate-lead titanate (PMN-PT) and lead lanthanum zirconate titanate (PLZT).

The elongation of electrostrictive materials is related quadratically [117] to the applied electric field [116]:

$$\varepsilon = \text{const} \cdot E^2 \tag{3.2}$$

where ε is strain and E is electric field strength. The relative percentual elongation for PMN-PT is 0.1% or 1000 $\mu\varepsilon$, but this is achieved under a field strength of 2 MV/m. Typical strains for special electrostrictors is in the range of 0.02–0.08%.

3.2.3 *Magneto- and Electrostrictive Materials in Vibration Control*

A mechanically amplified MS actuator for low frequency (1-10 Hz) vibration damping applications is suggested in [14], where the achievable displacement is rated between 0.5–4 mm and the force between 0.5–6 kN. Commercial actuator prototypes are also available; examples of such actuators are featured in Fig. 3.10[7] [20].

From the point of vibration control, MS actuators may deliver a high force output with high frequency [18, 73]. The underlying dynamics is a complex combination of electrical, mechanical and magnetic phenomena, which is further complicated by

7 Courtesy of the CEDRAT Group.

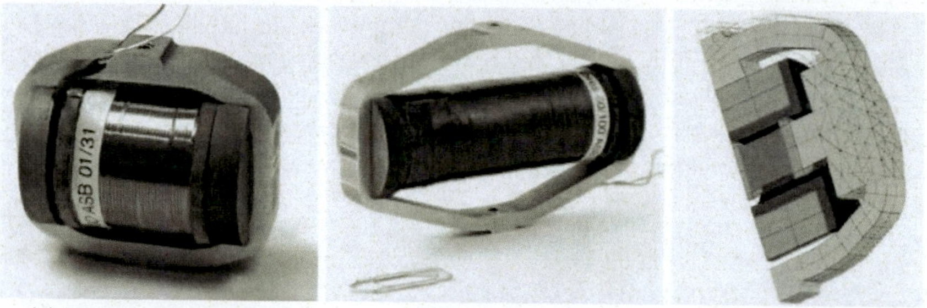

(**a**) Miniature MS actuator (**b**) MS actuator (**c**) FEM analysis

Fig. 3.10 Prototypes of different magnetostrictive actuators are shown in (**a**) and (**b**), while (**c**) features a FEM simulation of a deformed magnetostrictive actuator [20]

the nonlinear hysteretic behavior of Terfenol-D [18]. The linear properties of MS actuators hold only under the following assumptions [18]:

- low driving frequency
- reversible magnetostriction without power loss
- uniform stress and strain distribution

Under these assumptions the magnetomechanical equations are [27]:

$$\mathbf{S} = \mathbf{s}_H \sigma + \mathbf{g}\mathbf{H_m} \tag{3.3}$$

$$\mathbf{B_m} = \mathbf{g}\sigma + \mu_\sigma \mathbf{H_m} \tag{3.4}$$

where $\mathbf{S}$ is strain, σ is stress, $\mathbf{s}_H$ is mechanical compliance at constant applied magnetic field strength $\mathbf{H}$, $\mathbf{g}$ is the magnetic cross-coupling coefficient, μ_σ is magnetic permeability at constant stress and $\mathbf{B_m}$ is magnetic flux within the material.

Piezoelectricity is in fact a subclass of electrostrictive materials. However, while electrostrictive materials are nonlinear, piezoelectric materials behave linearly, which is an important feature for control applications. Moreover, electrostriction is not a reversible effect; unlike magnetostriction or piezoelectricity, the material does not generate an electric field upon the application of a mechanical deformation. Another important feature of electrostrictive materials is that they do not reverse the direction of the elongation with a reversed electric field [109]—note the quadratic dependence in (3.2). Therefore, electrostrictive transducers must operate under a biased DC electric field [116]. In comparison with piezoelectric materials, electrostrictive materials demonstrate a smaller hysteresis [109].

Braghin et al. introduces a model of magnetostrictive actuators for active vibration control in [18]. The authors propose a linear model for MS actuators, which is suitable for control design below the 2 kHz frequency range. This simple linearized numerical model has provided a good match with the experimental result for an inertial type of MS actuator. Such a linearization is not only important for the design of traditional feedback control systems, but is also essential for real-time model predictive control using MS actuators. Despite the complicated coupling, hysteresis and nonlinearity the static actuation displacement of MS actuators remains linear [73]. A linear SDOF system is the basis for the further analysis of the behavior of an MS actuator in a work by Li et al. as well [58].

The vibration suppression of composite shells using magnetostrictive layers is discussed in a work by Pradhan et al. [90]. The author formulated a theoretical model for composite shells and found that magnetostrictive layers should be placed further away from the neutral plane. In addition, thinner MS layers produced better damping. The MS actuator-based vibration damping of a simply supported beam is discussed by Moon et al. [73], where the experimental setup shows a significant reduction of vibrations in comparison with the scenario without control.

The use of electrostrictive actuation in vibration control is relatively uncommon. Sonar projectors are the typical field of use for electrostrictive actuators [89]; however, this does not concern vibration attenuation rather generating acoustic waves. An electrostrictive actuator has been utilized by Tzou et al. in [117] for the control of cantilever vibrations. Tzou et al. achieved only minimal damping under control when compared to the free response without actuation.

3.3 Magneto- and Electrorheological Fluids

Fluids based on the magnetorheological (MR) and electrorheological (ER) effects contain suspended particles, which upon the application of a magnetic/electric field align themselves to form columns within the fluid. The columns of suspended particles create an obstacle for fluid flow, thus increasing the overall net viscosity of the system. Damping systems using MR or ER fluids cannot supply energy to the controlled system, but may change the damping properties according to a controller strategy.

3.3.1 Magnetorheological Fluids

Magnetorheological fluids (MR) contain micrometer-sized particles in a dielectric carrier fluid [105]. The carrier fluid is usually a type of oil, while the particles are manufactured from multi-domain, magnetically soft materials such as metals and alloys [5]. MR fluids respond to an applied magnetic field. Upon application of a magnetic field, the MR fluid changes viscosity up to the point where it can be considered as a viscoelastic solid. It is worth noting that this behavior can be regarded as semi-active. This is because it is needed to use an external field to induce classical coupling. There is also the lack of reciprocal effect. The controlled system will remain conservative from the mechanical point of view and can only dissipate energy [92].

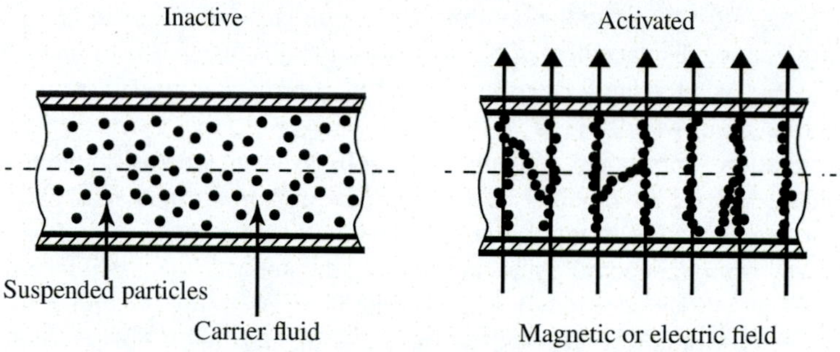

Fig. 3.11 Inactive MR fluid with suspended particles without an applied magnetic field is shown on the *left*, while the particles aligned to columns according to magnetic flux lines are featured on the *right*. ER fluids behave very similarly, but the material is activated by a strong electric field

Magnetorheological fluids are different from ferrofluids mainly in the size of the suspended particles. Brownian motion keeps the nanometer-sized particles constantly suspended in ferrofluids. The micrometer-sized particles in MR fluids are not subject to this phenomenon, therefore they eventually settle in the fluid because of the density difference of the carrier and the particles.

A typical magnetorheological fluid contains 20–40% by volume, 3–10 μm diameter iron particles [61]. These particles are suspended in a carrier liquid which can be mineral oil, synthetic oil or even water and may contain proprietary additives [5]. Additives prevent the suspended particles from settling due to the gravitational effect, decrease wear and modify viscosity.

When a magnetic field is applied to such a two-phased suspension, the randomly placed particles align themselves along the flux lines as illustrated in Fig. 3.11. In this way the flow of liquid is restricted in a direction perpendicular to the magnetic flux. The yield stress of magnetorheological fluids is typically 20–50 times higher than electrorheological fluids [91]. MR fluids are temperature stable, need low power supply and are able to react in a fully reversible fashion within milliseconds. The type of the aligned particle chains may vary if the MR liquid is activated in microgravity, ultimately changing its viscosity properties. Figure 3.12[8] shows the particles aligned to spikes in Earth gravity on the left, while the particle columns are broader in space.

3.3.2 Electrorheological Fluids

Electrorheological fluids (ER) differ from magnetorheological in the form of control coupling: while MR is controlled indirectly through the application of a magnetic field, ER is controlled through a direct electric field [105]. ER fluids contain

[8] Courtesy of NASA.

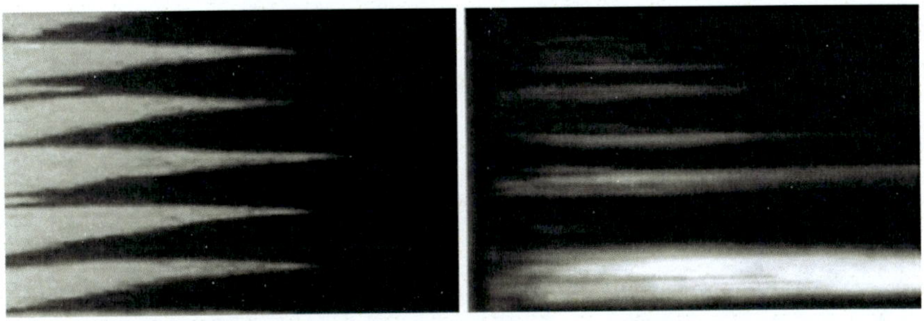

(a) Earth gravity **(b)** Microgravity

Fig. 3.12 Video microscope images of magnetorheological (MR) fluids. The MR fluid forms columns or spiked structures on Earth, while on the *right* broader columns are formed in micro-gravity environment aboard the International Space Station (ISS) [82]

non-conducting particles up to 50 μm in diameter and similar to MR fluids have a reaction time measurable in milliseconds. When we describe the consistency of the activated MR material from a practical point of view, we may refer to it as a kind of viscoelastic solid while the activated ER fluid resembles the consistency of a gel-like substance. In addition to the composition of the ER fluid, its behavior is greatly dependent on the geometry such as the size and distance of the plates acting as electrodes. This is similar to the design of capacitors, where capacitance in addition to the type of dielectric material used is influenced by electrode spacing and geometry.

The behavior of electrorheological fluids can be described by two alternative theories [104]: electrostatic theory and the interfacial tension theory. Electrostatic theory assumes a behavior similar to the MR fluids: the particles align in accordance to the electrical field to form chain-like structures. According to the interfacial tension theory, the ER liquid consists of three phases: the suspension liquid, the particles and another liquid within the particles. In the inactive state, this liquid is contained within the dispersed particles. When the liquid is activated, the electric field drives the liquid within the particles to a certain side through electroosmosis—and because of that neighboring particles start to bind and formulate chains on a macroscopic level. While no conclusion has been reached, it is possible that different types of ER materials exist which behave according to different principles.

3.3.3 Magneto- and Electrorheological Materials in Vibration Control

There are three possible operation modes for MR and ER fluid based semi-active damping devices. These modes are [105]:

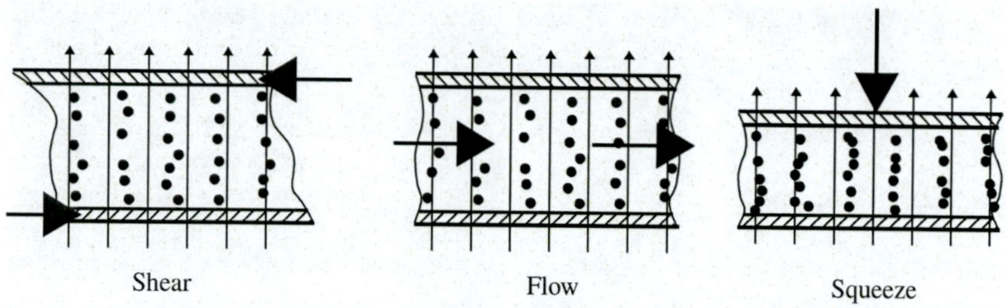

Fig. 3.13 Different operation modes for an MR fluid-based damping device. The figure shows shear, flow and squeeze modes (from *left* to *right*). Thick arrows denote movement directions, while thin arrows represent magnetic flux lines

- shear
- flow
- squeeze

To get an idea of these different operation modes, let us imagine a situation when the MR fluid is placed between two parallel plates. The magnetic field is acting perpendicular to these plates; therefore, the particles align themselves perpendicularly as well. In shear mode, one plate is moving in relation with each the other in a direction perpendicular to the magnetic flux and the particle columns. In flow mode, the plates are stationary and the fluid is moving due to a pressure difference. In the squeeze-flow mode, the plates are moving in relation to each other, however this time the movement is carried out along the flux lines and the particle columns. Operation modes of different MR and ER fluid based devices are illustrated in Fig. 3.13. The vibration damping performance of flow and squeeze mode semi-active insulation mounts with ER fluid actuators is discussed by Hong and Choi in [41], concluding that the squeeze mode actuator is more effective than the flow mode.

Mathematically, it is somewhat difficult to represent the behavior of MR and ER materials. The reason for this is that the model must account for hysteresis and static friction. The yield stress of the fluid is also variable: in the presence of a magnetic (electric) field the MR (ER) fluid acts as a solid, up to the point where the shear stress is reached (yield point).

The most common mathematical description is the so-called Bingham plastic model, which utilizes variable yield strength τ_Y. This depends on the strength of the applied magnetic field H. There is a point where the increased magnetic field H does not have an effect on the yield, the MR fluid is magnetically saturated. The flow is governed by the equation [87]:

$$\tau = \tau_Y(H) + \eta_v \dot{\gamma}_s \tag{3.5}$$

where shear stress is expressed by τ, shear strain by γ_s and fluid viscosity by η_v. Below the yield stress, the material exhibits viscoelastic properties, which may be described by:

$$\tau = G\gamma_s \tag{3.6}$$

where G is the complex material modulus. The Bingham plastic model is a good approximation for the activated MR and ER fluids with the respective substitutions for τ_Y, η_v, γ_s. In reality, the dynamic behavior of MR and ER fluids differs from the Bingham plastic model, for example in the absence of magnetic field, because the fluid is slightly non-Newtonian and its behavior is temperature dependent.

The above-mentioned types of semi-smart materials also have a few drawbacks: namely that they demonstrate high density (weight), good quality fluids are expensive and are prone to sedimentation after prolonged use. Particle sedimentation is controlled by surfactants; recent studies also used nanowire particles to decrease the sedimentation effect [83]. Sedimentation in ER fluids is prevented by using particles of densities comparable to the density of the fluid phase. Another disadvantage is that both MR and ER fluids have low shear strength; this is in the range of 100 kPa for MR [5]. Shear strength in both MR and ER can be improved by fluid pressurization [71, 122] while it can be further increased in MR using elongated particles [120].

In addition to sedimentation and low shear–stress, ER dampers require a high potential electric field to operate which is in the order of 1 kV for an electrode separation and fluid thickness of 1 mm.

From the control point of view, the dynamic behavior of MR and ER dampers is highly nonlinear, creating a considerable difficulty when designing a control algorithm. Controllers such as positive position and velocity feedback, acceleration feedback based LQ control, have been successfully utilized to generate control voltage to MR dampers. The operating voltage in ER fluids is close to the breakdown voltage of the material, therefore a constrained controller such as MPC is highly recommended to prevent actuator failure and increase lifetime while maintaining maximum actuating efficiency.

Flow mode MR dampers are used to control linear vibrations. Rotational motion can be controlled by using a shear mode damper, while a squeeze flow setup mode is useful for vibration damping platforms. The relatively large size of MR dampers prevents their use in certain application fields, such as submarines [5]. The use of MR dampers in earthquake control is favored because of low power requirements and high reliability [128]. Magnetorheological dampers are often utilized for the vibration damping of vehicle suspensions and civil engineering structures [5, 57, 67, 121]. MR dampers have been also suggested for the semi-active damping of washing machine vibrations in [103].

In addition to experimental vibration damping applications [42, 50, 107], ER materials have been successfully utilized in clutches [72] and brakes, machine tool vibration control in [123], engine suspensions in [125] and others. They have been considered to use in haptic displays and to create bulletproof vests. An ER fluid based composite sandwich material is suggested in [62]. The authors investigated the modal properties of such a system under different electric loads and arrived at the conclusion that the resonant frequencies shift up and peaks decrease with an increasing field strength. Such a composite smart structure could be very interesting for the aviation and space industry. Yet other applications for vibration control through MR-or ER-based dampers are featured in [17, 25, 134, 110].

3.4 Piezoelectric Materials

Piezoelectricity is the ability of certain materials to generate an electric charge in response to an applied stress. If the material is not short-circuited, the charge induces a voltage. This is the so-called *direct piezoelectric effect*. The piezoelectric effect is reversible, meaning that applied voltage generates mechanical stress and deformation. This is known as the *converse piezoelectric effect*. Piezoelectric materials are common in everyday appliances, such as sound reproduction instruments [96], igniters, optical assemblies and other devices. Due to the unique properties of piezoelectric materials, their application is also widespread in active vibration control and its related fields.

The usage of piezoelectric transducers as both sensors and actuators is demonstrated in Fig. 3.14[9] with the Aerostructures Test Wing (ATW) experimental device designed and used at NASA's Dryden Flight Research Center. The ATW is equipped with an array of piezoelectric patches, used in sensor mode to estimate aeroelastic flutter effects on the wing [77]. The wing is intentionally driven to mechanical failure, in order to assess flutter effect. This experiment is carried out by mounting the ATW to a testbed aircraft, but the larger piezoelectric patches are used as actuators to excite the structure to induce structural failure artificially. The figures and the experiment not only demonstrate the dual usage of piezoelectric transducers, but also point out the dramatic effects and real power of mechanical systems driven through piezoceramics.

3.4.1 The Piezoelectric Effect and Materials

A piezoelectric material mounted with electric terminals is illustrated in Fig. 3.15a where no strain change ($\varepsilon_{in} = 0$) is induced in the absence of deformation and no voltage ($V_{out} = 0$ V) can be measured on its terminals. Upon the application of deformation, the strain change induces charge within the material and a measurable voltage on the output. The piezoelectric material acts as a strain or deformation sensor. This direct piezoelectric effect is illustrated in Fig. 3.15b graphically. An initial situation identical to the previous case is illustrated in Fig. 3.15c, where no strain or voltage can be measured on the material and its terminals. Here however, the material works as actuator, taking advantage of the converse piezoelectric effect as illustrated in Fig. 3.15d. Upon the application of an input voltage to the terminals, a deformation thus a strain change is induced in the material. Note that the strain change $\varepsilon_{out} \neq 0$ is maintained only as long as there is an input voltage on the

[9] Courtesy of NASA.

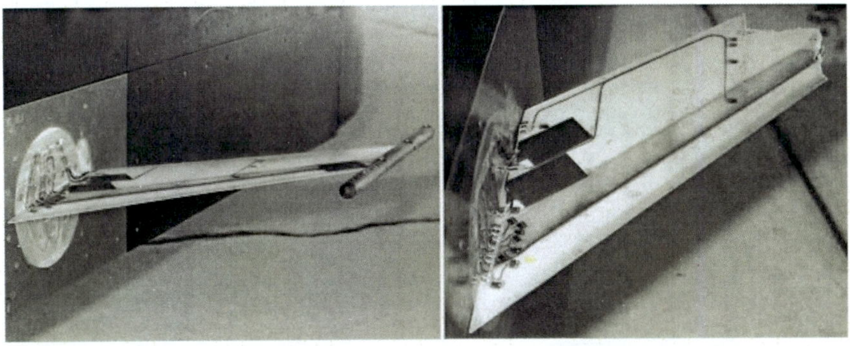

(a) Wing before failure **(b)** Wing after failure

Fig. 3.14 An aerostructures test wing (ATW) is pictured in (**a**), while the same device is shown in (**b**) after intentional failure. Piezoelectric transducers are mounted on the surface and visible on the image [75, 76]

terminals. Moreover, the strain change can be induced with DC voltage change as well. In the direct case, only dynamic strain changes are detected. In other words, steady-state deformations $\varepsilon_{out} \neq 0$ are not included in the output signal V_{out}, only the strain changes are represented.

The direct and converse piezoelectric effect is due to the electric (piezoelectric) domains or dipoles within the material [33]. In fact, the piezoelectric effect is a special case of the electrostrictive effect, where the randomly oriented electric domains in all dielectric materials orient themselves along the flux lines of an applied electric field, causing slight deformations. Although there are naturally occurring piezoelectric materials (quartz,topaz, Rochelle salt, cane sugar), today's applications use engineered polycrystalline materials such as lead-zirconate-titanate (PZT), lead titanate (PT), lead magnesium niobate-lead titanate (PMN-PT); more recently macrofiber composites (MFC) and ammonium dihydrogen phosphate (ADP); or the environmentally friendly lead-free ceramics such as bismuth sodium titanate (BNT), tungsten-bronze (TB), and others [26, 65, 94, 133]. PZT is probably the most commonly utilized piezoelectric material utilized in commercially available transducers. Piezoelectric transducers are relatively cheap to manufacture, and are available in an array of shapes and sizes. The material itself is hard and brittle. By the process of *poling* or *polarization*, the dominant geometry of the piezoelectric material can be set at the manufacturing stage. Both the direct and the converse piezoelectric effects will dominate this *polarization* direction.

The amount of deformation depends on the type of piezoelectric material in use. In addition to the material type, the size, geometry, positioning and the amount of applied voltage are also contributing factors in the achievable deformation. For PZT the percentual elongation is typically 0.1%. In addition to precision systems and nanopositioning, there are numerous examples where piezoelectric transducers can be used directly with or without mechanical amplification. For example [38] considers the use of PZT wafers in cyclic and collective pitch control of UAV

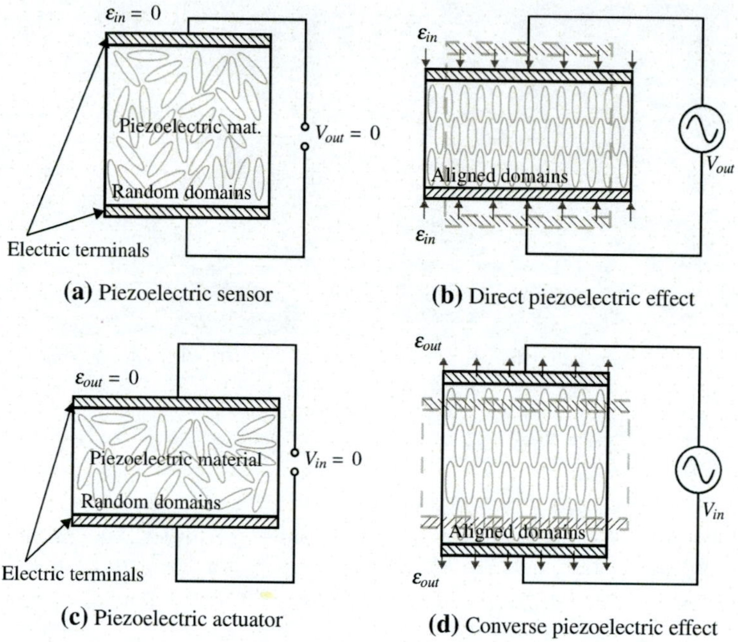

Fig. 3.15 The direct piezoelectric effect, or in other words the piezoelectric material in sensor mode, is illustrated in (**a**) and (**b**). After the application of a deformation to the material causing a strain change, voltage can be measured on the terminal. The reverse of this effect is illustrated in (**c**) and (**d**) where after the application of a voltage, the material deforms

helicopter rotor blades. Forces generated by converse piezoelectricity are enormous, in the order of tens of meganewtons but the small displacement is the reason why this is not so obvious.

From the viewpoint of control, piezoelectric materials have a linear response to the applied voltage [70, 91]. Since piezoceramics have a ferroelectric nature, they can show signs of hysteretic behavior. In fact, the hysteresis of PZT materials is relatively significant and is in the order of 10–15% [91]. In practical terms, *hysteresis* means that the output displacement of the piezoelectric actuator not only depends on the applied voltage, but also depends on how that voltage was applied previously. Hysteresis manifests mainly with high voltage drives, although in scenarios where the actuator is driven by low voltage amplifiers this effect may be practically neglected [32, 48, 70]. The presence of hysteresis may have an effect on stability and closed-loop performance [70, 129]. This type of actuator nonlinearity is mainly important in control systems where the piezoelectric transducers are used to position an object precisely, for example AFM scanning probes [56, 111]. Such quasi-static application fields may utilize a closed-loop compensation of hysteresis, but the use of such techniques in combination with dynamic systems is discouraged, since it may lead to instability [91]. The hysteresis effect in piezoceramic materials and techniques overcoming it are discussed in the book by Moheimani and Fleming [70]. In addition to hysteresis, piezoelectric materials may also demonstrate *creep* [21, 95, 129], which is a tendency of the material to slowly move or deform under the effect of a permanent voltage input. Methods aimed at overcoming hysteresis and creep in piezoelectric materials are discussed in [21, 37, 48] and others. The simulations and experiments presented in the upcoming chapters neglect the effect of hysteresis and creep in the piezoelectric transducers, because they are driven by low voltages in a dynamic application scenario.

3.4.2 Piezoelectric Transducers in Vibration Control

The converse piezoelectric effect may be readily utilized in active vibration control as a source of actuation force. At the same time, the direct piezo effect allows to use piezoelectric materials as sensors as well. The availability, price and electro-mechanical properties of piezoelectric transducers set these devices at the forefront of vibration control applications. Their usage in the field of AVC continues to be popular among both engineering practitioners and researchers.

Because experimental studies aimed at the active vibration control of flexible beams predominantly use piezoceramics as actuating elements, the AVC demonstration device introduced in the upcoming Chap. 5 will also feature piezoelectric actuators. The price range, effectiveness and simplicity of these devices and moreover the possibility to integrate them into active structures renders them as an ideal option for this laboratory application [112, 115]. As a complete review of all possible aspects of the use of piezoceramics in AVC is not in the scope of this work, we will leave the exhausting analysis of modeling techniques, placement optimization aspects and other details to our more experienced colleagues. The available literature on the application of piezoelectric materials, transducer design, placement strategy, types of actuators and their use in vibration control is very broad, we may recommend the excellent books by Fuller et al. [33] and many others [9, 45, 87].

The use of piezoelectric patches for the vibration control of cantilever beams [64, 97, 113, 114, 127] and plates [16, 59, 63] is especially popular in the literature and there is an abundance of publications on the topic of these demonstrative examples. Piezoelectric patches have been suggested for use on systems with dynamics similar to vibrating beams, such as on robotic manipulators arms [98, 124, 132] and satellite boom structures [74]. Further real-life examples of piezoceramics in vibration control are scanning position tables, scanning probe microscopes such as the atomic force microscope (AFM), magnetic force microscope (MFM), micropositioning platforms [29, 60], or the vibration attenuation of civil engineering structures [102]. Less common engineering applications of piezoelectric actuators in AVC are for example the vibration control of grinding machines proposed by Albizuri et al. [2], vibration suppression in gantry machines as proposed by Stöppler in [106], vibration damping of car body structures [52] and gearboxes [40], and aeroelastic wing control [108].

Fig. 3.16 A range of commercial piezoelectric transducers in different shape and size variations, contrasted to the 1 Euro coin for size reference

Figure 3.16 illustrates a range of commercially produced piezoelectric transducers. A wide selection of transducer shape and size configurations is currently available on the market, moreover they can be manufactured according to the needs of the customer. The transducers shown in the figure come in a pre-packaged form with the necessary electric leads bonded on the surface, equipped with a protective foil and a connection terminal. The longer transducer pictured at the bottom (marked as QP45N) and the transducer on the right (marked as QP25N) contain two layers of piezoceramics. These two layers can be used either with the same input signal to achieve larger actuation force or one layer can be utilized as an actuator while the other as sensor to achieve near perfect collocation.

3.4.3 Mathematical Description of the Piezoelectric Effect

The relative expansion or strain of a piezoelectric element is proportional to the applied field strength. It may be calculated very simply, by using the following formula [88]:

$$\varepsilon = \frac{\Delta l}{l_0} d_{ij} E \tag{3.7}$$

$$E = \frac{U}{w_s} \tag{3.8}$$

where d_{ij} is the piezoelectric constant of the material, E is the electric field strength, U the applied voltage and finally w_s is the thickness of the piezoelectric material in the poling direction. The electric behavior of the material is described by

$$\mathbf{D_e} = \varepsilon_\sigma \mathbf{E_e} \tag{3.9}$$

where $\mathbf{D_e}$ is the electric displacement, ε_σ is permittivity and $\mathbf{E_e}$ is the electric field strength. Similarly, the mechanical properties are described by Hooke's law:

$$\mathbf{S} = \mathbf{s}_E \sigma \tag{3.10}$$

where $\mathbf{S}$ is strain, $\mathbf{s}_E$ is the compliance matrix and σ is stress. When we combine these two equations, we obtain the coupled piezoelectric equations in the strain-charge form [33, 87]:

$$\mathbf{S} = \mathbf{s}_E \sigma + \mathbf{d}^T \mathbf{E_e} \tag{3.11}$$

$$\mathbf{D_e} = \mathbf{d}\sigma + \varepsilon_\sigma \mathbf{E_e} \tag{3.12}$$

where $\mathbf{d}$ represents the piezoelectric coupling constants. Subscripts $_E$ indicate zero (or constant) electric field and $_\sigma$ zero (or constant) stress field. Superscript ^{T} denotes matrix transposition. Equation (3.11) may be used to express relations for a piezoelectric actuator and similarly (3.12) is used to describe the behavior of a piezoelectric sensor. For additional clarity, these equations can be expressed for a poled piezoelectric ceramic like PZT in the following form:

$$\begin{bmatrix} S_1 \\ S_2 \\ S_3 \\ S_4 \\ S_5 \\ S_6 \end{bmatrix} = \begin{bmatrix} s_{11}^E & s_{12}^E & s_{13}^E & 0 & 0 & 0 \\ s_{12}^E & s_{11}^E & s_{13}^E & 0 & 0 & 0 \\ s_{13}^E & s_{13}^E & s_{33}^E & 0 & 0 & 0 \\ 0 & 0 & 0 & s_{44}^e & 0 & 0 \\ 0 & 0 & 0 & 0 & s_{44}^e & 0 \\ 0 & 0 & 0 & 0 & 0 & s_{66}^e \end{bmatrix} \begin{bmatrix} \sigma_1 \\ \sigma_2 \\ \sigma_3 \\ \sigma_4 \\ \sigma_5 \\ \sigma_6 \end{bmatrix} + \begin{bmatrix} 0 & 0 & d_{31} \\ 0 & 0 & d_{31} \\ 0 & 0 & d_{31} \\ 0 & d_{15} & 0 \\ d_{15} & 0 & 0 \\ 0 & 0 & 0 \end{bmatrix} \begin{bmatrix} E_1 \\ E_2 \\ E_3 \end{bmatrix} \tag{3.13}$$

$$\begin{bmatrix} D_1 \\ D_2 \\ D_3 \end{bmatrix} = \begin{bmatrix} 0 & 0 & 0 & 0 & d_{15} & 0 \\ 0 & 0 & 0 & d_{15} & 0 & 0 \\ d_{31} & d_{31} & d_{33} & 0 & 0 & 0 \end{bmatrix} \begin{bmatrix} \sigma_1 \\ \sigma_2 \\ \sigma_3 \\ \sigma_4 \\ \sigma_5 \\ \sigma_6 \end{bmatrix} + \begin{bmatrix} \varepsilon_{11} & 0 & 0 \\ 0 & \varepsilon_{11} & 0 \\ 0 & 0 & \varepsilon_{33} \\ 0 & 0 & 0 \end{bmatrix} \begin{bmatrix} E_1 \\ E_2 \\ E_3 \end{bmatrix} \tag{3.14}$$

The previous equation makes use of the crystal symmetry, thus showing only the relevant elements in the matrices. Naturally, for other fundamental types of piezoelectric materials these matrices are in a different form. The elements d_{ij} of the matrix d express the coupling between the electric field in the i direction and the strain in the j direction. In practice, due to crystal symmetries, the coupling matrix d has only a few non-zero elements.

3.4.4 FEM Formulation for Piezoelectric Transducers

The previously introduced Eqs. (3.11) and (3.12) provide the basis for the finite element (FE) formulation of an engineering problem involving a structure with piezoelectric actuators. The approach is similar to the modeling of lumped parameter mechanical systems introduced in Sect. 2.4, with both the mechanical and the electrical terms in the equation of motion. Analogies between the structural damping and dielectric loss, furthermore in between the stiffness / permittivity / coupling are made use of when formulating the underlying set of equations expressed in the following matrix form:

$$\begin{bmatrix} \mathbf{M} & \mathbf{0} \\ \mathbf{0} & \mathbf{0} \end{bmatrix} \begin{bmatrix} \ddot{\mathbf{q}} \\ \ddot{\mathbf{p}} \end{bmatrix} + \begin{bmatrix} \mathbf{B_d} & \mathbf{0} \\ \mathbf{0} & \mathbf{B_p} \end{bmatrix} \begin{bmatrix} \dot{\mathbf{q}} \\ \dot{\mathbf{p}} \end{bmatrix} + \begin{bmatrix} \mathbf{K_s} & \mathbf{K_z} \\ \mathbf{K_z}^T & \mathbf{K_p} \end{bmatrix} \begin{bmatrix} \mathbf{q} \\ \mathbf{p} \end{bmatrix} = \begin{bmatrix} \mathbf{f}_e \\ \mathbf{L_e} \end{bmatrix} \tag{3.15}$$

where $\mathbf{B_d}$ is responsible for structural damping, $\mathbf{B_p}$ for dielectric loss and $\mathbf{M}$ is a mass matrix. Terms $\mathbf{q}$ and $\mathbf{p}$ and their first and second order derivatives express the structural and electrical degrees of freedom. $\mathbf{K_p}$ is anisotropic permittivity, $\mathbf{K_s}$ is the anisotropic stiffness and finally $\mathbf{K_z}$ and is responsible for coupling—or simply stated, the piezoelectric effect.

The above formulation, familiar from lumped parameter vibrating systems is formulated for piezoelectrics using the stress–charge form of the piezoelectric equations, which are given by

$$\sigma = \mathbf{c}_E \mathbf{S} - \mathbf{e_p} \mathbf{E_e} \tag{3.16}$$

$$\mathbf{D_e} = \mathbf{e_p}^T \mathbf{S} + \varepsilon_S \mathbf{E_e}$$

where $\mathbf{e_p}$ are the piezoelectric coupling coefficients in the stress–charge form, $\mathbf{c}_E$ contains stiffness coefficients under constant electric field and ε_S is the electric permittivity matrix under constant stain. As we can see, this equation differs considerably from (3.11) and (3.12), which are given in the strain–charge form. Other representations of the linear piezoelectric equations are also possible, as we can express the constitutive equations in stress–charge, strain–charge, strain–voltage and stress–voltage forms.

Although the FEM representation of the piezoelectric effect is based on the stress–charge form using the $\mathbf{e_p}$ coupling term, the piezoelectric matrix $\mathbf{d}$ can also be directly used and defined in most FEM packages. For example in the ANSYS package, $\mathbf{d}$ is defined in the strain–charge matrix form. This is then internally converted into the piezoelectric stress–charge matrix, using the strain matrix at a constant temperature. Conversion between the strain charge to stress charge is calculated according to [44]:

$$\varepsilon_S = \varepsilon_\sigma - \mathbf{d} \mathbf{s}_E{}^{-1} \mathbf{d}^T \tag{3.17}$$

$$\mathbf{e_p} = \mathbf{d} \mathbf{s}_E{}^{-1} \tag{3.18}$$

$$\mathbf{c_E} = \mathbf{s}_E^{-1} \tag{3.19}$$

where subscripts $_\sigma$, $_S$, $_E$ indicate that the values were evaluated under a constant stress, strain and electric field. The elements of the anisotropic elasticity matrix may also be expressed by the following terms:

$$\mathbf{s}_E = \begin{bmatrix} \frac{1}{E_x} & -\frac{\nu_{xy}}{E_x} & -\frac{\nu_{xz}}{E_x} & 0 & 0 & 0 \\ 0 & \frac{1}{E_y} & 0 & 0 & 0 & 0 \\ 0 & 0 & \frac{1}{E_z} & 0 & 0 & 0 \\ 0 & 0 & 0 & \frac{1}{G_{xy}} & 0 & 0 \\ 0 & 0 & 0 & 0 & \frac{1}{G_{yz}} & 0 \\ 0 & 0 & 0 & 0 & 0 & \frac{1}{G_{xz}} \end{bmatrix} \qquad G_{xy} = \frac{1}{2[s_{11}^E] - [s_{11}^E]} \tag{3.20}$$

where ν_{xz} is Poisson's ratio in a given direction.

The piezoelectric constitutive equations of (3.16) can also be compacted to a matrix term:

$$\begin{bmatrix} \sigma \\ \mathbf{D} \end{bmatrix} = \begin{bmatrix} \mathbf{c}_E & \mathbf{e_p} \\ \mathbf{e_p}^T & -\varepsilon_S \end{bmatrix} \begin{bmatrix} \mathbf{S} \\ -\mathbf{E} \end{bmatrix} \tag{3.21}$$

After applying the variational principle and finite element discretization, one will arrive at the form introduced by (3.15). The damping matrix $\mathbf{B_p}$ responsible for the dielectric loss will contain the negative element dielectric damping matrix, $\mathbf{K_p}$ will contain the negative element dielectric permittivity coefficient matrix and $\mathbf{K_z}$ the piezoelectric matrix [3].

A finite element formulation of a vibrating beam may be created using the Euler Bernoulli model, where the beam section with piezo patches is considered to be laminated and the rest isotropic [36]. The numerical model of Gaudenzi et al. contains not only the mechanic and piezoelectric parts - but also the electronic components like the voltage amplifier.

A practical guide to simulating the electromechanical behavior of a transversally vibrating beam equipped with piezoelectric transducers is given in Appendix. A for those interested. The text lists an ANSYS code with a detailed description of the commands and the process of creating a working model.

3.5 Electrochemical Materials

Electroactive polymers (EAP) are artificial polymer-based smart materials reacting to applied electric current through a change in size. The coupling is electro-mechanical just like in the case of piezoceramic materials. While the normal piezoelectric materials may exhibit large forces, their percentual elongation is typically in the order of mere 0.1%. However, for certain types of EAP this can be more than 300% [11, 131]. Electroactive polymers can be divided into two fundamental classes:

- dielectric
- ionic

3.5.1 Dielectric EAP

In dielectric or "dry" EAP, actuation is caused by electrostatic forces [10]. Dielectric EAP materials work like a capacitor: due to the introduced electrical field the capacitance changes and the actuator compresses in thickness and expands in area. Typically, dielectric EAP actuators exhibit large strains, albeit they require a large actuating voltage in the range of hundreds or even thousands of volts. This type of EAP requires no electrical power to keep the actuator at the desired position.

The most common type of dielectric EAP material is polyvinylidene fluoride (PVDF), which is a part of the ferroelectric polymer family.[10] Electrostatic fields in the range of 200 MV/m can induce strains of 2%, which exceeds the capability of piezoceramics. In addition to the large actuating voltages, the applied electric field in PVDF is also very close to the breakdown potential of the material. Ferroelectric polymers are not to be confused with ferroceramics. The piezoelectric effect is linear and reversible, however electrostriction in PVDF is only one way and nonlinear.

Electro-statically stricted polymer (ESSP) actuators are made from polymers with low elastic stiffness and high dielectric constants. These polymers are then subjected to an electrostatic field to induce deformation. The most common physical configuration of ESSP are rope-like longitudinal or bending type actuators. A very large actuation potential in the range of 100 V/μm is required to induce strains in the order of 10–200%. Moreover, these excessive voltages are very close to the breakdown potential of the material, therefore in practice they have to be lowered. Thin actuators of less than 50 μm are used to decrease the required actuation voltage. Similar to the polarization limits of piezoceramics, different types of EAP also require constraints on the input voltage. An effective way to include these constraints into the controller (while maintaining the stability of the resulting nonlinear control law and full performance optimality) is the application of the model predictive control strategy.

Electro-viscoelastic elastomers (EVE) are a type of dielectric EAP that are closely related to magnetorheological fluids. In fact, before a curing process, EVE behave analogously to MR fluids. The difference between the two types of materials is introduced with a curing cycle, when an electric field is applied to fix in the position of the polar phase in the elastomer. EVE materials can be used as an alternative to MR fluids in vibration control systems [10].

According to the experiments by Palakodeti and Kessler, the relative increase in area of dielectric EAP is roughly linearly dependent on the applied electric field [84]. The highest actuator efficiency has been measured at the largest prestrain and smallest frequency values, while a frequency of 20 Hz caused an efficiency drop to 25%.

---

[10] Due to the inherent similarities with piezoceramic materials, sometimes polyvinylidene fluoride (PVDF) is regarded to be a piezoelectric material. The piezoelectric effect is reversible, however the actuating effect in PVDF is only one way.

3.5.2 Ionic EAP

In ionic EAP, actuation is caused by the diffusion and subsequent displacement of ions. The driving chemical reaction in ionic EAP is a change from an acid to a base environment, in other words the ionic EAP materials have fixed anions and mobile cations [7]. This reaction can be stimulated electrically with embedded electrodes [19]. Upon the application of electric current, the cathode becomes basic while the anode acidic—the cations begin to migrate towards the electrodes. Unfortunately, the response time is slow due to the need of the ions to diffuse through the material, and the electrodes degrade very rapidly. For example in the work of Calvert et al. the swelling of an ionic EAP gel structure to twice its original size took approximately 20 min, while the process could only be repeated 2–3 times due to the chemical degradation of electrodes [19]. Depending on thickness and the kinetics of the chemical reaction, other materials can exhibit actuation changes measured in milliseconds to minutes [10]. While much lower actuation voltages are needed in ionic EAP, larger electrical current is required to start and maintain the ionic flow within the material. The disadvantage is that power is also required to keep the actuator at a given position. Actuators made from ionic EAP are predominantly bending type structures [11]. Due to its fundamental working principle, ionic EAP are also referred to as "wet" EAP [10].

Perfluorinated ion exchange membrane platinum composite (IPMC) treated with an ionic salt and deposited with electrodes on both sides is a common type of ionic EAP material often cited in the literature and utilized in academic research. The ionic base polymer used to create IPMC is typically 200 μm thick [55], and it is available under the market names Nafion or Flemion [4, 31]. IPMC actuated with low voltages in the range of 1–10 V demonstrates large displacements in the sub 0.1 Hz frequency range. Unfortunately, the displacement response decreases very rapidly with increased frequencies [10]. An application of voltages above 1 V induces electrolysis; causing degradation, heat and gas generation. The schematic of the electro-mechanical behavior of an IPMC-based ionic EAP actuator is illustrated in Fig. 3.17.

Carbon nanotubes are another emerging type of ionic EAP actuators. As the name implies, these materials consists of two narrow sheets of carbon nanotubes bonded together through an electrically insulated adhesive layer. This composite structure is then immersed in an electrolyte. Carbon nanotube EAP can achieve strains in the order of 1% [10]. In addition to the usual disadvantages of wet EAP materials, carbon nanotube-based EAP are expensive and their mass production is difficult as well.

As of today, the commercial availability and acceptance of EAP materials is still very limited, most of the EAP research projects utilize custom made materials and actuators. The main limitations of EAP are low actuation forces, lack of robustness and no well-established commercial material types with guaranteed properties. Ionic EAP requires protective coatings to maintain moisture continuously within the material [8]. Bending type structures are the most convenient to manufacture and while the shape change is very easily induced, the actuating forces are weak.

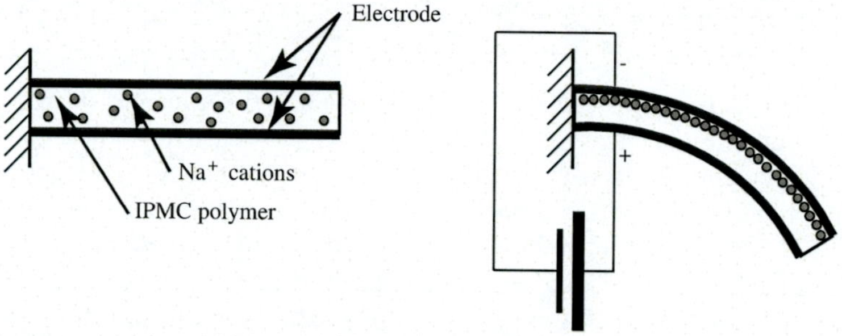

Fig. 3.17 Mechanical behavior of an IPMC based ionic EAP actuator

3.5.3 EAP in Vibration Control

Generally, the use of EAP in active vibration control applications cannot be recommended. This is due to the above-mentioned disadvantages such as the lack of robustness or the technical readiness, availability or acceptance of EAP materials. In addition to that, the effective bandwidth and internal chemical degradation or electrode corrosion makes the use of EAP in vibration control somewhat limited. Recent interest of the scientific and engineering community gives hope for the further progress of the development of electroactive polymers. A more mature EAP technology is certainly a good basis for practical applications in active vibration damping.

Passive structural vibration damping via EAP materials is possible and limited research exists in the field. The damping properties of carbon nanotube EAP enforced structures are for example discussed in [93], while other researchers are examining the passive vibration damping provided by deposited layer of electroactive polymers [66]. Only a handful of academic publications deal with the use of EAP as possible actuators in vibration control. Space mission vibration control through IPMC-based ionic EAP has been suggested without detailed elaboration in [53] by Krishen et al. This is of course only possible with very low bandwidth actuation of large and flexible structures in the sub Hertz range.

An earlier work discusses IPMC-based active vibration control in more detail: the tip of a flexible link has been stabilized by Bandopadhya et al. using ICMP patches controlled by distributed PD controllers [7]. Here EAP is considered instead of piezoceramics because of the lower actuation voltage and the lack of brittleness. It is interesting to note that according to Bandopadhya et al. in addition to the voltage and bending curvature the voltage bending moment relationship remains linear for the considered IPMC material. Moreover, experimental results suggest that, if piezo patches are used as sensors on an IPMC actuated bimorph, the generated voltage is proportional to the tip deflection [8]. The experimental evaluation in this work utilizes a 2 Hz excitation signal, it however demonstrates only a 1 second time response showing the vibration in a higher order mode around 50 Hz. It is unclearwhether

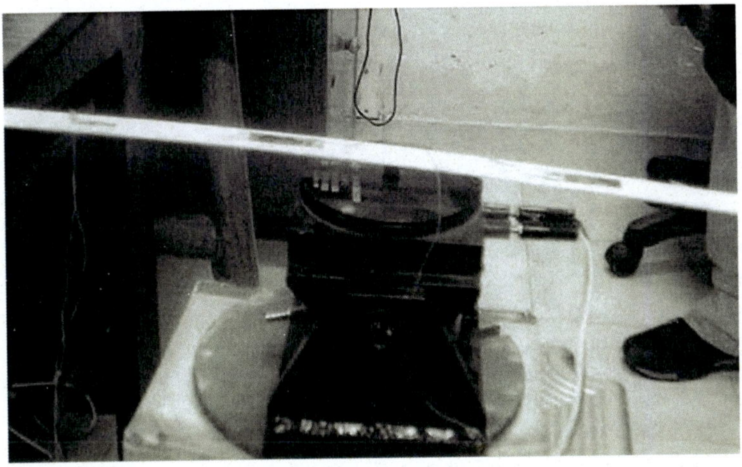

Fig. 3.18 Experimental vibration control setup using electroactive polymer actuators [7, 8]

the vibration damping is due to the active effect of the EAP enabled beam or the damping would even be present with activated EAP patches in a semi-active fashion. The experimental vibration control setup used by Bandopadhya et al., which features electroactive polymer actuators is shown[11] in Fig. 3.18.

Attempts to model the electro-mechanical dynamics of EAP actuators mathematically have been discouraged by the complexity of the task and the several unknown factors such as the humidity in ionic EAP, static prestrain, manufacturing irregularities and other environmental factors. Other challenges in the modeling of EAP include nonlinearity,[12] large mechanical compliance (mismatch between the properties of the material and electrodes), hysteresis and non-homogeneity of the material resulting from the manufacturing process [12]. It is more of an experimental approach, however we must note the work of Lavu et al. in which the authors developed an observer Kalman filter identification (OKID) based state-space multi-model for IPMC taking into account the relative humidity around the actuator [55].

3.6 Other Types of Materials and Actuators

There are several other types of smart materials that have not been explicitly discussed here. These include chemically activated polymers, light-activated materials, magnetically or thermally activated polymers and gels [11].

If an electrically conductive material is subjected to a time-changing magnetic field, eddy currents are formed in the conductor. The eddy currents circulate within the

---

11 Courtesy of Bishakh Bhattacharya.

12 As some researchers have pointed out through experimental tests, models could be linearized as EAP behaves linearly under certain conditions [8, 84].

conductor creating an inside magnetic field which is then mutually interacting with the outside magnetic field thus creating dynamic forces between the conductor and the field. Utilizing this effect, an electromagnet placed in the vicinity of a conductor can be used as an actuator to attenuate vibrations. According to Sodano and Inman, the first use of an active eddy current actuator for vibration suppression is demonstrated in [100]. Eddy current dampers are suggested for the vibration control of a cantilever beam earlier in [6], while Gospodarič et al. used a pair of electromagnets to control vibrations in a ferromagnetic cantilever beam in [39] as well. In a more traditional approach, a mechanical structure can be actuated by simply placing an electromagnet close to its surface and using the magnetic forces between the ferromagnetic materials or permanent magnets and the actuating coil as means of actuation. In this case, the coupling between the electric and magnetic effect is indirect, and we cannot speak of smart materials. For example, a cantilever beam is actuated by an electromagnet placed underneath the beam tip in the work of Fung et al. [34].

This chapter has been mainly concerned with the use of smart materials as actuators or sensors. However, vibrating mechanical systems are often actuated through traditional actuators such as linear motors or hydraulic devices. An electrodynamic shaker is directly utilized as an actuator for the vibration control of a robotic arm by Dadfarnia et al. in [30]. A two degree of freedom mass-spring-damper demonstration device is actuated using a hydraulic piston in a work by VandenBroeck et al. [118, 119].

References

1. Active Materials Laboratory, UCLA (2010) Magnetostriction and magnetostrictive materials. Online, http://aml.seas.ucla.edu/research/areas/magnetostrictive/mag-composites%/Magnetostriction%20and%20Magnetostrictive%20Materials.htm
2. Albizuri J, Fernandes M, Garitaonandia I, Sabalza X, Uribe-Etxeberria R, Hernández J (2007) An active system of reduction of vibrations in a centerless grinding machine using piezoelectric actuators. Int J Mach Tools Manuf 47(10):1607–1614. doi:10.1016/j.ijmachtools.2006.11.004, http://www.sciencedirect.com/science/article/B6V4B-4MR1K43-1/2/aa1014dd203b27a44c75cef37a2adf09
3. Ansys Inc (2009) Release 12.1 Documentation for ANSYS. Ansys Inc. / SAS IP Inc., Canonsburg
4. Asahi Glass (2010) Flemion: a fluoropolymer ion-exchange membrane which protects global environment through outstanding technologies. Tokyo. Online, http://www.agc.com/english/csr/environment/products/positive_seihin5.html
5. Aslam M, Xiong-Liang Y, Zhong-Chao D (2006) Review of magnetorheological (MR) fluids and its applications in vibration control. J Marine Sci Appl 5(3):17–29
6. Bae JS, Kwak MK, Inman DJ (2005) Vibration suppression of a cantilever beam using eddy current damper. J Sound Vib 284(3–5):805–824. doi:10.1016/j.jsv.2004.07.031, http://www.sciencedirect.com/science/article/B6WM3-4F1J8KP-D/2/f6ceedf67a74bb2aadd57e99f8bea787
7. Bandopadhya D, Bhogadi D, Bhattacharya B, Dutta A (2006) Active vibration suppression of a flexible link using ionic polymer metal composite. In: 2006 IEEE conference on robotics, automation and mechatronics, pp 1–6. doi:10.1109/RAMECH.2006.252638

8. Bandopadhya D, Bhattacharya B, Dutta A (2007) An active vibration control strategy for a flexible link using distributed ionic polymer metal composites. Smart Mater Struct 16(3):617. http://stacks.iop.org/0964-1726/16/i=3/a=008
9. Bandyopadhyay B, Manjunath T, Umapathy M (2007) Modeling, control and implementation of smart structures: A FEM-State Space Approach, 1st edn. Springer, Berlin
10. Bar-Cohen Y (2000) Electroactive polymers as artificial muscles —capabilities, potentials and challenges, 1st edn, NTS Inc., Chap 8.11, pp 936–950. Handbook on Biomimetics
11. Bar-Cohen Y (2002) Electro-active polymers: current capabilities and challenges. In: Proceedings of the SPIE smart structures and materials symposium, San Diego, pp 4695–4702
12. Bar-Cohen Y, Sheritt S, Lih SS (2001) Electro-active polymers: current capabilities and challenges. In: Proceedings SPIE 8th annual international symposium on smart structures and materials, Newport, pp 4329–4343
13. Bars R, Colaneri P, de Souza CE, Dugard L, Allgöwer F, Kleimenov A, Scherer C (2006) Theory, algorithms and technology in the design of control systems. Annu Rev Control 30(1):19–30. doi:10.1016/j.arcontrol.2006.01.006, http://www.sciencedirect.com/science/article/B6V0H-4K7WTXH-1/2/705feea29964327e5fd83768ae0f99e2, 2005 IFAC Milestone Reports
14. Bartlett P, Eaton S, Gore J, Metheringham W, Jenner A (2001) High-power, low frequency magnetostrictive actuation for anti-vibration applications. Sens Actuators A 91(1–2):133–136. doi:10.1016/S0924-4247(01)00475-7, http://www.sciencedirect.com/science/article/B6THG-4313YT1-14/2/851ac15043a568a17313eef3042d685f, Third European Conference on Magnetic Sensors & Actuators.
15. Baz A, Imam K, McCoy J (1990) Active vibration control of flexible beams using shape memory actuators. J Sound Vib 140(3):437–456. doi:10.1016/0022-460X(90)90760-W, http://www.sciencedirect.com/science/article/B6WM3-494TCC5-NS/2/7f451727d377dc9c69f873e2c0d85665
16. Birman V, Adali S (1996) Vibration damping using piezoelectric stiffener-actuators with application to orthotropic plates. Compos Struct 35(3):251–261, doi:10.1016/0263-8223(96)00011-6. http://www.sciencedirect.com/science/article/B6TWP-3VS925W-1/2/a20cdc4439f3d1f8f958988ad0b645f3
17. Bouzidane A, Thomas M (2008) An electrorheological hydrostatic journal bearing for controlling rotor vibration. Comput Struct 86(3–5):463–472. doi:10.1016/j.compstruc.2007.02.006, http://www.sciencedirect.com/science/article/B6V28-4NDVGTG-1/2/32b829850e8db109469179cdb7d7d4f6, Smart Structures
18. Braghin F, Cinquemani S, Resta F (2010) A model of magnetostrictive actuators for active vibration control. Sens Actuators A (in press) Corrected Proof:–, doi:10.1016/j.sna.2010.10.019, http://www.sciencedirect.com/science/article/B6THG-51F25N5-4/2/f5cf46980d38877c74a3c4d34fbd894d
19. Calvert P, O'Kelly J, Souvignier C (1998) Solid freeform fabrication of organic-inorganic hybrid materials. Mater Sci Eng C 6(2–3):167–174. doi:10.1016/S0928-4931(98)00046-0, http://www.sciencedirect.com/science/article/B6TXG-3W0G7X7-B/2/6975c02114a4ef48788129ec9b336c73
20. CEDRAT Group (2009) Magnetostrictive actuator prototypes and FEM simulation. Meylan Cedex. Online, http://www.cedrat.com/en/technologies/actuators/magnetic-actuators-motors.html
21. Changhai R, Lining S (2005) Hysteresis and creep compensation for piezoelectric actuator in open-loop operation. Sens Actuators A 122(1):124–130. doi:10.1016/j.sna.2005.03.056, http://www.sciencedirect.com/science/article/pii/S0924424705001512, sSSAMW 04 - Special Section of the Micromechanics Section of Sensors and Actuators based on contributions revised from the Technical Digest of the 2004 Solid-State Sensor, Actuator and Microsystems Workshop
22. Chen Q, Levy C (1999) Vibration analysis of flexible beam by using smart damping structures. Composites Part B: Eng 30:395–406

23. Choi SB, Hwang JH (2000) Structural vibration control using shape memory actuators. J Sound Vib 231(4):1168–1174. doi:10.1006/jsvi.1999.2637, http://www.sciencedirect.com/science/article/B6WM3-45CWW19-HC/2/a61f7091ea2a3e8325d72991c1da1b04
24. Choi SB, Han YM, Kim JH, Cheong CC (2001) Force tracking control of a flexible gripper featuring shape memory alloy actuators. Mechatronics 11(6):677–690, doi:10.1016/S0957-4158(00)00034-9, http://www.sciencedirect.com/science/article/B6V43-43MMSYG-6/2/cec92f6a51314ef45e91c5caf9e5859e
25. Choi SB, Hong SR, Sung KG, Sohn JW (2008) Optimal control of structural vibrations using a mixed-mode magnetorheological fluid mount. Int J Mech Sci 50(3):559–568. doi:10.1016/j.ijmecsci.2007.08.001, http://www.sciencedirect.com/science/article/B6V49-4PD4XHC-1/2/c491dc4a4a881e38b0e20ceef7206dec
26. Choy S, Jiang X, Kwok K, Chan H (2010) Piezoelectric and dielectric characteristics of lead-free BNKLBT ceramic thick film and multilayered piezoelectric actuators. Ceram Int 36(8):2345–2350. doi:10.1016/j.ceramint.2010.07.030, http://www.sciencedirect.com/science/article/B6TWH-50P9H3H-T/2/b62d4dee5b14f6a8678c3fc747ae42e8
27. Clark A, Savage H, Spano M (1984) Effect of stress on the magnetostriction and magnetization of single crystal Tb.27Dy.73Fe2. IEEE Trans Magn 20(5):1443–1445. doi:10.1109/TMAG.1984.1063469
28. Corbi O (2003) Shape memory alloys and their application in structural oscillations attenuation. Simul Modell Pract Theory 11(5–6):387–402. doi:10.1016/S1569-190X(03)00057-1, http://www.sciencedirect.com/science/article/B6X3C-49097BD-1/2/70ef399b2b05b43d6182b568e028b58c, Modeling and Simulation of Advanced Problems and Smart Systems in Civil Engineering
29. Cunningham M, Jenkins D, Clegg W, Bakush M (1995) Active vibration control and actuation of a small cantilever for applications in scanning probe instruments. Sens Actuators A 50(1–2):147–150. doi:10.1016/0924-4247(96)80099-9, http://www.sciencedirect.com/science/article/B6THG-3YVM62R-15/2/ea100dfeea242e7471472799494a5b93
30. Dadfarnia M, Jalili N, Liu Z, Dawson DM (2004) An observer-based piezoelectric control of flexible cartesian robot arms: theory and experiment. Control Eng Pract 12(8): 1041–1053. doi:10.1016/j.conengprac.2003.09.003, http://www.sciencedirect.com/science/article/B6V2H-49WMRJ4-4/2/eb147ee81930b34eb0aba81e90b3a711, Special Section on Emerging Technologies for Active Noise and Vibration Control Systems
31. DuPont (2010) Nafion membranes. Wilmington. Online, http://www2.dupont.com/Automotive/en_US/products_services/fuelCell/nafion.html
32. Fuller C (1990) Active control of sound transmission/radiation from elastic plates by vibration inputs: I. analysis. J Sound Vib 136(1):1–15. doi:10.1016/0022-460X(90)90933-Q, http://www.sciencedirect.com/science/article/pii/0022460X9090933Q
33. Fuller CR, Elliott SJ, Nelson PA (1996) Active Control of Vibration, 1st edn. Academic Press, San Francisco
34. Fung RF, Liu YT, Wang CC (2005) Dynamic model of an electromagnetic actuator for vibration control of a cantilever beam with a tip mass. J Sound Vib 288(4–5):957–980. doi:10.1016/j.jsv.2005.01.046, http://www.sciencedirect.com/science/article/B6WM3-4G4N5VD-1/2/fc3710f0625ef69f19d16c8778a63e58
35. Ganilova O, Cartmell M (2010) An analytical model for the vibration of a composite plate containing an embedded periodic shape memory alloy structure. Compos Struct 92(1):39–47. doi:10.1016/j.compstruct.2009.06.008, http://www.sciencedirect.com/science/article/B6TWP-4WMDHPH-1/2/5c0b8fba0f2e05dad5142ddfcbc48f32
36. Gaudenzi P, Carbonaro R, Benzi E (2000) Control of beam vibrations by means of piezoelectric devices: theory and experiments. Compos Struct 50:373–379
37. Gaul L, Becker J (2009) Model-based piezoelectric hysteresis and creep compensation for highly-dynamic feedforward rest-to-rest motion control of piezoelectrically actuated flexible structures. Int J Eng Sci 47(11–12):1193–1207. doi:10.1016/j.ijengsci.2009.07.006, http://www.sciencedirect.com/science/article/pii/S0020722509001219, Mechanics, Mathematics

and Materials a Special Issue in memory of A.J.M. Spencer FRS—In Memory of Professor A.J.M. Spencer FRS

38. Giannopoulos G, Santafe F, Vantomme J, Buysschaert F, Hendrick P (2006) Smart helicopter blade using piezoelectric actuators for both cyclic and collective pitch control. Multifunctional Structures / Integration of Sensors and Antennas 11(1)
39. Gospodaric B, Voncina D, Bucar B (2007) Active electromagnetic damping of laterally vibrating ferromagnetic cantilever beam. Mechatronics 17(6):291–298. doi:10.1016/j.mechatronics.2007.04.002, http://www.sciencedirect.com/science/article/B6V43-4NVSWV6-1/2/5c4672945cfa9b81238f0b1cb8a8eb13
40. Guan YH, Lim TC, Shepard WS (2005) Experimental study on active vibration control of a gearbox system. J Sound Vib 282(3–5):713–733. doi: 10.1016/j.jsv.2004.03.043, http://www.sciencedirect.com/science/article/B6WM3-4DHXFPJ-4/2/e98625c6c04fd1f5bb5712eb31806f54
41. Hong S, Choi S, Lee D (2006) Comparison of vibration control performance between flow and squeeze mode ER mounts: experimental work. J Sound Vib 291(3–5):740–748, doi:10.1016/j.jsv.2005.06.037, http://www.sciencedirect.com/science/article/B6WM3-4H16P3S-C/2/f98734a231c88a1ec1b43024a2a32f2e
42. Hong SR, Choi SB, Han MS (2002) Vibration control of a frame structure using electro-rheological fluid mounts. Int J Mech Sci 44(10):2027–2045. doi:10.1016/S0020-7403(02)00172-8, http://www.sciencedirect.com/science/article/B6V49-47BX3RX-4/2/53a10ce8cbf8dfa679c34e04beb688e4
43. Ibrahim R (2008) Recent advances in nonlinear passive vibration isolators. J Sound Vib 314(3–5):371–452. doi:10.1016/j.jsv.2008.01.014, http://www.sciencedirect.com/science/article/B6WM3-4S0R6TJ-3/2/8168db91488e18ca41869d56de24ca53
44. Efunda Inc (2007) Constitutive transforms of piezo materials. Sunnyvale. Website, available: http://www.efunda.com/materials/piezo/piezo_math/
45. Inman DJ (2006) Vibration with control. Wiley, Chichester
46. Inman DJ (2007) Engineering Vibrations, 3rd edn. Pearson International Education (Prentice Hall), Upper Saddle River
47. Janke L, Czaderski C, Motavalli M, Ruth J (2005) Applications of shape memory alloys in civil engineering structures: overview, limits and new ideas. Mater Struct 38:578–592. doi:10.1007/BF02479550
48. Janocha H, Kuhnen K (2000) Real-time compensation of hysteresis and creep in piezoelectric actuators. Sens Actuators A 79(2):83–89. doi:10.1016/S0924-4247(99)00215-0, http://www.sciencedirect.com/science/article/pii/S0924424799002150
49. John S, Hariri M (2008) Effect of shape memory alloy actuation on the dynamic response of polymeric composite plates. Composites Part A 39(5):769–776. doi:10.1016/j.compositesa.2008.02.005, http://www.sciencedirect.com/science/article/B6TWN-4RV7YMP-1/2/a8402bbfb476e507253cf32aea87cfb8
50. Jung WJ, Jeong WB, Hong SR, Choi SB (2004) Vibration control of a flexible beam structure using squeeze-mode ER mount. J Sound Vib 273(1–2):185–199. doi:10.1016/S0022-460X(03)00478-4, http://www.sciencedirect.com/science/article/B6WM3-49DFFMM-1/2/1255ad59eca53b0c021632de61aef0b8
51. Knight GPM, UCLA Active Materials Lab (2011) Magnetostrictive materials background. Available http://aml.seas.ucla.edu/research/areas/magnetostrictive/overview.htm
52. Kozek M, Benatzky C, Schirrer A, Stribersky A (2011) Vibration damping of a flexible car body structure using piezo-stack actuators. Control Eng Pract 19(3):298–310. doi:10.1016/j.conengprac.2009.08.001, http://www.sciencedirect.com/science/article/B6V2H-4X3MR4Y-2/2/3ef1d868e70c2b6f10fd9412f9c8c1de, Special Section: IFAC World Congress Application Paper Prize Papers
53. Krishen K (2009) Space applications for ionic polymer-metal composite sensors, actuators, and artificial muscles. Acta Astronautica 64(11–12):1160–1166.

doi:10.1016/j.actaastro.2009.01.008, http://www.sciencedirect.com/science/article/B6V1N-4VM2K65-3/2/f8b0b2d64f274154a5eb59da52fbf524
54. Lau K, Zhou L, Tao X (2002) Control of natural frequencies of a clamped-clamped composite beam with embedded shape memory alloy wires. Compos Struct 58(1):39–47. doi:10.1016/S0263-8223(02)00042-9, http://www.sciencedirect.com/science/article/B6TWP-45XTP9W-N/2/07b9a065ac866d8869a4240deb918851
55. Lavu BC, Schoen MP, Mahajan A (2005) Adaptive intelligent control of ionic polymermetal composites. Smart Mater Struct 14(4):466. http://stacks.iop.org/0964-1726/14/i=4/a=002
56. Lee J, Oh Y, Kim T, Choi M, Jo W (2007) Piezoelectric and electromechanical properties of relaxor ferroelectric Pb(Mg1/3Nb2/3)O3(65%)-PbTiO3(35%) thin films observed by scanning force microscopy. Ultramicroscopy 107(10–11):954–957. doi:10.1016/j.ultramic.2007.02.039, http://www.sciencedirect.com/science/article/B6TW1-4NN6TNC-2/2/a15081aabe1d05ba3faaa00f5797e41d. In: Proceedings of the 8th international conference on scanning robe microscopy, sensors and nanostructures
57. Lee JH, Su RK, Lee PK, Lam LC (2002) Semi-active damping device for vibration control of buildings using magnetorheological fluid. In: Anson M, Ko J, Lam E (eds) Advances in Building Technology, Elsevier, Oxford, pp 969–976, doi:10.1016/B978-008044100-9/50122-4, http://www.sciencedirect.com/science/article/B858K-4PCJRKH-47/2/1c6a74db22e114e2cbddec5d173950f8
58. Li H, Liu S, Wen F, Wen B (2007) Study on dynamic of giant magnetostrictive material transducer with spring of nonlinear stiffness. J Mech Sci Technol 21:961–964. doi:10.1007/BF03027077, http://dx.DOI.org/10.1007/BF03027077
59. Li YY, Cheng L, Li P (2003) Modeling and vibration control of a plate coupled with piezoelectric material. Compos Struct 62(2):155–162. doi:10.1016/S0263-8223(03)00110-7, http://www.sciencedirect.com/science/article/B6TWP-48R1WVK-1/2/f0788ece03ae40a5874f11852e927842
60. Liu YT, Fung RF, Huang TK (2004) Dynamic responses of a precision positioning table impacted by a soft-mounted piezoelectric actuator. Precis Eng 28(3):252–260, doi:10.1016/j.precisioneng.2003.10.005, http://www.sciencedirect.com/science/article/B6V4K-4BMCGD0-1/2/a0f54a4367d9a4fa037b99ba4762a3b9
61. LORD Corporation (2006) Magneto-Rheological (MR) Fluid. LORD Corporation, Cary
62. Lu H, Meng G (2006) An experimental and analytical investigation of the dynamic characteristics of a flexible sandwich plate filled with electrorheological fluid. Int J Adv Manuf Technol 28:1049–1055. doi:10.1007/s00170-004-2433-8, http://dx.DOI.org/10.1007/s00170-004-2433-8
63. Luo Y, Xie S, Zhang X (2008) The actuated performance of multi-layer piezoelectric actuator in active vibration control of honeycomb sandwich panel. J Sound Vib 317(3–5):496–513. doi:10.1016/j.jsv.2008.03.047, http://www.sciencedirect.com/science/article/B6WM3-4SJR2GN-1/2/04c4aad317afe74e20e6f5810f689674
64. Mahmoodi SN, Craft MJ, Southward SC, Ahmadian M (2011) Active vibration control using optimized modified acceleration feedback with adaptive line enhancer for frequency tracking. J Sound Vib 330(7):1300–1311. doi:10.1016/j.jsv.2010.10.013, http://www.sciencedirect.com/science/article/B6WM3-51D894K-1/2/25e8ef1bcadb5fd2aa078de4d678c7f4
65. Maleki M, Naei MH, Hosseinian E, Babahaji A (2011) Exact three-dimensional analysis for static torsion of piezoelectric rods. Int J Solids Struct 48(2):217–226. doi:10.1016/j.ijsolstr.2010.09.017, http://www.sciencedirect.com/science/article/B6VJS-513F90C-5/2/96a4ce72adde5d18a1c509ab880cb797
66. Malinauskas A (2001) Chemical deposition of conducting polymers. Polymer 42(9):3957–3972. doi:10.1016/S0032-3861(00)00800-4, http://www.sciencedirect.com/science/article/B6TXW-42C0RR9-1/2/e8084cbb0f228b86a5cc9d061a340e22
67. McManus SJ, St. Clair KA, Boileau P, Boutin J, Rakheja S (2002) Evaluation of vibration and shock attenuation performance of a suspension seat with a semi-active magnetorheological fluid damper. J Sound Vib 253(1):313–327.

doi:10.1006/jsvi.2001.4262, http://www.sciencedirect.com/science/article/B6WM3-45Y1C16-N/2/33a165ac8f2fe7d8fad7bd83d9484957
68. Memory-Metalle GmbH (2010) Infosheet no. 5: The memory effects—an introduction. White paper, Memory-Metalle GmbH, Weil am Rhein. http://www.memory-metalle.de/html/03_knowhow/PDF/MM_05_introduction_e.pdf, available online (1 page)
69. MIDÉ (2007) Shape memory alloy starter kit—reference manual. MIDÉ Technology, Medford
70. Moheimani S, Fleming AJ (2006) Piezoelectric transducers for vibration control and damping. Springer, London
71. Monkman GJ (1995) The electrorheological effect under compressive stress. J Phys D: Appl Phys 28(3):588. http://stacks.iop.org/0022-3727/28/i=3/a=022
72. Monkman GJ (1997) Exploitation of compressive stress in electrorheological coupling. Mechatronics 7(1):27–36. doi:10.1016/S0957-4158(96)00037-2, http://www.sciencedirect.com/science/article/B6V43-3WDCFB7-3/2/00d7c7757dd73812cfe88867f704ba25
73. Moon SJ, Lim CW, Kim BH, Park Y (2007) Structural vibration control using linear magnetostrictive actuators. J Sound Vib 302(4–5):875–891. doi:10.1016/j.jsv.2006.12.023, http://www.sciencedirect.com/science/article/B6WM3-4N2M6HH-5/2/417522adfca8640acfa76e890ae0533c
74. Moshrefi-Torbati M, Keane A, Elliott S, Brennan M, Anthony D, Rogers E (2006) Active vibration control (AVC) of a satellite boom structure using optimally positioned stacked piezoelectric actuators. J Sound Vib 292(1–2):203–220. doi:10.1016/j.jsv.2005.07.040, http://www.sciencedirect.com/science/article/pii/S0022460X050 05171
75. NASA Dryden Flight Research Center (NASA-DFRC) (2001) The Aerostructures Test Wing (ATW)— after intentional failure. Image ID: EC01-0124-1
76. NASA Dryden Flight Research Center (NASA-DFRC) (2001) The Aerostructures Test Wing (ATW)— before failure. Image ID: EC01-0086-4
77. NASA Dryden Flight Research Center (NASA-DFRC) (2001) The Aerostructures Test Wing (ATW) experiment (description). Online, http://nix.larc.nasa.gov/info;jsessionid=6f0l4hj7bt39u?id=EC01-0086-4\&orgid=7
78. NASA Glenn Research Center (NASA-GRC) (2008) Shape Memory Alloy (SMA) Demonstration Hardware. Image ID: C-2008-02698
79. NASA Glenn Research Center (NASA-GRC) (2008) Shape Memory Alloy (SMA) Demonstration Hardware. Image ID: C-2008-02707
80. NASA Langley Research Center (NASA-LaRC) (1996) High Displacement Actuator (HDA). Image ID: EL-1996-00133
81. NASA Langley Research Center (NASA-LaRC) (2000) F-15 model in the 16 foot transonic tunnel. Image ID: EL-2000-00147
82. NASA Marshall Space Flight Center (NASA-MSFC) (2002) Comparison of magnetorheological fluids on earth and in space. Image ID: MSFC-0700441
83. Ngatu GT, Wereley NM, Karli JO, Bell RC (2008) Dimorphic magnetorheological fluids: exploiting partial substitution of microspheres by nanowires. Smart Mater Struct 17(4):045,022, http://stacks.iop.org/0964-1726/17/i=4/a=045022
84. Palakodeti R, Kessler M (2006) Influence of frequency and prestrain on the mechanical efficiency of dielectric electroactive polymer actuators. Mater Lett 60(29–30):3437–3440. doi:10.1016/j.matlet.2006.03.053, http://www.sciencedirect.com/science/article/B6TX9-4JN2J11-2/2/bb86365e0ad88fd27e7dad13bd4d5ac0
85. Park JS, Kim JH, Moon SH (2004) Vibration of thermally post-buckled composite plates embedded with shape memory alloy fibers. Compos Struct 63(2):179–188, doi:10.1016/S0263-8223(03)00146-6, http://www.sciencedirect.com/science/article/B6TWP-48Y6PS8-6/2/d70ea3b2717f54027d999c7fe92da11f
86. Phillips DJ, Hyland DC, Collins CG (2002) Real-time seismic damping and frequency control of steel structures using nitinol wire. In: Proceedings of SPIE 2002, vol 4696, pp 176–185

87. Piefort V (2001) Finite element modelling of piezoelectric structures. PhD thesis, Université Libre de Bruxelles
88. Piezosystem-Jena (2007) Piezoline theory. Available: http://www.piezojena.com/files.php4?dl_mg_id=229\&file=dl_mg_1195142143.pdf\&SID=125nfb5prkpt35d2k9cops3021
89. Pilgrim SM (2001) Electrostrictive ceramics for sonar projectors. In: Buschow KHJ, Cahn RW, Flemings MC, Ilschner B, Kramer EJ, Mahajan S, Veyssicre P (eds) Encyclopedia of materials: science and technology, Elsevier, Oxford, pp 2738–2743. doi:10.1016/B0-08-043152-6/00488-5, http://www.sciencedirect.com/science/article/B7NKS-4KF1VT9-MR/2/9a3a6454bb6c54bb64719b3036d2789b
90. Pradhan S (2005) Vibration suppression of FGM shells using embedded magnetostrictive layers. Int J Solids Struct 42(9–10):2465–2488, doi:10.1016/j.ijsolstr.2004.09.049, http://www.sciencedirect.com/science/article/B6VJS-4F6SSGN-1/2/b6f9e2e6ffc65bfc0c4af5083e37df0b
91. Preumont A (2002) Vibration control of active structures, 2nd edn. Kluwer Academic Publishers, Dordrecht
92. Preumont A, Seto K (2008) Active control of structures, 3rd edn. Wiley, Chichester
93. Rajoria H, Jalili N (2005) Passive vibration damping enhancement using carbon nanotube-epoxy reinforced composites. Compos Sci Technol 65(14):2079–2093. doi:10.1016/j.compscitech.2005.05.015, http://www.sciencedirect.com/science/article/B6TWT-4GHBPN0-5/2/a67d954050aac7829a56e3e4302c8ef6
94. Ren TL, Zhao HJ, Liu LT, Li ZJ (2003) Piezoelectric and ferroelectric films for microelectronic applications. Mater Sci Eng B 99(1–3):159–163. doi:10.1016/S0921-5107(02)00466-X, http://www.sciencedirect.com/science/article/B6TXF-47P92FB-4/2/4f139e078631af2acf841b7db5d37316, Advanced electronic-ceramic materials. Proceedings of the 8th IUMRS international conference on electronic materials (IUMRS-ICEM2002), Symposium N
95. Richter H, Misawa EA, Lucca DA, Lu H (2001) Modeling nonlinear behavior in a piezoelectric actuator. Precis Eng 25(2):128–137. doi:10.1016/S0141-6359(00)00067-2, http://www.sciencedirect.com/science/article/pii/S0141635900000672
96. Rossing TD, Moore RF, Wheeler PA (2001) The science of sound. 3rd edn. Addison Wesley, San Francisco
97. Schlacher K, Kugi A, Irschik H (1998) $\mathcal{H}_\infty$-control of random structural vibrations with piezoelectric actuators. Comput Struct 67(1–3):137–145. doi:10.1016/S0045-7949(97)00165-X, http://www.sciencedirect.com/science/article/B6V28-3VKTNM7-J/2/faecf09ede8e0d56452351d4d30bc45b
98. Shin HC, Choi SB (2001) Position control of a two-link flexible manipulator featuring piezoelectric actuators and sensors. Mechatronics 11(6):707–729. doi:10.1016/S0957-4158(00)00045-3, http://www.sciencedirect.com/science/article/B6V43-43MMSYG-8/2/868a8129a5636d164e6aa1a89358b8fb
99. Sitnikova E, Pavlovskaia E, Wiercigroch M, Savi MA (2010) Vibration reduction of the impact system by an SMA restraint: numerical studies. Int J Non Linear Mech 45(9):837–849. doi:10.1016/j.ijnonlinmec.2009.11.013, http://www.sciencedirect.com/science/article/B6TJ2-4Y4PVRH-1/2/6c55c60479b498920a5272d281d5fd5d, dynamics, control and design of nonlinear systems with smart structures
100. Sodano HA, Inman DJ (2007) Non-contact vibration control system employing an active eddy current damper. J Sound Vib 305(4–5):596–613. doi:10.1016/j.jsv.2007.04.050 http://www.sciencedirect.com/science/article/B6WM3-4P2J38T-3/2/a75ecce7ed7841e00499a50d077bd23c
101. Song G, Ma N, Li HN (2006) Applications of shape memory alloys in civil structures. Eng Struct 28(9):1266–1274. doi:10.1016/j.engstruct.2005.12.010, http://www.sciencedirect.com/science/article/B6V2Y-4JRM0BH-1/2/ccac39dd197a641451117034df623bd1
102. Song G, Sethi V, Li HN (2006) Vibration control of civil structures using piezoceramic smart materials: a review. Eng Struct 28(11):1513–1524,

doi:10.1016/j.engstruct.2006.02.002, http://www.sciencedirect.com/science/article/B6V2Y-4JKRTRW-3/2/73299aa5fc196cb9802cbc961b896402

103. Spelta C, Previdi F, Savaresi SM, Fraternale G, Gaudiano N (2009) Control of magnetorheological dampers for vibration reduction in a washing machine. Mechatronics 19(3):410–421. doi:10.1016/j.mechatronics.2008.09.006, http://www.sciencedirect.com/science/article/B6V43-4TT1G22-1/2/3d8e5bd1cc63e7181272ef848f15508c
104. Stangroom JE (1983) Electrorheological fluids. Phys Technol 14(6):290. http://stacks.iop.org/0305-4624/14/i=6/a=305
105. Stanway R (2004) Smart fluids: current and future developments. Mater Sci Technol 20(8): 931–939
106. Stöppler G, Douglas S (2008) Adaptronic gantry machine tool with piezoelectric actuator for active error compensation of structural oscillations at the tool centre point. Mechatronics 18(8):426–433, doi:10.1016/j.mechatronics.2008.03.002, http://www.sciencedirect.com/science/article/B6V43-4SC5PJV-2/2/ca3951fdc7496b096233f1bd02e0898a
107. Su YX, Duan BY, Wei Q, Nan RD, Peng B (2002) The wind-induced vibration control of feed supporting system for large spherical radio telescope using electrorheological damper. Mechatronics 13(2):95–110. doi:10.1016/S0957-4158(01)00042-3, http://www.sciencedirect.com/science/article/B6V43-46WPHMS-2/2/eca7cd44909e99a1f8c6ad76a4fd4f19
108. Suleman A, Costa AP (2004) Adaptive control of an aeroelastic flight vehicle using piezoelectric actuators. Comput Struct 82(17–19):1303–1314. doi:10.1016/j.compstruc.2004.03.027, http://www.sciencedirect.com/science/article/B6V28-4CPD4P5-1/2/5a9c6f2a79e0f4e978f43d2b6ed45b93, computational mechanics in Portugal
109. Suleman A, Burns S, Waechter D (2004) Design and modeling of an electrostrictive inchworm actuator. Mechatronics 14(5):567–586. doi:10.1016/j.mechatronics.2003.10.007, http://www.sciencedirect.com/science/article/B6V43-49YD001-1/2/9b6a81729c9e0f3c7f59ce4bafbe5c2a
110. Sung KG, Han YM, Cho JW, Choi SB (2008) Vibration control of vehicle ER suspension system using fuzzy moving sliding mode controller. J Sound Vib 311(3–5): 1004–1019. doi:10.1016/j.jsv.2007.09.049, http://www.sciencedirect.com/science/article/B6WM3-4R2H1TN-4/2/b3a297765c3ac7767b2d64fda7a6a3d7
111. Tabak F, Disseldorp E, Wortel G, Katan A, Hesselberth M, Oosterkamp T, Frenken J, van Spengen W (2010) MEMS-based fast scanning probe microscopes. Ultramicroscopy 110(6):599–604. doi:10.1016/j.ultramic.2010.02.018, http://www.sciencedirect.com/science/article/B6TW1-4YJCKXY-1/2/4f5b9ba5875b8066d7cb20174f05ad61, 11th International scanning probe microscopy conference
112. Takács G, Rohal'-Ilkiv B (2009) Implementation of the Newton-Raphson MPC algorithm in active vibration control applications. In: Mace BR, Ferguson NS, Rustighi E (eds) Proceedings of the 3rd international conference on noise and vibration: emerging methods, Oxford
113. Takács G, Rohal'-Ilkiv B (2009) MPC with guaranteed stability and constraint feasibility on flexible vibrating active structures: a comparative study. In: Hu H (ed) Proceedings of the 11th IASTED international conference on control and applications, Cambridge
114. Takács G, Rohal'-Ilkiv B (2009) Newton-Raphson based efficient model predictive control applied on active vibrating structures. In: Proceedings of the European control control conference, Budapest
115. Takács G, Rohal'-Ilkiv B (2009) Newton-Raphson MPC controlled active vibration attenuation. In: Hangos KM (ed) Proceedings of the 28th IASTED international conference on modeling, identification and control, Innsbruck
116. TRS Technologies (2010) Electrostrictive materials. State College. Online, http://www.trstechnologies.com/Materials/electrostrictive_materials.php
117. Tzou H, Chai W (2007) Design and testing of a hybrid polymeric electrostrictive/piezoelectric beam with bang-bang control. Mech Syst Sig Process 21(1):417–429. doi:10.1016/j.ymssp.2005.10.008, http://www.sciencedirect.com/science/article/B6WN1-4HR75KY-1/2/73701e5908a2ea598fa7bec1ce093563

118. Van den Broeck L, Diehl M, Swevers J (2009) Time optimal MPC for mechatronic applications. In: Proceedings of the 48th IEEE conference on decision and control, Shanghai, pp 8040–8045
119. Van den Broeck L, Swevers J, Diehl M (2009) Performant design of an input shaping prefilter via embedded optimization. In: Proceedings of the 2009 American control conference, St-Louis, pp 166–171
120. Vereda F, de Vicente J, Hidalgo-Álvarez R (2009) Physical properties of elongated magnetic particles: Magnetization and friction coefficient anisotropies. ChemPhysChem 10:1165–1179
121. Wahed AKE, Sproston JL, Schleyer GK (2002) Electrorheological and magnetorheological fluids in blast resistant design applications. Mater Des 23(4):391–404. doi: 10.1016/S0261-3069(02)00003-1, http://www.sciencedirect.com/science/article/B6TX5-450HD50-1/2/0da443f054d99983150525d47bf17aeb
122. Wang H, Zheng H, Li Y, Lu S (2008) Mechanical properties of magnetorheological fluids under squeeze-shear mode. In: Society of photo-optical instrumentation engineers (SPIE) conference series, presented at the Society of Photo-optical Instrumentation Engineers (SPIE) conference, vol 7130. doi:10.1117/12.819634
123. Wang M, Fei R (1999) Chatter suppression based on nonlinear vibration characteristic of electrorheological fluids. Int J Mach Tools Manuf 39(12):1925–1934. doi:10.1016/S0890-6955(99)00039-5, http://www.sciencedirect.com/science/article/B6V4B-3X7N8GJ-7/2/6cc38d51af69b4fbb0aa1135681b5356
124. Wei JJ, Qiu ZC, Han JD, Wang YC (2010) Experimental comparison research on active vibration control for flexible piezoelectric manipulator using fuzzy controller. J Intell Rob Syst 59:31–56, doi:10.1007/s10846-009-9390-2, http://dx.DOI.org/10.1007/s10846-009-9390-2
125. Williams E, Rigby S, Sproston J, Stanway R (1993) Electrorheological fluids applied to an automotive engine mount. J Non-Newtonian Fluid Mech 47:221–238. doi: 10.1016/0377-0257(93)80052-D, http://www.sciencedirect.com/science/article/B6TGV-44V49DV-75/2/a6f4db8ffcb810f6167c845a984dd93f
126. Williams K, Chiu G, Bernhard R (2002) Adaptive-passive absorbers using shape-memory alloys. J Sound Vib 249(5):835–848. doi:10.1006/jsvi.2000.3496, http://www.sciencedirect.com/science/article/B6WM3-4576DS3-2N/2/63e7f46640d919db867f8b1e391f4c4c
127. Wills AG, Bates D, Fleming AJ, Ninness B, Moheimani SOR (2008) Model predictive control applied to constraint handling in active noise and vibration control. IEEE Trans Control Syst Technol 16(1):3–12
128. Yan G, Sun B, Lü Y (2007) Semi-active model predictive control for 3rd generation benchmark problem using smart dampers. Earthq Eng Eng Vib 6:307–315. doi:10.1007/s11803-007-0645-2, http://dx.DOI.org/10.1007/s11803-007-0645-2
129. Yeh TJ, Ruo-Feng H, Shin-Wen L (2008) An integrated physical model that characterizes creep and hysteresis in piezoelectric actuators. Simul Modell Pract Theory 16(1):93–110. doi:10.1016/j.simpat.2007.11.005, http://www.sciencedirect.com/science/article/pii/S1569190X07001396
130. Yongsheng R, Shuangshuang S (2007) Large amplitude flexural vibration of the orthotropic composite plate embedded with shape memory alloy fibers. Chin J Aeronaut 20(5):415–424. doi:10.1016/S1000-9361(07)60063-6, http://www.sciencedirect.com/science/article/B8H0X-4R5R8KJ-5/2/dfb4a7094008193f73ce3269cb319dbe
131. Yuse K, Guyomar D, Kanda M, Seveyrat L, Guiffard B (2010) Development of large-strain and low-powered electro-active polymers (EAPs) using conductive fillers. Sens Actuators A. In Press, Accepted Manuscript. doi:10.1016/j.sna.2010.08.008, http://www.sciencedirect.com/science/article/B6THG-50RVNJK-5/2/4002a5f80bb3323c2a1a5af618b089ca
132. Zhang X, Lu J, Shen Y (2003) Active noise control of flexible linkage mechanism with piezoelectric actuators. Comput Struct 81(20):2045–2051. doi:10.1016/S0045-7949(03)00230-X, http://www.sciencedirect.com/science/article/B6V28-49036KY-2/2/418991ffedba1e4e78d0f90c263b465e

133. Zhou C, Liu X, Li W, Yuan C, Chen G (2010) Structure and electrical properties of Bi0.5 (Na, K)0.5 TiO_3-BiGao3 lead-free piezoelectric ceramics. Curr Appl Phys 10(1):93–98. doi:10.1016/j.cap.2009.05.004, http://www.sciencedirect.com/science/article/B6W7T-4WBC1T8-1/2/8caea987dd9c442dea9061d54474df0e
134. Zhu C (2005) A disk-type magneto-rheological fluid damper for rotor system vibration control. J Sound Vib 283(3-5):1051–1069. doi:10.1016/j.jsv.2004.06.031, http://www.sciencedirect.com/science/article/B6WM3-4F4H9R2-1/2/48abebbf8d1230fcd80eee7d19fe52fa

Chapter 4
Algorithms in Active Vibration Control

Significant research effort has been devoted to explore the means and properties of active vibration attenuation in the last two to three decades. The complex discipline of active vibration control (AVC) is now leaving the realm of experimental applications and slowly starting to appear in advanced commercial products. In addition to the hardware components in AVC, such as the actuators performing changes in system dynamics and sensors providing feedback information about vibration levels, an essential unit of this feedback system is the control strategy itself.

The field of engineering concerned with the design of control strategies and algorithms for dynamic systems is called *control engineering*. Active vibration control heavily relies on the recent theoretical results of control engineering. Generic strategies commonly used for dynamic plants ranging from missile cruise control to washing machines are also applicable to AVC systems. These strategies can be as simple as multiplying a feedback signal with a constant gain and supplying it back to the actuator; or may be complex online optimization based on ideas such as the model predictive control (MPC) algorithm. The aim of this chapter is thus to present a few essential control algorithms which are commonly used in active vibration control and to provide examples of their utilization in academic studies. It is by no means our goal to replicate intricate details and results of control theory such as robustness or uncertainty. It is not our ambition to list and describe every possible control strategy either. The current chapter serves mainly as a taste of strategies utilized for AVC other than MPC. The reader interested in general aspects of control engineering is referred to feedback control textbooks such as the book on the basics of classical control theory by Antsaklis and Michel [10]; on state-space control by Williams and Lawrence [118] and others works [9, 50, 73, 74].

The first section is concerned with classical control methods applied to active vibration attenuation. These methods are often referred to as position or velocity feedback where the system simply uses a fixed gain to multiply position, velocity or acceleration signal in order to compute an input signal, which is in turn supplied to the actuator. Section 4.2 discusses proportional-integral-derivative (PID) controllers, which are very commonly used in all areas of engineering, including AVC.

G. Takács and B. Rohal'-Ilkiv, *Model Predictive Vibration Control*,
DOI: 10.1007/978-1-4471-2333-0_4, © Springer-Verlag London Limited 2012

Fig. 4.1 An excellent example of the possibilities offered by active vibration control: a modified F/A-18A test aircraft with active aeroelastic wings [76] is shown with accelerometers visible on the wing surfaces. In addition, the actuators and sensors the performance of active systems depends on the choice and design of control algorithms as well (Courtesy of NASA)

The next two sections review slightly more advanced controller strategies: Sect. 4.3 deals with linear quadratic control, while Sect. 4.4 considers $\mathscr{H}_\infty$ control—both being basic optimization-based methods. Of these two methods, linear quadratic control is particularly interesting, since it is used both extensively and effectively in AVC. Moreover, it can be regarded as a basis for the model predictive control approach. The last major section of this chapter reviews the more exotic control approaches that are exciting and potentially powerful, albeit seldom used in the area of active vibration control due to several practical limitations. These methods are referred to as soft computing approaches and we will cover the essential basics of neural networks, genetic algorithms and fuzzy control here. The former two approaches rely on ideas borrowed from nature, that is the working principles of the nervous system and evolutionary processes. The latter, fuzzy control utilizes the idea of fuzzy sets where complex dynamics can be controlled using common sense and trivial statements. The reason why we have decided to briefly cover these less common methods is their potential to control and model highly nonlinear and hysteretic dynamics, such as magnetorheological (MR) dampers. Finally, Sect. 4.6 will mention some of the alternative algorithms which can be used in active vibration control in a nutshell.

This book is first and foremost concerned with the application of model predictive control to vibration attenuation, therefore this control strategy will be introduced in a comprehensive detail starting from the next part—Part II. As it has been mentioned before, the current chapter is only concerned with algorithms other than MPC; hence we will only give brief description in order to familiarize the reader with the idea. Model predictive control is an advanced control algorithm where optimal inputs are calculated based on predictions given by an internal plant model. Model predictive control belongs to the broader family of algorithms based on optimal control. Unlike in the case of for instance linear quadratic (LQ) control, MPC does not only generate optimal inputs but it also actively respects various system constraints. As with every conceivable real plant, vibration control systems have inherent limitations—such as actuator saturation and others. These limitations render the system nonlinear, thus precautions are to be made to guarantee the stability of the control system. As it will be elaborated later, the application of model predictive control to vibration attenuation can be a non-trivial task, mainly because of the high sampling rates necessary because of fast system dynamics.

Books on the topic of vibration mechanics and vibration control start to adapt to the new trend of AVC and set aside chapters on control theory. In addition to the material provided by this chapter, an excellent treatment of vibration control concepts for a single degree of freedom vibrating system is given in the recent book by Benaroya and Nagurka [13]. For those with minimal or no background in control theory, parts[1] of this publication may give a fast yet still very thorough discussion on classical transfer function-based control strategy, and state-space control related to the problem of vibration. Moreover, there is an abundance of excellent literature discussing control theory from the active vibration control viewpoint such as the classical book on AVC by Fuller et al. [42] or Inman [55], Preumont [96] and others [49, 56, 97].

Figure 4.1 illustrates a heavily modified F/A-18A test aircraft equipped with actuators and accelerometers to create a complex control system that is capable of altering the aerodynamic properties of the plane. Just as in the case of this aircraft with active aeroelastic wings—or any other control system—the proper choice of hardware in AVC is not the only important aspect of the design. The overall effectiveness and safety of the system also depends on a fitting control architecture.

4.1 Classical Feedback Methods

When the vibration signal measured by the sensors is simply amplified by a gain and fed back to the actuators, we may classify this type of feedback control system as a *classical feedback* method. To demonstrate the concept mathematically, let us consider a vibrating system described as a continuous state-space system or in other words given by a set of first order differential equations [32]:

---

1 See Chap. 6 in [13].

$$\begin{aligned} \dot{\mathbf{x}}(t) &= \mathbf{A}\mathbf{x}(t) + \mathbf{B}\mathbf{u}(t) \\ \mathbf{y}(t) &= \mathbf{C}\mathbf{x}(t) + \mathbf{D}\mathbf{u}(t) \end{aligned} \tag{4.1}$$

where **A** is the state matrix, **B** is the input matrix and **C** is the output matrix. Because the term $\mathbf{Du}(t)$ represents direct input–output feedthrough, it is omitted from our representation[2]. Although it is more common to represent system models in classical control theory by continuous (Laplace domain) or discrete (Z-domain) transfer functions, the state-space representation will be preferred here, since the model predictive controllers (MPC) introduced in upcoming chapters will also utilize a state-space model.

Vibration control literature often uses to the terms *position feedback* and *velocity feedback* [55, 97]. These two very common control methods are a part of classical feedback-based vibration control. The idea in *direct* position or velocity feedback methods is in fact very simple: in direct position feedback, the position signal is amplified by a gain and fed back to the force actuators, while in direct velocity feedback the sensor output is differentiated, amplified and fed back to the force actuators. According to this underlying idea, we may define the input in direct position feedback as [42]:

$$\mathbf{u}(t) = -\mathbf{K}\mathbf{y}(t) \tag{4.2}$$

where **K** is the feedback gain matrix. In case the signal is based on velocity measurement, we have a velocity feedback and the input is given by [96, 97]:

$$\mathbf{u}(t) = -\mathbf{K}\dot{\mathbf{y}}(t) \tag{4.3}$$

Furthermore, it is also possible to utilize the acceleration measurement and formulate control input as

$$\mathbf{u}(t) = -\mathbf{K}\ddot{\mathbf{y}}(t) \tag{4.4}$$

but the use of a second order filter to generate a force proportional to the output of that filter is also a possible strategy when using direct acceleration feedback.

Note that unlike in the case of linear quadratic (LQ) control, here the input is not calculated using the state $\mathbf{x}(t)$, instead the output $\mathbf{y}(t)$ and its derivatives are utilized. Considering the case of direct position feedback given by (4.2) and substituting the output equation the input will be rendered to

$$\mathbf{u}(t) = -\mathbf{K}\mathbf{C}\mathbf{x}(t) \tag{4.5}$$

where the input term can be now substituted back to the original state equation in (4.1) to get

$$\dot{\mathbf{x}}(t) = (\mathbf{A} - \mathbf{B}\mathbf{K}\mathbf{C})\mathbf{x}(t) \tag{4.6}$$

2 Matrix D is omitted when accelerometers are not used for measuring output.

In case the sensors and actuators are co-located, we may assume that $\mathbf{C} = \mathbf{B}^T$, rendering the output equation to $\mathbf{y}(t) = \mathbf{B}^T \mathbf{x}(t)$.

In direct velocity feedback, two very common approaches exist. These are [113]:

- constant amplitude velocity feedback (CAVF)
- constant gain velocity feedback (CGVF)

In constant amplitude velocity feedback (CAVF), the gain of the i-th actuator is the opposite of that of the i-th sensor. This is mathematically denoted as:

$$\mathbf{u}(t) = -\mathbf{K}\,\mathrm{sign}(\dot{\mathbf{y}}(t)) \tag{4.7}$$

One may easily see that this control law is nonlinear and discontinuous. As the name implies, the feedback voltage amplitude in CAVF is constant. The feedback gain matrix used in this approach is defined by a diagonal matrix of the individual constant amplitudes $\bar{A}_i$:

$$\mathbf{K} = [0\ \mathrm{diag}(\bar{A}_1\ \bar{A}_2\ \ldots\ \bar{A}_i\ \ldots \bar{A}_N)] \tag{4.8}$$

where $i = 1 \ldots N$ is the number of actuating points. In constant gain velocity feedback the driving voltage of the i-th actuator is given by the relation introduced earlier in (4.2). The gain matrix here is simply a diagonal matrix of individual actuator gains, creating a linear continuous controller:

$$\mathbf{K} = [0\ \mathrm{diag}(K_1\ K_2\ \ldots\ K_i\ \ldots\ K_N)] \tag{4.9}$$

where $i = 1 \ldots N$ is the number of actuating points, and K_i are the associated gains.

Direct position feedback is not to be confused with the different concept of *positive position feedback* (PPF) [37, 45]. While in direct position/velocity/acceleration feedback the use of a position, velocity or acceleration sensor in combination of force actuator is implied, positive position feedback assumes a strain sensor in combination with a strain actuator [96]. The best-known combination of such a sensor and actuator pair are the commonly used piezoceramic transducers. The essential idea behind positive position feedback is to use the position signal in combination with a second order filter to generate the output for the strain actuator. The second order filter shall have an increased damping, ultimately attenuating vibrations in the closed-loop system. Preumont suggests the use of such a system in a decentralized manner, with co-located sensor and actuator pairs, thereby establishing stability [96]. An advantage of the PPF approach is that it can be designed based on an experimental transfer function, without a deeper analytical knowledge of the structure [55]. For a SISO system, the PPF control input is based on the strain signal and its transfer function can be given by [55, 96, 97]:

$$G(s) = \frac{-K}{s^2 + 2\zeta_{\mathrm{fil}}\omega_{\mathrm{fil}}s + \omega_{\mathrm{fil}}^2} \tag{4.10}$$

which combined by the error signal $e = r - y = -y$ and the negative gain $-K$ gives an overall positive feedback. The variables ζ_{fil} and ω_{fil} denote the damping and the frequency of the filter tuned to the mode, which is to be damped. If the output of this second order filter is marked by $\mathbf{v}_{\text{fil}}$, the input from the sensors is the displacement y, then for a MDOF system the controller can be described in terms of the filter equation and the output equation [55, 96, 97]:

$$\begin{aligned} \ddot{\mathbf{v}}_{\text{fil}}(t) + \xi_{\text{fil}}\dot{\mathbf{v}}_{\text{fil}}(t) + \Lambda_{\text{fil}}\mathbf{v}_{\text{fil}}(t) &= \mathbf{y}(t) \\ \mathbf{u}(t) &= \mathbf{K}\mathbf{v}_{\text{fil}}(t) \end{aligned} \tag{4.11}$$

where ξ_{fil} is a diagonal matrix containing the terms $2\zeta_{\text{fil}_i}\omega_{\text{fil}_i}$ on its main diagonal, similarly Λ_{fil} is a diagonal matrix containing the squares of the filter frequencies $\omega^2_{\text{fil}_i}$ on its main diagonal, for each individual filter i. The filter in (4.11) may be augmented by a rectangular matrix $\mathbf{E}_{\text{fil}}$ that allows using more filters than actuators, thereby allowing to damp more modes than the number of available actuators:

$$\begin{aligned} \ddot{\mathbf{v}}_{\text{fil}}(t) + \xi_{\text{fil}}\dot{\mathbf{v}}_{\text{fil}}(t) + \Lambda_{\text{fil}}\mathbf{v}_{\text{fil}}(t) &= \mathbf{E}_{\text{fil}}\mathbf{y}(t) \\ \mathbf{u}(t) &= \mathbf{E}^T_{\text{fil}}\mathbf{K}\mathbf{v}_{\text{fil}}(t) \end{aligned} \tag{4.12}$$

Given co-located sensors with the dynamics given by $\mathbf{y} = \mathbf{B}^T\mathbf{x}$ and system dynamics described by the equation $\mathbf{M}\ddot{\mathbf{q}}+\mathbf{B_d}\dot{\mathbf{q}}+\mathbf{K_s}\mathbf{q} = \mathbf{B}\mathbf{u}$, we may couple the PPF controller with the system to obtain [55]:

$$\begin{bmatrix} \mathbf{M} & 0 \\ 0 & 1 \end{bmatrix}\begin{bmatrix} \ddot{\mathbf{q}} \\ \ddot{\mathbf{v}}_{\text{fil}} \end{bmatrix} + \begin{bmatrix} \mathbf{B_d} & 0 \\ 0 & \xi_{\text{fil}} \end{bmatrix}\begin{bmatrix} \dot{\mathbf{q}} \\ \dot{\mathbf{v}}_{\text{fil}} \end{bmatrix} + \begin{bmatrix} \mathbf{K_s} & -\mathbf{KB} \\ -\mathbf{KB}^T & \Lambda_{\text{fil}} \end{bmatrix}\begin{bmatrix} \mathbf{q} \\ \mathbf{v}_{\text{fil}} \end{bmatrix} = \begin{bmatrix} 0 \\ 0 \end{bmatrix} \tag{4.13}$$

A review of the stability properties of direct velocity and acceleration feedback, moreover the stability of PPF assuming co-located sensors and actuators is given in [96], while the necessary and sufficient condition for the asymptotic stability of (4.11) has been established by Fanson and Caughey in [37] based on Lyapunov's direct method. Since both the augmented mass matrix and the augmented damping matrix are positive definite, the stability of the closed-loop system in (4.13) will only depend on the positive definiteness of the augmented stiffness matrix [55].

The control action in classical feedback control is realized through the manipulation of the closed-loop system poles by the feedback gain matrix **K**. If a well defined **K** is used, the original lightly damped poles of the open-loop system **A** are transformed into the better damped poles of the closed-loop system $(\mathbf{A} - \mathbf{BKC})$ [113]. Unlike in the case of optimization-based approaches such as linear quadratic control or $\mathscr{H}_\infty$ control, here an indirect computation of the feedback matrix **K** is employed. We may use several well established methods for the computation of **K** such as direct experimentation, strategies based on the pole-zero representation of the system (root-locus, pole-placement) or frequency domain methods (such as the Nyquist method) [113]. Although finding a direct fixed static output-feedback seems simple and intuitive enough, this fundamental control engineering problem is relatively challenging in terms of computational complexity [15, 16]. For example, the essential technique of pole-placement is NP-hard even for linear time-invariant systems [40].

A comparison of the use of classical feedback control methods with optimal control for vibration attenuation is discussed by Vasques and Rodrigues in [113]. A delayed position feedback is utilized for the vibration control of a flexible link manipulator in [59], while others employ position feedback for similar systems as well [102]. The counter-phased sound signal is employed to attenuate sound in enclosures by Lee et al. in [77], while a similar counter-phase signal is based on a gain scheduled observer in state-space in [17]. Optical tracking on satellites is ensured by AVC using a transfer function representation in [83]. Seismic activity is attenuated on a model using positive acceleration feedback in [99], while rotor vibrations are damped using PPF in [2]. Other works utilizing position feedback-based vibration control systems are [26, 53, 62, 71, 95, 106, 114].

Velocity or strain rate-based classical feedback controllers are very commonly implemented in vibration control applications as well. Aircraft tail vibrations are damped based on velocity feedback strategies in several works [8, 33, 34]. Other examples of such controllers are presented in [18, 106, 107, 119, 132]. A modified acceleration feedback-based method is applied for a cantilever beam in [84].

An even simpler on-off type controller is used in [130] for the control of rotor vibrations. A variation of this is referred to as *bang–bang* control. Here the controller switches between two extreme states depending on the position, velocity or the combination of the two. Such a controller has been utilized for example by Tzou et al. in [112] for cantilever vibration control.

Stability in systems controlled through classical feedback methods is guaranteed through the usual stability tests known in classical continuous or discrete controls. Moreover, Preumont states that stability in such systems can be guaranteed through the perfect physical collocation of sensors and actuators [96].

In addition to active systems, semi-active vibration damping gained some interest because of its simple electronics and hardware realization [81]. In the case of semi-active systems the obvious advantage is that A/D and D/A converters, voltage or charge amplifiers, microcontrollers are not needed; therefore making product integration simpler and economic. This method takes advantage of a fact that a circuit—using piezoelectric transducers and other simple electronic components like resistors and capacitors—may be tuned analogously to a vibration absorber. The disadvantage of this method is that one absorber may be tuned only to damp one vibration mode. Semi-active state switched resistive circuits with simple control logic are the next iteration of this concept. For example in case that the displacement and velocity at a specific point satisfies Eq. (4.14), the circuit will be switched to open circuit state. In all other cases, it will be closed circuit. Such a system including the optimal placement and resistance is discussed in [81]. The state switching law can be expressed by:

$$q(x, y, t)\dot{q}(x, y, t) \geq 0 \tag{4.14}$$

where q is a displacement depending on coordinates x, y and time t; and its derivation is velocity with the same parameters.

4.2 Proportional-Integral-Derivative Controllers

Proportional-integral-derivative (PID) controllers are widely used in industrial practice. In the absence of knowledge about the exact underlying process PID is a good controller choice, since its tuning parameters can be translated into physical changes on the closed-loop system dynamics. However, using advanced modeling and control approaches one may develop strategies offering much more than PID does. Among others, the disadvantages of the PID control strategy are that even though the controller has been tuned very carefully, it still does not guarantee the best possible control course. This situation can be remedied with the use of optimization-based algorithms such as LQ. Moreover, as with every real control system, the inputs and often the outputs are constrained as well. Such constraints are in practice implemented using saturation limits; this however introduces a level of nonlinearity in the control law. The nonlinearity of the law means that the proofs of stability and optimality no longer apply. Constraint handling even with guaranteed stability is successfully solved by the use of model predictive control.

The position or velocity-based classical feedback methods, where the measured signal is simply multiplied with a fixed gain are in fact not so distant from a PID controller or its variants. The similarity is clearer in the case if the controlled pant is a one degree of freedom vibrating system. PID controllers from the active vibration control viewpoint are reviewed in a very intuitive way by Fuller et al. in [42]. Let us now imagine a one DOF vibrating system where the amplitudes are measured and given as $q(t)$. We may devise a controller for this system, which calculates the control signal from the position coordinate $q(t)$, the velocity coordinate $\dot{q}(t)$ and the acceleration value $\ddot{q}(t)$. In case we could measure these values independently, an input to the actuators $u(t)$ could be calculated by these signals multiplied by three independent gains [42]:

$$u(t) = g_d q(t) + g_v \dot{q}(t) + g_a \ddot{q}(t) \tag{4.15}$$

where g_d, g_v and g_a are gains for the displacement, velocity and acceleration component. Due to physical limitations and practical and economic considerations not all signals can be measured. Let us therefore imagine that we can only measure the velocity signal $\dot{q}(t)$, and the acceleration is computed using numerical derivation, while the displacement is computed using a numerical equivalent of integration. In this case, we could formulate our problem as

$$u(t) = g_d \int_0^t q(t)dt + g_v \dot{q}(t) + g_a \frac{d}{dt}\dot{q}(t) \tag{4.16}$$

This in fact would be nothing else than a simple continuous PID controller. Generally, the displacement gain g_d is called the integral gain (K_i) in control engineering, since it is associated with the integral action. Similarly, the gain g_v associated with the unchanged signal is known as the proportional gain (K_p) while g_a as the derivative gain (K_d).

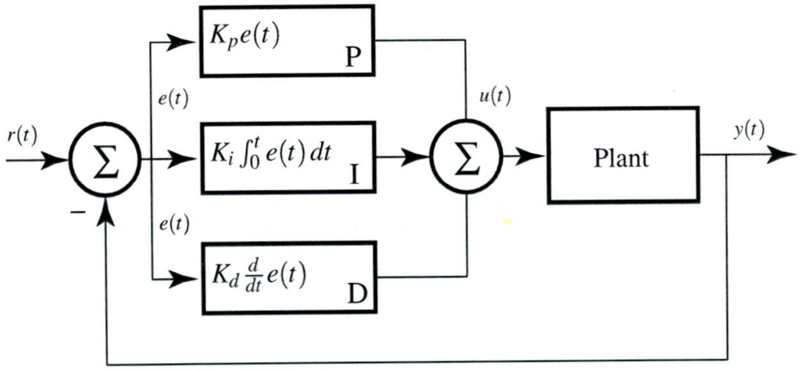

Fig. 4.2 Block algebra scheme of the inner workings of a PID controller

We will now review the principles of a generic PID controller designed for a single-input and single-output (SISO) system. Instead of the representation used in (4.15) and (4.16) focusing on vibrating systems, we will use notation known from control engineering. As with other feedback controllers, the first step in a PID algorithm is to calculate an error value $e(t)$ which is a difference between the desired reference setting $r(t)$ and the actual measured output $y(t)$:

$$e(t) = r(t) - y(t) \tag{4.17}$$

The sum of the error value itself $e(t)$, its time integral $\int_0^t e(t)dt$ and its derivative $\frac{d}{dt}e(t)$ multiplied by individual tuning constants creates the input $u(t)$ to the controlled plant. This in fact defines a PID controller. Mathematically we can express this as

$$u(t) = K_p e(t) + K_i \int_0^t e(t)dt + K_d \frac{d}{dt} e(t) \tag{4.18}$$

which is the so-called *ideal form* of a PID regulator. As the name implies, the first term is the proportional term where the error is multiplied by the proportional gain K_p. The second term in (4.18) is the integral term, which is multiplied by the integral gain K_i. This is followed by the derivative term, multiplying the error derivative by the derivative constant K_d. A block algebra scheme of this process is featured in Fig. 4.2.

We may try to imagine the meaning of the three components of a PID controller by relating the error and the practical interpretation of the integration and derivation operations:

- The proportional term P is related to the *current* error.
- The integral term I is related to the history of errors or the *past*, since the integral expresses the area under the error curve or in discrete terms the sum of all errors.

- The derivative term D is related to the *future* of the error, since a derivative expresses the rate of change or slope of the error curve, creating a kind of prediction about its upcoming trend.

It is not always necessary to use every component of the PID controller. By setting the appropriate tuning constants to zero, we can create an array of controllers missing one or two components of the original PID. In practice, however, only the following combinations are used: P, I, PI, PD. Another common notation expressing a PID controller is its so-called standard form:

$$u(t) = K_p\left(e(t) + \frac{1}{T_i}\int_0^t e(t)dt + T_d\frac{d}{dt}e(t)\right) \tag{4.19}$$

where K_p is the proportional constant and T_i and T_d are the integral and derivative time constants. We may also express a PID controller as a Laplace transform, more suited to numerical simulations in software prototyping environments such as Matlab/Simulink:

$$G(s) = K_p + \frac{K_i}{s} + K_d s = \frac{K_d s^2 + K_p s + K_i}{s} \tag{4.20}$$

where $G(s)$ is the continuous transfer function of the PID controller and s is the Laplace operator. An alternative transfer function of a PID controller is given by [75]:

$$G(s) = K_p\left(1 + \frac{1}{T_i s} + \frac{T_d s}{1 + \frac{T_d}{N}s}\right) \tag{4.21}$$

where the additional term $1 + T_d/Ns$ is a low pass filter introduced on the derivative action. As it has been already noted in (4.16), a PID controller implemented on a vibrating mechanical system can be interpreted as an analogy of velocity feedback, where position and acceleration measurement is estimated by numerical methods.

Let us now briefly return to the three independent proportional gain formulation of (4.15) and investigate how a closed-loop vibrating system will change if we implement such a simple controller. Remember that for this example we will assume that all three values of displacement, velocity and acceleration can be directly measured and the controller is a sum of these three proportional values. The vibrating mechanical system shall be represented by a one degree of freedom system in (2.108) for which the transfer function is defined as:

$$H(s) = \frac{Q(s)}{F_e(s)} = \frac{1}{ms^2 + bs + k} \tag{4.22}$$

where $Q(s)$ is the vibration amplitude and F_e is the external disturbance in the Laplace domain. Here $H(s)$ represents the dynamics of the vibrating system in *open-loop* that is, without a controller. Furthermore, let us perform a Laplace transform on the simple proportional control law given by (4.15) to get:

$$U(s) = g_d Q(s) + g_v Q(s)s + g_a Q(s)s^2 \tag{4.23}$$

and finally obtain the transfer function of the control law [42]:

$$G(s) = \frac{U(s)}{Q(s)} = g_d + g_v s + g_a s^2 \tag{4.24}$$

where $Q(s)$ is the measured position and $U(s)$ is the controller input in the Laplace domain. To calculate the *closed-loop* response of this system, we must consider the direct path from the disturbance to the displacement ($H(s)$) and divide it by the indirect path ($1+H(s)G(s)$) which contains the controller as well. After substituting for $H(s)$ and $G(s)$ we will get [42]:

$$F(s) = \frac{Q(s)}{F_e(s)} = \frac{H(s)}{1+H(s)G(s)} = \frac{1}{(m+g_a)s^2 + (b+g_v)s + (k+g_d)} \tag{4.25}$$

which is a transfer function describing the new, controlled relationship between excitation $F_e(s)$ and vibration $Q(s)$. One may easily see that there is a direct and physically interpretable connection between the individual gains g_d, g_v and g_a which help to create the new modified mass, stiffness and damping properties of the system:

$$F(s) = \frac{Q(s)}{F_e(s)} = \frac{1}{m's^2 + b's + k'} \tag{4.26}$$

where m', b' and k' are the modified closed-loop mass, damping and stiffness values.

The above discussion is valid to systems with no delays. Unfortunately, delays are always present in control systems and are caused by imperfect sensor or actuator dynamics. The digital sampling process itself may introduce delays into the closed-loop system as well. These delays may cause that the damping properties will change dramatically if the excitation frequency is much higher than the resonance frequency of the system. We can model the dynamics of a controller similarly to (4.22), which also takes delays into account by [42]:

$$G(s) = \frac{U(s)}{Q(s)} = e^{-\tau_d s}(g_d + g_v s + g_a s^2) \tag{4.27}$$

where τ_d is delay and $e^{-\tau_d s}$ models this delay in the Laplace domain. Let us assume that the delay is small and then the frequency response can be expressed by:

$$e^{-j\omega\tau_d} \approx 1 - j\omega\tau_d \tag{4.28}$$

which is valid for $\omega\tau_d << 1$. Now the closed-loop frequency response of this system can be expressed similarly to (4.26) by equivalent mass, damping and stiffness terms:

$$F(s) = \frac{Q(s)}{F_e(s)} = \frac{1}{\omega^2 m'' + \omega b'' + k''} \tag{4.29}$$

where the new equivalent effective mass m'', effective damping b'' and effective stiffness k'' terms can be expressed by [42]:

$$m'' = m + g_a - \tau_d g_v \tag{4.30}$$

$$b'' = b + g_v - \tau_d g_d + \omega^2 \tau_d g_a \tag{4.31}$$

$$k'' = k + g_d \tag{4.32}$$

If we compare k'' and k' we can see that the delay has no impact on the effective stiffness. For lightly damped systems the term $\tau_d g_v$ is small when compared to the mass m, therefore its impact on the effective mass is minimal. The effective damping is however greatly influenced by both the delay τ_d and the frequency ω. Let us assume changing the effective mass and stiffness twice to their relative magnitude under displacement and acceleration feedback [42]. For a lightly damped system, the term $\tau_d g_d$ is comparable to b if the delay is small compared to the period of the natural resonant frequency on the system. On the other hand, for frequencies ω over the damped natural frequency ω_d the term $\omega^2 \tau_d g_a$ becomes comparable to b. With a displacement or acceleration-based feedback even a small delay may dramatically alter the effective damping or even render the system unstable. That is why in classical feedback control (see Sect. 4.1) velocity-based feedback is preferred. Velocity feedback will not alter effective mass, stiffness or damping properties of the system significantly if unmodeled delay is introduced into the closed-loop system.

Although it is possible to create a purely continuous-time PID controller, it is more common to implement it in a digital control system. For this it is necessary to replace the integral term with its discrete-time equivalent, summing:

$$\int_0^t e(t)dt \approx \sum_{i=1}^{k-1} e(i)T_s \tag{4.33}$$

It is also necessary to compute differences numerically, instead of symbolic differentiation:

$$\frac{de(t)}{dt} \approx \frac{e(k) - e(k-1)}{T_s} \tag{4.34}$$

for a sampling time T_s. The resulting controller will be suitable for discrete-time application. A discrete PID controller is sometimes referred to as a PSD controller, exchanging the integral term with summation. A so-called velocity form of a discrete PID (PSD) controller can be expressed as:

$$\begin{aligned} u_k =& u_{(k-1)} \\ &+ K_p \left[\left(1 + \frac{T_s}{T_i} + \frac{T_d}{T_s} \right) e(k) + \left(-1 - \frac{2T_d}{T_s} \right) e(k-1) + \frac{T_d}{T_s} e(k-2) \right] \end{aligned} \tag{4.35}$$

or alternatively we may write [12]:

$$u_k = K_p \left[e(k) + \frac{T_s}{T_i} \left(\frac{e(0) + e(k)}{2} + \sum_{i=1}^{k} e(i) \right) + \frac{T_d}{T_s} \left((e(k) - e(k-1) \right) \right] \tag{4.36}$$

where T_s is the discrete sampling period. The discrete-time PID controller may be expressed after Z-transformation in the Z-domain by [50]:

$$G(z) = \frac{U(z)}{E(z)} = K_p + K_i \frac{z}{z-1} + K_d \frac{z-1}{z} \tag{4.37}$$

There are different methods to tune a PID controller. One of the most widely used is a simple iterative trial and error process. Other methods include Ziegler-Nichols, Cohen-Coon, iterative response shaping and others. Note that, as it has been previously implied, the use of a PID controller neither guarantees stability of the control loop nor is it optimal in any sense. The available literature on PID controllers is extensive, therefore we will not discuss the control engineering aspects and details of this method. The reader shall refer to the relevant publications on the topic.

PID controllers are utilized by Fung et al. in [43] to control the vibrations of a flexible beam actuated through an electromagnet. Semi-active suspensions can be also controlled via PID [35]. The use of the PID strategy for earthquake-induced vibration control in civil engineering structures is suggested by Carotti and Lio; and Guclu in [21, 47]. Yet another possible application of PID in vibration control is for the AVC of flexible link mechanisms as described in [62] and other similar vibration control systems [6, 60, 111].

4.3 Linear Quadratic Control

Linear quadratic control belongs to the broader family of algorithms based on optimal control. In optimal control, a cost function indicating a performance index is chosen which is then minimized to obtain an optimal input $u(k)$ [55].

Let us consider a continuous, linear time-invariant state-space system as defined by (4.1). The cost function in the continuous linear quadratic optimal control problem can be chosen to be quadratically dependent on the control input and the state or output response:

$$J = \frac{1}{2} \int_{t_0}^{t_f} \left(\mathbf{x}^T(t) \mathbf{Q} \mathbf{x}(t) + \mathbf{u}^T(t) \mathbf{R} \mathbf{u}(t) \right) dt + \frac{1}{2} \mathbf{x}^T(t_f) \mathbf{P}_f \mathbf{x}(t_f) \tag{4.38}$$

where $\mathbf{Q}$ is a state weighting matrix, $\mathbf{R}$ is an input weighting matrix and $\mathbf{P}_f$ is a terminal weighting matrix. All these weighting factors or penalty matrices can be chosen by the control engineer to fine-tune the behavior of the controller, according to the particular needs of the plant.

A linear quadratic (LQ) regulator (LQR) is a special case of the generic linear quadratic control problem. Contrary to the general case described above, the weighting matrices in the LQR problem are constant. Moreover, the control horizon t_f is assumed to approaching infinity. The matrix $\mathbf{Q}$ is positive semidefinite, while matrix $\mathbf{R}$ is positive definite. The generic LQ optimal control problem is expressed as the minimization of the following cost function [42, 97]:

$$J = \frac{1}{2}\int_{t_0}^{\infty} \left(\mathbf{x}^T(t)\mathbf{Q}\mathbf{x}(t) + \mathbf{u}^T(t)\mathbf{R}\mathbf{u}(t)\right)\mathrm{d}t \tag{4.39}$$

We may interpret the above formulation as an attempt to minimize the overall control energy measured in a quadratic form. In fact, the LQR controller is an automated way to find an optimal fixed feedback matrix. The final control law then assumes the form of a constant matrix state feedback gain in the form [10, 97]:

$$\mathbf{u}(t) = -\mathbf{K}\mathbf{x}(t) \tag{4.40}$$

rendering the continuous-time state-space representation in (4.1) to:

$$\dot{\mathbf{x}}(t) = (\mathbf{A} - \mathbf{B}\mathbf{K})\mathbf{x}(t) \tag{4.41}$$

which is the closed-loop state equation of the continuous system with an LQ fixed feedback law.

The matrix gain $\mathbf{K}$ can be expressed as [10]:

$$\mathbf{K} = \mathbf{R}^{-1}\mathbf{B}^T\mathbf{P} \tag{4.42}$$

where $\mathbf{P}$ is the solution of the differential Ricatti equation given as [42, 118]:

$$\dot{\mathbf{P}} = -\mathbf{P}\mathbf{A} - \mathbf{A}^T\mathbf{P} + \mathbf{P}\mathbf{B}\mathbf{R}^{-1}\mathbf{B}^T\mathbf{P} - \mathbf{Q} \tag{4.43}$$

which for the infinite horizon LQR problem is replaced by the so-called algebraic Ricatti equation (ARE) defined as

$$\mathbf{0} = -\mathbf{P}\mathbf{A} - \mathbf{A}^T\mathbf{P} + \mathbf{P}\mathbf{B}\mathbf{R}^{-1}\mathbf{B}^T\mathbf{P} - \mathbf{Q} \tag{4.44}$$

For a discrete time-invariant state-space system, we may define the LQR controller as the fixed matrix feedback gain $\mathbf{K}$, which minimizes the following infinite horizon cost function [50]:

$$J = \sum_{k=0}^{\infty}\left(\mathbf{x}_k^T\mathbf{Q}\mathbf{x}_k + \mathbf{u}_k^T\mathbf{R}\mathbf{u}_k\right) \tag{4.45}$$

The output voltage at the actuators is then:

$$\mathbf{u}_k = -\mathbf{K}\mathbf{x}_k \tag{4.46}$$

and the discrete linear time-invariant state-space system will be rendered to

$$\mathbf{x}_{k+1} = (\mathbf{A} - \mathbf{B}\mathbf{K})\mathbf{x}_k = \boldsymbol{\Phi}\mathbf{x}_k \tag{4.47}$$

where $\boldsymbol{\Phi}$ expresses the state dynamics of the closed-loop system controlled through a fixed LQ gain. The LQR feedback gain may be calculated from

$$\mathbf{K} = (\mathbf{R} + \mathbf{B}^T\mathbf{P}\mathbf{B})^{-1}\mathbf{B}^T\mathbf{P}\mathbf{A} \tag{4.48}$$

where $\mathbf{P}$ is the solution of the discrete-time algebraic Ricatti equation (DARE) defined by

$$\mathbf{P} = \mathbf{Q} + \mathbf{A}^T\left(\mathbf{P} - \mathbf{P}\mathbf{B}\left(\mathbf{R} + \mathbf{B}^T\mathbf{P}\mathbf{B}\right)^{-1}\mathbf{B}^T\mathbf{P}\right)\mathbf{A} \tag{4.49}$$

LQ controllers are extensively used both in general industrial applications and in vibration control. To list some of the applications, the LQ strategy has been suggested for the active vibration control of buildings during an earthquake [85, 97], semi-active control for vehicle suspensions and mounts platforms [24, 39, 60], in a hybrid feedforward-feedback setup for active noise control in [66], in optical drives [22] and in numerous other academic studies aimed at vibration control [1, 43, 51, 52, 59, 67, 88].

4.4 $\mathscr{H}_2$ and $\mathscr{H}_\infty$ Control

Just like the previously introduced linear quadratic control scheme, *H-infinity* or as it is commonly denoted in the literature $\mathscr{H}_\infty$ (H_∞) controllers are a subclass of optimization-based control methods too. $\mathscr{H}_\infty$ control methods are often utilized with the aim to create a robust and stabilizing control system [72]. Similarly to LQ controllers, one of the biggest disadvantage of $\mathscr{H}_\infty$ controllers is their general inability to handle constraints, such as saturation limits or naturally occurring process constraints. The advantage of $\mathscr{H}_\infty$ controllers is the possibility to control multi-variable systems and robust control formulations.

In essence, the $\mathscr{H}_\infty$-optimization of control systems is based on the minimization of the peak value of closed-loop frequency functions [72]. Let us consider a simple problem involving a SISO system plant with a disturbance, which is illustrated on Fig. 4.3. Here we have a plant $\mathscr{P}(s)$ and a controller $\mathscr{H}_\infty(s)$. The output of the system is denoted by $y(s)$, while the system is also subjected to an outside disturbance $v(s)$. The reference value is denoted by $w(s)$, let us keep it at zero for now. We can denote the Laplace transform of the plant output as:

$$y(s) = v(s) - \mathscr{P}(s)\mathscr{H}_\infty(s)y(s) \tag{4.50}$$

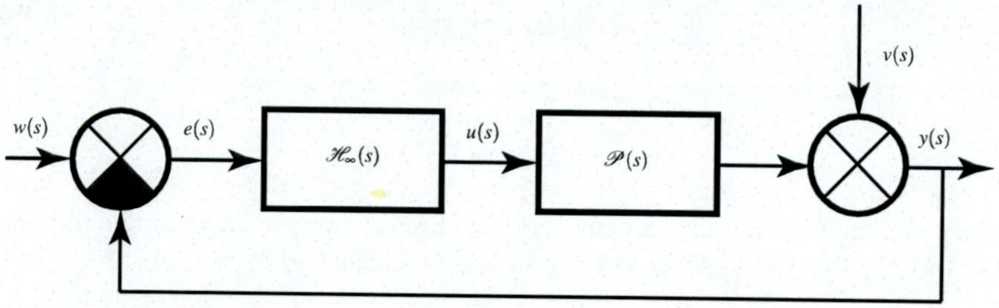

Fig. 4.3 Formulating the sensitivity function in $\mathscr{H}_\infty$ control: general SISO control loop with a reference and outside disturbance

from this it follows that

$$y(s) = \mathscr{S}(s)v(s) \tag{4.51}$$

where $\mathscr{S}(s)$ is the so-called sensitivity function according to

$$\mathscr{S}(s) = \frac{1}{1 + \mathscr{P}(s)\mathscr{H}_\infty(s)} \tag{4.52}$$

or in matrix terms

$$\mathscr{S}(s) = [\mathbf{I} + \mathscr{P}(s)\mathscr{H}_\infty(s)]^{-1} \tag{4.53}$$

and we may also define the *complementary sensitivity* function

$$\mathscr{T}(s) = [\mathbf{I} + \mathscr{P}(s)\mathscr{H}_\infty(s)]^{-1}\mathscr{P}(s)\mathscr{H}_\infty(s) \tag{4.54}$$

The sensitivity function characterizes the sensitivity of the system output to disturbances and its value is in the ideal case $\mathscr{S}(s) = 0$. One may regard the sensitivity function as a performance indicator, similarly to the *cost* function that is used in LQ control or in MPC control as well. A low sensitivity function value implies a low tracking error, thus ultimately increasing the controller performance.

Alternatively for no outside disturbance $v(s) = 0$ but for a given reference tracking $w(s) \neq 0$ we may define that [31]:

$$\begin{aligned} y(s) &= \mathscr{P}(s)\mathscr{H}_\infty(s)[w(s) - y(s)] \\ &= [\mathbf{I} + \mathscr{P}(s)\mathscr{H}_\infty(s)]^{-1}\mathscr{P}(s)\mathscr{H}_\infty(s)w(s) \\ &= \mathscr{T}w(s) \end{aligned} \tag{4.55}$$

Similarly to the output of the plant, for the control error we may define

$$\begin{aligned} e(s) &= [w(s) - y(s)] \\ &= [\mathbf{I} - \mathscr{T}(s)]w(s) \\ &= \mathscr{S}(s)w(s) \end{aligned} \tag{4.56}$$

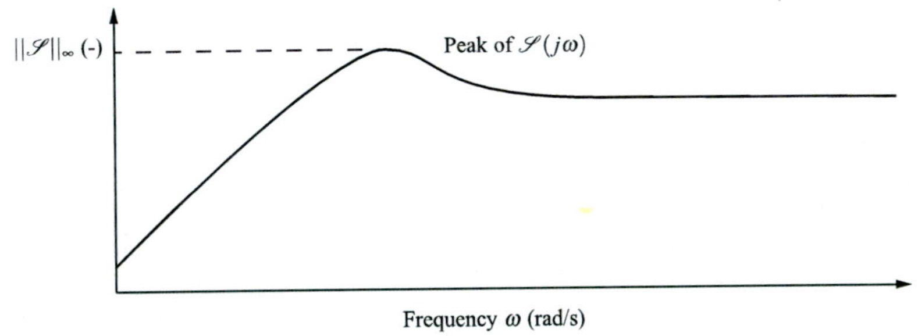

Fig. 4.4 Peak value $||\mathscr{S}||_\infty$ of the sensitivity function $\mathscr{S}(j\omega)$

Our aim is to make the closed-loop feedback system stable and find a controller $\mathscr{H}_\infty(s)$ which minimizes the peak value of the sensitivity function in the frequency domain: $\mathscr{S}(j\omega)$ [124, 125]. The peak value of the sensitivity function can be defined as the infinite norm of the function $\mathscr{S}(j\omega)$ given by:

$$||\mathscr{S}||_\infty = \max_\omega |\mathscr{S}(j\omega)| \tag{4.57}$$

The maximal or peak value of the sensitivity function in the frequency domain as defined by (4.57) is graphically illustrated in Fig. 4.4.

Although the maximum of the absolute value of the sensitivity function is a very intuitive way to define its peak, it is not always logically feasible. This is because for some functions the peak value is not assumed at all for a finite frequency ω. Instead, one may replace the maximum with supremum or at least an upper bound, so (4.57) will change to:

$$||\mathscr{S}||_\infty = \sup_\omega |\mathscr{S}(j\omega)| \tag{4.58}$$

We would like to minimize the peak of the sensitivity function, since in case this peak is small, then so is the magnitude of $\mathscr{S}$ for all frequencies. This implies that the disturbances are attenuated uniformly well over the whole frequency range [72]. The minimization of $||\mathscr{S}||_\infty$ is a worst-case optimization procedure, since it minimizes the effect of the worst disturbance on the output. For a physical vibrating system this may be understood as minimizing the effect of a harmonic disturbance on the closed-loop controlled system in its resonance—that is where the sensitivity function $|\mathscr{S}|$ has its peak value.

In order to provide a mathematically more detailed interpretation of $\mathscr{H}_\infty$ controllers in general, let us define what the $\mathscr{H}_\infty$ norm means: if $\mathscr{H}_\infty$ is a space of matrix-valued functions bounded in the right-half of the complex space, the value of the $\mathscr{H}_\infty$ norm is the maximal singular value of the function over that space [97]. In other words, the $\mathscr{H}_\infty$ norm is the maximal gain in any direction and frequency for a SISO system, or as it has been previously pointed out, the maximal magnitude of the frequency response.

Let us now define a controlled plant $\mathscr{P}(s)$ that has two inputs: $\mathbf{w}(s)$ is the reference signal containing disturbances, while $\mathbf{u}(s)$ is the controlled input to the plant.

The plant has two outputs as well, namely $\mathbf{e}(s)$, which is the error signal we aim to minimize and $\mathbf{y}(s)$ which is the measurable plant output. Unlike in the previous case our system is not SISO anymore, but MIMO therefore the variables $\mathbf{u}(s)$, $\mathbf{e}(s)$, $\mathbf{y}(s)$ and $\mathbf{w}(s)$ are vectors while $\mathcal{H}_\infty$ and $\mathcal{P}$ are matrices of transfer functions.

For this system we desire to find a matrix $\mathcal{H}_\infty$ (or essentially a feedback matrix $\mathbf{K}$), which will generate the optimal input $\mathbf{u}(s)$ based on the measured signal $\mathbf{y}(s)$—see Fig. 4.5 for illustration. This augmented system can be described by [46, 58, 104]:

$$\begin{bmatrix}\mathbf{e}(s)\\ \mathbf{y}(s)\end{bmatrix} = \mathcal{P}(s)\begin{bmatrix}\mathbf{w}(s)\\ \mathbf{u}(s)\end{bmatrix} = \begin{bmatrix}\mathcal{P}_{11}(s) & \mathcal{P}_{12}(s)\\ \mathcal{P}_{21}(s) & \mathcal{P}_{22}(s)\end{bmatrix}\begin{bmatrix}\mathbf{w}(s)\\ \mathbf{u}(s)\end{bmatrix} \tag{4.59}$$

$$\mathbf{u}(s) = \mathcal{H}_\infty(s)\mathbf{y}(s) \tag{4.60}$$

We may express the dependence of error $\mathbf{e}(s)$ on the reference $\mathbf{w}(s)$ by a term very similar to (4.56) using the sensitivity function to express the error based on the reference. For this, we substitute (4.60) into (4.59) and separate the matrix expression into two equations to get:

$$\mathbf{e}(s) = \mathcal{P}_{11}(s)\mathbf{w}(s) + \mathcal{P}_{12}(s)\mathcal{H}_\infty(s)\mathbf{y}(s) \tag{4.61}$$

$$\mathbf{y}(s) = \mathcal{P}_{21}(s)\mathbf{w}(s) + \mathcal{P}_{22}(s)\mathcal{H}_\infty(s)\mathbf{y}(s) \tag{4.62}$$

Expressing $\mathbf{y}(s)$ from the second equation yields

$$\mathbf{y}(s) = [\mathbf{I} + \mathcal{P}_{22}(s)\mathcal{H}_\infty(s)]^{-1}\mathcal{P}_{21}(s)\mathbf{w}(s) \tag{4.63}$$

which after substituting into the first equation yields

$$\mathbf{e}(s) = \mathcal{P}_{11}(s) + \mathcal{P}_{12}(s)\mathcal{H}_\infty(s)[\mathbf{I} - \mathcal{P}_{22}(s)\mathcal{H}_\infty(s)]^{-1}\mathcal{P}_{21}(s)\mathbf{w}(s) \tag{4.64}$$

$$= F_\ell(\mathcal{P}, \mathcal{H}_\infty)\mathbf{w}(s) \tag{4.65}$$

where the operator F_ℓ is known as the lower linear fractional transformation and it expresses the sensitivity function.

The objective of $\mathcal{H}_\infty$ control for the system defined above is to find such a feedback matrix $\mathcal{H}_\infty(s)$, which minimizes the lower linear fractional transformation or the F_ℓ part of (4.64) according to the $\mathcal{H}_\infty$ norm. The same definition also applies for $\mathcal{H}_2$ control. The infinity norm for a general MIMO system can be expressed as the peak value of the largest singular value taken as a function of frequency [46, 97]:

$$||F_\ell(\mathcal{P}, \mathcal{H}_\infty)||_\infty = \sup_\omega \bar{\sigma}_s(F_\ell(\mathcal{P}, \mathcal{H}_\infty)(j\omega)) \tag{4.66}$$

where $\bar{\sigma}_s$ is the maximal singular value of the matrix $F_\ell(\mathcal{P}, \mathcal{H}_\infty)(j\omega)$.

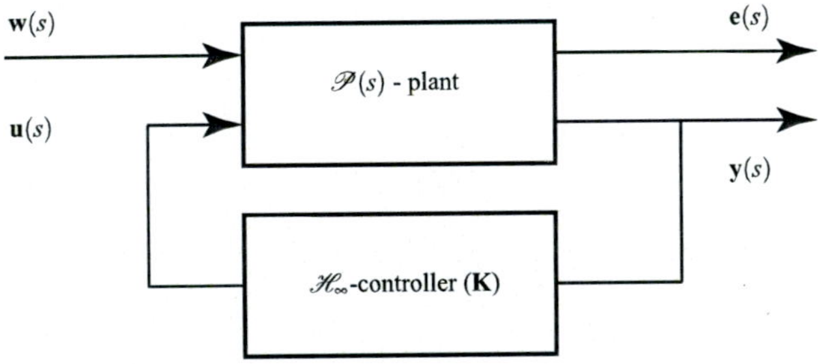

Fig. 4.5 Controlled plant and $\mathscr{H}_\infty$ controller

$\mathscr{H}_\infty$ control is utilized in [14] to control the vibration of rotor blades in a helicopter individually. Time-invariant but linear nature of the forward helicopter flight is solved through gain scheduling of the $\mathscr{H}_\infty$ control laws. Other applications of $\mathscr{H}_\infty$ based vibration control systems are for example active seats for the automotive industry or spacecrafts [109], active magnetic suspensions for rotors [58], active noise control [20] and active seismic vibration control in buildings [63, 97].

4.5 Soft Computing Approaches

The use of control approaches based on genetic algorithms, artificial neural networks and fuzzy control is fairly atypical for active vibration control. The reason for this is that *soft computing* control systems are rather suited for plants and phenomena, which are difficult if not impossible to model using exact mathematical, respectively numerical approaches. However, the dynamic behavior of vibrating mechanical systems can be easily characterized using ordinary or partial differential equations. This process then results in transfer function or state-space based models. By the aid of these models, exact *hard* control rules can be formulated.

Direct vibration control through genetic algorithms, neural networks or fuzzy control is rare. These somewhat "exotic" methods may however be utilized to tune more traditional controllers or to define the physical size or distribution of sensors and actuators. Other vibration control related applications in which the above-mentioned control methods are useful are the ones with large actuator hysteresis or other significant nonlinearities, such as magnetorheological dampers. The following sections will briefly characterize these *soft computing* methods.

4.5.1 Neural Networks

Artificial neural networks (ANN) mimic the behavior of biological neural networks found in nature by using programming constructs that resemble neurons and their

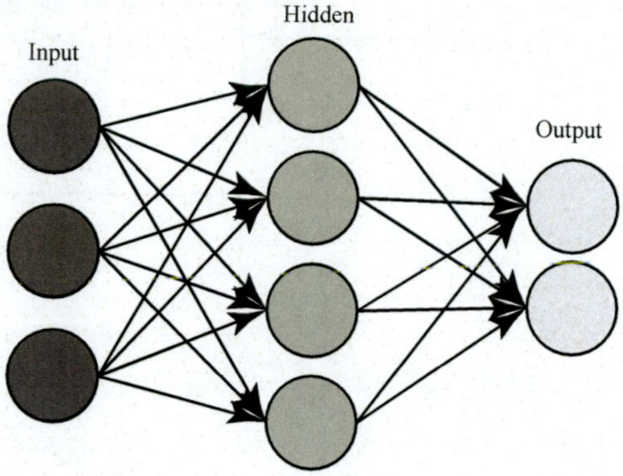

Fig. 4.6 Schematic illustration of a simple feedforward artificial neural network. The *nodes* denote programming modules mimicking neurons in a biological neural network. Synapses or connections between the individual neurons are denoted by *arrows*

interconnections. Just as in nature, the structure of an ANN changes and adapts according to the inflowing information emulating the learning process.

Biological neurons are replaced by nodes[3] in an artificial neural network and they are represented by the shaded circle in Fig. 4.6. The simplest ANN has three layers, as denoted in Fig. 4.6 consisting of an input, a hidden and an output layer. The input nodes or neurons send data via synapses to the second hidden layer, which in turn sends data to the output layer via other synapses. The synapses are denoted as arrows on the figure and in practice they store weighting parameters used to manipulate the transferred data.

For practical reasons, real life implementations of artificial neural networks rely on statistical and signal processing ideas more heavily than exact biological principles. However, what ANN and a real biological neural network have in common is their adaptive, distributed, nonlinear and parallel processing nature.

Let us represent the neural network with a function $f(x)$, which takes x as its input. The function $f(x)$ is a composition of other functions g_i which in turn may be a composition of yet other sets of function. This functional dependency is represented in Fig. 4.7. The dependency of functions can be interpreted in a so-called functional view, which is predominantly associated with optimization tasks. If we assume the set of functions g_i to be a vector $\mathbf{g} = [g_1\ g_2\ \dots\ g_i\ \dots\ g_n]$, then from the functional view the input x is transformed into a three-dimensional vector $\mathbf{h}$ which is in turn transformed into the two-dimensional vector $\mathbf{g}$ and finally to f. Another equivalent view of the artificial neural network is the so-called probabilistic view, which is commonly used in the context of graphical models.

The neural networks represented in Figs. 4.6 and 4.7 are of the feedforward type, without cycles. It is possible to include cycles in ANN, in that case we are talking about a recurrent network.

The use of neural networks in magnetorheological (MR) damper-based semi-active control systems is justified by the large hysteretic and nonlinear behavior of MR dampers. An MR damper actuated semi-active vehicle suspension that is

[3] Nodes are also referred to as neurons, processing elements or units.

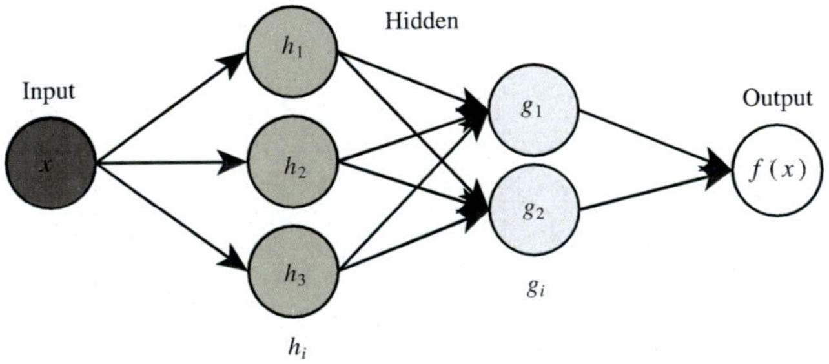

Fig. 4.7 Schematic illustration of the dependency of an artificial neural network. Input data x is mapped through a series of functions h_i and g_i to the final neural network $f(x)$

indirectly controlled by artificial neural networks has been proposed by Zapaterio et al. in [126]. A neural network is used as an inverse model of the MR damper: the desired force acts as an input, which is used to calculate the voltage needed to generate that force. The voltage level is then input into a controller acquired via traditional methods [126]. A neural network approach is used for the control of semi-active vehicle suspensions by Eski at al. in [35] as well. The control system is contrasted to a traditional PID controller in simulation. Eski et al. combine a PID controller with a novel ANN-based dynamics predictor.

Chen et al. combine different artificial neural network methods to attenuate acoustic signals with a voice-coil actuator in [23]. The ANN methods provide means to tune the parameters of traditional transfer function-based controllers adaptively. Adaptive vibration control is implemented similarly using ANN in [68], where the authors suggest that an ANN-based adaption method can be computationally less intensive than traditional adaptation methods. ANN has been used in [133] as well to create models for a predictive controller-based vibration flexible link manipulator vibration suppression system.

Neural networks are used to suppress vibrations in a permanent magnet linear motor in [123], rotor system in [3] and in other vibration control applications [7, 27, 64, 121, 122, 127, 129].

4.5.2 Genetic Algorithms

Similar to the artificial neural networks presented previously, genetic algorithms (GA) mimic nature's behavior. Instead of emulating the working principles of a nervous system, genetic algorithms copy the evolutionary selection process. In fact, genetic algorithms belong to the larger class of evolutionary algorithms and are often utilized in optimization and search problems.

The candidate solutions of a GA problem are represented by the individuals[4] and these individuals carry encoded genetic information represented by chromosomes.[5] The population of such chromosome carrying individuals is the genetic algorithm itself, which is gradually evolving toward an optimal solution through several generations. Naturally, in GA the genetic information is represented by binary or other type of strings instead of the DNA. We may describe the steps of a genetic algorithm in a simplified manner:

- initialization
- selection
- reproduction
- termination

At the initialization stage, a population of individuals with random genetic information is generated. Typically, a population consists of several hundreds or thousands of individual "creatures", covering the range of all possible solutions. It is also possible to insert individuals with possible optimal genetic material, so to aid the speed and succession of the selection process.

Just as in nature, the fitter individual survives. In the next stage of the genetic algorithm, a sub-set of the original population is selected based on *fitness* to survive and allowed to reproduce. Naturally, the fitness function is a measure of solution quality and is based on the desired type of solution, what is better for the individual changes according to the problem type. The selection process also contains a random element, so genetic information from individuals with a smaller fitness level can also enter the next generation. This helps to diversify the population.

The individuals surviving the selection process can reproduce to create the successive generation. This selection process also emulates the natural selection process. The genetic information of the "parents" is combined by genome crossover and a degree of randomness is also introduced by mutation. The process is repeated until a population with the desired size is created and the process continues with the next iteration of the selection process. With each new generation, a pool of genetic material is created which is different from that of the previous generation.

The genetic algorithm is usually terminated after a pre-set number of generations has evolved, or is terminated based on the fitness of the population.

Genetic algorithms for complex problems require extensive computation time. The computation of a complex fitness function for each individual in the population is the main limiting factor of GA. Another drawback is that the GA tends to converge toward local optima, instead of the global optimum. Certain techniques exist to diversify the population and prevent this, but no guarantee for the global optimum can be given.

The schematic representation of a simple genetic algorithm is featured in Fig. 4.8. The shaded circles represent the individuals, while the column of circles is the actual generation. The genetic information here is the color of the circle, which of course

---

[4] Also referred to as phenotypes or creatures.

[5] Also referred to as genomes, genotypes and strings.

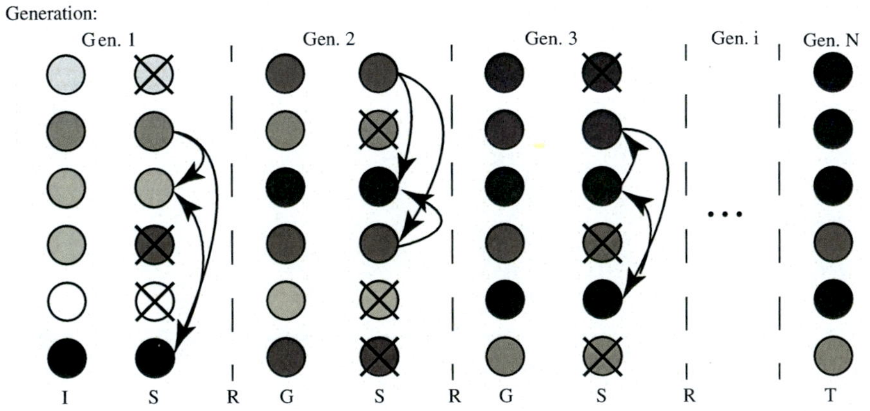

Fig. 4.8 Simplified schematic representation of the progression of a genetic algorithm. I is for initialization, S is for selection, R for reproduction G denotes the next generation and T is for termination

could be represented very easily by a binary string. The fitness function here is the darkest shade, we can state that the fittest is the darkest individual because it could hide well against a dark background and thus survive to pass on its genes. At an initialization stage (I) a population is generated with random genetic information, after this selection takes place (S). Selection includes fitness evaluation, where the lighter shades are removed from the population and some random "deaths" also occur—the unfavorable mutations are selected against. The rest of the population may reproduce (R) and the new generation appears (G). The reproduction happens through the crossing of the genetic information of the parents and possibly random mutations. The favorable mutation is more likely to survive and reproduce. After a satisfactory population fitness or generation number is reached, the algorithm is terminated (T).

It is clear that the nature of GA is more suited to supplementary optimization in vibration control, such as the optimal placement of actuators and sensors. It is possible to use GA as an adaptation feature, augmenting the function of other control systems. The direct utilization of GA in vibration control is not recommended because of the possible computational burden or the occurrence of local minima. Despite of its limitations, GA has been used in active vibration control applications.

The most popular way to utilize GA in the field of vibration control is to perform a geometric design optimization and therefore passively reach a better vibration response [65, 89]. A certain application of this principle is the optimal sizing and placement of actuators and sensors for active control [19, 86, 100, 120]. Tuning parameter optimization for traditional control systems can be carried out with the help of GA as well [5, 100, 120].

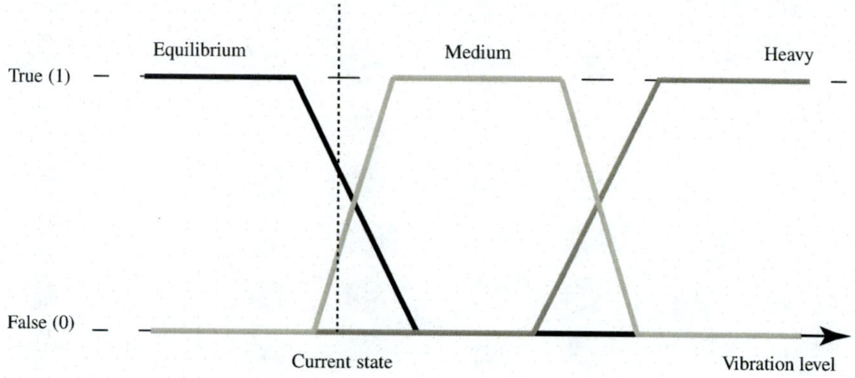

Fig. 4.9 Example fuzzy rules describing the measured vibration levels

4.5.3 Fuzzy Control

Fuzzy control allows creating intricate nonlinear controllers, based on a set of simpler heuristic laws. These heuristic control laws may come from the experience of an engineer, common sense actions or may be a result of extensive mathematical simulation and optimization.

Fuzzy controllers are based on fuzzy logic, derived from fuzzy set theory. In contrast to binary logic where a statement can have either true (1) or false (0) values, in fuzzy logic the statements can assume values in between these two extremes [128]. Fuzzy controllers may use so-called *linguistic variables* to describe the control laws [87]. For example, instead of assigning certain acceleration values to the vibration of a mechanical system, in fuzzy control we can replace these by terms like "in equilibrium", "medium vibrations" and "heavily vibrating". Control laws and functions can be associated with these linguistic terms to create a fuzzy controller. Figure 4.9 illustrates a simple set of three rules, describing the vibration level of a mechanical structure. Let us take a look at the dashed line which represents the measure of our current vibration level. It is certainly not heavily vibrating and close to the equilibrium—the value of this statement is about 60% true or 0.6. The actual level is also a little into medium vibration levels, the value of the statement that we have medium vibrations is about 30% or 0.3. In linguistic terms, we can say that our structure is slightly vibrating.

Similarly, it is possible to associate the actuator actions with an analogous set of rules and linguistic descriptors [87]. Let us imagine three different rules for an actuator: "no action", "medium action" and "intense action". Let us now formulate a set of rules based on these vibration levels and actuator actions, for example in linguistic terms we may logically define:

- *if* vibration *is* in equilibrium *then* take no action
- *if* vibration *is* medium *then* take medium action

- *if* vibration *is* heavy *then* take intense action

The shape of the membership functions featured in Fig. 4.9 may be completely changed or altered by the designer, and one may use various logical statements and operators in addition to the *if-is-then* construct [94]. The example illustrated above is very simple, but it is always possible to add more rules and insert other logical twists and turns into the control law. In contrast to genetic algorithms or neural networks, the fuzzy control laws can be interpreted in a way that a human operator or designer can easily understand. For those interested in the theoretical basics of fuzzy sets, fuzzy logic and the design of fuzzy controllers, books by Michels et al. [87] and others [36, 94, 128] can be recommended. An interesting connection is made between multi-parametric programming based MPC (MPMPC) and the control of systems described modeled by a set of fuzzy laws by Kvasnica et al. in [69, 70], where the explicit minimum-time MPC controllers are proposed for Takagi-Sugeno fuzzy systems.

Fuzzy control is more suited to direct vibration control than for example artificial neural networks or genetic algorithms. Although it is possible to use fuzzy control to tune the parameters in classical controllers [80], fuzzy control systems may be used alone for vibration attenuation. Fuzzy control is combined with a (moving) sliding mode controller by Sung et al. in [110], while a fuzzy control-based vehicle suspension is suggested in [108] by Sun and Yang. The performance of a fuzzy controller is contrasted to a simple PD controller in by Guclu and Yazici in [48] for the active control of earthquake-induced vibrations. Fuzzy control is also suggested for the use on space borne robotic manipulator arms in [116].

Fuzzy control is often utilized for the active vibration control of active and semi-active vehicle suspension systems [82, 110] because of its ability to emulate and control the highly hysteretic and nonlinear behavior of MR dampers. The fuzzy strategy is also employed in civil engineering [48, 91] and in other works [23, 29, 54, 79, 92, 117, 129].

4.6 Other Approaches

The creativity of the human mind is limitless, and this is also true for designing control strategies which can be used in active vibration control. Minor or major alterations of algorithms introduced previously are abundant in the academic literature. Furthermore, several works discuss the combination of two methods, for example using soft computing techniques to turn classical methods into more advanced adaptive or robust control systems. Here we will list briefly some of the approaches used in AVC that have not been explicitly mentioned before.

Sliding mode control (SMC) applies a state switching strategy to alter the dynamics of the control system. The control law is not a continuous function of time, instead it is a nonlinear system of alternate structures, which are switched based on the current state [11]. The main advantage of SMC is robustness, moreover if *bang-bang* control is required; the SMC strategy can be even optimal.

Examples of state-space representation-based control laws other than the ones presented here are dynamic response shaping, eigenvalue placement and minimum energy control [118]. In a different state-space-based approach Bohn et al. utilizes an observer to attenuate engine-induced vibrations in [17]. This observer is used to reconstruct the original disturbance signal, which is then fed back with a negative sign as a control input. Due to the ever changing speed of the engine, the observer gains are scheduled based on a speed signal.

Optimization-based control methods may take different aspects of the vibration engineering task into account, such as the minimization of deflections [98], velocities, accelerations [115], maximizing resonant frequencies [103] and minimizing vibration or acoustic energy [41]. Optimization may be used as a tool to find the ideal placement and number of sensors and actuators offline and in combination with traditional controllers online [19, 44, 90, 103]. Moreover, as it has been previously mentioned, various optimization-based methods can be used offline to tune the parameters of traditional controllers based on a cost function [41, 115]. Examples of direct optimization-based vibration control approaches in addition to the ones presented here can be found in [25, 78, 98] and other works.

Feedback loops are not the only way to control vibrating systems; numerous studies use the feedforward approach to attenuate mechanical disturbances. In fact, feedforward is often preferred over feedback in active noise cancellation systems [42, 105]. If there is no information available about the disturbance acting on the system, a feedback loop must be used. In case the type and character of the disturbance is known a priori, feedforward may be a very good choice. Such scenarios include periodic oscillations caused by rotating machines [93], or structures where a sensor may be placed in between the transmission path of the source and the primary point of actuation. In the case of feedforward control, sensors are not used to directly affect the response of the controller, they are employed as a type of adaptive measure to tune the feedforward controller and monitor its performance instead [42]. One of the common strategies in feedforward control is the use of the so-called filtered-x LMS algorithm, which is used to tune a FIR filter adaptively. Feedback control-based on the $\mathscr{H}_\infty$ method is contrasted to feedforward control in the work of Seba et al. [101] for a car engine vibration attenuation system. Feedforward-based vibration control has been applied to a single-link manipulators in [4, 38], while noise attenuation and control applications for windows [57], loudspeakers [131], heating ventilation and air conditioning (HVAC) [30] and other systems [28, 61] are also very common.

References

1. Agrawal BN, Bang H (1996) Adaptive structures for large precision antennas. Acta Astronaut 38(3):175–183. doi:10.1016/0094-5765(96)00062-8, http://www.sciencedirect.com/science/article/B6V1N-3VTW8Y7-3/2/a53f7c4acb3ee1541568e0db4062d985

2. Ahmed B, Pota H (2011) Dynamic compensation for control of a rotary wing UAV using positive position feedback. J Intell Rob Syst 61:43–56. doi: 10.1016/0094-5765(96)00062-8, http://10.1007/s10846-010-9487-7,
3. Al-Nassar YN, Siddiqui M, Al-Garni AZ (2000) Artificial neural networks in vibration control of rotor-bearing systems. Simul Pract Theory 7(8):729–740. doi:10.1016/S0928-4869(00)00004-5
4. Alam M, Tokhi M (2008) Designing feedforward command shapers with multi-objective genetic optimisation for vibration control of a single-link flexible manipulator. Eng Appl Artif Intell 21(2):229–246. doi:10.1016/j.engappai.2007.04.008, http://www.sciencedirect.com/science/article/B6V2M-4P0N8W7-1/2/0151e11caeeaab40012fcffe7059861b
5. Alam M, Tokhi M (2008) Hybrid fuzzy logic control with genetic optimisation for a single-link flexible manipulator. Eng Appl Artif Intell 21(6):858–873. doi:10.1016/j.engappai.2007.08.002, http://www.sciencedirect.com/science/article/B6V2M-4PPNM5K-1/2/e0cab5d17ae1330863c846be1692e6c8
6. Allaire PE, Lewis DW, Knight JD (1983) Active vibration control of a single mass rotor on flexible supports. J Franklin Inst 315(3):211–222. doi:10.1016/0016-0032(83)90025-X, http://www.sciencedirect.com/science/article/B6V04-45D9SMR-M/2/62024de7918cc7b0b23d9703691ab67a
7. Alli H, Uçar A, Demir Y (2003) The solutions of vibration control problems using artificial neural networks. J Franklin Inst 340(5):307–325. doi:10.1016/S0016-0032(03)00036-X, http://www.sciencedirect.com/science/article/B6V04-48XD30R-1/2/6aea0a5487a31fa2ed46dc803323b4f5
8. Amer Y, Bauomy H (2009) Vibration reduction in a 2DOF twin-tail system to parametric excitations. Commun Nonlinear Sci Numer Simul 14(2):560–573. doi:10.1016/j.cnsns.2007.10.005, http://www.sciencedirect.com/science/article/B6X3D-4PYP723-2/2/b9d5375168fadb0b4e67857e92948bfc
9. Antsaklis PJ, Michel AN (2005) Linear systems, 2nd edn. Birkhäuser, Boston (originally published by McGraw-Hill, Englewood Cliffs, 1997)
10. Antsaklis PJ, Michel AN (2007) A Linear systems Primer. Birkhäuser, Boston (originally published by McGraw-Hill, Englewood Cliffs, 1997)
11. Bandyopadhyay B, Janardhanan S (2004) Discrete-time sliding mode control: a multirate output feedback approach. In: Thoma M, Morari M (eds) Hybrid systems: computation and control, lecture notes in control and information sciences. Springer, Berlin
12. Belavý C (2009) Teória Automatického Riadenia II: Návody na cvičenia, 1st edn. Slovenská vysoká škola technická v Bratislave: Strojnícka Fakulta, Bratislava, (Theory of automatic control ii: seminar guide) in Slovak language
13. Benaroya H, Nagurka ML (2010) Mechanical vibration: analysis, uncertainities and control, 3rd edn. CRC Press, Taylor & Francis Group, Boca Raton
14. Bittanti S, Cuzzola FA (2002) Periodic active control of vibrations in helicopters: a gain-scheduled multi-objective approach. Control Eng Pract 10(10):1043–1057. doi:10.1016/S0967-0661(02)00052-7, http://www.sciencedirect.com/science/article/B6V2H-45KSPJJ-3/2/9647861ce849d131c7d4b90cdb964751
15. Blondel V, Tsitsiklis JN (1996) NP-hardness of some linear control design problems. SIAM J Control Optim 35:2118–2127
16. Blondel VD, Tsitsiklis JN (2000) A survey of computational complexity results in systems and control. Automatica 36(9):1249–1274. doi:10.1016/S0005-1098(00)00050-9, http://www.sciencedirect.com/science/article/pii/S0005109800000509
17. Bohn C, Cortabarria A, Härtel V, Kowalczyk K (2004) Active control of engine-induced vibrations in automotive vehicles using disturbance observer gain scheduling. Control Eng Pract 12(8):1029–1039. doi:10.1016/j.conengprac.2003.09.008, http://www.sciencedirect.com/science/article/B6V2H-49Y3VWS-1/2/dd7bcefd1618f3820896ddbd6dce7430, in special section on emerging technologies for active noise and vibration control systems

18. Braghin F, Cinquemani S, Resta F (2010) A model of magnetostrictive actuators for active vibration control. Sens Actuators A (in press). Corrected proof. doi: 10.1016/j.sna.2010.10.019, http://www.sciencedirect.com/science/article/B6THG-51F25N5-4/2/f5cf46980d38877c74a3c4d34fbd894d
19. Bruant I, Gallimard L, Nikoukar S (2010) Optimal piezoelectric actuator and sensor location for active vibration control, using genetic algorithm. J Sound Vib 329(10):1615–1635. doi:10.1016/j.jsv.2009.12.001, http://www.sciencedirect.com/science/article/pii/S0022460X09009869
20. Camino J, Arruda J (2009) $\mathscr{H}_2$ and $\mathscr{H}_\infty$ feedforward and feedback compensators for acoustic isolation. Mech Syst Sig Process 23(8):2538–2556. doi:10.1016/j.ymssp.2009.04.006, http://www.sciencedirect.com/science/article/B6WN1-4W7J0YN-2/2/918091cd3d7b23193d5b3637eb2342ce

21. Carotti A, Lio G (1991) Experimental active control: bench tests on controller units. Eng Struct 13(3):242–252. doi:10.1016/0141-0296(91)90036-C, http://www.sciencedirect.com/science/article/B6V2Y-4829VWB-CG/2/4414a8cb4321f4e346ca04468e610264
22. Chang CS, Liu TS (2007) LQG controller for active vibration absorber in optical disk drive. IEEE Trans Magn 43(2):799–801. doi:10.1109/TMAG.2006.888417
23. Chen K, Chou C, Chang S, Liu Y (2008) Intelligent active vibration control in an isolation platform. Appl Acoust 69(11):1063–1084. doi: 10.1016/j.apacoust.2007.06.008, http://www.sciencedirect.com/science/article/B6V1S-4PMYXPB-1/2/55a19d9f917a53bef57682bb9d03dac2
24. Choi SB, Hong SR, Sung KG, Sohn JW (2008) Optimal control of structural vibrations using a mixed-mode magnetorheological fluid mount. Int J Mech Sci 50(3):559–568. doi:10.1016/j.ijmecsci.2007.08.001, http://www.sciencedirect.com/science/article/B6V49-4PD4XHC-1/2/c491dc4a4a881e38b0e20ceef7206dec
25. Chu CL, Wu BS, Lin YH (2006) Active vibration control of a flexible beam mounted on an elastic base. Finite Elem Anal Des 43(1):59–67. doi:10.1016/j.finel.2006.07.001, http://www.sciencedirect.com/science/article/pii/S0168874X06001144
26. Creasy M, Leo D, Farinholt K (2008) Adaptive positive position feedback for actively absorbing energy in acoustic cavities. J Sound Vib 311(1–2):461–472. doi: 10.1016/j.jsv.2007.09.013, http://www.sciencedirect.com/science/article/B6WM3-4R2HKR0-3/2/e9d3c9817e3b4c302a861a4a3bb6fcb1
27. Darus IM, Tokhi M (2005) Soft computing-based active vibration control of a flexible structure. Eng Appl Artif Intell 18(1):93–114. doi: 10.1016/j.engappai.2004.08.017, http://www.sciencedirect.com/science/article/B6V2M-4DFT21W-2/2/0e01e702eebed40a2e2dbd2925feed5c
28. Davari P, Hassanpour H (2009) Designing a new robust on-line secondary path modeling technique for feedforward active noise control systems. Signal Process 89(6):1195–1204. doi:10.1016/j.sigpro.2009.01.003, http://www.sciencedirect.com/science/article/pii/S0165168409000103
29. de Abreu GLCM, Ribeiro JF (2002) A self-organizing fuzzy logic controller for the active control of flexible structures using piezoelectric actuators. Appl Soft Comput 1(4):271–283. doi:10.1016/S1568-4946(02)00020-0, http://www.sciencedirect.com/science/article/B6W86-454T5BG-1/2/a19272d9ab350a5ebca766ecab7b2192
30. de Callafon R, Zeng J, Kinney C (2010) Active noise control in a forced-air cooling system. Control Eng Pract 18(9):1045–1052. doi:10.1016/j.conengprac.2010.05.007, http://www.sciencedirect.com/science/article/pii/S0967066110001243
31. De Cuyper J, Swevers J, Verhaegen M, Sas P (2000) $\mathcal{H}_\infty$ feedback control for signal tracking on a 4 poster test rig in the automotive industry. In: 25th international conference on noise and vibration engineering, Leuven, pp 61–68
32. Doyle J, Francis B, Tannenbaum A (1992) Feedback control theory, 2nd edn. Macmillan Publishing, New York
33. Eissa M, Bauomy H, Amer Y (2007) Active control of an aircraft tail subject to harmonic excitation. Acta Mech Sin 23:451–462. doi:10.1007/s10409-007-0077-2, http://10.1007/s10409-007-0077-2
34. El-Badawy AA, Nayfeh AH (2001) Control of a directly excited structural dynamic model of an F-15 tail section. J Franklin Inst 338(2–3):133–147. doi:10.1016/S0016-0032(00)00075-2, http://www.sciencedirect.com/science/article/B6V04-42HNMDV-3/2/e3bf6f797834c8e8638324be88fb78f7
35. Eski I, Yıldırım S (2009) Vibration control of vehicle active suspension system using a new robust neural network control system. Simul Modell Pract Theory 17(5):778–793. doi:10.1016/j.simpat.2009.01.004, http://www.sciencedirect.com/science/article/B6X3C-4VHSDJ4-1/2/d2fe946695b369279d2e1229f15a61bd
36. Espinosa J, Vandewalle J, Wertz V (2005) Fuzzy logic, identification and predictive control. Advances in industrial control. Springer, London

37. Fanson JL, Caughey TK (1990) Positive position feedback control for large space structures. AIAA J 28(4):717–724. doi:10.2514/3.10451
38. Feliu V, Pereira E, Díaz IM, Roncero P (2006) Feedforward control of multimode single-link flexible manipulators based on an optimal mechanical design. Rob Autom Syst 54(8):651–666. doi:10.1016/j.robot.2006.02.012, http://www.sciencedirect.com/science/article/pii/S0921889006000649, Morphology, Control and Passive Dynamics
39. Fischer D, Isermann R (2004) Mechatronic semi-active and active vehicle suspensions. Control Eng Pract 12(11):1353–1367. doi:10.1016/j.conengprac.2003.08.003, http://www.sciencedirect.com/science/article/B6V2H-49V1CR4-2/2/0dd89d1b7760e7303a32b5bdd2cbbf9b , Mechatronic Systems
40. Fu M (2004) Pole placement via static output feedback is np-hard. IEEE Trans Autom Control 49(5):855–857. doi:10.1109/TAC.2004.828311
41. Fuller C (1990) Active control of sound transmission/radiation from elastic plates by vibration inputs: I. analysis. J Sound Vib 136(1):1–15. doi:10.1016/0022-460X(90)90933-Q, http://www.sciencedirect.com/science/article/pii/0022460X9090933Q
42. Fuller CR, Elliott SJ, Nelson PA (1996) Active control of vibration, 1st edn. Academic Press, San Francisco
43. Fung RF, Liu YT, Wang CC (2005) Dynamic model of an electromagnetic actuator for vibration control of a cantilever beam with a tip mass. J Sound Vib 288(4–5):957–980. doi:10.1016/j.jsv.2005.01.046, http://www.sciencedirect.com/science/article/B6WM3-4G4N5VD-1/2/fc3710f0625ef69f19d16c8778a63e58
44. Gao W, Chen JJ, Ma HB, Ma XS (2003) Optimal placement of active bars in active vibration control for piezoelectric intelligent truss structures with random parameters. Comput Struct 81(1):53–60. doi:10.1016/S0045-7949(02)00331-0, http://www.sciencedirect.com/science/article/pii/S0045794902003310
45. Goh CJ, Caughey TK (1985) On the stability problem caused by finite actuator dynamics in the collocated control of large space structures. Int J Control 41(3):787–802. doi:10.1080/00207178508961163, http://10.1080/00207178508961163
46. Green M, Limebeer D (1995) Linear robust control. Prentice Hall, Englewood Cliffs
47. Guclu R (2006) Sliding mode and PID control of a structural system against earthquake. Math Comput Modell 44(1–2):210–217. doi: 10.1016/j.mcm.2006.01.014, http://www.sciencedirect.com/science/article/B6V0V-4JP9FV5-1/2/0900f85ba6e764d746c054ac040aff77 (Advances in business modeling and decision technologies, pp 1–95)
48. Guclu R, Yazici H (2008) Vibration control of a structure with ATMD against earthquake using fuzzy logic controllers. J Sound Vib 318(1–2):36–49. doi:10.1016/j.jsv.2008.03.058, http://www.sciencedirect.com/science/article/B6WM3-4SM0XJT-1/2/fe8f6a66297ad6e12f0791a83e4eed36
49. Hatch MR (2000) Vibration simulation using MATLAB and ANSYS, 1st edn. Chapman and Hall/CRC, Boca Raton
50. Hellerstein JL, Diao Y, Parekh S, Tilbury DM (2004) Feedback control of computing systems. Wiley/IEEE Press, Hoboken
51. Ho CC, Ma CK (2007) Active vibration control of structural systems by a combination of the linear quadratic Gaussian and input estimation approaches. J Sound Vib 301(3–5):429–449. doi:10.1016/j.jsv.2005.12.061, http://www.sciencedirect.com/science/article/B6WM3-4MV19X0-1/2/39db74e66a9494e834cdab9f0da4b886
52. Hong SR, Choi SB, Han MS (2002) Vibration control of a frame structure using electro-rheological fluid mounts. Int J Mech Sci 44(10):2027–2045. doi:10.1016/S0020-7403(02)00172-8, http://www.sciencedirect.com/science/article/B6V49-47BX3RX-4/2/53a10ce8cbf8dfa679c34e04beb688e4
53. Hu Q (2009) A composite control scheme for attitude maneuvering and elastic mode stabilization of flexible spacecraft with measurable output feedback. Aerosp Sci Technol 13(2–3):81–91. doi:

10.1016/j.ast.2007.06.007, http://www.sciencedirect.com/science/article/B6VK2-4P96269-2/2/5fbc47249fdd3f1963c5ba856f071c55

54. Huang SJ, Man RJ (1996) Active vibration control of a dynamic absorber using fuzzy algorithms. Mechatronics 6(3):317–336. doi:10.1016/0957-4158(95)00081-X, http://www.sciencedirect.com/science/article/B6V43-3WRJ1S0-5/2/d0cec849f4495c8788cf58ae50bcd708
55. Inman DJ (2006) Vibration with control. Wiley, Chichester
56. Inman DJ (2007) Engineering vibrations, 3rd edn. Pearson International Education (Prentice Hall), Upper Saddle River
57. Jakob A, Möser M (2003) Active control of double-glazed windows, Part I: feedforward control. Appl Acoust 64(2):163–182. doi:10.1016/S0003-682X(02)00070-1, http://www.sciencedirect.com/science/article/pii/S0003682X02000701
58. Jastrzebski RP, Hynynen KM, Smirnov A (2010) $\mathscr{H}_\infty$ control of active magnetic suspension. Mech Syst Sig Process 24(4):995–1006. doi: 10.1016/j.ymssp.2009.10.008, http://www.sciencedirect.com/science/article/B6WN1-4XJP3XR-2/2/51b0222180b2610516135c196f226b0e
59. Jnifene A (2007) Active vibration control of flexible structures using delayed position feedback. Syst Control Lett 56(3):215–222. doi: 10.1016/j.sysconle.2006.10.005, http://www.sciencedirect.com/science/article/B6V4X-4MJC1V9-1/2/5fe33b4788d9ca97d9a9938bc7742194
60. Jung WJ, Jeong WB, Hong SR, Choi SB (2004) Vibration control of a flexible beam structure using squeeze-mode ER mount. J Sound Vib 273(1–2):185–199. doi:10.1016/S0022-460X(03)00478-4, http://www.sciencedirect.com/science/article/B6WM3-49DFFMM-1/2/1255ad59eca53b0c021632de61aef0b8
61. Kamman JW, Naghshineh K (1999) A comparison of open-loop feedforward and closed-loop methods for active noise control using volume velocity minimization. Appl Acoust 57(1):29–37. doi: 10.1016/S0003-682X(98)00043-7, http://www.sciencedirect.com/science/article/pii/S0003682X98000437
62. Kang B, Mills JK (2005) Vibration control of a planar parallel manipulator using piezoelectric actuators. J Intell Rob Syst 42:51–70. doi:10.1007/s10846-004-3028-1, http://10.1007/s10846-004-3028-1
63. Karimi H, Zapateiro M, Luo N, Rossell J (2010) Feedback vibration control of a base-isolated building with delayed measurements using $\mathscr{H}_\infty$ techniques. In: American control conference (ACC), 2010, pp 750–55
64. Kawabe H, Tsukiyama N, Yoshida K (2006) Active vibration damping based on neural network theory. Mater Sci Eng A 442(1–2):547–550. doi:10.1016/j.msea.2006.02.234, http://www.sciencedirect.com/science/article/B6TXD-4KPFKNH-2/2/51634002bdd85fe7ee55df4b6b28e7e4, Proceedings of the 14th international conference on internal friction and mechanical spectroscopy
65. Keane AJ (1995) Passive vibration control via unusual geometries: the application of genetic algorithm optimization to structural design. J Sound Vib 185(3):441–453. doi: 10.1006/jsvi.1995.0391, http://www.sciencedirect.com/science/article/B6WM3-45R8DN4-16/2/ef1e97179ca4c87ba2111dd2da839fd5
66. Kim I, Kim YS (2009) Active vibration control of trim panel using a hybrid controller to regulate sound transmission. Int J Precis Eng Manuf 10:41–47. doi:10.1007/s12541-009-0007-2
67. Krishnaswamy K, Rajamani R, Woo J, Cho Y (2005) Structural vibration control for broadband noise attenuation in enclosures. J Mech Sci Technol 19:1414–1423. doi:10.1007/BF03023900, http://10.1007/BF03023900
68. Kumar R, Singh S, Chandrawat H (2007) MIMO adaptive vibration control of smart structures with quickly varying parameters: neural networks vs classical control approach. J Sound Vib 307(3–5):639–661. doi:10.1016/j.jsv.2007.06.028, http://www.sciencedirect.com/science/article/B6WM3-4PJ6BP9-1/2/69b80beb8f5338317e59823d40598c23

69. Kvasnica M, Herceg M, Čirka Ľ, Fikar M (2009) Time-optimal control of Takagi-Sugeno fuzzy systems. In: Proceedings of the 10th European control conference, Budapest, pp 916–921
70. Kvasnica M, Herceg M, Čirka Ľ, Fikar M (2011) Explicit minimum-time controllers for fuzzy systems. In: Selected topics on constrained and nonlinear control. Preprints, STU Bratislava-NTNU Trondheim, pp 287–292
71. Kwak MK, Heo S (2007) Active vibration control of smart grid structure by multiinput and multioutput positive position feedback controller. J Sound Vib 304(1–2):230–245. doi:10.1016/j.jsv.2007.02.021, http://www.sciencedirect.com/science/article/B6WM3-4NH6N96-2/2/ca7b43602b9d052e388f4b2a28f1ebae
72. Kwakernaak H (1993) Robust control and $\mathscr{H}_\infty$-optimization-tutorial paper. Automatica 29(2):255–273. http://doc.utwente.nl/29962
73. Kwakernaak H, Sivan R (1972) Linear optimal control systems. Wiley-Interscience/Wiley, New York
74. Landau ID, Zito G (2006) Digital control systems: design, identification and implementation. Communications and control engineering. Springer, London
75. Landau ID, Constantinescu A, Rey D (2005) Adaptive narrow band disturbance rejection applied to an active suspension–an internal model principle approach. Automatica 41(4):563–574. doi:10.1016/j.automatica.2004.08.022, http://www.sciencedirect.com/science/article/B6V21-4FB3X55-3/2/28887440b73dcde4fdbaefe4d507e857
76. Landis T, NASA Dryden Flight Research Center (NASA-DFRC) (2001) Full scale dynamic model of the EOS-AM1 satellite. Image ID: EC01-0288-2
77. Lee J, Kim J, Cheong C (1999) Piezoelectric smart structures for noise reduction in a cabin. J Mech Sci Technol 13:451–458. doi:10.1007/BF02947714, http://10.1007/BF02947714
78. Lim CW (2008) Active vibration control of the linear structure with an active mass damper applying robust saturation controller. Mechatronics 18(8):391–399. doi:10.1016/j.mechatronics.2008.06.006, http://www.sciencedirect.com/science/article/pii/S0957415808000913
79. Lin J, Liu WZ (2006) Experimental evaluation of a piezoelectric vibration absorber using a simplified fuzzy controller in a cantilever beam. J Sound Vib 296(3):567–582. doi:10.1016/j.jsv.2006.01.066, http://www.sciencedirect.com/science/article/B6WM3-4K0FG0H-2/2/e4fad7e52e98cf46123aa869cf780b65
80. Lin LC, Lee TE (1997) Integrated PID-type learning and fuzzy control for flexible-joint manipulators. J Intell Rob Syst 18:47–66. doi:10.1023/A:1007942528058, http://10.1023/A:1007942528058
81. Lin Q, Ermanni P (2004) Semi-active damping of a clamped plate using PZT. Int J Solids Struct 41:1741–1752
82. Liu SJ, Huang ZH, Chen YZ (2004) Automobile active suspension system with fuzzy control. J Central South Univ Technol 11:206–209. doi:10.1007/s11771-004-0042-1, http://10.1007/s11771-004-0042-1
83. Luo T, Hu Y (2002) Vibration suppression techniques for optical inter-satellite communications. In: IEEE 2002 international conference on communications, circuits and systems and west sino expositions, vol 1, pp 585–589. doi:10.1109/ICCCAS.2002.1180687
84. Mahmoodi SN, Craft MJ, Southward SC, Ahmadian M (2011) Active vibration control using optimized modified acceleration feedback with adaptive line enhancer for frequency tracking. J Sound Vib 330(7):1300–1311. doi:10.1016/j.jsv.2010.10.013, http://www.sciencedirect.com/science/article/B6WM3-51D894K-1/2/25e8ef1bcadb5fd2aa078de4d678c7f4
85. Marzbanrad J, Ahmadi G, Jha R (2004) Optimal preview active control of structures during earthquakes. Eng Struct 26(10):1463–1471. doi:10.1016/j.engstruct.2004.05.010,http://www.sciencedirect.com/science/article/B6V2Y-4CYNR00-1/2/271b4c49fa053fb1a95d5df632c701c8

86. Mehrabian AR, Yousefi-Koma A (2011) A novel technique for optimal placement of piezoelectric actuators on smart structures. J Franklin Inst 348(1):12–23. doi:10.1016/j.jfranklin.2009.02.006, http://www.sciencedirect.com/science/article/B6V04-4VTCM9T-1/2/1d68ecf523d642a7246481a506f3edab, International symposium on mechatronics and its applications 2007
87. Michels K, Klawonn F, Kruse R, Nurnberger A (2006) Fuzzy control: fundamentals, stability and design of fuzzy controllers, studies in fuzziness and soft computing, vol 200. Springer, Berlin
88. Moon SJ, Lim CW, Kim BH, Park Y (2007) Structural vibration control using linear magnetostrictive actuators. J Sound Vib 302(4–5):875–891. doi:10.1016/j.jsv.2006.12.023, http://www.sciencedirect.com/science/article/B6WM3-4N2M6HH-5/2/417522adfca8640acfa76e890ae0533c
89. Moshrefi-Torbati M, Keane AJ, Elliott SJ, Brennan MJ, Rogers E (2003) Passive vibration control of a satellite boom structure by geometric optimization using genetic algorithm. J Sound Vib 267(4):879–892. doi:10.1016/S0022-460X(03)00192-5, http://www.sciencedirect.com/science/article/B6WM3-48NJ208-4/2/d00d9d286c87c83da9f2a01bba7d9209
90. Moshrefi-Torbati M, Keane A, Elliott S, Brennan M, Anthony D, Rogers E (2006) Active vibration control (AVC) of a satellite boom structure using optimally positioned stacked piezoelectric actuators. J Sound Vib 292(1–2):203–220. doi:10.1016/j.jsv.2005.07.040, http://www.sciencedirect.com/science/article/pii/S0022460X05005171
91. Ok SY, Kim DS, Park KS, Koh HM (2007) Semi-active fuzzy control of cable-stayed bridges using magneto-rheological dampers. Eng Struct 29(5):776–788. doi:10.1016/j.engstruct.2006.06.020, http://www.sciencedirect.com/science/article/B6V2Y-4KM46VD-4/2/1c85c3a0d12e30e2d5afddaa590f7059
92. Park HW, Yang HS, Park YP, Kim SH (1999) Position and vibration control of a flexible robot manipulator using hybrid controller. Rob Autom Syst 28(1):31–41. doi:10.1016/S0921-8890(99)00027-5, http://www.sciencedirect.com/science/article/B6V16-3X9YY2M-4/2/991e70955258e7604c6775467c5eea35, Robotics applications at FLINS'98
93. Pasco Y, Robin O, Bélanger P, Berry A, Rajan S (2011) Multi-input multi-output feedforward control of multi-harmonic gearbox vibrations using parallel adaptive notch filters in the principal component space. J Sound Vib (in press). Corrected proof. doi:10.1016/j.jsv.2011.06.008, http://www.sciencedirect.com/science/article/pii/S0022460X11004858
94. Passino KM, Yurkovich S (1998) Fuzzy control. Addison-Wesley, Berkley
95. Pradhan S (2005) Vibration suppression of FGM shells using embedded magnetostrictive layers. Int J Solids Struct 42(9–10):2465–2488. doi:10.1016/j.ijsolstr.2004.09.049, http://www.sciencedirect.com/science/article/B6VJS-4F6SSGN-1/2/b6f9e2e6ffc65bfc0c4af5083e37df0b
96. Preumont A (2002) Vibration control of active structures, 2nd edn. Kluwer Academic Publishers, Dordrecht
97. Preumont A, Seto K (2008) Active control of structures, 3rd edn. Wiley, Chichester
98. Rashidi M (1990) A computational strategy for active control of dynamic systems via minimizing the displacement magnitudes of dominant harmonics of vibration. Math Comput Modell 14:410–412. doi:10.1016/0895-7177(90)90217-B, http://www.sciencedirect.com/science/article/pii/089571779090217B
99. Ríos-Gutiérrez M, Silva-Navarro G (2010) Suppression of mechanical vibrations in a building like structure by means of a piezoelectric patch actuator and positive acceleration feedback. In: 2010 7th international conference on electrical engineering computing science and automatic control (CCE), pp 452–457. doi:10.1109/ICEEE.2010.5608581
100. Roy T, Chakraborty D (2009) Optimal vibration control of smart fiber reinforced composite shell structures using improved genetic algorithm. J Sound Vib 319(1–2): 15–40. doi:10.1016/j.jsv.2008.05.037, http://www.sciencedirect.com/science/article/B6WM3-4T0X2NT-1/2/6e02883f5e6352192210eb9b36700538

101. Seba B, Nedeljkovic N, Paschedag J, Lohmann B (2005) $\mathscr{H}_\infty$ feedback control and Fx-LMS feedforward control for car engine vibration attenuation. Appl Acoust 66(3):277–296. doi:10.1016/j.apacoust.2004.07.015, http://www.sciencedirect.com/science/article/B6V1S-4DTKD2W-1/2/d413b004e2a2e14e9df7fcf75f2df02f
102. Shan J, Liu HT, Sun D (2005) Slewing and vibration control of a single-link flexible manipulator by positive position feedback (PPF). Mechatronics 15(4): 487–503. doi:10.1016/j.mechatronics.2004.10.003, http://www.sciencedirect.com/science/article/B6V43-4DR87K7-4/2/2dd311fdd61308e1415cd45c1edc3076
103. Simões-Moita JM, Correia VMF, Martins PG, Soares CMM, Soares CAM (2006) Optimal design in vibration control of adaptive structures using a simulated annealing algorithm. Compos Struct 75(1–4):79–87. doi:10.1016/j.compstruct.2006.04.062, http://www.sciencedirect.com/science/article/pii/S0263822306001206, thirteenth international conference on composite structures-ICCS/13
104. Skogestad S, Postlethwaite I (2005) Multivariable feedback control: analysis and design, 2nd edn. Wiley, Chichester
105. Snyder SD (2000) Active noise control primer. Modern acoustics and signal processing. Springer/AIP Press, New York
106. Song G, Qiao PZ, Bibienda WK, Zhou GP (2002) Active vibration damping of composite beam using smart sensors and actuators. J Aerosp Eng 15(3):97–103
107. Sun D, Mills JK, Shan J, Tso SK (2004) A PZT actuator control of a single-link flexible manipulator based on linear velocity feedback and actuator placement. Mechatronics 14(4):381–401. doi:10.1016/S0957-4158(03)00066-7, http://www.sciencedirect.com/science/article/B6V43-49DN5K4-1/2/fa21df547f182ad568cefb2ddf3a6352
108. Sun J, Yang Q (2007) Automotive suspension system with an analytic fuzzy control strategy. In: IEEE international conference on vehicular electronics and safety, 2007. ICVES, pp 1–4. doi:10.1109/ICVES.2007.4456375
109. Sun W, Li J, Zhao Y, Gao H (2010) Vibration control for active seat suspension systems via dynamic output feedback with limited frequency characteristic. Mechatronics (in press). Corrected proof. doi:10.1016/j.mechatronics.2010.11.001, http://www.sciencedirect.com/science/article/B6V43-51KH6DW-1/2/9f06f9d31ca4a47bf3b8e034ba8c6150
110. Sung KG, Han YM, Cho JW, Choi SB (2008) Vibration control of vehicle ER suspension system using fuzzy moving sliding mode controller. J Sound Vib 311(3–5): 1004–1019. doi:10.1016/j.jsv.2007.09.049, http://www.sciencedirect.com/science/article/B6WM3-4R2H1TN-4/2/b3a297765c3ac7767b2d64fda7a6a3d7
111. Tso SK, Yang TW, Xu WL, Sun ZQ (2003) Vibration control for a flexible-link robot arm with deflection feedback. Int J Non Linear Mech 38(1):51–62. doi:10.1016/S0020-7462(01)00040-3, http://www.sciencedirect.com/science/article/B6TJ2-46BSCBF-5/2/db9a6ea06f0106fae187a067a96b1888
112. Tzou H, Chai W (2007) Design and testing of a hybrid polymeric electrostrictive/piezoelectric beam with bang-bang control. Mech Syst Sig Process 21(1):417–429. doi:10.1016/j.ymssp.2005.10.008, http://www.sciencedirect.com/science/article/B6WN1-4HR75KY-1/2/73701e5908a2ea598fa7bec1ce093563
113. Vasques C, Rodrigues JD (2006) Active vibration control of smart piezoelectric beams: Comparison of classical and optimal feedback control strategies. Comput Struct 84(22–23):1402–1414. doi:10.1016/j.compstruc.2006.01.026, http://www.sciencedirect.com/science/article/B6V28-4K4219V-1/2/fc83fdc87b19e200d95c2b596f8f0201, Composite adaptive structures: modelling and simulation
114. Šolek P (2009) Numerical analyses of piezoelectric elements, 1st edn. Slovenská technická univerzita v Bratislave, Nakladateľstvo STU, Bratislava
115. Wang Z, Chen S, Han W (1999) Integrated structural and control optimization of intelligent structures. Eng Struct 21(2):183–191. doi:10.1016/S0141-0296(97)90158-9, http://www.sciencedirect.com/science/article/pii/S0141029697901589

116. Wei JJ, Qiu ZC, Han JD, Wang YC (2010) Experimental comparison research on active vibration control for flexible piezoelectric manipulator using fuzzy controller. J Intell Rob Syst 59:31–56. doi:10.1007/s10846-009-9390-2, http://dx.doi.org/10.1007/s10846-009-9390-2
117. Wenzhong Q, Jincai S, Yang Q (2004) Active control of vibration using a fuzzy control method. J Sound Vib 275(3-5):917–930. doi: 10.1016/S0022-460X(03)00795-8, http://www.sciencedirect.com/science/article/B6WM3-49P82Y8-3/2/4041c663559fb530f34deadda058c82d
118. Williams RL II, Lawrence DA (2007) Linear state-space control systems. Wiley, Hoboken
119. Yaman M, Sen S (2007) Vibration control of a cantilever beam of varying orientation. Int J Solids Struct 44(3–4):1210–1220. doi:10.1016/j.ijsolstr.2006.06.015, http://www.sciencedirect.com/science/article/B6VJS-4K6KB0P-6/2/ec9c328d3a430cb47cf393bb4917a950
120. Yang Y, Jin Z, Soh CK (2005) Integrated optimal design of vibration control system for smart beams using genetic algorithms. J Sound Vib 282 (3–5):1293–1307. doi:10.1016/j.jsv.2004.03.048, http://www.sciencedirect.com/science/article/B6WM3-4DJBPM1-6/2/944b2e30a1b99c969b56adbf527d9b1c
121. Yau J (2009) Vibration control of maglev vehicles traveling over a flexible guideway. J Sound Vib 321(1–2):184–200. doi:10.1016/j.jsv.2008.09.030, http://www.sciencedirect.com/science/article/B6WM3-4TWSWP3-1/2/c2ef06bef3677e1ed29b82857a322d58
122. Yildirim S (2004) Vibration control of suspension systems using a proposed neural network. J Sound Vib 277(4–5):1059–1069. doi:10.1016/j.jsv.2003.09.057, http://www.sciencedirect.com/science/article/B6WM3-4BM6CCP-4/2/0db857f0580d634772e8d782485e76bf
123. Yousefi H, Hirvonen M, Handroos H, Soleymani A (2008) Application of neural network in suppressing mechanical vibration of a permanent magnet linear motor. Control Eng Pract 16(7):787–797. doi:10.1016/j.conengprac.2007.08.003, http://www.sciencedirect.com/science/article/B6V2H-4R003K1-1/2/42098496ccc03cdc28602bd04bc4858e
124. Zames G (1979) Optimal sensitivity and feedback; weighted seminorms, approximate inverses, and plant invariant schemes. In: Proceedings 17th of the Allerton conference, pp 744–752
125. Zames G (1981) Feedback and optimal sensitivity: model reference transformations, multiplicative seminorms, and approximate inverses. IEEE Trans Autom Control 26(2):301–320. doi:doi:10.1109/TAC.1981.1102603
126. Zapateiro M, Luo N, Karimi H, VehíJ (2009) Vibration control of a class of semiactive suspension system using neural network and backstepping techniques. Mech Syst Sig Process 23(6):1946–1953. doi:10.1016/j.ymssp.2008.10.003, http://www.sciencedirect.com/science/article/B6WN1-4TTMJRM-1/2/b6b45074716201902e0b01b664ebbeb9, special issue: Inverse Problems
127. Zhang CL, Mei DQ, Chen ZC (2002) Active vibration isolation of a micro-manufacturing platform based on a neural network. J Mater Process Technol 129 (1–3): 634–639. doi:10.1016/S0924-0136(02)00671-4, http://www.sciencedirect.com/science/article/B6TGJ-46V46C0-4P/2/8e8228760a4ac6759cef159e6fcb7606
128. Zhang H, Liu D (2006) Fuzzy modeling and fuzzy control. Control engineering. Birkhäuser, Boston
129. Zheng K, Zhang Y, Yang Y, Yan S, Dou L, Chen J (2008) Active vibration control of adaptive truss structure using fuzzy neural network. In: Control and decision conference. CCDC 2008. Chinese, pp 4872–4875. doi:10.1109/CCDC.2008.4598254
130. Zhu C (2005) A disk-type magneto-rheological fluid damper for rotor system vibration control. J Sound Vib 283(3–5):1051–1069. doi:10.1016/j.jsv.2004.06.031, http://www.sciencedirect.com/science/article/B6WM3-4F4H9R2-1/2/48abebbf8d1230fcd80eee7d19fe52fa
131. Zhu H, Rajamani R, Dudney J, Stelson K (2003) Active noise control using a distributed mode flat panel loudspeaker. ISA Trans 42(3):475–484. doi:10.1016/S0019-0578(07)60148-7, http://www.sciencedirect.com/science/article/pii/S0019057807601487

132. Zilletti M, Elliott SJ, Gardonio P (2010) Self-tuning control systems of decentralised velocity feedback. J Sound Vib 329(14):2738–2750. doi:10.1016/j.jsv.2010.01.024, http://www.sciencedirect.com/science/article/B6WM3-4YCGKVX-1/2/81a53368279e8e5c8664ee835d4b4985
133. Zmeu K, Shipitko E (2005) Predictive controller design with offline model learning for flexible beam control. In: Proceedings of the 2005 international conference on physics and control, pp 345–350. doi:10.1109/PHYCON.2005.1514005

Chapter 5
Laboratory Demonstration Hardware for AVC

This chapter introduces the design and construction of a simple experimental laboratory device that models a problem class in active vibration control (AVC) applications. The laboratory setup serves as a test bed to verify and compare the stabilized model predictive control algorithms featured in the experiments in the upcoming chapters. This laboratory device models the dynamic behavior of lightly damped flexible structures with piezoelectric actuation. A clamped cantilever beam with bonded piezoelectric actuation, fixed at one end and free at the other serves this purpose.

Those who are familiar with the general construction aspects of experimental laboratory hardware or are not particularly interested in the implementation details of the demonstration device featured in the upcoming discussion, may go ahead and after reading a brief introduction of the device in Sect. 5.1 skip the rest of this chapter. For researchers inexperienced with the construction of laboratory hardware, the initial task of familiarizing with equipment and assessing priorities may seem a little overwhelming. The majority of this chapter is thus an aid to the reader who attempts to build, identify and model such or similar laboratory devices aimed at active vibration control.

The chapter begins with a brief introduction of the experimental system, providing a summary of its components and features (Fig. 5.1). This introductory section also features a functional diagram of the control and measurement chain and a scaled engineering drawing of the vibrating blade and its base. The following section concentrates on the experimental identification of the blade dynamics. The state-space models identified in Sect. 5.2 are later used in the MPC algorithms to generate predictions of the system dynamics. In addition to a feedback control model, the sensing models for the piezoelectric patch and a capacitive proximity probe are also evaluated.

The following section is devoted to introducing the properties of the laboratory device and its components. Characteristics such as piezoelectric actuator linearity, step response and frequency response, along with a noise and disturbance analysis are introduced here. Section 5.3 ends with a short analysis on alternative sensing methods to provide feedback to the AVC controller strategy.

G. Takács and B. Rohal'-Ilkiv, *Model Predictive Vibration Control*,
DOI: 10.1007/978-1-4471-2333-0_5, © Springer-Verlag London Limited 2012

Fig. 5.1 The image illustrates an array of BNC connectors on a terminal, used for supplying analog inputs and outputs to the experimental demonstration hardware

Design aspects and cantilever dimension considerations have been largely influenced by a series of finite element analyses performed parallel to the development of the physical model. Section 5.4 introduces some of the results of the FEM simulations. In addition to this, a more complete overview of the modeling process is given in Appendix A with ANSYS code listing. Appendix A may serve as an outline to those interested in the FEM analysis of smart structures with piezoelectric transducers.

Finally, the individual components of the laboratory device are discussed in depth in Sect. 5.5. Here, different hardware components are introduced such as the piezoelectric patches, beam material and the supporting base, real-time control hardware and others. This section is recommended to those readers who are not familiar with the components and equipment used in the laboratory verification tests of AVC systems.

5.1 Experimental Device

To test the performance of various MPC algorithms with constraint feasibility and stability guarantees on lightly damped vibrating active structures, a small-scale laboratory model has been created. *Lightly damped vibrating structures* or, as they are often referred to *flexible systems*, can be regarded as a specific engineering problem in active vibration control and are treated as such in numerous academic works [2, 3, 6, 15, 16, 26, 32, 38, 41, 46, 53, 54, 63, 80, 87]. The laboratory device considered in this work is also a lightly damped flexible system and it consists of a clamped beam with a free end which is equipped with actuators and sensors to form an active structure.

5.1.1 The Cantilever Beam as a Dynamic Model for a Class of Real-Life Applications

The clamped cantilever beam may be a very simple demonstration hardware, however it fully models the dynamic behavior of a whole array of real-life applications from the control engineering point of view. Such simple demonstration devices and basic benchmark problems with minimal damping can be effectively used as dynamic models of more complicated lightly damped structures [16].

In relation to stabilized model predictive control the piezoelectrically actuated clamped cantilever beam and the class of examples it represents have the following common and important physical properties:

- fast dynamics
- light damping
- actuator-disturbance asymmetry

As it will be later thoroughly analyzed, these physical properties create a set of requirements, which render the implementation of stabilized MPC very difficult. The conditions for MPC implementation which directly arise from the physical properties above are:

- fast sampling
- long settling time
- large regions of attraction

The fast sampling time can be prohibitive for many MPC implementations because the computational time of the online optimization process cannot exceed sampling periods. The second condition of the long settling times arising from the lightly damped nature of these systems has a simple logical explanation: the horizon of an MPC controller even without stability guarantees shall cover the expected duration of dynamic effects. What is even more important, the third and final condition creates an especially unfavorable implementation requirement on stabilized MPC. The long settling times and the actuator-disturbance asymmetry warrant a region of attraction encompassing all possible states arising from the disturbances and can steer them into the target set. These concepts and possible issues will be evaluated in detail in the upcoming chapters.

Fig. 5.2 From the control engineering point of view a clamped cantilever beam has similar dynamic properties to helicopter rotor blades. The photograph shows an experimental smart rotor structure in a whirl tower test [10]

There are several examples of engineering structures for which active vibration attenuation systems have been considered and which have comparable dynamic properties to a simple active cantilever beam demonstration hardware. Flexible manipulator arms damped with piezoelectric actuators are a very common example of such lightly damped mechanical systems and appear in the academic literature very often [12, 35, 52, 69, 96, 106, 119, 121, 124]. Other examples of lightly damped mechanical structures with fast dynamics are helicopter rotor beams with active vibration attenuation [9, 62, 71] or wing surfaces in fixed wing aircraft [4, 23, 24]. Figure 5.2 illustrates such an experimental smart rotor structure[1] equipped with an active

1 Courtesy of The Boeing Company.

vibration control system.[2] Current space structures such as solar panels [40, 61, 72], antenna systems [1] and future light constructions for space applications such as support frames for interferometric applications [84, 91] are other examples of the general class of dynamic behavior which can be demonstrated by clamped active beams.

5.1.2 Brief Device Description

The beam is made of aluminum with one end clamped and fixed to a base and the other allowed to vibrate freely. The demonstration device is illustrated in Fig. 5.3 along with the actuators and sensors used in the active vibration control scheme. The beam dimensions are 550 $\times$ 40 $\times$ 3 mm and it is made of EN AW 1050A commercially pure aluminum. The aim of the stabilized predictive control system is to *minimize the vibrations at the beam tip*, when subjected to an external excitation. The dynamics of the beam have been assumed to be completely linear, in order to enable the implementation of linear model predictive control. In addition to the obvious advantages of using a linear prediction model, the assumed linear nature of the device is important when performing experimental testing such as frequency response tests [50].

After assessing the properties of different smart materials commonly employed in active structures, piezoelectric transducers have been chosen as actuating elements. Piezoelectric actuators are commonly used in lightly damped structures as they can be integrated within the original structure to form a composite; moreover, these materials are relatively cheap. Unlike shape memory alloys or electrochemical materials, the FEM simulation support of piezoelectric materials is also well established. The actuating elements in this smart structure are single layer MIDÉ QP16n piezoelectric transducers. These transducers are electrically connected counter-phase and receive the same electric signal from a MIDÉ EL-1225 operational amplifier. The polarization voltage of the transducers is given by the manufacturer as $\pm$120 V, which shall be later incorporated into the MPC law as system constraints.

Accelerometers are used widely in academic studies as a feedback method [29, 31, 94, 123], however a contact-less strategy has been preferred in this case. In this demonstration setup, the relative tip displacement is measured by a contact-free sensing method. Solutions like laser Doppler vibrometry are excellent for such a laboratory application; however, the price range of such devices is excessive. For the average laboratory user laser triangulation is a suitable measuring method with an acceptable precision and equipment price. Beam tip deflections in this device are therefore measured using a Keyence LK-G82 industrial grade laser triangulation system. The Keyence LK-G3001V central processing unit provides analog outputs

[2] See Fig. 1.4. on p. 9 for another photograph depicting the same smart helicopter rotor structure with AVC.

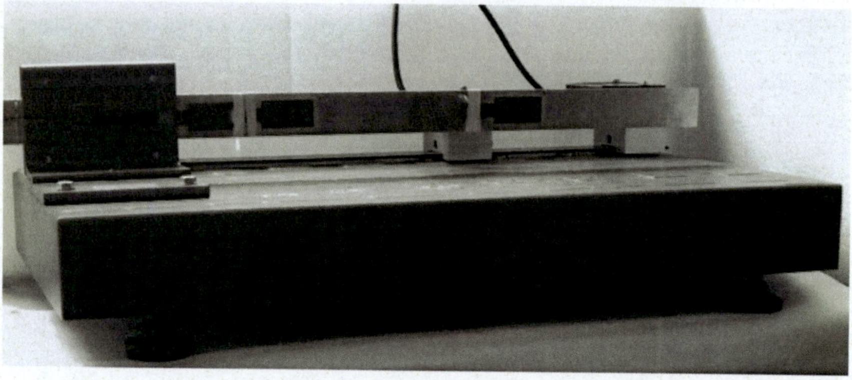

(a) Side view

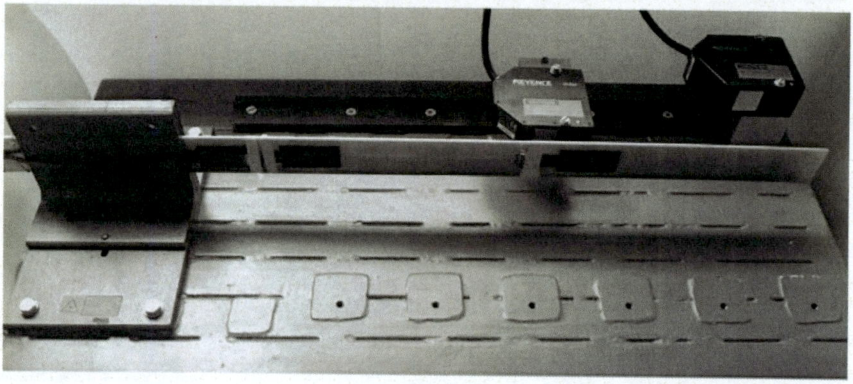

(b) Partial top view

Fig. 5.3 The active vibration control demonstration device is shown from a side (**a**) and a partial top view (**b**). The aluminum beam actuated by piezoceramic transducers is held in place by a clamp while the displacements at the end are measured by a laser triangulation sensor

to the A/D input of the measurement card. The signal from this triangulation system is used for both system identification and direct control feedback.

A laboratory measurement card with a terminal is also necessary to implement a control system on an active vibrating structure. The deciding factors upon selecting the card are amongst others a number of analog inputs and outputs, sampling rate, software prototyping suite compatibility and resolution. A National Instruments PCI-6030E measurement card with a maximal 100 kS/s sampling speed and 16-bit resolution is installed in a personal computer, providing A/D and D/A conversion to the controller. The controller itself is implemented using the xPC Target application platform.

In this work the dimensioning and positioning of the piezoelectric strips has been mostly influenced by iterative FE modeling. Conventions used in similar applications have been taken into consideration as well [33, 64, 67, 118]. The base and

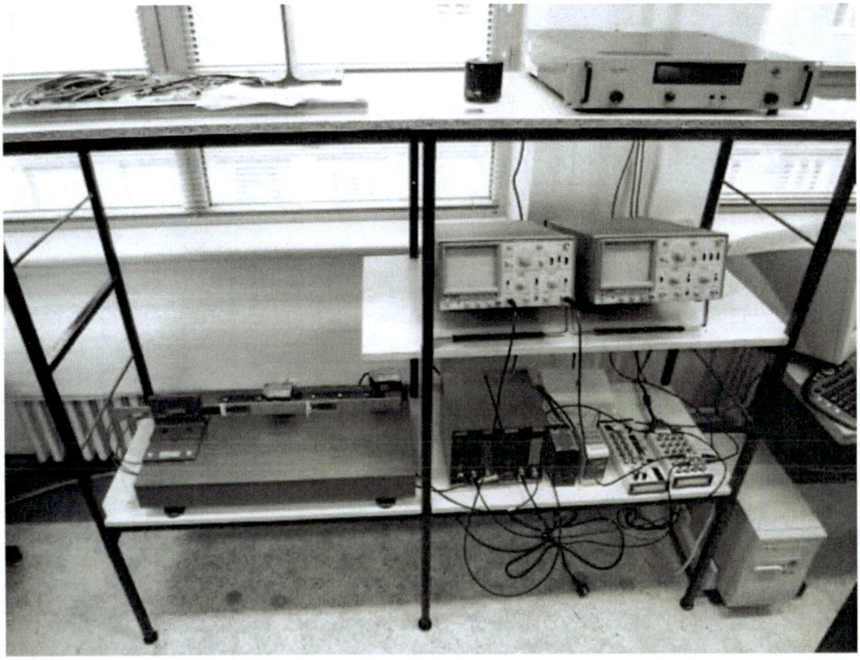

Fig. 5.4 Complete experimental setup for testing efficient MPC in vibration attenuation applications (excluding the Target PC used for real-time control and Host PC used for controller development)

measurement mount was modified to suit the need and dimensions of the active beam and the optical measuring system. A more detailed treatment of the individual hardware components and the justification of particular choices is given in the final section of this chapter, that is Sect. 5.5 for those interested. The complete assembly of the AVC experimental device is shown in Fig. 5.4 alongside supplementary instruments such as amplifiers for piezoceramic actuators, amplifier for a magnetodynamic shaker, oscilloscopes and others.

5.1.3 Functional Scheme of the Device

Figure 5.5 illustrates the functional scheme of the laboratory device. The hardware components introduced in detail in Sect. 5.5 serve together in the experimental setup considered for implementing effective predictive algorithms for active vibration suppression.

Controllers are tested and developed on a computer, from which are loaded via Ethernet onto a PC running a real-time control kernel. This PC contains the data acquisition card, which is connected onto a BNC cable terminal through a

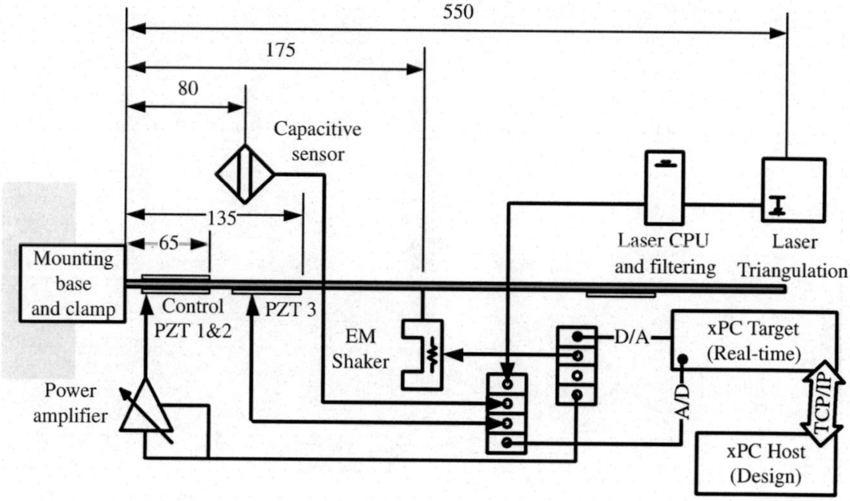

Fig. 5.5 Simplified functional scheme of the experimental laboratory device. The capacitive sensor is only used in alternative feedback testing experiments while the laser sensor is used for evaluating model predictive control algorithms

proprietary cable. The cable terminal has two analog outputs, of which one drives the piezoelectric actuators through a voltage amplifier. The second analog output drives an additional piezoelectric patch, which may generate disturbance signals in certain measurement or test configurations. In other configurations, this output drives an electrodynamic shaker through an amplifier.

The feedback signal in the experimental verification tests of model predictive vibration control comes from the laser triangulation head, connected to a proprietary CPU performing various measurement processing tasks. This CPU then provides the measurement card via the cable terminals an analog input of the displacement signal. Only the laser triangulation head close to the free end of the beam is utilized.

Since the terminal has several more free input terminals, it is possible to utilize a voltage signal from a piezoelectric patch in sensor mode. A capacitive proximity sensor is also placed near the beam surface. The PZT patch-based and capacitive sensor-based feedback systems have been only used to compare the damping effect of these feedback schemes under identical control laws. The MPC tests utilize direct laser readouts. Both the capacitive proximity sensor and the piezo patch reconstruct the deflection, which is to be approximated at the tip of the beam using a process model, while the measurement itself is carried out in different places due to placement and other practical issues.

As the MPC controlled beam has been subjected to different disturbance scenarios, an electrodynamic shaker has also been in use. This shaker is driven through an amplifier, which receives an analog signal from the control computer through the measurement card. The physical connection of the shaker was only left in place when the shaker was utilized. Other tests like initial deflection or piezo-induced disturbance

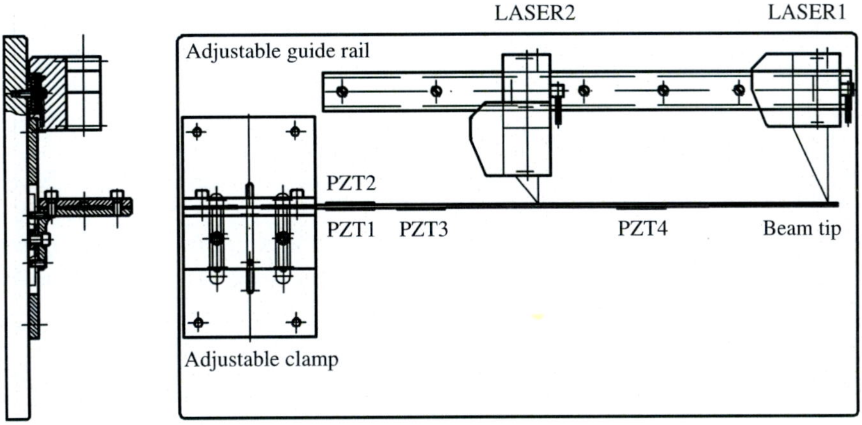

Fig. 5.6 Position and designation of PZT patches and laser heads on the experimental hardware. Patches PZT1 and PZT2 are used as input, PZT3 as disturbance (select experiments), while only the second laser head is utilized for feedback measurements

have not been performed with a connected shaker. To preserve clarity, monitoring, support and other non-essential equipment are not shown on the functional scheme presented.

5.1.4 PZT Transducer Configuration and Usage

As shown in Fig. 5.6, the piezoelectric transducer designated as PZT1 is placed 12 mm from the clamped edge. This transducer is used in actuator mode. On the opposite surface, symmetrically from PZT1 is a transducer designated as PZT2, which is also utilized as an actuator. Actuators PZT1 and PZT2 are electrically connected in counter-phase to maximize bending force. Some studies prefer to use co-located piezoelectric patches for actuation and sensing, for example [22, 31, 118] and others. In that case, the beam tip displacement feedback is reconstructed from the strictly dynamic voltage output of the patch working in sensor mode.

There are two additional piezo wafers bonded to the surface. Both of them may be easily reconfigured to either actuator or sensor mode. Most of the experiments performed in this work did not make use of these, therefore they have been short-circuited to prevent electro-mechanical interference with the structure.

The experimental demonstrator considered throughout this work assumes single point sensing and single point actuation. Although this SISO system is very simple in terms of modeling and represents only a basic and limited experimental scenario, due to the complexity of the online optimization task the practical application of MPC is already very burdensome. The traditional dual-mode infinite horizon quadratic programming-based MPC (QPMPC) formulation was chosen as a baseline method, acting as the best possible achievable performance benchmark, but with the highest real-time computational need. In order to allow the implementation of QPMPC, the complexity of the plant and the resulting model had to be seriously limited. This is the reason why all experiments assume a SISO system representation. The SISO plant has one enormous drawback when one attempts to contrast constrained MPC algorithms to simple saturated LQ controllers. The optimality advantage of MPC over hard-saturated methods is the most apparent with MIMO systems, SISO systems are not the best choice for this purpose. Since the main goal of this book is not to prove the advantages of MPC but to explore the possibility to apply MPC to vibration damping systems, the SISO representation was deemed sufficient.

The modal and harmonic analysis performed in Sects. 5.4.2 and 5.4.3 provided an insight into the behavior of the structure, and the proper placement of the remaining PZT patches. For the placement of transducer designated as PZT3 the deciding factor was the proximity of the clamped end. Therefore, PZT3 was placed as close to the other patches as physically possible. Cabling and electrical connectors permitted to position the edges of PZT2 and PZT3 25 mm from each other.

The transducer patch designated as PZT3 in the figure has been used as a source of mechanical disturbance in some experiments, involving the testing of controllers. The mechanical disturbance in static mode is too small for practical controller evaluation, just as a random signal fed into PZT3 would produce negligible displacement—at least for this application.

Transducer PZT4 is not used in any of the experiments in this work, although it has been added for its potential to provide a dynamic feedback to the controller instead of the triangulation device. Considering its use as a sensor, the vibration nodes and anti-nodes of the different modes (mainly the first three transversal) showed one satisfactory spot for PZT4. It was necessary to avoid an anti-node or else the output voltage from the patch would be very low. Another consideration for a piezo patch in sensor mode is to place it in the spots of maximal strain, for maximal output signal level. The edge of PZT4 starts 320 mm from the clamped end, and is approximately in the two thirds of the beams total length.

At the time of the mathematical system identification process, piezo patches PZT3 and PZT4 were short-circuited. This is necessary to minimize their effect on the structure. The patches are connected through a factory provided cable to the monitoring instruments and to the amplifiers or measuring card terminals.

5.2 Identification Procedure

The aim of the system identification procedure for the AVC demonstrator is to obtain a mathematical model of the controlled structure in a state-space form. This model can be then utilized in the implemented predictive control algorithm to evaluate future system outputs both in simulation and in experiments. A separate state-space model may be used for estimating the tip deflection using a piezoelectric transducer in sensor mode or a capacitive proximity sensor.

In case one aims to cover a wide bandwidth of disturbances in a predictive AVC system, it is necessary to create a prediction model, which does include the resonant modes located in that bandwidth. Initially a high order prediction model has been considered which covers the bandwidth of 0–500 Hz, thus including five measurable transversal modes of the AVC demonstrator. Because of the implementability issues of stabilized MPC with higher model orders, later only simpler model orders were considered with the corresponding longer sampling intervals. These models are second order and include only the dynamic behavior of the first measurable transversal mode, however as the dynamic response of lightly damped vibrating structures is dominated by the first mode, this proved to be sufficient.

In addition to experimental identification or a first-principle analysis [37], it is possible to create a state-space model using the amplitude and phase information acquired for example from FEM analysis [36]. Due to the uncertainty of the quality of the FEM model, the control schemes featured in this work have been obtained experimentally.

Subspace identification methods were used to create a state-space model from the response data. The System Identification Toolbox developed by MathWorks is an up-to-date instrument intensively used in different aspects of control engineering, including vibration control [65, 100, 120]. All experimental identification procedures in this work, including post-processing and identification of the measurement data have been carried out in the System Identification Toolbox [115].

5.2.1 Control Model

To obtain an input–output relationship between the driving voltage and the tip vibration, the piezoelectric transducers PZT1 and PZT2 were driven by a chirp signal. The excitation function was generated in a Matlab / Simulink scheme, where the amplitude of the signal going into the measuring card was ± 5 V. This signal was then fed through the power amplifiers with a $20\times$ gain to reach a ± 100 V peak voltage level on the actuators, well under the acceptable ± 100 V RMS maximal permissible actuator potential. The high voltage level is a necessity due to the need to minimize signal to noise ratio in the measurement, especially in between resonant frequencies.

The laser triangulation sensors have a limited precision for a given deflection span, therefore the bandwidth from 0 to 500 Hz was divided into separate parts accordingly. At each partial measurement, the laser properties were set based on the expected maximal deflection and frequency. To improve precision, the amplitude and voltage ratio was modified. Digital low-pass filtering was enabled to improve precision. The measurements were carried out according to the data presented in Table 5.1.

Sampling rate was set at 0.0002 s, which is 5000 Hz. This allows a sufficient frequency resolution even at the higher end of the required bandwidth. To reach the resonant amplitudes at the eigenfrequencies, one needs to leave sufficient time for the

Table 5.1 Measurement settings of the laser triangulation system

Pass no.	1	2	3	4	5
Freq. span (Hz)	0.01–7	7–9	9–45	45–55	55–500
Amp. gain (mm/V)	0.1	1.5	0.2	0.5	0.25
Low pass filter (Hz)	10	10	100	100	1000
No. of samples (× 1E6)	1.5	3	1	3	5

measurement. If the chirp signal passes through the resonances too fast, the resulting maximal vibrational amplitudes will be lower than expected.

This brings an additional factor into consideration—that is the high number of measurement points to be stored in real-time. The xPC Target system provides the capability to use file scopes, with a storage capacity only limited by the target machines RAM. The computer used for the original measurements had 300 MB RAM, which enabled to store approximately 14 million samples along with the time series data. The resulting data file was retrieved and transformed into a Matlab usable format using xPC target specific Matlab commands [114].

The partial measurement files were combined into one result file using Matlab. The resulting raw time series data contained 13.5 million samples. Manipulation with a data vector of this size has been difficult on a personal computer conforming to today's standard specifications.[3]

The raw data file was loaded into the Matlab System Identification Toolbox. Post-processing of the measurements included detrending and removing signal means. To convert the time series measurements into frequency domain, a fast Fourier transformation (FFT) was performed on the measured data set, up to the half of the sampling bandwidth: that is 2500 Hz. This data file was then filtered using a low-pass band filter to cut off unnecessary frequencies and to reduce the amount of working data.

Since the amount of data points in the working file still prohibited the practical use of identification routines, the frequency response and the spectrum was estimated using spectral analysis with frequency-dependent resolution returning. The estimation procedure has been performed with a logarithmic resolution of 2000 frequencies ranging from 1 to 500 Hz. The size of the original data set has been reduced also cutting down the implied computational load, while still containing enough information to create a high quality model. The resulting processed measurement was thus suitable for direct identification.

After comparing singular values and taking into consideration the required bandwidth, a 12th order model has been chosen. The identification routine utilized a subspace iteration method [70], implemented as default in the System Identification toolbox [115].

As previously noted, due to the specifics of stabilized MPC on lightly damped systems, the later stages of controller development and testing required much lower model orders. These have been created by the above-described procedure.

[3] See the description of the *host* PC in Sect. 5.5.4.1.

The measurement data has been resampled to lower rates in order to prevent oversampling and unnecessarily small sampling intervals. This also produced a desirable side effect, namely that the model could be now identified directly from the frequency domain data—instead of spectral estimate. A second order model acquired by this technique is given by the following relation:

$$\mathbf{A} = \begin{bmatrix} 0.867 & 1.119 \\ -0.214 & 0.870 \end{bmatrix} \quad \mathbf{B} = \begin{bmatrix} 9.336\text{E}-4 \\ 5.309\text{E}-4 \end{bmatrix} \tag{5.1}$$

$$\mathbf{C} = [-0.553 \quad -0.705]$$

where the output of the model is beam tip deflection in millimeters, and input is direct actuator voltage. Sampling in this case is 100 Hz, or 0.01 s. The larger order model containing measurable dynamic response up to the fifth transversal bending mode, or approximately 500 Hz can be described by the discrete, linear-time-invariant system sampled by 5000 Hz (0.0002 s) with matrices **A**, **B** and **C**.

Simulation and measured response comparison of vibration attenuation systems is ideally performed in the frequency domain. The particular method of obtaining frequency domain measurements unfortunately prohibits the direct comparison of the results throughout the whole considered bandwidth. The reason for this is the size of the raw frequency domain measurement file: comparison on computers conforming today's standards and using the System Identification Toolbox is not a viable option. However, the results of partial bandwidth measurements and matching simulation output are shown in Fig. 5.7a. In this case, the response of a second order system is compared with the simulation results up to a partial bandwidth of 20 Hz.

To evaluate the match between large order model and measurement data, spectrum estimate of the measured frequency response is utilized. The System Identification Toolbox then also provides a numerical indicator of precision—a percentage match between the model and measurement data. One also has to take into account the planned practical implementation and use of the state-space model to validate the results. In this case, the match of the model in the proximity of the first resonant frequency is more important than the general correspondence in the whole bandwidth. In the process of fine-tuning the identification algorithm, fitting the model to measurement data within the high frequency portions of the signal was of less importance. Graphical results of this comparison in are shown in Fig. 5.7b.

Contrasting measurement results in the time domain is not a good indicator of the model accuracy. The main reason for this is the fact that in static mode, the beam tip deflection is minimal, only measurable in tenths of millimeters—the practical use will utilize a dynamic excitation of the piezoelectric transducers anyways. Outside disturbances have a substantial effect on the measurements. The physical device is also placed in an environment, where the effect of road traffic or people moving within the laboratory is clearly indicated in the vibration of the beam tip. Direct comparison of the model response and the real system measurements to a pulse signal is shown

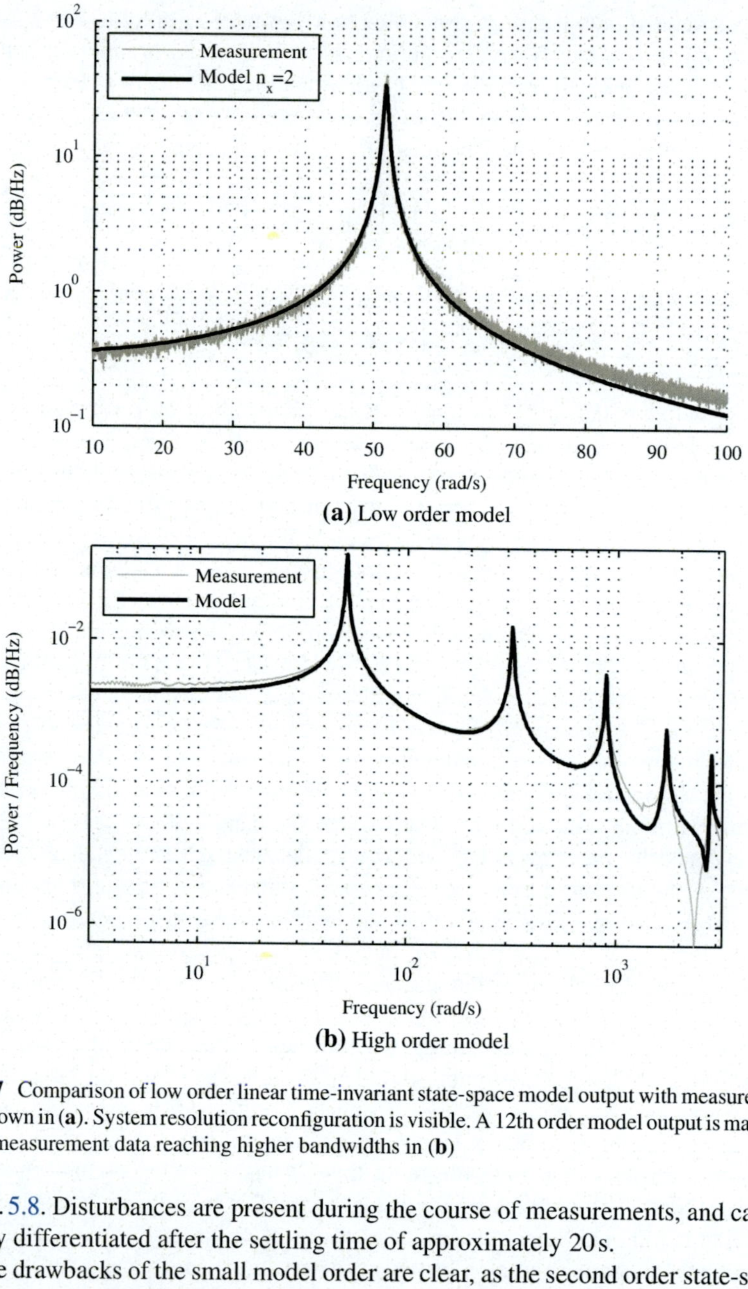

Fig. 5.7 Comparison of low order linear time-invariant state-space model output with measurement data shown in (**a**). System resolution reconfiguration is visible. A 12th order model output is matched to the measurement data reaching higher bandwidths in (**b**)

in Fig. 5.8. Disturbances are present during the course of measurements, and can be clearly differentiated after the settling time of approximately 20 s.

The drawbacks of the small model order are clear, as the second order state-space model cannot represent system dynamics above the first resonance. This is clear from Fig. 5.9 where the deflection response of a second order model to a chirp signal input is compared with the measured beam tip deflection response. The frequency of the signal is increasing and passing through the first, second and third transversal

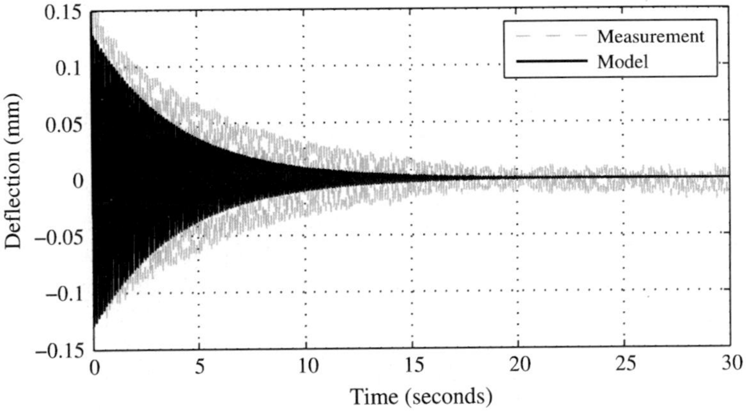

Fig. 5.8 Comparing measurement data and model response to an impulse generated by a piezoelectric transducer. Note the added mechanical noise present in the laboratory

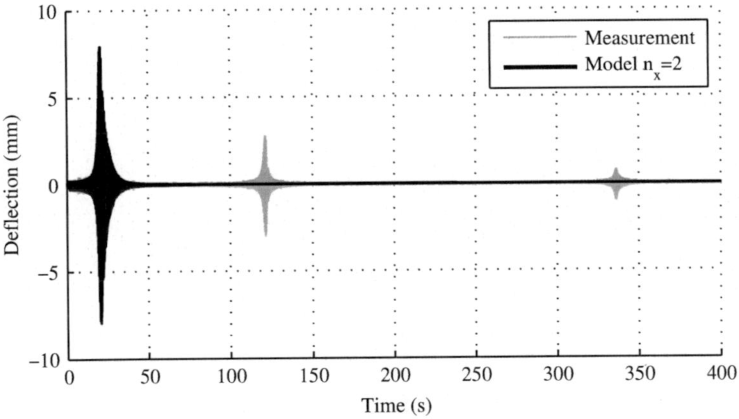

Fig. 5.9 Comparison of the response of the second order model and the real structure to a chirp signal with increasing frequency

resonant mode as it can be seen from the measured structural response. However, the low order model can only represent the dynamic response only up to the first resonant frequency.

5.2.1.1 The Experimental System as a Point Mass

It is possible to approximate the dynamic behavior of the smart cantilever beam demonstrator as a vibrating point mass system. The properties of this equivalent system have been measured as follows:

- *mass* $m = 0.1782$ kg
- *spring constant* $k = 470$ N/m
- *damping constant* $b = 0.08344$ Ns/m
- *linear force conversion* $k_f = 0.585\,\text{E}{-3}$ N/V. This is for one actuator, therefore the total is $2 \times k_f$.

This model has its output in meters instead of millimeters as featured in the models identified from the experimental data. The equivalent point mass system has not been utilized in the MPC control schemes featured here. The point mass equivalent of the demonstrator has been used as an example for a moving horizon observer (MHO) study performed by Poloni et al. in [90].

5.2.2 Capacitive Sensor Feedback Model

The capacitive sensor model takes the proximity sensor voltage as input, and outputs the estimated beam tip position in millimeters for the controller. The sensor feedback model has been identified experimentally, using a time domain data set. The beam has been subjected to manual excitation, while the voltages coming from the capacitive sensor were acquired along with the laser measured beam tip positions.

The sensor feedback models featured in this work are second order linear time-invariant systems, explicitly including only the first mode of vibration. Sampling time for both the capacitive and piezoelectric sensor feedback model are $T_s = 0.01$ s.

This data set was then filtered, detrended and a suitable portion used for identification. The state-space model of the capacitive feedback measurement process has been identified using an iterative prediction-error minimization method introduced in [70].

Akaike Final Prediction Error (FPE) criterion for this model has been calculated to be 0.0432 (–). After numerous trials, the model in (5.2) has been selected as best for the purposes of the feedback model experiment featured later in Sect. 5.3.4. The model validation process proved to yield a satisfactory match, while the transient and frequency response of the model was also adequate.

$$\mathbf{A} = \begin{bmatrix} 1.229 & -0.216 \\ 2.048 & 0.190 \end{bmatrix} \quad \mathbf{B} = \begin{bmatrix} -2.086E{-2} \\ -7.674E{-2} \end{bmatrix} \tag{5.2}$$

$$\mathbf{C} = [97.28 \quad -1.159]$$

5.2.3 Piezoelectric Sensor Feedback Model

The measurement model based on the output voltage of the piezoelectric strip is a single input, single output state-space model as well, and has been created according to [111]. It takes voltages as its input and outputs the beam tip deflection position estimates for the controller. This feedback model has also been identified using a time domain data set, while subjected to pseudo-random manual excitation. Beam tip deflections have been recorded along with sensor voltage output.

The data set was then filtered, detrended, mean values removed and divided into parts for system identification and model validation. The mathematical model of the feedback based on piezoelectric wafer signals has been identified using a subspace iteration method [70].

FPE criterion for this model has been calculated to be 0.0091 (–). Comparing the measured and estimated model output for different portions of the data set, and by judging the quality of the transient response, frequency response and model residuals; the model in (5.3) has been chosen to assess the feedback control quality of PZT patched based sensing in Sect. 5.3.4.

$$\mathbf{A} = \begin{bmatrix} 0.987 & 0.144 \\ -0.274 & 0.009 \end{bmatrix} \quad \mathbf{B} = \begin{bmatrix} 3.959E{-2} \\ 1.851E{-1} \end{bmatrix} \tag{5.3}$$

$$\mathbf{C} = [34.72 \quad -1.3595]$$

5.3 Device Properties

This section introduces experiments, where select properties of the *smart* structure are characterized. These include actuator linearity and other characteristics, ambient mechanical noise present during the trials, feedback routes alternative to the direct laser triangulation, electrical noise and certain other mechanical aspects of the device.

The experimentally verified AVC demonstrator properties featured in this section do not contribute to the model predictive vibration control of the structure directly, albeit they form the basis of many assumptions featured later. The linearity of the piezoelectric actuators justifies the linear time-invariant model used to generate dynamics predictions. The noise test featured in this section explains some of the minor irregularities in the MPC verification results. The simple modal tests verify the general accuracy of the FE model, while the evaluation of the feedback quality with the piezoelectric and capacitive sensor points out the effectiveness of AVC with simpler and cheaper hardware components ready for mass deployment.

5.3.1 Actuator and Sensor Characteristics

5.3.1.1 Piezoelectric Actuator Linearity

Actuator linearity has been verified in experiment. A simple test has been designed, where the beam tip displacement was measured subject to an increasing voltage load on the actuators. A ramp signal provided the elevating voltage level, starting from

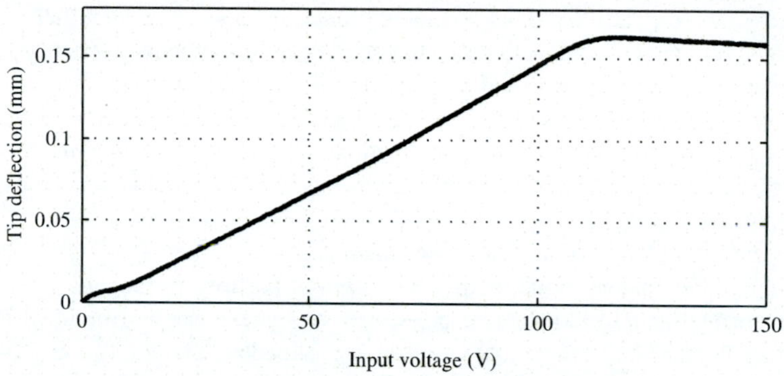

Fig. 5.10 Beam tip deflection subject to increasing static excitation voltage on the piezoelectric actuators. The response is practically linear in the normal working voltage range

zero up to exceeding the allowable maximal 120 V. Going beyond the recommended allowable voltage levels was not possible because of the built-in safety features in the power amplifier. The ramp signal increased voltage inputs steadily, allowing sufficient time for the system to settle in the desired position.

Expected deflections were comparable to the usual background mechanical disturbance levels in the laboratory, therefore precautions were made to minimize their effect. Although the beam under actuation is less prone to the causes of outer disturbances, filtering has been used at the measurement stage. The laser triangulation system CPU was set to engage an 18 byte[4] moving average operation on the samples. Measurement sampling has been set to 50 μs, data logging to 0.1 s. To increase measurement precision, the analog output of the laser system has been established at 0.1 V/mm instead of the usual 1.5 V/mm permitting deflections over only under 1 mm.

The results of this trial are indicated in Fig. 5.10. The horizontal axis presents increasing voltage levels, the vertical shows corresponding deflections measured at the beam tip. As is evident from the figure, the dependence of deflections at the beam tip is linear on the supplied voltage. This suggests that on the voltage range of interest, a linear time-invariant state-space model sufficiently emulates input–output behavior. In the region over approximately 120 V the safety features of the amplifier engage, and do not allow further increase in the deflection. The lower portion of the graph shows a little deviation from the linear characteristics, but this can be attributed to the effects of averaging on a system where the initiation of actuation caused some minor transient vibrations.

Maximal static deflection achievable with 120 V supplied to the piezoelectric actuators connected counter-phase is only around 0.15 mm. This fact is especially interesting, given that the vibrating blade under control is capable of surpassing a magnitude higher deflections very effectively, if no further force effect is present.[5]

---

[4] That is 262 144 data points.

[5] See the initial deflection, frequency domain and other experiments on the controlled system, as described in Sects. 12.2, 12.3.2, 12.4.

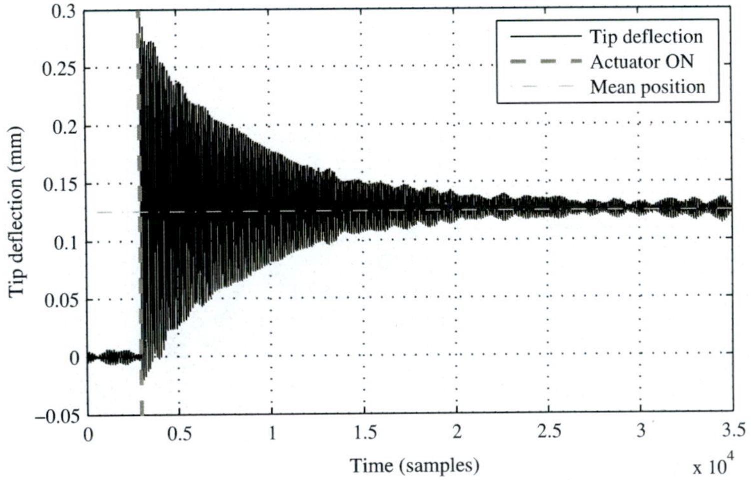

Fig. 5.11 Beam tip deflection subject to a step change of actuator voltage from 0 into –120 V. The change in position is instantaneous, however causes vibrations which settle very slowly due to the small damping ratio

5.3.1.2 Step Response of the Piezoelectric Actuator

Actuator step response has been measured in experiment. Figure 5.11 shows beam tip displacement measured in millimeters, sampled by a 0.001 s interval. Digital filtering has been enabled to remove frequencies above 1000 Hz. The two counter-phase connected actuators PZT1 and PZT2 have been supplied with an instantaneous step change of voltage potential from 0 to –120 V. This is visible in the evolution of states after 3 s, or 3E3 samples.

Reaction of the beam tip to the change in input is practically immediate. However, due to the highly under-damped nature of the physical structure, vibrations prevail after several seconds the change in position occurs. The initial position changes to the peak maximum of approximately 0.3 mm and then eventually settles to the static position of 0.13 mm. Repeated tests also indicated static positions in the vicinity of 0.15 mm, this being subject of disturbances, filtering methods and others. According to this, static gain can be approximated only as 1 μm/V.

Even maximal allowable voltage levels produce deflections at the beam tip only in the order of fractions of millimeters. This could suggest that piezoelectric actuation will be ineffective in the damping of orders of magnitude larger deflections.[6] However, as controlled actuation experiments[7] will prove, piezoelectric transducers can be very effective even when the disturbance is much higher than maximal static deflections. This could be attributed to the fact that, unlike the displacement or

[6] Presuming initial disturbance is removed and the system is able to settle on its own or under control.

[7] see Sect. 12.1 for LQ, Sect. 12.2 and others for NRMPC and MPMPC.

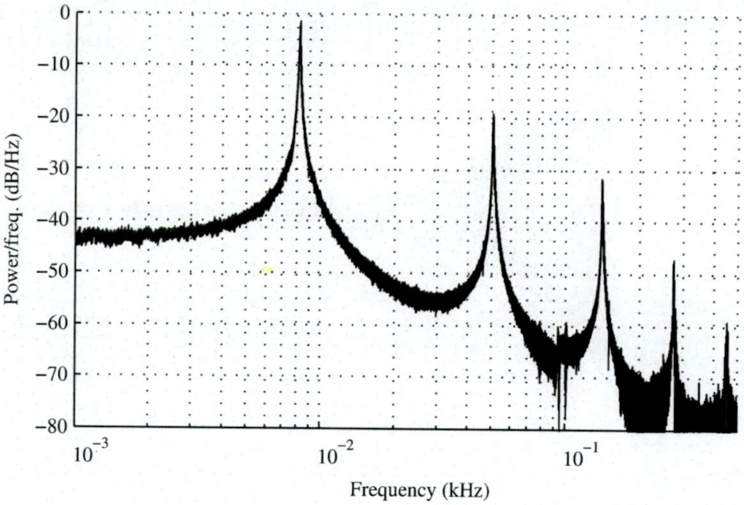

Fig. 5.12 Single-sided power spectral density of beam deflections, showing five transversal vibration modes in the given bandwidth. The two actuators were excited by a chirp signal of ±120V peak amplitude

deformation, the force exerted by piezoelectric actuators can be very high. Also, the dynamic response indicates a large range of possible deflections at the first resonant frequency.

Vibrations are most visible in the section without actuation, this is due to the mechanical noise and disturbances constantly present in the laboratory (see 5.3.2 for details.) Unwanted disturbance also prevails after the beam tip has settled in its static position.

5.3.1.3 Frequency Characteristics of the Piezoelectric Actuator

In this application, a dynamic test of actuator properties holds more information about the real-system behavior than a static trial. Figure 5.12 shows single-sided amplitude spectra of beam deflections, subject to an excitation by a chirp signal of ±120 V peak amplitude. This chirp signal ranged from 0.1 Hz up to 500 Hz in 20 min. Measurement sampling rate has been set to 0.0002 s (5000 Hz) and the results were stored in a 6E6 data point time domain file. The time domain information has been later converted into the frequency domain.

Finite element tests indicate seven resonances in the bandwidth of interest. Two of those are not measurable since they are twisting and lateral modes. There are two small spikes visible near the frequency range of approximately 100 Hz, this could be theoretically caused by the non-measurable vibration modes.

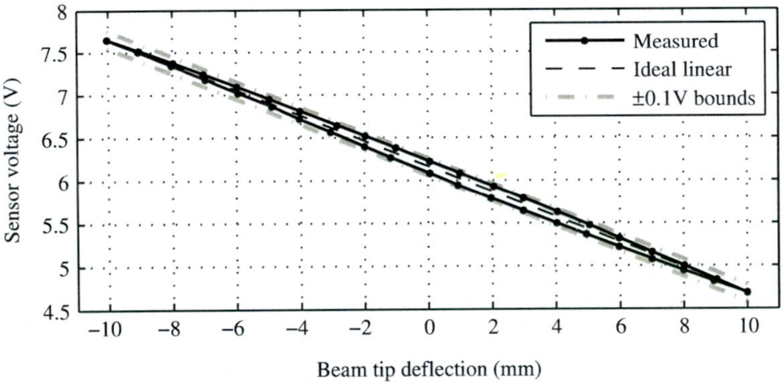

Fig. 5.13 Static characteristics of the capacitive feedback setup, related to beam tip deflection. Dashed line denotes the ideal linear relation, while the solid line indicates actual measurements

The remaining five bending modes are well visible in Fig. 5.12. Higher frequency modes produce only small vibration amplitudes, therefore, noise is more dominant in the higher frequency ranges of the measurement. The first bending mode is quite significant—reaching deflections up to ±15 mm, which materializes as an extraordinary violent movement of the structure. Results of such and similar dynamic actuator tests were utilized in creating an input–output state-space model of the structure behavior, as described in Sect. 5.2.

5.3.1.4 Linearity of the Capacitive Proximity Sensor

To assess linearity and hysteresis of the capacitive sensor in relation to deflections measured at the beam tip, the end of the cantilever has been deflected –10 mm away from equilibrium, slowly repositioned to 10 mm and finally back to its original position. Laser triangulation has been used to measure deflections, while sensor voltage output has been recorded.

Figure 5.13 presents static characteristics of the capacitive sensor in relation to cantilever beam tip deflections. The ideal linear relationship is denoted with a dashed thin line, while the 0.1 V sensor linearity deviation bounds are also indicated for reference.

In addition to device specific built-in irregularities, linearity deviation may also be attributed to material changes. The rated sensing distance is based on a standard target, presumably steel [99]. Effective sensing distance therefore must be reduced. Moreover, the beam surface at the point of measurement is only parallel to the sensor area if the beam is resting in its equilibrium position. With increasing deflections measured at the tip, the beam and sensing surface angle elevates as well.

While there is a slight deviation from the ideal linear response, this irregularity does not prevent its efficient use as feedback source in vibration control applications. Experimentally identified mathematical feedback models could include effects caused by linearity deviations, or it is possible to correct them if necessary.

5.3.2 *Noise and Disturbances*

Ambient mechanical disturbance and electrical noise may affect measurements in a laboratory environment. A series of experiments have been designed to approximate the real magnitude and nature of these unwanted effects on experimental results. Disturbances of both electrical and mechanical nature have been analyzed.

5.3.2.1 Mechanical Disturbances and Noise

Undesired mechanical noise and disturbances are always present in the laboratory and affect measurements in various ways. Controller trials with large deflections are practically not altered by mechanical noise, since its levels are small comparable of those induced during the trials. Other tests like the static step response of the actuators or random excitation trials with small amplitudes may be influenced by unwanted mechanical noise.

Sources of this mechanical disturbance are diverse, and are expected given the under-damped nature of the structure. Convective heat, movement of air is constantly present in the laboratory. Heavy traffic on adjacent streets can also influence certain types of tests, and is well measurable on the free blade vibration. Activity of persons in the laboratory is probably the most influential of all disturbances. Movement, moving of objects and talking can all show up in measurements.

Placement of the active structure on the laboratory shelf system may increase the source of mechanical disturbance as well. For measurements involving low range of tip displacements, it would be beneficial to place the device on the ground or a much more rigid support base. Because the device is used for educational purposes and needs frequent physical reconfiguration, it was always used on the shelf system visible in Fig. 5.4.

Figure 5.14 shows vibration of the uncontrolled, freely moving beam tip in the absence of deliberately introduced disturbances. This time domain measurement has been sampled by 0.0002 s (5000 Hz). A 5000 Hz low pass filter has been enabled in the laser triangulation system CPU. The lower level vibration is generally present and affecting measurements. It is caused by the combination of traffic, rotation of power amplifier cooling fans, air convection and others. The higher level vibration contains the additional factor of persons moving around the laboratory.

As is clearly visible from the figure, levels of mechanical noise easily approach that of the static deflection possible through piezoelectric actuation. Some measurements therefore include the effect of unwanted mechanical disturbance. Noise has another implication: there is no state of absolute equilibrium, therefore the beam is in constant motion. Any controller has to take this into account by implementing a dead-zone, to prevent the amplification of ambient mechanical disturbance.

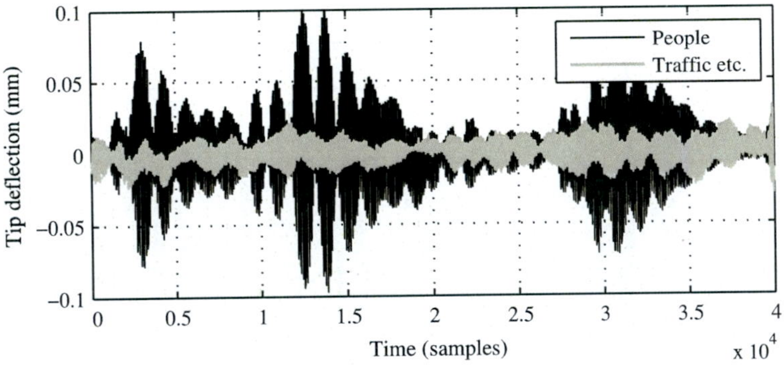

Fig. 5.14 Free, uncontrolled beam tip deflection in the time domain, subject to mechanical noise and disturbance caused by convective heat, road traffic, equipment ventilators and other factors. Movement and conversation of persons in the room increases undesired vibrations significantly

5.3.2.2 Electric Noise

Measurement noise generated by electrical properties of the system is not significant and in agreement with the measurement card specifications as given by the manufacturer. Three experiments have been performed, each measuring electric potential on the terminal used to acquire feedback signal. The analog input has been set to measure a symmetric bipolar voltage of ±10 V. Sampling rate has been established as 0.01 s, filtering was set to 5000 Hz low pass. Time domain signals have been recorded with the input terminal short circuited to create zero potential. This has been performed in three diverse cases: at the cable terminal only, cable terminal with equipment connected and the third possibility included power amplifiers in operation.

Initially a larger noise level has been expected with the cabling and equipment connected to the measurement card. Electromagnetic noise generated by the high voltage amplifiers suggested a significant pollution of the measurement signal. However, this turned out to be a false assumption: all three experiments produced identical results. Figure 5.15 shows the zero potential measurements in all three cases.[8] The peak signals range from 0.3 to 0.9 mV with a mean value of 0.6 mV. Manufacturer lists a 1.147 mV absolute accuracy at the given scale, with a 723 μV single point resolution [82]. A 16-bit resolution would indicate 65,536 voltage levels differentiated at a given level,[9] which at 20 V span would give a 30 mV detail. We may state that this noise may be expected and is according to the hardware specifications provided by the manufacturer.

---

8 Not actually plotted. The three different experiments produced random noise with the same specifications.

9 Actual quantization may differ significantly, possibly encoding potential polarity and others. For example a 10 V span can be divided into 1.5 mV portions.

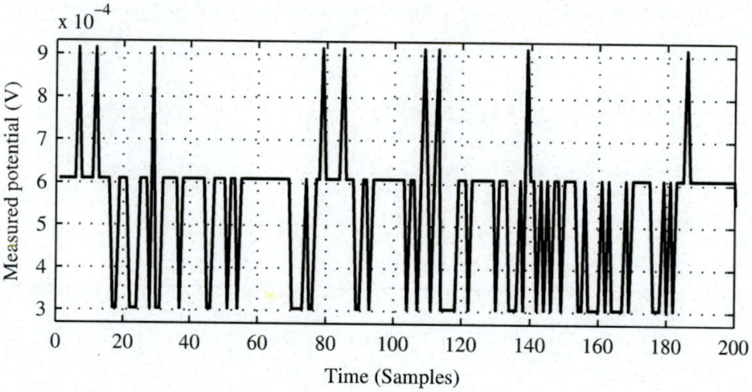

Fig. 5.15 Measured electric potential in Volts. The figure shows inherent electrical and systematical noise in agreement with the measurement card specifications

5.3.3 Mechanical Properties

Several ad hoc tests were carried out to explore select mechanical properties of the AVC demonstrator system to aid the design process of the hardware and the controller itself. At the time of performing these trials, the smart structure had only one piezoelectric transducer bonded to its surface, and the FEM model corresponded to this state. Simple experimental approximations investigated the real resonant frequency and vibration amplitude levels, which helped to improve on the numerical model.

5.3.3.1 Resonance Frequencies

A test has been carried out to compare eigenfrequencies obtained by finite element analysis and experimentally on the physical system. The optical measurement system was set to take readings of the beam tip vibration. The actuator was excited with a ± 100 V sine wave, using a sine wave generator. Although excitation amplitude is irrelevant to the results, this corresponds to the loading used in the FEM model harmonic tests. The frequency of the actuating sine wave has been varied around the values expected from the simulations, until the highest amplitude was reached. The frequency and the corresponding highest vibration amplitude were noted. This experiment has been performed when only one of the piezoelectric transducers was bonded onto the beam surface, marked as PZT1 in Fig. 5.6.

Although this approach provides only an approximation of the resonance frequencies and amplitudes of the physical system, it is sufficient enough to validate the finite element model. The results show good agreement with the FE simulations. Frequencies estimated and measured are shown in Table 5.2. Comparison of the modal frequencies obtained by the finite element method and experimental approximation is shown in Table 5.2.

Table 5.2 Comparison of experimentally approximated resonant frequencies with FEM results, considering one piezoelectric transducer on the surface

Mode	1	2	4	6	7
Freq. FEM (Hz)	8.2	50.6	139.8	271.9	449.2
Freq. Exp. (Hz)	8.2	50.7	140.9	275.5	455.1

Table 5.3 Comparison of experimentally measured tip vibration amplitudes to FEM model results considering one piezoelectric transducer bonded to the surface

Mode	1	2	4
Amplitude FEM (mm)	12.4	3.2	1.1
Amplitude Experimental (mm)	12.3	3.0	1.0

5.3.3.2 Resonant Amplitudes

It is also worth comparing maximal average tip displacement amplitudes with simulation results. Table 5.3 lists these results for each bending vibration mode. Modes 3 and 5 are not shown, since irrelevant to the control application. Only the first three bending modes are listed, since most of the vibration energy is concentrated in lower bandwidths. In addition, measurement of higher vibration amplitudes proved to be unreliable because of the ambient mechanical noise and disturbances present.

After adjusting properties of the FE model, tests showed a good agreement with the simulation results. If one subjects the piezo actuator to the maximal voltage in static mode, the beam tip displacement is indistinguishable from equilibrium with the human eye. Measurement in an uncontrolled environment may also be problematic. Vibration in the first three transversal modes is visible, the fourth and the fifth modes are only measurable.

5.3.3.3 Impact Response

Another preliminary experiment was carried out with one piezoelectric actuator bonded to the beam surface. Since an impact hammer and the required signal processing and sensing instrumentation have not been available, a real modal analysis has not been carried out. A rather improvised method was used to gain some insight to the modal properties, namely the eigenfrequencies of the system.[10]

An optical displacement and vibration measuring instrument was utilized as a means of sensing the vibration amplitude. The amplitude signal was saved to the intermediary buffer of the sensors CPU with 65,536 data points. Several tests were carried out by impacting the beam base. Different sampling frequency settings were utilized, of which 5 kHz produced the most satisfactory results.

10 Accelerometers and amplifiers registering both excitation and response.

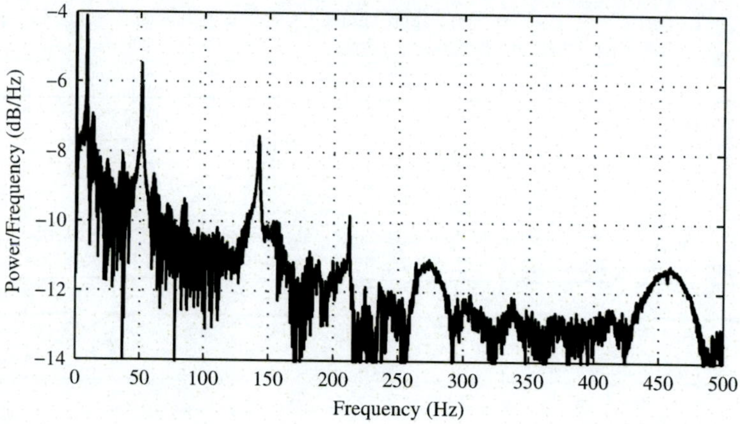

Fig. 5.16 Logarithmic single-sided amplitude spectrum of beam tip vibrations. The beam base has been subjected to an unmeasured impact

Table 5.4 Modal frequencies with multiple piezo transducers bonded onto the active structure

Mode	1	2	4	6	7
Freq. Exp. (Hz)	8.25	51.03	141.9	275.7	456.9

The raw time series data was downloaded from the CPU buffer into a computer using proprietary software [57]. The time series data were imported into Matlab and transformed using a standard fast Fourier transformation, to obtain single-sided amplitude spectra of the beam tip vibration. The result of one of these tests is presented in Fig. 5.16.

Although the amplitude values are shown in the figure, they act only as a relative reference. The logarithmic amplitude spectrum chart shows most of the bending modes clearly visible. The acquired vibration amplitude spectrum is very noisy and not suitable for system identification. This ad hoc experiment has been designed for informative purposes only, to gain insight into the eigenfrequencies of the AVC demonstrator structure. Resonant frequencies of the experimental device with all the piezoelectric transducers and sensors in place are shown in Table 5.4.

5.3.4 Capacitive and Piezoelectric Sensor-Based Feedback

Feedback models have been created for both a piezoelectric patch in sensor mode and a capacitive proximity probe. These feedback sources alternative to the more precise laser triangulation device were used to compare the damping performance of the same LQ strategy under different sensing schemes. Despite both the PZT and the capacitive probe-based feedback strategies provided comparable vibration attenuation levels to the direct laser one, the experiments shown in the upcoming chapters assume feedback from the optical device.

The feedback model has been validated experimentally both in time and frequency domain. Although model estimate sampling is set at the period of $T_s = 0.01$ s, output measurements have been adjusted in the case of the frequency domain measurement to a rate of $T_s = 0.0002$ s in order to capture higher frequency effects.

5.3.4.1 Position Estimates in Time and Frequency Domain

The measured and estimated beam tip deflections were compared in a test involving an initial deflection 10 mm away from equilibrium. The beam has been displaced and released to vibrate under saturated LQ control without further outside force interaction. The results of this test are indicated in Fig. 5.17.

As is evident from Fig. 5.17a, both model estimates have a difficulty correctly assessing slow displacement changes. In addition to the fact that the piezoelectric patch is only usable in dynamic mode, slow frequency changes cannot be readily detected. However, as the beam is released just before the 1100 sample time mark, the estimates become more accurate—in fact indistinguishable from the laser triangulation measurement taken as reference. Figure 5.17b shows the unprocessed voltage output from the capacitive and piezoelectric sensors.[11]

Tip displacement estimates have been compared to the true laser-based reference under a wide-band mechanical excitation as well. The shaker received a chirp excitation signal in the frequency range of 0–500 Hz, time span of 200 s.

Figure 5.18 indicates the single-sided amplitude power spectra of laser measured and piezoelectric sensor and capacitive proximity sensor-based tip displacement estimates. Laser reference indicated with the solid black line on the figure shows the measured response the numbered resonant nodes.[12] As it is expected from a second order tip deflection estimate model, deformations are assessed correctly only in the vicinity of the first resonant mode.

Frequencies above approximately 15 Hz are not covered by the tip position estimate models, as their order and sampling period does not allow this. The repeating and reoccurring peaks in the capacitive and piezoelectric feedback data are merely artifacts of the FFT process on a data set lacking high frequency components, often referred to as spectral leakage.

Higher order position estimate models could explicitly include higher resonant modes, but would also require faster sampling periods. Additionally, simpler models can perform very well even at higher frequencies.

The capacitive sensor has bandwidth limitations, although it is hard to assess this from the given experiment as the vibration ranges decrease with increasing

[11] Laser reference output is directly proportional to the measured value, in this case, there is a 1.5 mm/V gain.

[12] Modes (3) and (5) are twisting modes and cannot be measured nor controlled with the sensor/actuator configuration assumed throughout this book.

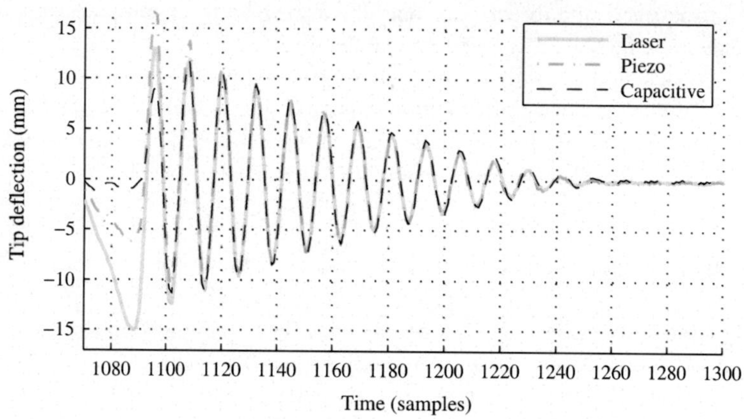

(a) Measured and model estimated beam tip deflection

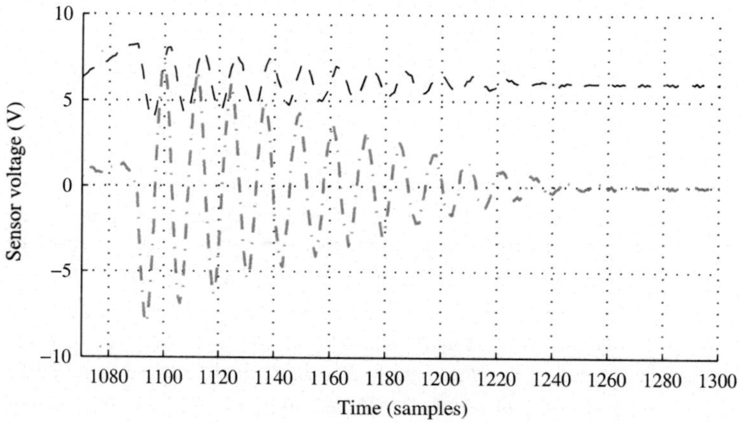

(b) Capacitive and piezoelectric sensor voltages

Fig. 5.17 Comparison of measured and estimated beam tip deflection is shown in (**a**), while output voltages for the piezoelectric patch and capacitive proximity sensor for the same experiment are featured in (**b**)

frequency. Furthermore, slow- or near-DC vibrations cannot be detected through the piezoelectric sensor feedback, due to the physical nature of the hardware.

5.3.4.2 Effect on Damping Performance

We will now introduce experiments performed in order to determine the effect of different feedback schemes on damping performance. Control sampling is set to $T_s = 0.01$ s.

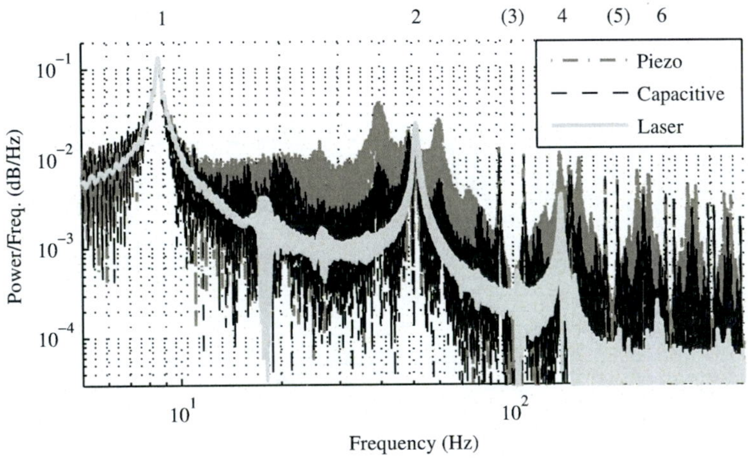

Fig. 5.18 Single-sided amplitude spectrum of measured and estimated beam tip deflections. Numbers denote laser measured amplitude peaks due to corresponding structural vibration modes

The control method used throughout the experiments comparing different feedback schemes was a simple saturated linear quadratic controller. Saturation levels were set to ±120 V according to the safety limits specified by the manufacturer, preventing the depolarization of piezoelectric material. The LQ controller has been calculated using a state penalty matrix $\mathbf{Q} = \mathbf{C}^T\mathbf{C}$ and an input penalty $\mathbf{R} = r = 1\mathrm{E}-4$. The state controller gain $\mathbf{K}$ is according to relation (5.4).

$$\mathbf{K} = [12.97 - 125.50] \tag{5.4}$$

The beam tip has been set to an initial position of 10 mm away from equilibrium, and then controlled responses have been recorded without further structural interaction. Beam tip deflections in all scenarios have been measured using laser triangulation, but the feedback control schemes utilized different sensors. Saturated LQ control with direct LQ feedback provides a very effective damping performance, comparable to constrained MPC control for this SISO system.[13] Settling times are reduced by an order of magnitude, demonstrating the effectiveness of both the actuators and the control system applied to the AVC demonstrator.

Figure 5.19a shows the evolution of beam tip vibrations to this type of excitation while control voltage supplied to the actuators is indicated in Fig. 5.19b. Both position feedback estimating methods are considered here, with direct laser measurement acting as reference.

As it is clear from the results, there is no substantial difference in the damping performance when estimates are used as feedback. Measured beam tip deflections

13 An MPC control would provide significantly better performance than LQ given a MIMO vibration control system, where the real performance of the linear quadratic controller would be degraded due to the saturation limits not included in the original optimization task.

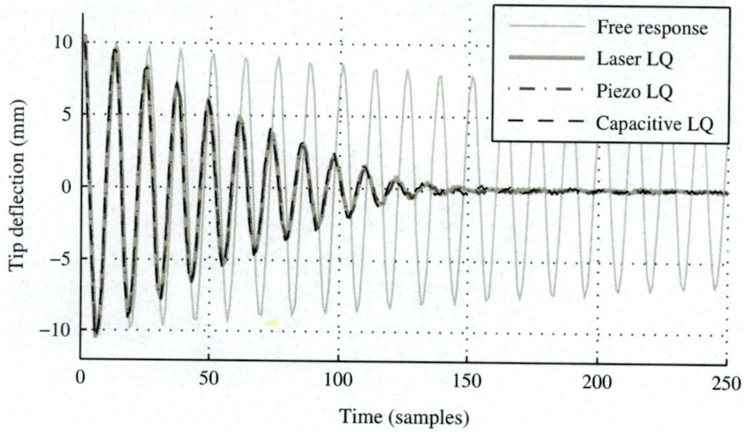

(a) Laser measured beam tip deflections

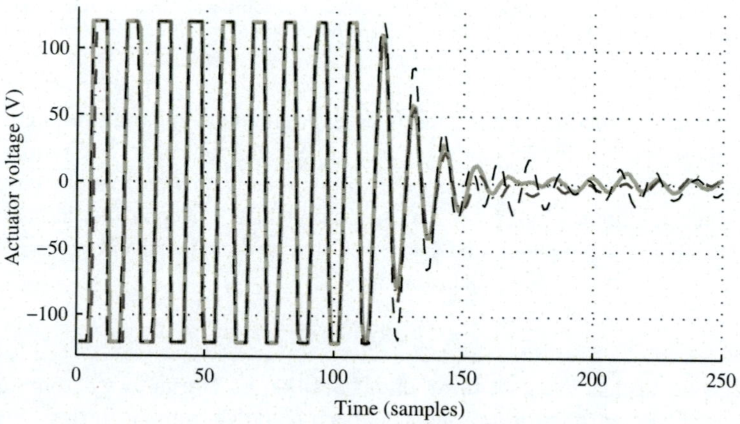

(b) Controller voltage signal to actuators

Fig. 5.19 Comparison of direct linear quadratic position feedback-based control with piezoelectric and capacitive sensor estimates in an initial deflection test is shown in (**a**), while corresponding controller voltage outputs are presented in (**b**)

under LQ control with piezoelectric sensor feedback or capacitive proximity sensor feedback are practically identical to directly measured laser feedback.

The evolution of controller voltages in Fig. 5.19b demonstrates no considerable deviation from the reference; response is virtually identical up to time sample 120. Slight deviations after this time sample are attributed to the fact that these are three separate measurements with the beam subjected to outside sources of error at each time. These errors are compensated by the controller, therefore to be noted only in the voltage output but not on the beam deflections.

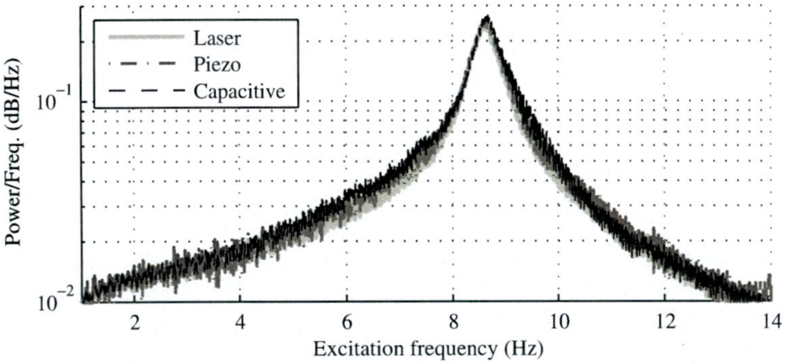

(a) Laser measured amplitude spectra

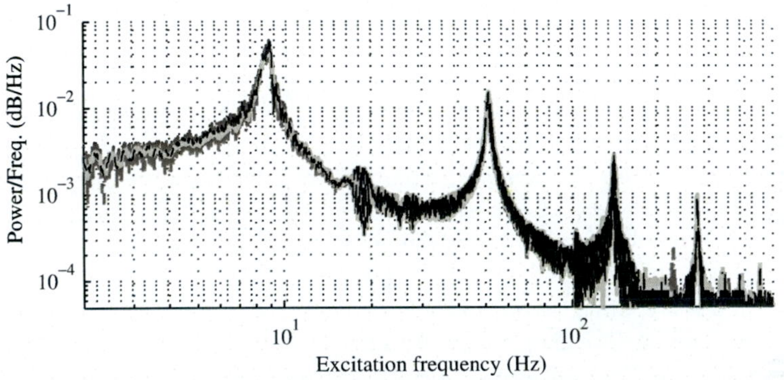

(b) Response to wide band excitation

Fig. 5.20 Laser measured single-sided power density spectra of beam tip deflections in the region of the first vibration mode are indicated in (**a**), while (**b**) shows damping behavior in the first three dominant structural modes

All three different control schemes have been evaluated in a frequency domain experiment. The beam has been excited using an electromagnetic exciter, utilizing identical hardware and software setup—except for the feedback loop configuration.

Figure 5.20 indicates the directly measured single-sided amplitude spectra of beam tip vibrations, showing all considered feedback control schemes. As it is expected, Fig. 5.20a demonstrates that the estimated position-based feedback schemes provide a nearly indistinguishable damping performance to direct feedback in the region of the first structural vibration frequency.

It has been demonstrated previously that capacitive proximity sensor and piezoelectric sensor-based second order feedback models are limited to give a correct deflection estimate only in the vicinity of the first resonant frequency. Despite this fact, all three control schemes provide a good damping performance even when excited by higher frequency mechanical disturbances. The measurement illustrated by Fig. 5.20b utilized second order system and measurement estimate models; however, it has been excited to frequencies exceeding 300 Hz. As the results clearly indicate, using low order tip deflection estimate models, the damping performance is practically indistinguishable from that using direct tip position measurements.

5.4 FEM Analysis

With the advent of cheaper computing platforms and more sophisticated software tools, finite element modeling and FEM-based analysis has affected the workflow of all engineering disciplines. Vibration analysis, particularly active vibration control is no exception as the dynamic analyses like the modal, harmonic, and transient aid the design of both active and passive vibration control systems. The correct assessment of resonance frequencies is essential in a well-designed PPF controller scheme, where the filter frequencies should closely match the natural frequencies of the structure [91]. In vibration control applications used in space flight the elevated acceleration during liftoff and microgravity in orbit cannot be measured beforehand, only approximated by numerical methods.

In order to improve the demonstration hardware, this work required to investigate the behavior of the AVC laboratory structure in different design versions. The material and dimensions of the beam and more importantly the placement and configuration of the sensors was also explored by means of FE simulations. Each of these design iterations required to take the modal, transient and frequency response of the system into account. The choice FE software to carry out simulations was ANSYS. This software provides readily available elements for multidisciplinary simulations. Amongst others, there are elements with piezoelectric properties suitable to prepare models of active structures.

ANSYS is widely used to analyze piezoelectric transducer actuated smart structures in vibration control [18, 22, 36, 103]. This commercial FE package can also be used to identify the open-loop dynamics of the controlled plant for AVC design; for example, Hong et al. uses ANSYS to identify modal parameters for an ER fluid controlled frame in [39]. Models of SMA materials have also been developed for ANSYS by Torra et al. in [116]. The integration of control strategies directly into the FEM simulation has an advantage of more precise models and wider disturbance/response scenarios, however with an increased computational load. Integration of a direct velocity feedback strategy with a FEM model has been presented by Malgaca in [75]. Nguyen and Pietrzko use ANSYS to simulate a PZT actuated semi-active R-L shunt circuit-based vibration controller in [85]. ANSYS is also used to aid the design of a vibration control system for high-rise buildings to suppress earthquake movement by Preumont and Seto in [92]. Other commercial finite element packages are also used in the simulation of vibration dynamics for AVC systems such as COMSOL Multiphysics which is particularly suited for vibration-related applications in MEMS [25, 60, 86, 107]. Yet another commercial FEM package is MSC/NASTRAN which is widely utilized for vibration and vibration control-related analysis as well [17, 76, 117].

This section gives an overview of the simulation results performed on the FE model of the vibration control demonstration device. For those interested, a complete explanation of the simulation steps is given along with an ANSYS APDL code listing in Appendix A.

Figure 5.21 shows the finite element model of the smart structure. All the piezoelectric actuators are explicitly modeled; their size and position correspond to the real device. The detailed image indicates the piezo wafer and the adhesive layer under the transducer. The ANSYS symbols indicate the point where the beam is clamped to the support base. Electric loading and the coupling between the relevant nodes of the piezoelectric transducers is also visible. Coupling the piezoelectric elements and the assignment of the voltage levels simulates the electrodes connected to the piezoceramic material.

5.4.1 Static Loading

Static loading tests were simulated by applying a 100 V electrical load to the actuating piezo element. The utilized piezo actuators would withstand a continuous ±120 V loading. To prevent depoling, the amplifiers have a cutoff voltage of ±100 V RMS.[14]

Initial tests only investigated the behavior of the standalone piezo wafer. In this stage the PZT transducer was not yet bonded to the cantilever beam. Static deflection proved to be within the expected range. This also corresponded to the results obtained by the simplified calculations, for example (3.7). Following these initial simulations, a more complex model was built, simulating the static deflection of the actual beam tip. The static tests provided a good basis for the upcoming more complex dynamic simulations.

The static step response of a similar cantilever structure has been used for control system design in [103, 104], utilizing the set of nodal displacements resulting from a unit step input of voltage potential to the actuator. The vibration suppression strategy has been based on the idea of lumped-input and distributed output control [47], where the structure dynamics have been separated to independent time and spatial domains based on the ANSYS nodal step responses. Then the time domain used the well-known PPF and PVF methods for control synthesis, while the beam response in the space domain has been computed as an approximation problem using the set of nodal step displacements. As the simulation results yielding merely sub-millimeter beam tip deflections suggest [103], it is not ideal to use step responses of structures actuated by piezoceramics as mathematical models in feedback control. Static simulations may misrepresent the real dynamic properties of the controlled structures—especially when the system is lightly damped, oscillating and underactuated. The use of methods based on a set of distributed step responses is rather suited for systems with non-oscillatory and slower dynamics, such as assessing temperature fields in metallurgical processes [48, 49].

---

[14] See (5.5.1) for more details on amplifier safety measures.

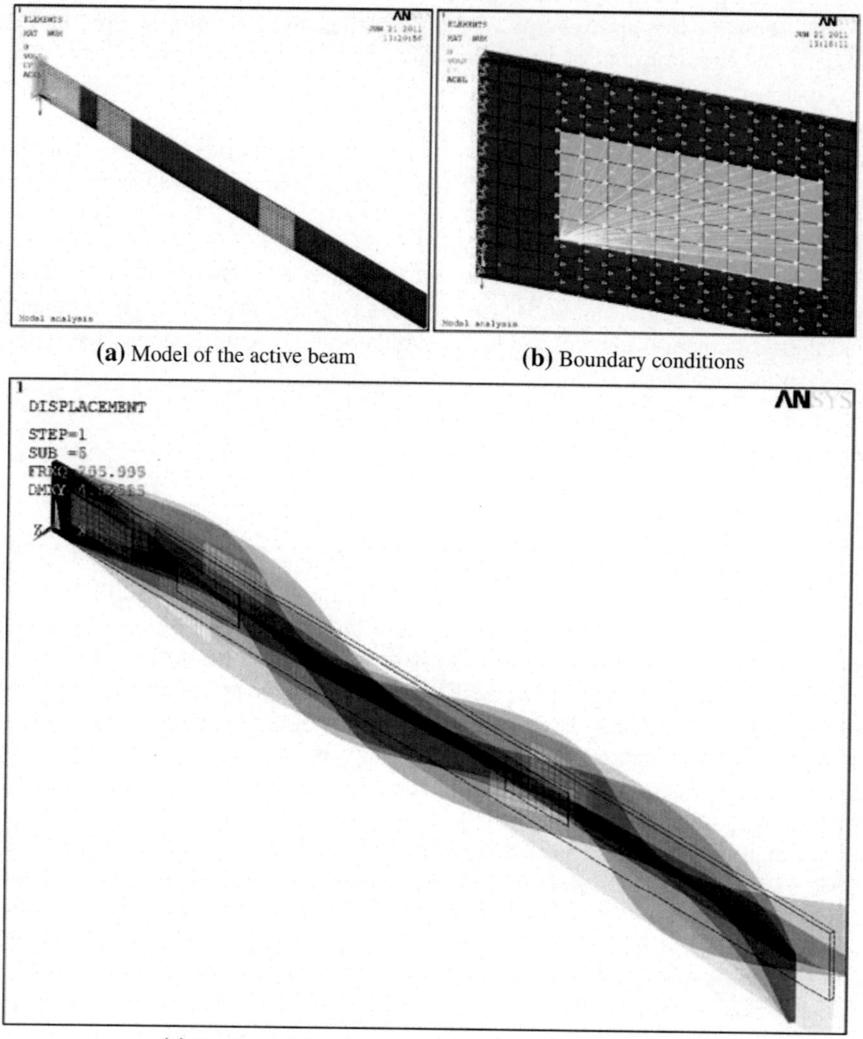

(**a**) Model of the active beam

(**b**) Boundary conditions

(**c**) Overlay of vibration modes and the undeformed structure

Fig. 5.21 Finite element model of the smart structure (**a**), showing details of the boundary conditions including displacement, voltage potential and gravity loads in (**b**). An overlay of various vibration modes and the original undeformed structure is illustrated in (**c**)

5.4.2 Modal Analysis

The modal analysis of the smart structure was the first step towards determining the experimental setup configuration. Frequency range investigated by the modal analysis corresponded to the originally planned control bandwidth, namely 0–500 Hz.

Table 5.5 FEM modal tests to determine ideal beam length

Mode	1	2	3	4	5	6	7
Length			Frequency (Hz)				
450	12.5	77.0	162.3	215.2	262.1	422.1	–
500	10.1	62.5	131.7	174.5	235.2	342.4	–
550	8.3	51.7	108.9	144.2	213.2	282.5	467.8
600	7.0	43.1	91.3	121.2	195.4	237.3	392.8
650	5.9	37.2	78.3	103.1	179.1	202.4	334.1

This however later proved to be too extensive, given the difficulties with implementing an MPC controller with a priori guarantees on the physical system.[15]

Block Lanczos, the ANSYS default mode extraction method was used to obtain the first 10 modes. The rest of the simulation parameters were corresponding to the program defaults. Several simulations were carried out which attempted to determine the cantilever dimensions. The aim was to obtain five bending modes within the originally planned control bandwidth. Moreover, these modes were required to have a desirable distribution within the examined frequency range.

Primarily only one piece of piezoelectric actuator was considered. The edge of the actuator was fixed at 20 mm from the clamped end. The width of the beam was planned to be 40 mm, as this is the most sensible for the given actuator width. The thickness of the beam was fixed to be 3 mm. Beam length was the only dimension remaining as a variable. As Table 5.5 shows, varying the beam length greatly influences the number and distribution of the first seven modes under 500 Hz.

Beams with lengths 450 and 500 mm showed only 4 bending vibration modes under 500 Hz. On the contrary, if the beam had a length of 650 mm it presented an extra mode within the planned control range. Taking all considerations into account, the beam length of 550 mm was chosen. This length provided five bending modes within the bandwidth, and these modes had a desirable distribution.

After the cantilever beam material, dimensions and the placement of one piezoelectric actuator[16] was assigned and fixed, the modal simulation showed the presence of seven modes within the considered frequency range. As planned, five of these modes were bending modes also detectable by observing the tip deflection. After visualizing the modes, the second mode showed a transversal movement and the fifth one a twisting behavior. The relevant modal frequencies using one piezoelectric actuator are listed in Table 5.6.

The first mode is a pure bending mode. The second mode shows one major node, which is located about 290 mm from the clamped base. There is also an anti-node approximately 50 mm from the beam tip. The fourth is the next relevant mode, this presents two nodes of vibration which are placed at one and two thirds of the total

[15] Considering the first three transversal vibration modes instead of five at design stage would be more suitable for this application. However, the issues regarding the size of the region of attraction and unexpected levels of NRMPC suboptimality were at this time unknown. See Sects. 11.1 and 11.4 for more details.

[16] Actuator marked as PZT1, see Sect. 5.1 for details.

Table 5.6 FE modal simulation results utilizing one piezoelectric actuator

Mode	1	2	3	4	5	6	7
Frequency (Hz)	7.8	48.5	101.2[a]	135.1	198.1	264.0[b]	435.9

[a] Sideways mode
[b] Twisting mode

Table 5.7 FE modal simulation results utilizing piezoelectric transducers corresponding to the real and finished experimental device

Mode	1	2	3	4	5	6	7
Frequency (Hz)	7.9	48.9	101.1[a]	135.5	204.0	264.0[b]	446.2

[a] Sideways mode
[b] Twisting mode

beam length. The sixth and seventh mode shows three, respectively four vibration nodes according to the FEM simulation.

After completing the FEM model by placing the rest of the piezoelectric transducers in their respective locations, the modal analysis was repeated. This time all the piezoelectric transducers were modeled in a short circuited state; meaning that there was zero potential in between their electric terminals. Table 5.7 lists the resonance frequencies of the smart structure with all the piezoelectric actuators in place. As is evident from the table, no significant frequency shifts occurred. There was no special effort taken to readjust the simulation model parameters to the real properties of the system. Higher resonance frequencies may be imprecise, although the first three eigenfrequencies correspond to measurements.

Figures 5.22 and 5.23 show plots of modal shapes for modes 1 through 6. All the images were acquired through an ANSYS FEM simulation, using the fully modeled experimental device with all transducers in place. A side view of the shape is presented on the left, while pictures on the right show an isometric three-dimensional angle of vibration modes. Mode shape is shaded with dark, while the undeformed geometry is indicated by contour lines. Note that the third mode in Fig. 5.22e, f deforms perpendicular to the measurement point, therefore it cannot be detected with the considered measurement system. Similarly, mode five in Fig. 5.23c, d is a twisting mode—not considered in the control application. There is one more vibration mode just under 500 Hz not depicted on the figures.

5.4.3 *Harmonic Analysis*

Harmonic analyses have also been carried out to test the frequency response between the actuating piezoelectric actuator and the beam tip at the point of interest. The number of sub-steps or the resolution of the analysis was adjusted according to the actual mesh density used in order to decrease the computational load. None of the default simulation parameters were changed—a harmonic analysis of a piezoelectric actuator enabled structure can be easily prepared in ANSYS. The obtained

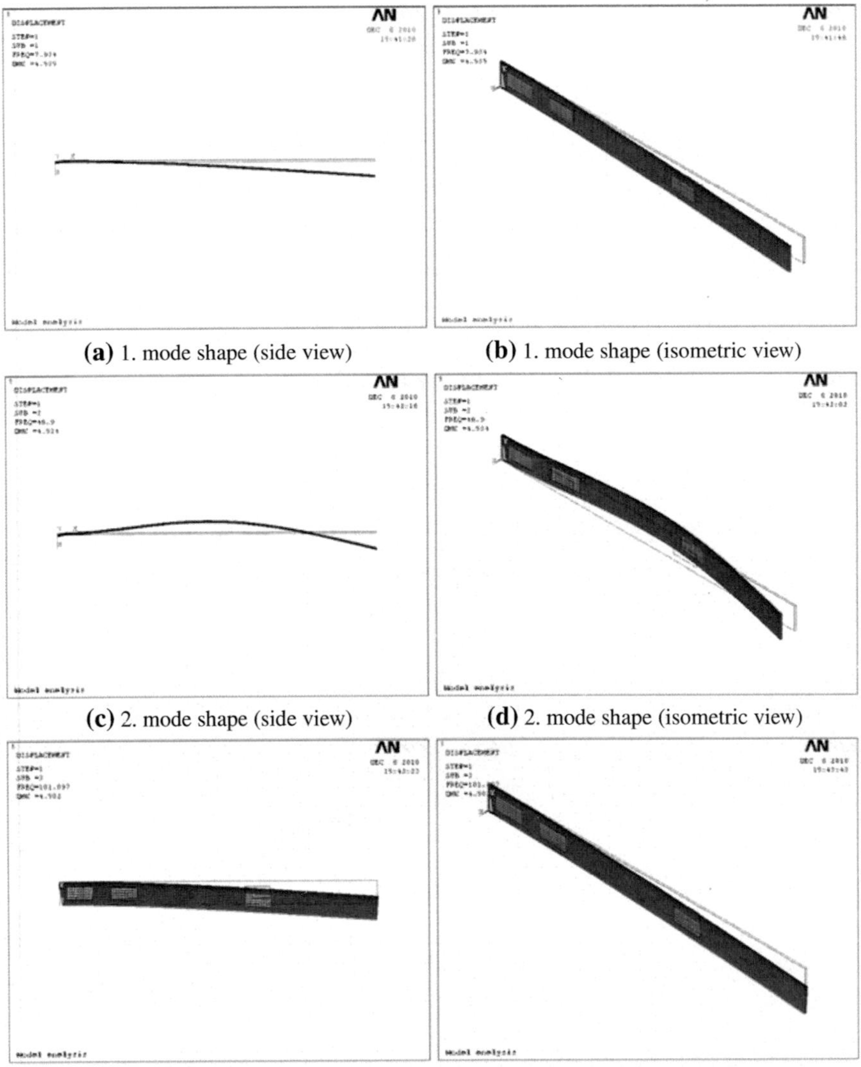

(a) 1. mode shape (side view)
(b) 1. mode shape (isometric view)
(c) 2. mode shape (side view)
(d) 2. mode shape (isometric view)
(e) 3. mode shape (front view)
(f) 3. mode shape (isometric view)

Fig. 5.22 FEM simulated vibration mode shapes 1. (**a**, **b**), 2. (**c**, **d**), 3. (**e**, **f**) of the active structure. Showing side (front for (**e**)) views on left and isometric on right. Note that mode 3 is perpendicular to the laser measurement direction, thus not considered in the control application

frequency responses have been subsequently exported to a data file, which in turn was utilized in Matlab for system identification. This system identification approach was later abandoned in favor of the simpler and more direct experimental identification procedure.

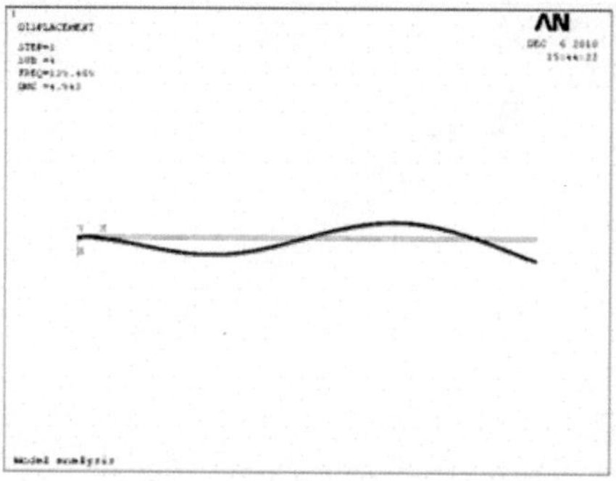

(a) 4. mode shape (side view)

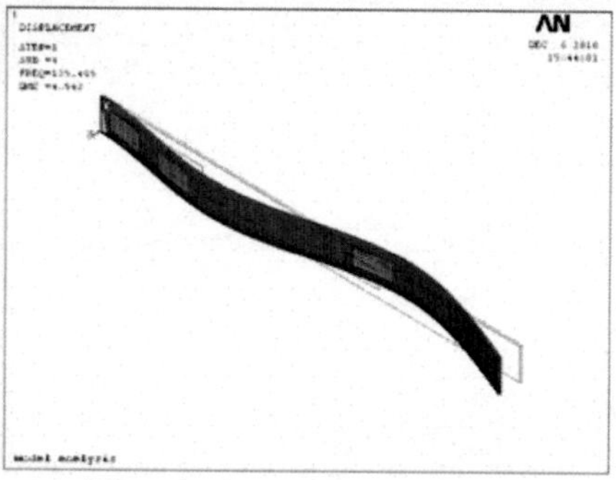

(b) 4. mode shape (isometric view)

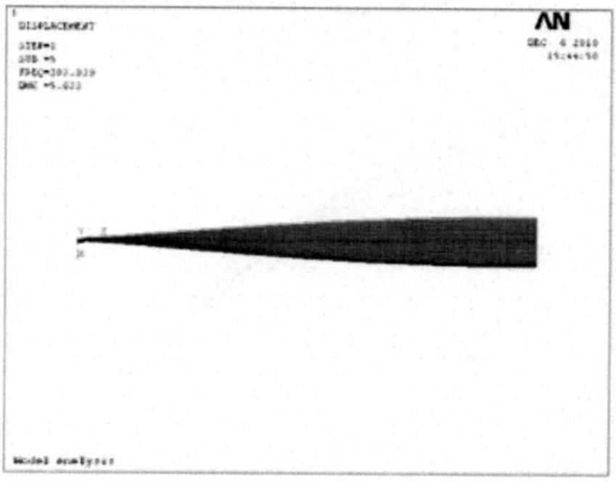

(c) 5. mode shape (side view)

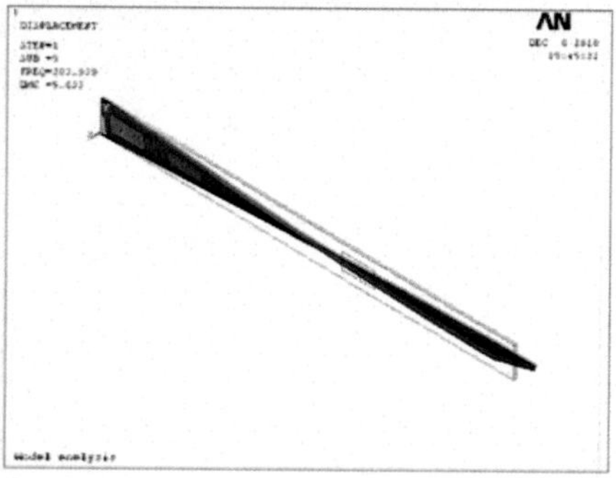

(d) 5. mode shape (isometric view)

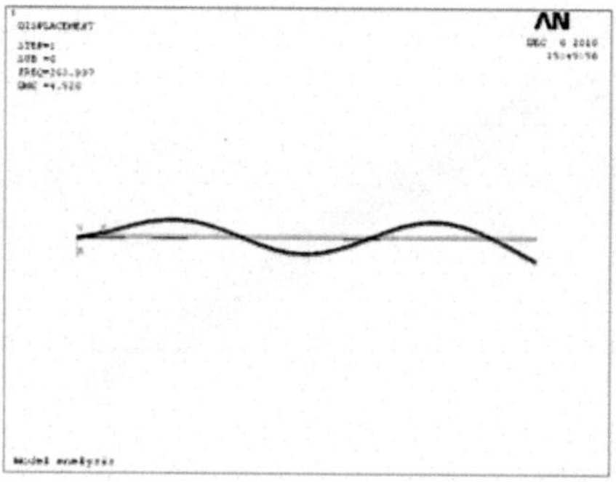

(e) 6. mode shape (side view)

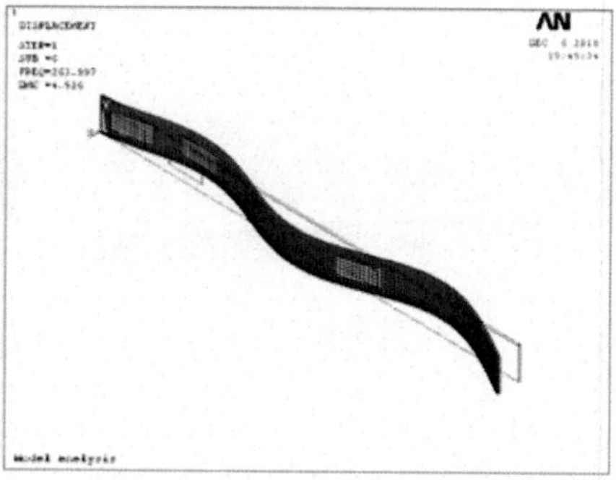

(f) 6. mode shape (isometric view)

Fig. 5.23 FEM simulated vibration mode shapes 4. (**a**, **b**) 5. (**c**, **d**) 6. (**e**, **f**) of the active structure. Showing side views on left and isometric on right. Note that mode 5 is a twisting mode, not considered in the control application

5.4.3.1 Transducer Placement

Several works deal with the optimal placement of piezoelectric transducers in active and semi-active vibration attenuation. Criterion on placement are diverse, usually include prevention of control spillover, controllability, observability and others

Table 5.8 Primary tests on actuator position with one piezoelectric transducer

Mode	1	2	4
Edge position (mm)		Vibration Amplitude (mm)	
0	20.70	1.49	0.38
4	15.50	2.37	0.34
8	15.09	3.24	0.29
12[a]	14.83	3.99	0.31
20	14.04	3.03	0.30
40	10.85	1.01	0.10
60	8.19	0.43	0.06

[a] Final position used on the physical device

[34, 93]. Unlike the sophisticated optimization-based placement approach utilized by many researchers, this work used a more straightforward method. The placement of actuators is based entirely on practical considerations: the first and foremost decision factor was to maximize beam tip displacement under the first resonant frequency, while ensuring measurable and visible vibration amplitudes at higher resonances at the same time.

The primary placement FEM simulations involved only one piezoelectric transducer in actuator mode. It was necessary to determine the position of the PZT wafer starting edge measured from the clamped end. Several simulations have been carried out all with a fixed carrier beam dimension of $550 \times 40 \times 3$ mm. The piezo transducer length was aligned to the symmetry line of the beam width. The main aim was to maximize the vibration amplitude at the beam tip in the first mode, although the amplitudes of the second and fourth nodes were not neglected either. Results from the preliminary harmonic excitation FEM analyses are summarized in Table 5.8.

Note that the vibration amplitudes in this initial analysis are somewhat different from those measured in the final device. This is simply because at the time of preparing these simulations the FEM model has not been readjusted according to the real device properties. The elastic modulus of the beam, PZT material properties and the damping constant has been assumed from table values and engineering judgment. In spite of this, the first resonant mode with the actual PZT placement (12 mm) gives a realistic estimate of the real amplitudes.

Amplitudes bear information about the relative change of tip displacement related to the actuator position. It is clear that for the first mode displacements, the closer we get to the fixed end with the actuator edge, the higher the tip vibration amplitude will be. It is less obvious but observable from the simulation data shown in Table 5.8, that for the second vibration mode there is a turning point in the amplitude values. Up to a certain position it seems to be increasing, then as we get closer to the fixed end it decreases again. Therefore, the placement of the first actuating piezo edge was determined to be 12 mm from the clamped end. This provides a tradeoff between first, second and fourth mode vibration amplitudes measured at the end. Because of the electrical connections on the piezo transducer, it would be impossible to get closer to the clamp.

5.4.4 Transient Analysis

A transient analysis has been performed on the FEM model of the active vibrating beam. The actuator voltage has been set to zero; the simulation denotes the mechanical response of the beam equipped with the piezoelectric actuators without control. An initial deflection has been simulated using an equivalent force effect applied at the beam tip at the first step. The force effect has been removed for the rest of the simulation. In the experiment the beam has been deflected 10 mm away from equilibrium, an equivalent force represented the same initial displacement.

As is demonstrated in Fig. 5.24, the mechanical response of the beam can be precisely simulated using FEM, even in the case of a composite aluminum beam equipped with piezoceramic actuators. We have to note, however, that a short-circuited set of piezoelectric patches do not have a significant effect on the uncontrolled mechanical behavior from the beam. The detailed response shows very little difference between the simulated and real deflections. The transient response simulation at first overestimates the effect of damping, then later underestimates when we compare it to the experimentally measured response. This is simply due to the imperfect damping model, as proportional damping has been considered in the simulations.

Another transient test (not shown) has been performed as well. Here the actuator voltage was set to a nominal value of 100 V and then back to zero at the next time step. The driving voltage was stepped up and down instead of ramping. The rest of the simulation parameters were similar to a general case transient analysis previously. The simulated transient deflection response to a change of voltage in the piezoelectric actuators has been in good agreement with experimental results. The setup of a transient response simulation is briefly discussed in Appendix A.

Figure 5.25 illustrates the transient response of the beam near the first resonant frequency as shown in the frequency domain. The FEM simulation results match the beam dynamics acquired through experiment.

5.4.5 Control Prototyping

Finite element analysis software usually do not contain tools to implement control strategies in dynamic simulations. The ANSYS software environment lacks this feature as well.[17] However, using its internal scripting language APDL, it is possible to include simple control strategies in transient simulations.

In order to evaluate the effect of a closed-loop control strategy applied to the AVC demonstrator and the actuation potential of the piezoceramics, a modified ANSYS transient analysis has been prepared. A linear quadratic controller has been implemented in the proprietary APDL language along with a simple state observer. The state-space model of the system has been considered the same as the direct laser-based system model in Sect. 5.3.4, while the fixed LQ gain **K** has been computed

[17] As of its current version: Release 13.0.

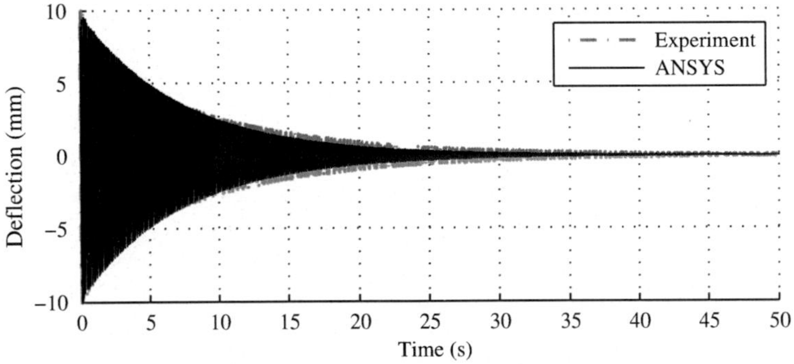

(a) Free transient response

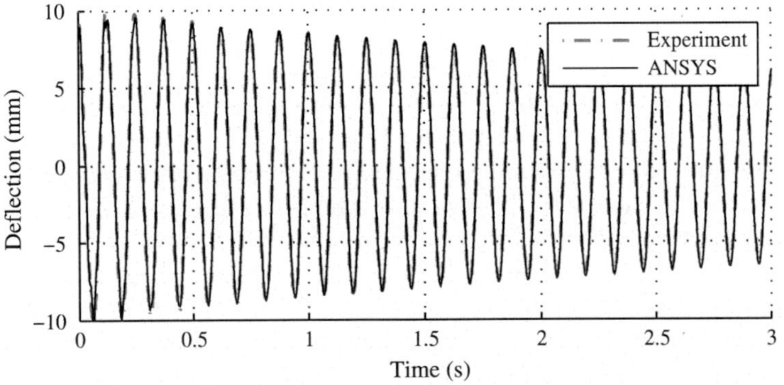

(b) Detail of the free transient response

Fig. 5.24 Comparison of the experimentally measured transient vibration response with the FEM simulation. The figures denote response to a 10 mm initial deflection without control and the resulting settling time well in excess of 30 s

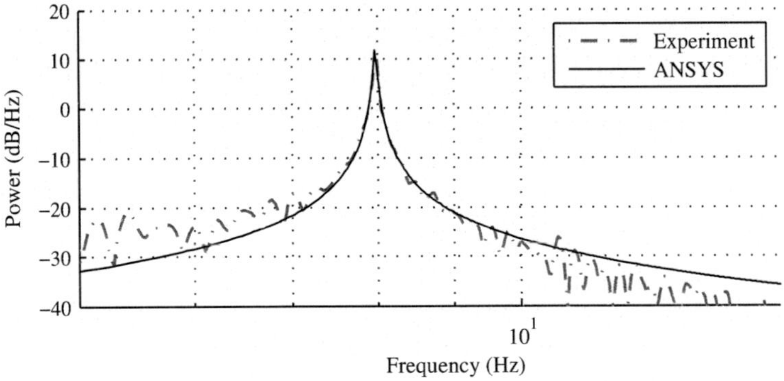

Fig. 5.25 The transient response of the beam in the vicinity of the first resonance

to be equal to (5.4). The experimental data featured earlier in Sect. 5.3.4 served as a basis of comparison for the ANSYS simulation results.

The transient simulation of the AVC demonstrator has been programmed using the ANSYS APDL macro language in the usual manner. The transient time steps have been however divided into steps within an APDL **DO* program cycle. The time steps between the subsequent transient analysis steps have been set equal to the sampling period of the controller and the observer, that is, $T_s = 0.01$ s. After a *SOLVE* command has been issued, the nodal displacement at the end of the beam is evaluated and used as an input to the state observer. The observer then outputs the estimated state, and the control input to the piezoceramic actuators is computed based on the fixed feedback matrix **K**. The control input value at the actual step is input as the voltage potential boundary condition for the next simulation step, and the cycle repeats.

Figure 5.26 illustrates the result of this ANSYS simulation. The experimental data of the beam subjected to a 10 mm initial displacement without control and with LQ control is contrasted to the ANSYS controlled transient simulation results on Fig. 5.26a. The controlled FEM simulation not only matches laboratory measurements well, but also indicates the considerable damping effect of the controlled piezoceramic transducers: the settling time has been reduced well in excess of an order of magnitude. Figure 5.26b shows the saturated control inputs supplied to the piezoceramic transducers. The control inputs do not converge to zero in the real measurements unlike in the ANSYS simulation, because of the unmodeled effects of ambient mechanical noise and minor disturbances in the laboratory.

5.5 Hardware Components

This section introduces the individual hardware components of the experimental AVC demonstrator device in detail. It is intended for the reader who is not familiar with the components and laboratory equipment used in the construction of piezoelectric actuated smart structures. This section may also give pointers to those who wish to build such a laboratory device. In addition to the technical description of the piezoelectric actuators, feedback measurement methods or the beam placement and a brief overview of their use in the academic literature is also given.

5.5.1 Piezoelectric Transducers

The individual piezoelectric actuators and sensors utilized on the AVC demonstrator laboratory device are identical. MIDÉ QuickPack QP16n transducers are used in actuator mode to apply a deformation to the beam thus deflecting the whole structure away from equilibrium and in sensor mode to generate a voltage proportional to the mechanical deformation. The product shows higher variance in capacitance, since it is intended for the mid-range market segment. The transducers come in a

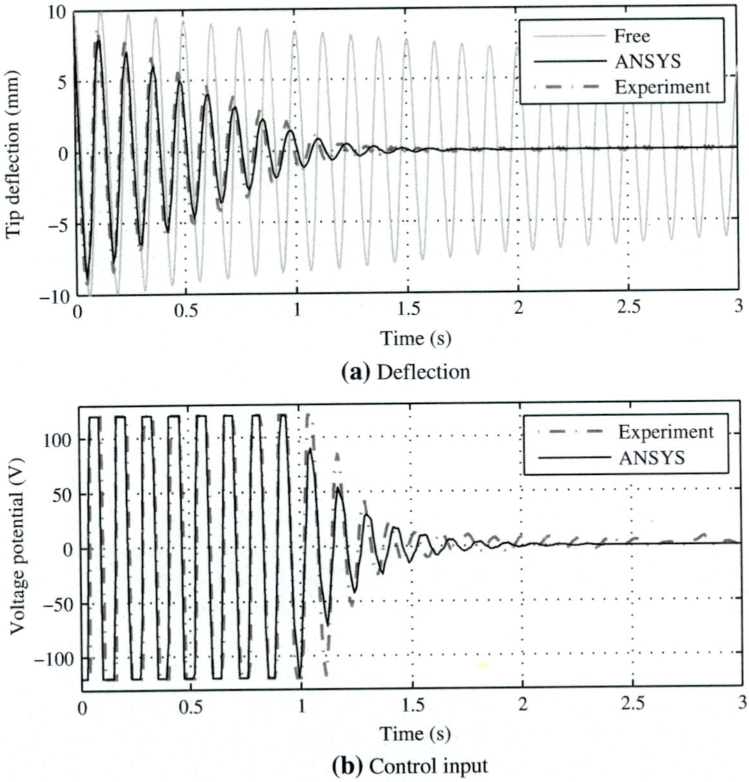

Fig. 5.26 Comparison of the experimentally measured transient vibration response under LQ control with the FEM simulation, showing the free response for reference. The figures denote response to a 10 mm initial deflection and the significantly reduced forced settling course

convenient pre-packaged form, along with the electrodes and a connector block. A QP16n transducer is shown in Fig. 5.27a with the electrical connections and protection film visible.

Similar to the experimental implementation featured in this book, other academic works dealing with the vibration damping of cantilever-like structures often utilize piezoelectric patches [7, 30, 64, 67, 74 118].

5.5.1.1 Piezoelectric Transducers in Actuator Mode

The type of transducer used in this work is suitable for strain actuation only [77]. It consists of one piezo wafer, made of PZT5A. The dimensions of the wafer are $45.9 \times 20.7 \times 0.25$ mm and the full-scale voltage applicable is ± 120 V. The nominal device capacitance is 137 nF. The manufacturer gives a ± 225 $\mu\varepsilon$ full-scale strain. The piezo wafer is surrounded by a protective film and comes with the electrodes bonded to the surface. The electrodes are connected to a socket, compatible with

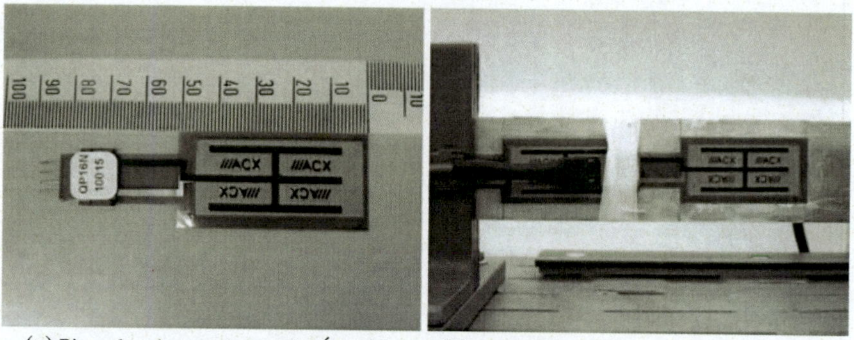

(a) Piezoelectric transducer MIDÉ QP16n **(b)** Transducers bonded on the beam surface

Fig. 5.27 Piezoelectric transducer MIDÉ QP16n in a factory-packaged state with connector and protective film (**a**) and transducers bonded directly onto the beam surface in (**b**)

the factory provided cables. The size of one PZT transducer in the protective film, without the electrical connection is $50 \times 25 \times 0.25$ mm.

The transducers are bonded to the beam surface using a two-component high-performance structural epoxy resin. After marking the correct position of the transducers, the surface was lightly sanded to allow better adhesion. The relevant parts were cleaned; grease was removed using isopropyl alcohol. The surfaces around the device were masked off using TEFLON tape. A thin layer of resin was carried on the surface. Finally, a pressure of approximately 100 kPa was applied on the actuator surface during the whole 24 h cure cycle [79]. An even more uniform bond may be acquired using a vacuum bonding technique, although the former method is considered to be satisfactory for the application in question.

Transducers bonded on the beam surface are shown in Fig. 5.27b. One of the symmetrically placed actuators close to the clamped end is visible here, along with the additional transducer acting either as a source of disturbance or as feedback sensor in certain experiments.

In the case of an actuator mode application, the PZT transducers are connected to the amplifier output terminal via CB-014 cables, while the same type of leads are used for the senor mode transducers. The cable is suitable for transducers with a single PZT wafer and has a BNC termination. The two inner pins on the connector are not in use with this type of transducer; signal is transmitted only through the outermost connectors.

5.5.1.2 Piezoelectric Patch in Sensor Mode

A piezoelectric patch identical to the actuators is used in sensor mode, in experiments analyzing differences between direct and estimated feedback. The edge of the piezoelectric sensor is located 80 mm away from the beam clamp, coinciding with the point where the capacitive proximity sensor is mounted. The axis of symmetry of the piezoelectric patch along its length coincides with that of the beam.

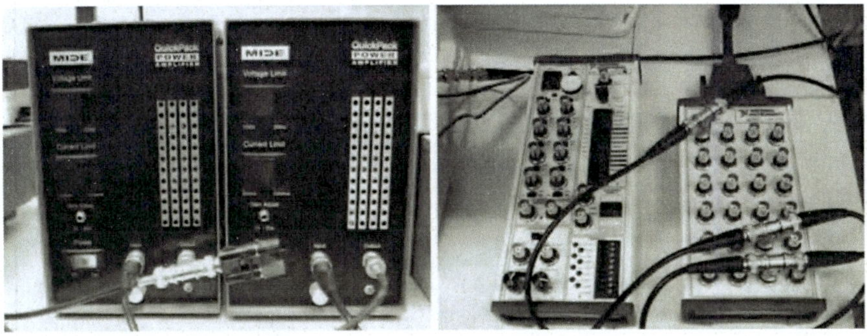

(a) Power amplifiers **(b)** Cable terminal

Fig. 5.28 MIDÉ EL-1225 high voltage inverted operational amplifier used to drive piezoelectric actuators (**a**) and the NI BNC-2111 cable terminal connecting outputs and inputs to the measurement card shown in (**b**) (*right*)

Initial tests demonstrated that the signal level coming from the wafer is more than satisfactory. This allows the sensor to be connected directly to the monitoring equipment. No voltage following circuit is required to drive the data acquisition device [105]. If there is no stress, a piezoelectric transducer is a capacitor in the electrical point of view. On the other hand, if there is stress present, the generated voltage is proportional to it.

Piezoelectric sensor voltage output level exceeds signal acquisition card specifications, therefore has to be attenuated. In order to do this a 100 kΩ resistor is connected in parallel with the piezoelectric sensor strip. In this way voltage output is matched to the expected tip deformation range of the cantilever beam.

5.5.1.3 Amplifiers

A power amplifier drives piezoelectric actuators through an applied voltage. The device used in the series of experiments featured in this work is a MIDÉ EL-1225 high voltage inverted operational amplifier, shown in Fig. 5.28a. The device accepts an input signal and amplifies it up to ±200 V peak. It is stable to drive large capacitive loads, like piezoelectric actuators. It provides an output bandwidth of 5 kHz with a 3 dB attenuation, with a continuous user adjustable gain of 20 V/V [78].

The device provides safety measures against overloading the actuators, since too much power and high voltage may damage the piezoelectric devices and the amplifier circuitry. There is a selectable ±100 or ±200 V voltage limit[18] to suit the depolarization voltage of the actuator. There is a 50 or 200 mA peak current limit as well, of which the lower one is suitable for devices with low power dissipation. Because the actuators described previously have a maximum admissible voltage of ±100 V,

[18] This safety limit is given in RMS, not peak values.

the lower safety settings were in effect throughout most of the experiments. The maximal output power of the amplifier is 20 W (RMS).

The signal of the output terminal is split using a BNC-T junction. It is connected to the actuator and an analog oscilloscope for convenient signal monitoring.

5.5.2 *Beam Material and Base*

Several material types have been considered for the cantilever beam. Plastics would be an interesting application for active vibration damping, although a plastic base would bring significant unwanted difficulties both in modeling and in control: plastics demonstrate a nonlinear, time dependent behavior and significant creep. In this case, classic Hookean engineering does not apply, thus complicating the FE simulation, modeling, mathematical identification and necessary control system considerations significantly.

5.5.2.1 Beam

Aluminum has been chosen for the cantilever beam material. There are several types of aluminum alloys commercially available in a sheet form. The most common is 6061, which contains magnesium and silicone as alloying elements. To maximize the vibration amplitudes, this application required a rather soft aluminum type, one with low elastic strength. Aluminum with the designation EN AW 1050A was chosen, as this is 99.5% pure commercially available material [89].

Mechanical properties are subject to great variation, the exact elastic strength is not a known parameter unless the sheet metal is purchased with a certificate. The cantilever has dimensions of $550 \times 40 \times 3$ mm ($L \times W \times H$), with an additional 200 mm length reserve. This allows the adjustment of system properties by changing the clamping position, thus changing modal frequencies. The possibility to change modal frequencies and thus the dynamic properties of the beam can be utilized when investigating controller robustness or simulate parametric vibration. The beam in question is shown mounted to its support base in Fig. 5.29a.

To avoid sideways vibration, which is not directly measured nor considered in this work, one could ideally choose the beam width to be the same as its height. Unfortunately, the width of the piezoelectric transducers prevented the significant reduction of the beam width; therefore, it was chosen to be 40 mm, slightly exceeding the width of the transducers.

5.5.2.2 Base

A heavy iron structure acts as the base of the mechanical part of the experimental setup. The clamp initially considered would fix the smart structure in a vertical position [27]. This means that the beam tip would vibrate alongside the Earth's

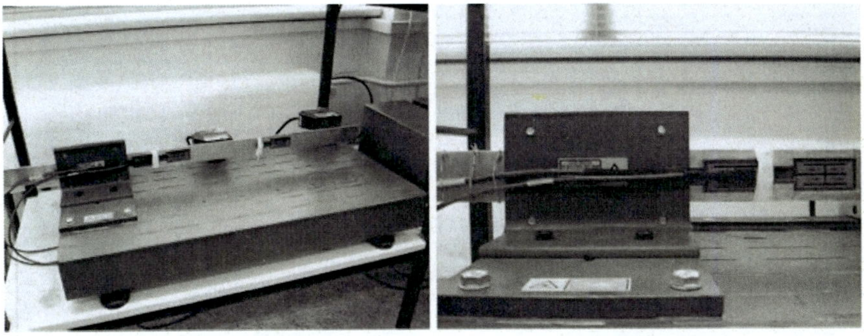

(a) Beam and support base **(b)** Adjustable clamp

Fig. 5.29 Beam and its support base construction (**a**) and the adjustable clamping system shown in (**b**)

gravitational field. The clamp was manufactured out of a section of a *HEB* iron construction profile. Although this setup would present some advantages, it proved to be unsatisfactory. The primary tests involved only one PZT actuator bonded to an aluminum beam, excited with the first modal frequency. Measurements showed asymmetric movement: the vibration amplitude was enlarged by the gravitational pull. This effect was deemed undesirable; therefore, the clamp design was changed.

The final clamp design shown in Fig. 5.29a was modified to hold the beam horizontally. This means that the beam tip vibrates perpendicular to the Earth's gravitational field. This prevents the problem of the gravitational inertia significantly effecting the measurements. The smart structure (aluminum beam with the bonded PZT transducers) is held in place with an adjustable clamp, as shown in Fig. 5.29b. Alongside the beam direction there is a guide rail for the optical head assembly. This allows accurate positioning of the laser heads. Several academic studies prefer the vertical positioning of the beam, such as [7, 33, 64, 67, 74, 96, 102, 118] and others. It is also possible to fix the beam in an upright position [5].

5.5.3 *Measurement of the Tip Displacement*

As is the case with every control application, active vibration cancelling systems need feedback and some real-time measure of vibration levels as well. A selection guide for strain, displacement, velocity and acceleration sensing instrument useful for vibration attenuation feedback is presented in [98], while different industrial sensors and actuators are discussed in [114].

Accelerometers are the most common means of acquiring a feedback signal to controllers [22, 88, 94]; albeit mounting or bonding the accelerometers can be impractical or impossible in certain situations. Despite recent advances in accelerometer miniaturization, these devices may still alter the mass and stiffness properties of

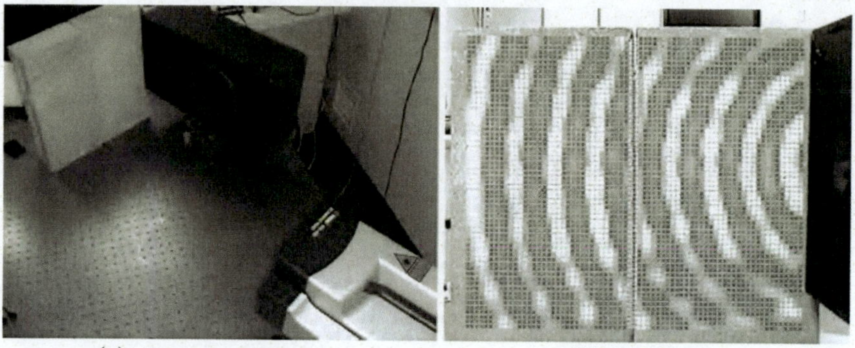

(**a**) Vibrometer and speaker (**b**) Sound field

Fig. 5.30 Measurement of the sound field emitted from a speaker using a Polytec PSV-400 scanning laser Doppler vibrometer is shown in (**a**), while (**b**) features the resulting measured sound field in the scan grid [42, 43]

mechanical systems. Other problems may arise by the bond between accelerometer and measured surface, and the presence of lead wires. Accelerometers provide an excellent measure of acceleration levels that can be integrated to gain velocity or position estimates, but they cannot sense static position changes.

Piezoelectric wafers, piezoresistive strips and other similar devices are commonly used as feedback sources in active vibration attenuation [55, 68, 101, 118]. These devices are cheap to manufacture and can be integrated into the controlled mechanical structure. This is an excellent option for a plethora of applications; however, it still requires altering the original structure. In addition, the presence of lead wires is not solved and piezoelectric sensors cannot provide a DC component of mechanical changes in the controlled structure.

Contact-free measurement of vibration levels is an excellent alternative to the formerly mentioned feedback sources, if the modification of the controlled structure is not desired or permitted. Mass, damping and stiffness properties are naturally not affected by the sensor or its lead wires, which is especially important in micro cantilever vibration damping in MEMS [73]. Laser Doppler vibrometry is an attractive choice for experimental verification tests because of its accuracy and it is often utilized in academic studies on vibrating cantilevers and other structures [7, 44, 45, 51, 73]. A laboratory measurement of the sound field emitted from a speaker by using a scanning laser Doppler vibrometer is illustrated[19] in Fig. 5.30 [42, 43]. The vibrometer scans through a predetermined grid of measurement points (Fig. 5.30a), and creates a visual representation of phase and magnitude at the given grid point at a particular time instant (Fig. 5.30b). Laser Doppler vibrometers are very valuable in vibration and sound-related research, nevertheless mainly suited to the laboratory environment because of their relatively bulky size and price range.

[19] Courtesy of Thomas Huber.

Industrial optical sensors based on laser triangulation are more suitable to practical use and product integration. Unfortunately, the size of triangulation sensors is still somewhat large, and the price range of such devices is too high for a mass produced commercial item. Industrial laser triangulation feedback sensor has been used for vibration damping in [109] and later in [108]. The use of laser triangulation is also suggested for road profile feedback in vehicles with active vibration control in [19].

An array of frequency selective capacitive vibration sensors has been presented in [95] and suggested for use in vibration measurements. Although the surface near silicon bulk microtechnology fabrication process permits low cost sensors, these devices still have to come into physical contact with the measured structure. Moreover, this device is only sensitive to selected frequency lines in the range of 1−10 kHz, thus not suitable for lightly damped structures with slow dynamics.

Ultrasonic sensors are relatively cheap and small, albeit not suitable for high-speed measurements. The fundamental working principle of ultrasonic sensors prevents their use in high sampling speed real-time vibration measurement and control applications. Cheap, industrial grade capacitive sensors may be suitable for vibration damping applications. While capacitive proximity sensors are commonly used in control engineering practice, their utilization in the field of vibration control is atypical but promising possible benefits such as low price and availability [110].

The measurement of the beam tip deflection is ensured by a contact-free state of the art laser distance measuring system for the AVC demonstrator. The measuring chain consists of a laser head, cabling and a proprietary processing unit. The distance is output to the processing unit as an analog scalable voltage value. Other types of signal output are also possible, although not utilized in this particular application. The analog voltage output from the processing unit is connected to the PCI measuring card via an appropriate BNC-BNC cable terminal.

5.5.3.1 Head Unit of the Triangulation System

The laser triangulation head utilized as a feedback source in this work is shown in Fig. 5.31a. It contains a linearized CCD sensor, which determines the distance from a reference point by means of triangulation. Position of the reflected light passes through the lenses and hits the CCD. The light on the CCD moves, as the position of the target changes. The sensor detects the peak value of the light quantity distribution of the beam spot for each pixel. This is identified as the target position [56]. Errors in pixel edges are reduced by a proprietary technology to ensure high accuracy.

The type of the triangulation head is a Keyence LK-G82 high stability, multi-purpose laser device, with one of the fastest sampling in its class with a 50 kHz sampling rate [3]. It is providing an accuracy of $\pm 0.05\%$ with the resolution of 0.2 μm. The measuring range of the head is 80 ± 15 mm. These properties make the triangulation device suitable for application as vibrometry equipment for this experimental device.

There is a second laser triangulation head present in the experimental configuration; a Keyence LG-G32 device. It has a slightly smaller range from the reference

(a) Laser triangulation head **(b)** Laser measurement CPU

Fig. 5.31 Keyence LK-G82 laser triangulation measurement head in (**a**) and Keyence LK-G3001V proprietary CPU processing and filtering signals from the heads shown in (**b**)

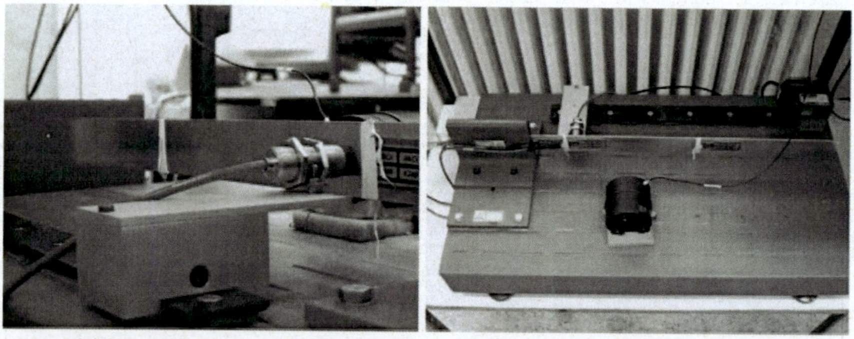

(a) Capacitive proximity sensor **(b)** Capacitive and PZT feedback setup

Fig. 5.32 The capacitive proximity sensor mounted in front of the beam is shown in (**a**) while the overall configuration of the feedback verification scheme is demonstrated in (**b**)

point: 30 $\pm$ 5 mm. The resolution of this device is higher and in the range of 0.05 μm. The second laser head provides additional measuring points in addition to evaluating solely the beam tip vibration. The feedback signal from the second head is not utilized in the control systems considered in this work.

5.5.3.2 Processing and Filtering Unit

The processing and filtering unit for the laser head is of the type Keyence LK-G3001V. The central processing unit amongst others provides the means to calibrate and set the measurement properties. It allows for scaling distances to analog voltage values to refine the resolution output. The triangulation system controller is shown in Fig. 5.31b.

It is possible to use different types of filters to eliminate unwanted measurement noise. The settings are administered via USB, utilizing a personal computer. The unit also contains a 65,000 point internal memory. This allows for proper system identification and validation tasks. The CPU is connected to a BNC-2111 cable terminal with BNC connectors. This allows fast reconfiguration of the system for different experiments.

It is worth noting that the measurement precision is mainly limited by the surroundings of the experimental setup. Since the long clamped aluminum beam is an under-damped structure, it presents vibration at all times. This is due to the small turbulence of air within the laboratory, temperature gradients and even the ongoing traffic on the street adjacent to the building.[20] Measurements that are more precise could be carried out in a sound isolated environment.

Realistically, in the surroundings available to us the beam tip vibrates with an approximate 0.025 mm amplitude, which can be as high as 0.15 mm with heavy traffic. This measurement disturbance shall be limited by filtering.

5.5.3.3 Repositioning of the Beam and the Measurement System

Laser heads 1 and 2 are mounted on an adjustable guide rail system. This allows convenient repositioning of the measurement system along the length of the beam. The current work uses only laser head referenced in Fig. 5.6 as LASER 1 for its feedback measurements. This measurement head is used at the end of the beam because it is capable of handling the deflections encountered at the first resonant frequency. The measurement head LASER 2 is currently unused, although it is possible to construct a feedback model with more than one measurement points.

An additional guide rail for the aluminum beam allows changing the reference position of the laser heads from the measuring system. This movement takes place along the axis parallel to the vibration direction and perpendicular to the length of the beam. The equilibrium position of the beam tends to drift under load, therefore the readjustment of the feedback reference is often necessary.

5.5.3.4 Capacitive Proximity Sensor

An 18 mm diameter Pepperl+Fuchs 924 Series industrial capacitive proximity sensor is utilized in this work to evaluate alternative feedback signals to the control system. Unless otherwise noted, the experiments have been carried out with the more precise laser triangulation system and its feedback. The capacitive sensor has been only in place to compare the LQ enabled vibration attenuation under different feedback schemes.

The sensing range of this capacitive sensor is listed as 2–5 mm and its linear voltage output is supplied from 1 to 9 V with a ±0.25 V linearity deviation. According

---

[20] See Sect. 5.3.2 for a measured mechanical noise and disturbance sample.

to manufacturer specifications, its sensitivity is 2.66 V/mm and the response time 1 V/ms.

If a capacitive proximity sensor is considered as feedback source in active vibration damping applications, the main limiting factor is its response time. At full 8 V output range this device provides a 125 Hz bandwidth, which is the worst-case scenario. As the sensing range decreases, the bandwidth also widens. If the maximum sensing range is set to the first vibration mode at the point of measurement, higher modes may be also measured since they are less dominant and thus the sensor bandwidth increases. Based on the bandwidth, capacitive sensors are suitable feedback sources for active vibration attenuation and control of lightly damped structures.

For the experimental setup in question, the expected beam tip deflections are ± 15 mm away from the equilibrium position, giving an overall 30 mm deformation range. Naturally, the range of the capacitive proximity sensor is much smaller; therefore, it has to be placed closer to the clamped end. A sensor range of ± 1.5 mm and proportional relation to beam length and deflection gives an ideal position of 55 mm measured from the fixed end. The effective sensing range is also decreased because the beam is made of aluminum instead of steel. Taking these facts into account the sensor has been mounted 80 mm from the clamped end, its zero equilibrium position fixed at 6 V. An expected full range bandwidth of 125 Hz is well sufficient for this application, since the first three measurable vibration modes are under this limit. The capacitive sensor is powered by 19 V direct current through a stabilized laboratory supply, within the specified nominal range.

5.5.4 *Real-Time Control Environment*

Identification, experimental measurement and controller testing have been carried out in real-time, using the combination of hardware and software tools described below.

5.5.4.1 Controller Development

The development, testing and final implementation of the control system along with mathematical identification tasks are carried out under the Matlab suite, also using Simulink. The computer running the Matlab suite is referred to as *host PC*. The host PC has been used for simulations described in Sect. 11.2.1, evaluating MPMPC controller calculation times. Although this computer takes no role in the real-time control of the vibrating beam, it is an integral part of the laboratory hardware.

The host PC conforms to current average specifications, having an AMD Athlon X2 DualCore 4400+ processor running at 2.00 GHz. A large, 2.93 GBytes of RAM is necessary for the identification process and other memory intensive tasks. This computer is connected to the PC performing real-time tasks via Ethernet, as described in

Sect. 5.5.4.2. The rest of the hardware components included in the host PC configuration are irrelevant to the application, therefore will not be listed.

5.5.4.2 Controller Application

In this work, the real-time controller is implemented under the xPC Target rapid prototyping system [112]. The xPC Target software prototyping system is a popular choice for the real-time implementation and verification of controllers for AVC both in academic studies and for industrial practitioners [20, 28, 59, 81]. Amongst a few others, dSPACE is also a well-known hardware platform for real-time controller implementation for active vibration control [11, 21, 24, 66]. Other specialized hardware and code development software is available from vendors such as National Instruments, Quanser and others.

The xPC Target solution uses a separate PC, often referred to as *target PC* as a hardware platform for the prototyping, testing and deployment of control and measurement software. The environment makes use of Simulink block schemes, including custom applications developed in C language [113]. After creating the desired block scheme on the host PC as described in Sect. 5.5.4.1, and adding appropriate input and output functions, a compiler creates an executable code. The executable is then loaded onto the target PC. The process will run under the xPC kernel in real-time only in case the calculations are tractable on the given hardware.

This laboratory device is controlled by a software prototyping system running on a target PC with an Intel Core 2 Duo (E6550) processor running at 2.33 GHz and 2024 MBytes of RAM. RAM size directly affects the number of measurement values possible to store online, while processor speed directly influences the execution times which can be used in real-time control.[21] Although not a requirement, the target PC has a hard drive which is used for loading the xPC kernel at startup. This hard drive is also used for saving measurement data, which is then downloaded to the host PC for further processing.

The target PC communicates with the host PC using an xPC Target 4.0 compatible networking interface, manufactured by 3COM. The communication between the host PC and target PC is ensured via Ethernet, using the TCP/IP protocol. The most important hardware component in the target PC is the data acquisition (commonly referred to as the measurement card). It is used for connection with the actuators performing the control moves and sensors providing feedback signal.

5.5.4.3 Data Acquisition

Data acquisition for the demonstration hardware featured in this work is carried out utilizing a PCI DAQ card. The device in question is a National Instruments

---

21 This computer has been used to compute the task execution times of various predictive controller featured in Sect. 12.5.

PCI-6030E[22] multifunction DAQ card. It is capable of providing sampling rates up to 100 kS/s with 16-bit resolution. There are 16 analog inputs and 2 analog outputs [82]. The card is capable of self-calibration, utilizing a high precision on board voltage source [83]. This ensures precise measurements at all times. The particular piece of equipment was chosen, because it is compatible with the hardware and software requirements of this application. The choice of measurement cards and boards mainly depends on the control software prototyping platform compatibility, resolution, precision and the number of input an output ports.

The data acquisition device is connected to a National Instruments BNC-2111 cable terminal with BNC connectors for the analog input and output channels. BNC-2111 is a shielded connector block providing 16 single ended analog input terminals, two analog output terminals and five digital I/O. It is compatible with the data acquisition card in question and delivers an external reference voltage for the analog outputs. Selection between ground referenced and floating source signals is ensured by a two-position switch. The cable terminal used throughout the experiments is shown in Fig. 5.28b.

5.5.5 *Electrodynamic Shaker and Amplifier*

An electro magneto-dynamic shaker has been utilized in select verifications tests as a source of mechanical disturbance. Shakers or modal testing equipment based on similar working principles are used as motive force generators in several applications, including mechanical impedance measurements, vibration testing and accelerometer calibration [50]. In the field of active vibration control, shakers can be used as a primary source of outside disturbance to verify and compare control systems. In addition to modal tests, shakers of different sizes and makes are widely utilized in academic studies and industrial settings to test active vibration control systems [8, 39, 74, 97].

5.5.5.1 Shaker

The electrodynamic shaker with a permanent field magnet utilized in this application has been manufactured by Brüel&Kjær. The miniature laboratory shaker, Type 4810 is capable of delivering a 10 N peak sine force, with a bandwidth of DC to 18 kHz [13]. Its first resonance is above the usable bandwidth, and is capable of delivering a bare table acceleration of 550 ms^{-2}. Maximal stroke is set at 6 mm. The shaker body with a visible electrical connection is shown in Fig. 5.33a. Radial flexure springs restrict the moving element, ensuring near perfect rectilinear motion. Laminated flexure springs minimize distortion due to resonances, by providing necessary damping to the system.

22 Formerly known as PCI-MIO-16XE-10.

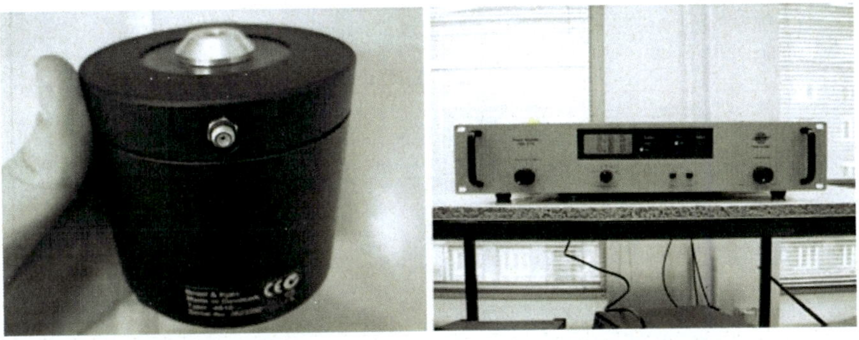

(**a**) Electrodynamic shaker (**b**) Shaker amplifier

Fig. 5.33 Brüel&Kjær Type 4810 electrodynamic shaker (**a**) and Brüel&Kjær Type 2718 general-purpose shaker amplifier shown in (**b**)

The object is attached to the shaker by a 10–32 UNF screw, commonly used with accelerometers. Although capable of modal testing, this application only uses the shaker as a source of disturbance, generated by the measurement setup. The shaker acquires its voltage source through an amplifier.

5.5.5.2 Amplifier

The amplifier driving the miniature shaker introduced in Sect. 5.5.5.1 has been manufactured by Brüel&Kjær and is shown on is shown in Fig. 5.33b. This general-purpose operational amplifier Type 2718 has been specifically designed for small vibration exciters, and drives shaker Type 4810 up to its full rating.

The device features a 75 VA output capability, continuously variable current limit from 1 to 5 A RMS and a 40 dB voltage gain [14]. This amplifier provides a flat frequency response in the bandwidth of 10–20 kHz (±0.5 dB).

Monitoring of output is ensured by a front panel display and information about distortion, temperature and current overload, stand-by status and power on. Monitoring of the output current and voltage is ensured by an appropriate output at the back of the amplifier using two BNC connectors.

The amplifier is delivering the high voltage signal to the mini shaker using an 10–32 UNF Microdot plug. The amplifier gains its signal from the measuring card, and its input is connected to the cable terminal using a BNC cable.

References

1. Agrawal BN, Bang H (1996) Adaptive structures for large precision antennas. Acta Astronautica 38(3):175–183. doi:10.1016/0094-5765(96)00062-8, http://www.sciencedirect.com/science/article/B6V1N-3VTW8Y7-3/2/a53f7c4acb3ee1541568e0db4062d985

2. Alam M, Tokhi M (2008) Designing feedforward command shapers with multi-objective genetic optimisation for vibration control of a single-link flexible manipulator. Eng Appl Artif Intell 21(2):229–246. doi:10.1016/j.engappai.2007.04.008, http://www.sciencedirect.com/science/article/B6V2M-4P0N8W7-1/2/0151e11caeeaab40012fcffe7059861b
3. Allen M, Bernelli-Zazzera F, Scattolini R (2000) Sliding mode control of a large flexible space structure. Control Eng Pract 8(8):861–871. doi:10.1016/S0967-0661(00)00004-6, http://www.sciencedirect.com/science/article/B6V2H-4118CFK-2/2/7c38da4cfa63c75479f00bdc58a9fa9e
4. Amer Y, Bauomy H (2009) Vibration reduction in a 2DOF twin-tail system to parametric excitations. Commun Nonlinear Sci Numer Simul 14(2):560–573. doi:10.1016/j.cnsns.2007.10.005, http://www.sciencedirect.com/science/article/B6X3D-4PYP723-2/2/b9d5375168fadb0b4e67857e92948bfc
5. Bae JS, Kwak MK, Inman DJ (2005) Vibration suppression of a cantilever beam using eddy current damper. J Sound Vib 284(3–5):805–824. doi:10.1016/j.jsv.2004.07.031, http://www.sciencedirect.com/science/article/B6WM3-4F1J8KP-D/2/f6ceedf67a74bb2aadd57e99f8bea78
6. Balas M (1978) Feedback control of flexible systems. IEEE Trans Autom Control 23(4):673–679. doi:10.1109/TAC.1978.1101798
7. Barrault G, Halim D, Hansen C (2007) High frequency spatial vibration control using method. Mech Syst Sig Process 21(4):1541–1560. doi:10.1016/j.ymssp.2006.08.013, http://www.sciencedirect.com/science/article/B6WN1-4M33W0V-2/2/d0c429b3c7910c838e4e1c39d7d042e6
8. Barrault G, Halim D, Hansen C, Lenzi A (2008) High frequency spatial vibration control for complex structures. Appl Acoust 69(11):933–944. doi:10.1016/j.apacoust.2007.08.004, http://www.sciencedirect.com/science/article/B6V1S-4R11Y2G-2/2/796db2eb20c42f55ab1a0ac73f869b45
9. Bittanti S, Cuzzola FA (2002) Periodic active control of vibrations in helicopters: a gain-scheduled multi-objective approach. Control Eng Pract 10(10):1043–1057. doi:10.1016/S0967-0661(02)00052-7, http://www.sciencedirect.com/science/article/B6V2H-45KSPJJ-3/2/9647861ce849d131c7d4b90cdb964751
10. Boeing Company (2004) Boeing-led team successfully tests SMART materials helicopter rotor. Online, http://www.boeing.com/news/releases/2004/photorelease/q2/ml
11. Bohn C, Cortabarria A, Härtel V, Kowalczyk K (2004) Active control of engine-induced vibrations in automotive vehicles using disturbance observer gain scheduling. Control Eng Pract 12(8):1029–1039. doi:10.1016/j.conengprac.2003.09.008, http://www.sciencedirect.com/science/article/B6V2H-49Y3VWS-1/2/dd7bcefd1618f3820896ddbd6dce7430, in Special Section on Emerging Technologies for Active Noise and Vibration Control Systems
12. Boscariol P, Gasparetto A, Zanotto V (2010) Model predictive control of a flexible links mechanism. J Intell Rob Syst 58:125–147. doi:10.1007/s10846-009-9347-5
13. Brüel & Kjær Inc (2008) Mini-shaker—type 4810. Product data sheet. Brüel & Kjær, Sound & Vibration Measurement A/S, Nærum. http://wwww.bksv.com/doc/bp0232.pdf
14. Brüel & Kjær Inc (2008) Operational amplifier—type 2718. Product data sheet. Brüel & Kjær, Sound & Vibration Measurement A/S, Nærum. http://wwww.bksv.com/doc/bp0232.pdf
15. Cavallo A, De Maria G, Leccia E, Setola R (1997) A robust controller for active vibration control of flexible systems. In: Proceedings of the 36th IEEE conference on decision and control, vol 2. pp 1355–1360. doi:10.1109/CDC.1997.657648
16. Chiang RY, Safonov MG (1991) Design of $\mathscr{H}_\infty$ controller for a lightly damped system using a bilinear pole shifting transform. In: American control conference, pp 1927–1928
17. Choi SB, Hong SR, Sung KG, Sohn JW (2008) Optimal control of structural vibrations using a mixed-mode magnetorheological fluid mount. Int J Mech Sci 50(3):559–568. doi:10.1016/j.ijmecsci.2007.08.001, http://www.sciencedirect.com/science/article/B6V49-4PD4XHC-1/2/c491dc4a4a881e38b0e20ceef7206dec

18. Chu CL, Wu BS, Lin YH (2006) Active vibration control of a flexible beam mounted on an elastic base. Finite Elem Anal Des 43(1):59–67. doi:10.1016/j.finel.2006.07.001, http://www.sciencedirect.com/science/article/pii/S0168874X06001144
19. Cigada A, Mancosu F, Manzoni S, Zappa E (2010) Laser-triangulation device for in-line measurement of road texture at medium and high speed. Mech Syst Sig Process 24(7):2225–2234. doi:10.1016/j.ymssp.2010.05.002, http://www.sciencedirect.com/science/article/B6WN1-502V6TK-1/2/721efad2686da854aa76ed846c28861d, special Issue: ISMA 2010
20. Cole MO, Wongratanaphisan T, Pongvuthithum R, Fakkaew W (2008) Controller design for flexible structure vibration suppression with robustness to contacts. Automatica 44(11):2876–2883. doi:10.1016/j.automatica.2008.03.022, http://www.sciencedirect.com/science/article/B6V21-4TNTN71-1/2/c5e859942698eab6c6b7f95e5dab2296
21. Daley S, Johnson FA, Pearson JB, Dixon R (2004) Active vibration control for marine applications. Control Eng Pract 12(4):465–474. doi:10.1016/S0967-0661(03)00135-7, http://www.sciencedirect.com/science/article/B6V2H-495051H-2/2/9b7cbd1e4f539a3b0c92c698ce1bad19 uKACC Conference Control 2002
22. Dong X, Meng G, Peng J (2006) Vibration control of piezoelectric smart structures based on system identification technique: numerical simulation and study. J Sound Vib 297:680–693
23. Eissa M, Bauomy H, Amer Y (2007) Active control of an aircraft tail subject to harmonic excitation. Acta Mech Sin 23:45–462. http://10.1007/s10409-007-0077-2 doi:10.1007/s10409-007-0077-2
24. El-Badawy AA, Nayfeh AH (2001) Control of a directly excited structural dynamic model of an F-15 tail section. J Franklin Inst 338(2–3):133–147. doi:10.1016/S0016-0032(00)00075-2, http://www.sciencedirect.com/science/article/B6V04-42HNMDV-3/2/e3bf6f797834c8e8638324be88fb78f
25. Falconi C, Mantini G, D'Amico A, Wang ZL (2009) Studying piezoelectric nanowires and nanowalls for energy harvesting. Sens Actuators B 139(2):511–519. doi:10.1016/j.snb.2009.02.071, http://www.sciencedirect.com/science/article/B6THH-4VWB16R-F/2/cc7ca2b1281134e75cf141e8ef942105
26. Friedman J, Khargonekar P (1995) Application of identification in $\mathscr{H}_\infty$ to lightly damped systems: two case studies. IEEE Trans Control Syst Technol 3(3):279–289. doi:10.1109/87.406975
27. Fung RF, Liu YT, Wang CC (2005) Dynamic model of an electromagnetic actuator for vibration control of a cantilever beam with a tip mass. J Sound Vib 288(4–5):957–980. doi:10.1016/j.jsv.2005.01.046, http://www.sciencedirect.com/science/article/B6WM3-4G4N5VD-1/2/fc3710f0625ef69f19d16c8778a63e58
28. Gani A, Salami M, Khan R (2003) Active vibration control of a beam with piezoelectric patches: real-time implementation with xPC target. In: Proceedings of 2003 IEEE conference on control applications. CCA 2003, vol 1. pp 538–544. doi:10.1109/CCA.2003.1223494
29. Gatti G, Brennan MJ, Gardonio P (2007) Active damping of a beam using a physically collocated accelerometer and piezoelectric patch actuator. J Sound Vib 303(3–5):798–813. doi:10.1016/j.jsv.2007.02.006, http://www.sciencedirect.com/science/article/B6WM3-4NH6N96-1/2/13ee638f653d035bca14ce9109e1cd96
30. Gaudenzi P, Carbonaro R, Barboni R (1997) Vibration control of an active laminated beam. Compos Struct 38(1–4):413–420. doi:10.1016/S0263-8223(97)00076-7, http://www.sciencedirect.com/science/article/B6TWP-3SP98BG-1F/2/4153da95a6c08e2ac7317fcc21271 Ninth International Conference on Composite Structures
31. Gaudenzi P, Carbonaro R, Benzi E (2000) Control of beam vibrations by means of piezoelectric devices: theory and experiments. Compos Struct 50:373–379
32. Gaudiller L, Hagopian JD (1996) Active control of flexible structures using a minimum number of components. J Sound Vib 193(3):713–741. doi:10.1006/jsvi.1996.0310, http://www.sciencedirect.com/science/article/B6WM3-45PVM9F-35/2/146b67a462e38c197fe3acde5d1df

33. Gospodaric B, Voncina D, Bucar B (2007) Active electromagnetic damping of laterally vibrating ferromagnetic cantilever beam. Mechatronics 17(6):291–298. doi:10.1016/j.mechatronics.2007.04.002, http://www.sciencedirect.com/science/article/B6V43-4NVSWV6-1/2/5c4672945cfa9b81238f0b1cb8a8eb
34. Halim D, Moheimani S (2003) An optimization approach to optimal placement of collocated piezoelectric actuators and sensors on thin plate. Mechatronics 13:27–47
35. Hassan M, Dubay R, Li C, Wang R (2007) Active vibration control of a flexible one-link manipulator using a multivariable predictive controller. Mechatronics 17(1):311–323
36. Hatch MR (2000) Vibration simulation using MATLAB and ANSYS, 1st edn. Chapman and Hall / CRC, Boca Raton
37. Hellerstein JL, Diao Y, Parekh S, Tilbury DM (2004) Feedback control of computing systems. Wiley / IEEE Press, Hoboken
38. Herrick DC (1980) Study of velocity output vibration suppression controllers with a multiloop root locus. In: 19th IEEE conference on decision and control including the symposium on adaptive processes, vol 19. pp 1088–1090. doi:10.1109/CDC.1980.271970
39. Hong SR, Choi SB, Han MS (2002) Vibration control of a frame structure using electro-rheological fluid mounts. Int J Mech Sci 44(10):2027–2045. doi:10.1016/S0020-7403(02)00172-8 http://www.sciencedirect.com/science/article/B6V49-47BX3RX-4/2/53a10ce8cbf8dfa679c34e04beb688e4
40. Hu Q (2009) A composite control scheme for attitude maneuvering and elastic mode stabilization of flexible spacecraft with measurable output feedback. Aerosp Sci Technol 13(2–3):81–91. doi:10.1016/j.ast.2007.06.007, http://www.sciencedirect.com/science/article/B6VK2-4P96269-2/2/5fbc47249fdd3f1963c5ba856f071c55
41. Hu QL, Wang Z, Gao H (2008) Sliding mode and shaped input vibration control of flexible systems. IEEE Trans Aersp Electron Syst 44(2):503–519. doi:10.1109/TAES.2008.4560203
42. Huber T (2011) Image showing the scan points on the two bricks. Website, http://physics.gac.edu/huber/acoustics/speaker_vibrometer_fields/10khz_tweeter_finergrid3_cropped.png
43. Huber T (2011) Photograph showing the speaker placed in front two bricks and the vibrometer scan head. Website, http://physics.gac.edu/huber/acoustics/speaker_vibrometer_fields/pic_0075.jpg
44. Huber TM (2011) Measurement of mode shapes of musical instruments using a scanning laser Doppler vibrometer. J Acoust Soc Am 129(4):2615–2615. doi:10.1121/1.3588688, http://link.aip.org/link/?JAS/129/2615/5
45. Huber TM, Mellema DC, Abell B (2009) Selective excitation of microcantilever array using ultrasound radiation force. J Acoust Soc Am 125(4):2635–2635. http://link.aip.org/link/?JAS/125/2635/5
46. Hubinský P (2010) Riadenie mechatronických systémov s nízkym tlmením, 1st edn. In: Slovenská technická univerzita v Bratislave, Nakladateľstvo STU, Control of mechatronic systems with low damping, Slovak language, Bratislava
47. Hulkó G, Belavý C, Belanský J, Szuda J, Végh P (1998) Modeling, control and design of distributed parameter systems: with demonstrations in Matlab, 1st edn. Publishing house of Slovak University of Technology, Bratislava
48. Hulkó G, Belavý C, Buček P, Ondrejkovič K, Zajíček P (2009) Engineering methods and software support for control of distributed parameter systems. In: ASCC 2009: 7th Asian control conference, pp 1432–1438
49. Hulkó G, Belavý C, Mészáros A, Buček P, Ondrejkovič K, Zajíček P (2009) Engineering methods and software support for modelling and design of discrete-time control of distributed parameter systems. Euro J Control 15(3–4):407–417
50. Inman DJ (2007) Engineering vibrations, 3rd edn. Pearson International Education (Prentice Hall), Upper Saddle River
51. John S, Hariri M (2008) Effect of shape memory alloy actuation on the dynamic response of polymeric composite plates. Composites Part A 39(5):769–776. doi:

10.1016/j.compositesa.2008.02.005, http://www.sciencedirect.com/science/article/B6TWN-4RV7YMP-1/2

52. Kang B, Mills JK (2005) Vibration control of a planar parallel manipulator using piezoelectric actuators. J Intell Rob Syst 42:51–70. doi:10.1007/s10846-004-3028-1
53. Kapucu S, Yıldırım N, Yavuz H, Bayseç S (2008) Suppression of residual vibration of a translating-swinging load by a flexible manipulator. Mechatronics 18(3):121–128. doi:10.1016/j.mechatronics.2007.10.007, http://www.sciencedirect.com/science/article/B6V43-4R8M99T-1/2/75f50dfb587306d4ed19a2be2d79471e
54. Karl W, Verghese G, Lang J (1994) Control of vibrational systems. IEEE Trans Autom Control 39(1):222–226. doi:10.1109/9.273372
55. Kermani MR, Moallem M, Patel RV (2004) Parameter selection and control design for vibration suppression using piezoelectric transducers. Control Eng Pract 12:1005–1015
56. Keyence Inc (2003) Displacement sensors technical guide. Keyence Inc., Osaka
57. Keyence Inc (2004) LK Navigator user's manual. Keyence Inc., for LK-G Series Setting and Support Software LK-H1W, Osaka
58. Keyence Inc (2006) LK-G series user's manual. Keyence Inc., Osaka
59. Kim SM, Wang S, Brennan MJ (2011) Dynamic analysis and optimal design of a passive and an active piezo-electrical dynamic vibration absorber. J Sound Vib 330(4):603–614. doi:10.1016/j.jsv.2010.09.004, http://www.sciencedirect.com/science/article/B6WM3-5172KF3-2/2/719330ee33502c8438c9b629587c52ca
60. Kursu O, Kruusing A, Pudas M, Rahkonen T (2009) Piezoelectric bimorph charge mode force sensor. Sens Actuat A 153(1):42–49. doi:10.1016/j.sna.2009.04.026, http://www.sciencedirect.com/science/article/B6THG-4W6Y34K-4/2/ae7628bbda5ac689d58eaec08bddb9
61. Kwak MK, Heo S (2007) Active vibration control of smart grid structure by multiinput and multioutput positive position feedback controller. J Sound Vib 304(1–2):230–245. doi:10.1016/j.jsv.2007.02.021, http://www.sciencedirect.com/science/article/B6WM3-4NH6N96-2/2/ca7b43602b9d052e388f4b2a28f1ebae
62. Lau K, Zhou L, Tao X (2002) Control of natural frequencies of a clamped-clamped composite beam with embedded shape memory alloy wires. Compos Struct 58(1):39–47. doi:10.1016/S0263-8223(02)00042-9, http://www.sciencedirect.com/science/article/B6TWP-45XTP9W-N/2/07b9a065ac866d8869a4240deb918851
63. Lee HK, Chen ST, Lee AC (1996) Optimal control of vibration suppression in flexible systems via dislocated sensor/actuator positioning. J Franklin Inst 333(5):789–802. doi:10.1016/0016-0032(96)00038-5, http://www.sciencedirect.com/science/article/B6V04-3VVCHR0-B/2/05cf1444f44d099257f07563677c080
64. Lewis B, Tran H (2007) Real-time implementation of a frequency shaping controller on a cantilever beam. Appl Numer Math 57(5–7): 778–790. doi: 10.1016/j.apnum.2006.07.017, http://www.sciencedirect.com/science/article/B6TYD-4KNM9VK-1/2/ec6 special Issue for the International Conference on Scientific Computing
65. Li M, Lim TC, Lee JH (2008) Simulation study on active noise control for a 4-T MRI scanner. Magn Reson Imaging 26(3):393–400. doi:10.1016/j.mri.2007.08.003, http://www.sciencedirect.com/science/article/B6T9D-4R8KT3W-2/2/2797c565f329cf6cd1e567eefb69607
66. Li YY, Cheng L, Li P (2003) Modeling and vibration control of a plate coupled with piezoelectric material. Compos Struct 62(2):155–162. doi:10.1016/S0263-8223(03)00110-7, http://www.sciencedirect.com/science/article/B6TWP-48R1WVK-1/2/f0788ece03ae40a5874f11852e9278
67. Lin J, Liu WZ (2006) Experimental evaluation of a piezoelectric vibration absorber using a simplified fuzzy controller in a cantilever beam. J Sound Vib 296(3):567–582. doi:10.1016/j.jsv.2006.01.066 http://www.sciencedirect.com/science/article/B6WM3-4K0FG0H-2/2/e4fad7e52e98cf46123aa869cf780b65
68. Lin J, Nien MH (2005) Adaptive control of a composite cantilever beam with piezoelectric damping-modal actuators/sensors. Compos Struct 70:170–176

69. Lin LC, Lee TE (1997) Integrated PID-type learning and fuzzy control for flexible-joint manipulators. J Intell Rob Syst 18:47–66. http://10.1023/A:1007942528058,10.1023/A:1007942528058
70. Ljung L (1999) System identification: theory for the user, 2nd edn. PTR Prentice Hall, Upper Saddle River
71. Lu H, Meng G (2006) An experimental and analytical investigation of the dynamic characteristics of a flexible sandwich plate filled with electrorheological fluid. Int J Adv Manuf Technol 28:1049–1055 10.1007/s00170-004-2433-8
72. Luo T, Hu Y (2002) Vibration suppression techniques for optical inter-satellite communications. In: IEEE 2002 international conference on communications, circuits and systems and west sino expositions, vol 1. pp 585–589. doi:10.1109/ICCCAS.2002.1180687
73. Mahmoodi S, Daqaq MF, Jalili N (2009) On the nonlinear-flexural response of piezoelectrically driven microcantilever sensors. Sens Actuators A 153(2):171–179. doi:10.1016/j.sna.2009.05.003, http://www.sciencedirect.com/science/article/B6THG-4W8TW31-1/2/4fc1a15727232bf3e6741b9f2618c61c
74. Mahmoodi SN, Craft MJ, Southward SC, Ahmadian M (2011) Active vibration control using optimized modified acceleration feedback with adaptive line enhancer for frequency tracking. J Sound Vib 330(7):1300–1311. doi:10.1016/j.jsv.2010.10.013, http://www.sciencedirect.com/science/article/B6WM3-51D894K-1/2/25e8ef1bcadb5fd2aa078de4d678c7f4
75. Malgaca L (2010) Integration of active vibration control methods with finite element models of smart laminated composite structures. Compos Struct 92(7):1651–1663. doi:10.1016/j.compstruct.2009.11.032, http://www.sciencedirect.com/science/article/B6TWP-4XY4K3R-1/2/39a31e42ad10d6e7aafe32a91352372a
76. Mehrabian AR, Yousefi-Koma A (2011) A novel technique for optimal placement of piezoelectric actuators on smart structures. J Franklin Inst 348(1):12–23. doi:10.1016/j.jfranklin.2009.02.006, http://www.sciencedirect.com/science/article/B6V04-4VTCM9T-1/2/1d68ecf523d642a7246481a506f3edab, mechatronics and its Applications
77. MIDÉ Technology Corporation (2007) QuickPack actuator catalog. MIDÉ Technology Corporation, Medford
78. MIDÉ Technology Corporation (2007) QuickPack power amplifier. MIDÉ Technology Corporation, Medford (Operator's Manual)
79. MIDÉ Technology Corporation (2008) Attaching the quickpack/powerAct transducer to a structure with epoxy. MIDÉ Technology Corporation, Medford (quick Pack Technical Notes)
80. Montazeri A, Poshtan J, Choobdar A (2009) Performance and robust stability trade-off in minimax LQG control of vibrations in flexible structures. Eng Struct 31(10):2407–2413. doi:10.1016/j.engstruct.2009.05.011, http://www.sciencedirect.com/science/article/B6V2Y-4WJ2CGY-1/2/66039e162a0fe4f3e4aa37dbae422b04
81. Moon SM, Cole DG, Clark RL (2006) Real-time implementation of adaptive feedback and feedforward generalized predictive control algorithm. J Sound Vib 294(1–2):82–96. doi:10.1016/j.jsv.2005.10.017, http://www.sciencedirect.com/science/article/B6WM3-4HYMY76-1/2/50d98047187533ebe9d3ea8310446e77
82. National Instruments Corporation (2005) NI 6030E/6031E/6032E/6033E Family Specifications. National Instruments Corporation, Austin. http://www.ni.com/pdf/manuals/370720c.pdf
83. National Instruments Corporation (2007) DAQ E series user manual. National Instruments Corporation, Austin. http://www.ni.com/pdf/manuals/370720c.pdf
84. Neat G, Melody J, Lurie B (1998) Vibration attenuation approach for spaceborne optical interferometers. IEEE Trans Control Syst Technol 6(6):689–700. doi:10.1109/87.726529
85. Nguyen C, Pietrzko S (2006) FE analysis of a PZT-actuated adaptive beam with vibration damping using a parallel R-L shunt circuit. Finite Elem Anal Des 42(14–15):1231–1239. doi:10.1016/j.finel.2006.06.003,http://www.sciencedirect.com/science/article/B6V36-4KGX810-1/2/ada2103f6ecf58789b588a65756803d5

86. Nouira H, Foltęte E, Brik BA, Hirsinger L, Ballandras S (2008) Experimental characterization and modeling of microsliding on a small cantilever quartz beam. J Sound Vib 317(1–2):30–49. doi:10.1016/j.jsv.2008.03.017, http://www.sciencedirect.com/science/article/B6WM3-4SGTM4T-1/2/6adaaffb834cdd20916ebdbc5d0b3592
87. O'Connor WJ (2006) Wave-echo control of lumped flexible systems. J Sound Vib 298(4–5):1001–1018. doi:10.1016/j.jsv.2006.06.010, http://www.sciencedirect.com/science/article/B6WM3-4KM46VY-3/2/b2b255aa52ecbc2dfe781e84207a479f
88. Petersen IR, Pota HR (2003) Minimax LQG optimal control of a flexible beam. Control Eng Pract 11:1273–1287
89. Podolan M (2007) Hliník a jeho zliatiny, ich porovnanie a dostupnosť na trhu. Tech. rep., IMC Slovakia s.r.o. In: Aluminum and its alloys: comparison and market accessibility, Slovak language. Považská Bystrica
90. Polóni T, Rohaľ-Ilkiv B, Johansen TA (2010) Damped one-mode vibration model state and parameter estimation via pre-filtered moving horizon observer. In: 5th IFAC symposium on mechatronic systems. Mechatronics 2010, IFAC, Boston, pp 24–31
91. Preumont A (2002) Vibration control of active structures, 2nd edn. Kluwer Academic Publishers, Dordrecht
92. Preumont A, Seto K (2008) Active control of structures, 3rd edn. Wiley, Chichester
93. Qiu Z, Zhang X, Wu H, Zhang H (2007) Optimal placement and active vibration control for piezoelectric smart flexible cantilever plate. J Sound Vib 301:521–543
94. Qiu ZC, Wu HX, Ye CD (2009) Acceleration sensors based modal identification and active vibration control of flexible smart cantilever plate. Aerosp Sci Technol 13(6):277–290. doi:10.1016/j.ast.2009.05.003, http://www.sciencedirect.com/science/article/B6VK2-4WB3NH7-2/2/e7bef32fa0e1ef301516f9b393ea8a97
95. Scheibner D, Mehner J, Brämer B, Gessner T, Dötzel W (2003) Wide range tuneable resonators for vibration measurements. Microelectron Eng 67–68:542–549
96. Shan J, Liu HT, Sun D (2005) Slewing and vibration control of a single-link flexible manipulator by positive position feedback (PPF). Mechatronics 15(4):487–503. doi:10.1016/j.mechatronics.2004.10.003, http://www.sciencedirect.com/science/article/B6V43-4DR87K7-4/2/2dd311fdd61308e1415cd45c1edc3076
97. Shen H, Qiu J, Ji H, Zhu K, Balsi M, Giorgio I, Dell'Isola F (2010) A low-power circuit for piezoelectric vibration control by synchronized switching on voltage sources. Sens Actuators A 161(1–2):245–255. doi:10.1016/j.sna.2010.04.012, http://www.sciencedirect.com/science/article/B6THG-5017HGW-1/2/a638dcd70899a318cd3871a5112021bb
98. Shieh J, Huber JE, Fleck NA, Ashby MF (2001) The selection of sensors. Prog Mater Sci 46(3–4):461–504. doi:10.1016/S0079-6425(00)00011-6,, http://www.sciencedirect.com/science/article/B6TX1-42JYVPK-G/2/b23a52757764ef929c1164920807d3d6
99. Siemens Industry Incorporated (2010) Capacitive proximity sensors: theory of operation. Munich, http://www.eandm.com/eandm/training/siemenscourses/snrs_3.pdf
100. Skullestad A, Hallingstad O (1998) Vibration parameters identification in a spacecraft subjected to active vibration damping. Mechatronics 8(6):691–705. doi:0.1016/S0957-4158(97)00051-2, http://www.sciencedirect.com/science/article/B6V43-3W18XD5-4/2/d6e2d2a478a77ff09e9f6f7dfe7fd503
101. Sloss J, Bruch J, Sadek I, Adali S (2003) Piezo patch sensor/actuator control of the vibrations of a cantilever under axial load. Compos Struct 62:423–428
102. Sodano HA, Inman DJ (2007) Non-contact vibration control system employing an active eddy current damper. J Sound Vib 305(4–5):596–613. doi:10.1016/j.jsv.2007.04.050, http://www.sciencedirect.com/science/article/B6WM3-4P2J38T-3/2/a75ecce7ed7841e00499a50d077bd23
103. Šolek P (2009) Numerical analyses of piezoelectric elements, 1st edn. Slovenská technická univerzita v Bratislave, Nakladateľstvo STU, Bratislava

104. Šolek P, Starek L, Hulkó G, Šedivý C, Cibiri Š (2005) Theoretical and experimental study of efficient suppression vibrations in a clamped square plate. Inženýrská Mechanika Engineering Mechanics 12(A1):277–284
105. Spangler R (2007) Piezo sensor technical note, 2nd edn. MIDÉ Technology Corporation, Medford
106. Sun D, Mills JK, Shan J, Tso SK (2004) A PZT actuator control of a single-link flexible manipulator based on linear velocity feedback and actuator placement. Mechatronics 14(4):381–401. doi:10.1016/S0957-4158(03)00066-7, http://www.sciencedirect.com/science/article/B6V43-49DN5K4-1/2/fa21df547f182ad568cefb2ddf3a6352
107. Tabak F, Disseldorp E, Wortel G, Katan A, Hesselberth M, Oosterkamp T, Frenken J, van Spengen W (2010) MEMS-based fast scanning probe microscopes. Ultramicroscopy 110(6):599–604. doi:10.1016/j.ultramic.2010.02.018, http://www.sciencedirect.com/science/article/B6TW1-4YJCKXY-1/2/4f5b9ba5875b8066d7cb20174f05ad61, 11th International Scanning Probe Microscopy Conference
108. Takács G, Rohal'-Ilkiv B (2009) Implementation of the Newton-Raphson MPC algorithm in active vibration control applications. In: Mace BR, Ferguson NS, Rustighi E (eds). Proceedings of the 3rd international conference on noise and vibration: emerging methods. Oxford
109. Takács G, Rohal'-Ilkiv B (2009) Newton-Raphson MPC controlled active vibration attenuation. In: Hangos KM (eds) Proceedings of the 28. IASTED international conference on modeling, identification and control, Innsbruck
110. Takács G, Rohal'-Ilkiv B (2010) Capacitive proximity sensor position feedback in active vibration control of lightly damped cantilevers. In: Shokin YI, Bychkov I, Potaturkin O (eds). Proceedings of the 3rd IASTED international multiconference ACIT-CDA. Novosibirsk, pp 692–700
111. Takács G, Rohal'-Ilkiv B (2010) Piezoelectric wafer based feedback in vibration control of lightly damped beams. In: Proceedings of the 9th international scientific—technical conference—Process control 2010, Kouty nad Desnou, p C43a
112. The Mathworks (2007) xPC Target 4. Software. The MathWorks Inc., Natick. http://www.mathworks.com/products/xpctarget/
113. The MathWorks (2008) xPC target for use with real-time workshop, 6th edn. The MathWorks Inc., Natick. http://www.mathworks.com/products/xpctarget/
114. The Mathworks (2009) How can I look at the data in a file created by the file scope block on the host machine in xPC Target 2.7.2 (R14SP2). The MathWorks Inc., Natick. Available: http://www.mathworks.de/support/solutions/data/1-1J7RAI.html?product=XP\&solution=1-1J7RAI
115. The Mathworks (2011) Matlab system identification toolbox v7.4.2 (R2011a). Software. The MathWorks Inc., Natick. Available: http://www.mathworks.com/help/toolbox/ident/
116. Torra V, Isalgue A, Martorell F, Terriault P, Lovey F (2007) Built in dampers for family homes via SMA: An ANSYS computation scheme based on mesoscopic and microscopic experimental analyses. Eng Struct 29(8):1889–1902. doi:10.1016/j.engstruct.2006.08.028, http://www.sciencedirect.com/science/article/B6V2Y-4MFKD84-1/2/8742fa675c346a7b34f395d9422cbc22
117. Wang W, Yang Z (2009) A compact piezoelectric stack actuator and its simulation in vibration control. Tsinghua Sci Technol 14(Suppl 2):43–48. doi:10.1016/S1007-0214(10)70029-8, http://www.sciencedirect.com/science/article/B7RKT-4YJ4CW2-9/2/11681292e718c72b12ae9969c514f4
118. Wills AG, Bates D, Fleming AJ, Ninness B, Moheimani SOR (2008) Model predictive control applied to constraint handling in active noise and vibration control. IEEE Trans Control Syst Technol 16(1):3–12
119. Wilson DG, Robinett RD, Parker GG, Starr GP (2002) Augmented sliding mode control for flexible link manipulators. J Intell Rob Syst 34:415–430. doi:10.1023/A:1019635709331
120. Xiaojin Z, Miao Z, Zhiyuan G, Zhiyan C (2010) Analysis of active vibration control for piezoelectric intelligent structures by ANSYS and MATLAB. In: 2010 international confer-

ence on computer application and system modeling (ICCASM), vol 4. pp V4-184 –V4-188. doi: 10.1109/ICCASM.2010.5619058
121. Yim W (1996) Modified nonlinear predictive control of elastic manipulators. In: Proceedings of the 1996 IEEE international conference on robotics and automation, vol 3. pp 2097–2102. 10.1109/ROBOT.1996.506180
122. Zhang P (2008) Sensors and actuators for industrial control. In: Industrial control technology, William Andrew Publishing, Norwich, pp 1–186. doi:10.1016/B978-081551571-5.50002-5 http://www.sciencedirect.com/science/article/B8M8S-4TRTX8X-4/2/7446382607dd6d46d9befe145fcee0
123. Zheng K, Zhang Y, Yang Y, Yan S, Dou L, Chen J (2008) Active vibration control of adaptive truss structure using fuzzy neural network. In: Control and decision conference, CCDC 2008. Chinese, pp 4872–4875. doi: 10.1109/CCDC.2008.4598254
124. Zmeu K, Shipitko E (2005) Predictive controller design with offline model learning for flexible beam control. In: Proceedings of the 2005 international conference on physics and control, pp 345–350. doi:10.1109/PHYCON.2005.1514005

Part II
Model Predictive Control

Chapter 6
Basic MPC Formulation

Model predictive control (MPC) is an advanced optimization-based control method that has been in use for applications with slow dynamics, such as petrochemical plants since the 1980s (Fig. 6.1). Unlike linear quadratic (LQ) control, in addition to providing an optimal control process, MPC offers the explicit handling of process constraints that arise from natural requirements, for example cost effectiveness, safety, actuator limits and others. In fact due to the active interest of industry, the early theoretical development of MPC has been influenced greatly by the requirements of corporate users. A review of the industrial applications for those interested is given in [35, 43–45]. The design of MPC controllers is nowadays supported by numerous off-the-shelf commercial packages [45]. These tools typically contain means for model identification, controller design, controller tuning and controller performance analysis and are intended for the industrial user without a deep knowledge of the theoretical aspects of MPC.

Control decisions in MPC are computed online using an internal model of the plant dynamics. The big advantage of MPC over other control strategies is that it can handle process constraints on an algorithmic level. Unfortunately, the inclusion of constraints renders the MPC law nonlinear, which has dramatic effects on its stability properties. Just as in the case of any other well-designed system employing an arbitrary control law, the closed-loop stability of constrained MPC needs to be investigated and if possible guaranteed as well.

Given an otherwise stable plant model, it is always possible to conceive a system state, which can render the MPC controller unstable. A constrained MPC control law (in its primal, online optimization-based form) does not have an explicit closed-loop form, therefore stability guarantees can be given only by applying additional constraints which have a task to ensure future feasibility and stability. The stability aspects of model predictive control will be discussed in the following chapter, that is in Chap. 7.

The aim of this chapter is to introduce model predictive control to the reader who has no or minimal prior knowledge of this advanced control strategy. For this reason, we are beginning our discussion from the essentials and build a controller

G. Takács and B. Rohal'-Ilkiv, *Model Predictive Vibration Control*,
DOI: 10.1007/978-1-4471-2333-0_6, © Springer-Verlag London Limited 2012

Fig. 6.1 The petrochemical industry was the first to recognize the merits of MPC and adopt it in everyday operations [22]

from the grounds up. Fundamental concepts such as prediction, cost and penalization are explained first in order to introduce the simplest possible MPC control law, which in the absence of constraints can be expressed in a closed form. By the end of this chapter, the reader shall be familiar with the theoretical fundamentals of the popular dual-mode formulation of the quadratic programming-based MPC strategy.

After an introduction of the underlying idea of model predictive control, a historic overview of the development of MPC is presented in the first section. Here the development from fixed feedback laws based on finite impulse responses up to the constrained and stable online optimization strategies used today are briefly reviewed. This is followed by a section discussing how we can predict the evolution of states based on a state-space model, and how all of this can be given in a compact matrix notation. Section 6.3 introduces the idea of the cost function, which can give a clear numerical measure of the performance of a control law formulating the basis of the optimization task in MPC. Following this, further building blocks of the predictive strategy are discussed, namely the penalization matrices that help to fine tune the behavior of the controlled system. The first working MPC control law is derived in Sect. 6.6, which thanks to the absence of constraints is just a fixed feedback matrix. As one of the main advantages of using MPC over classical methods is its ability to handle process constraints, the formulation of constraints is reviewed in the following section. Following this Sect. 6.8 finally arrives at the central element of the predictive strategy: the most common dual-mode constrained MPC law. This MPC formulation is evaluated online using the mathematical optimization tool called quadratic programming. The chapter ends with a short section discussing the idea of different predictive and control horizons while briefly examining the problem of tracking in state-space systems as well.

The style of presentation and the content of this chapter is aimed at the reader unfamiliar with model predictive control. The ultimate objective is to present constrained MPC based on quadratic programming in a straightforward manner, without distractions leading the reader off-course. This however requires omitting some aspects of predictive control from the explanation. The theoretical view on stability, feasibility and efficiency of MPC is discussed in the subsequent chapters. In case one is interested in MPC formulations based on transfer function models, impulse or step responses, we suggest to read one of the excellent books discussing the basics and more advanced concepts of model predictive control, such as the popular books by Maciejowski [34], Rossiter [48] or several other publications [6, 7, 26, 36].

6.1 The MPC Idea

In essence, an MPC controller is based on an iterative, finite horizon (constrained) optimization of a plant model. At each discrete sampling time (k) the plant is sampled and the actual state[1] $\mathbf{x}_k$ is measured or estimated using observers. The performance of the controller is expressed by a so-called *cost function*. Based on a *dynamic model* of the plant, this cost function is formulated in such a way that it expresses the performance of the MPC controller in the future, given a current plant state $\mathbf{x}_k$ and a sequence of future inputs $\mathbf{u}_k$. In other words, this *predicted cost function* gives a numerical indicator of the quality of control, assuming that the current plant state is influenced by a certain sequence of past inputs. The question is not what the performance of the controller will be, but rather what the sequence of inputs $\mathbf{u}_k$ is which will produce the best performance. To calculate the optimal sequence of inputs, one must *minimize the cost function* at each sampling interval using a numerical optimization algorithm. As in the case of most real plants inputs, outputs and states can be bounded by physical constraints, which can be easily incorporated into the numerical minimization task. Of the sequence of future inputs $\mathbf{u}_k$ only the first is applied, then the process is repeated based on brand new measured state information. This type of repeated measure-predict-optimize-apply cycle is called *receding horizon control*.

The model predictive control algorithm is schematically illustrated in Fig. 6.2. The structure of this scheme illustrates that essentially MPC is a form of a feedback control algorithm, where instead of a fixed feedback law a dynamic online optimization process determines inputs based on the actual measurements.

---

1 Note that a state $\mathbf{x}_k$ is a vector, however in our further discussions we will use a scalar notation x_k instead, see Sect. 6.2 in p. 237 for explanation.

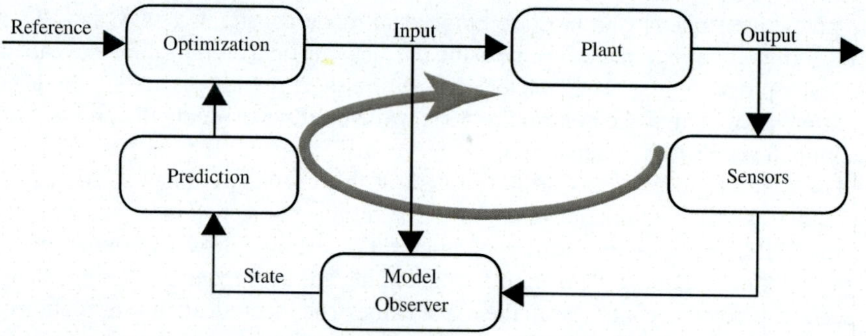

Fig. 6.2 Schematic representation of the model predictive control algorithm

A model predictive controller is based on the following concepts:

- using a mathematical model of the plant dynamics
- predicting future plant dynamics
- expressing process optimality by a cost function
- predicting cost of future plant dynamics
- cost function minimization (optimal control)
- receding horizon control

Figure 6.3 illustrates the concept of the receding horizon model predictive controller. At the time step (k) the controller measures or observes the current plant state from the outputs, denoted by the black dot. An optimal course of inputs is calculated, which is associated with a predicted output course. At time (k) however, only the first element of the sequence is applied to the plant. At the next step $(k+1)$ the whole process is repeated, shifting the horizon one step further.

An interesting view of predictive control is presented by Camacho and Bordons in [6, 7], where the control process is represented by the analogy of driving a motor vehicle. While model predictive control represents driving based on the information gathered by looking out the front windshield, classical feedback control is closer to looking out the *back window* or the rear-view mirror. A real driver uses an MPC-like approach in steering the car, since it looks forward and chooses an ideal action based on possible future outcomes, taking the real characteristics of the car into consideration. A hypothetical driver using a classical control engineering approach (e.g. PID) would only look out the back window, trying to steer the vehicle based on information about its past behavior. Moreover, our hypothetical driver would not take into consideration the real limits and boundaries of its vehicle: it would try to drive through a curve with a semi-truck, assuming it handles just like a sports car.

While this is an oversimplified approach, the analogy has more to offer. The driver of a car bases his or her judgments at the current time on predictions of the future. The driver is familiar with a mental image of the car, knows how it can accelerate, how fast it can stop and how it handles in general. This mental image of the car is

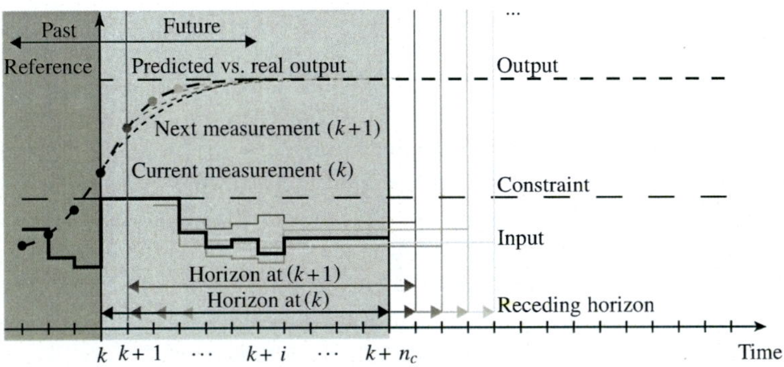

Fig. 6.3 Model predictive control demonstrating the receding horizon control idea—new state measurements (observations) are used to compute a new optimal sequence of inputs shifting the horizon forward at all times

substituted by a simplified internal *mathematical model* in MPC. In good visibility, a driver may see far ahead in the horizon, and thus go fast. If the visibility conditions are bad, the horizon in front of the driver is also short and one may easily misjudge the situation. The portion of the road one can see while driving is represented by the *prediction horizon* in MPC. Instability may occur in MPC if the prediction horizon is too short, this would be the equivalent of crashing the car because of entering an unknown curve too fast in low visibility conditions. A real driver continuously updates its decisions, it does not make a plan before starting the engine and stick to it by all means. Similarly, MPC continuously *updates* the decisions in real-time and it uses only the most recent one, then *repeats* this procedure. Since just like the real driver the MPC law updates its decisions at all times, the MPC horizon is *receding* forward in time. Of course, one of the most important aspects of driving is given by the essential requirement of not leaving the road and crashing the car. We can think of this as a type of *constraint*, along with other constraints such as the physical properties of the vehicle. With a limited portion of the road ahead, one may misjudge its current actions leaving the road: the trajectory and handling of the car becomes *unstable*, leading to dramatic consequences. Similarly, in MPC the mere presence of constraints can complicate the situation and affect the stability of the control course.

6.1.1 Historical Overview

Historically, we can differentiate between three independent lines of development of model predictive control [6, 25, 26, 51]:

- model predictive control
- generalized predictive control
- receding horizon control

The first general category is model predictive control (MPC) which encompasses for example model algorithmic control (MAC) based on finite impulse response (FIR) models and dynamic matrix control (DMC) which is based on finite step response (FSR) models. Successful application of MAC has been introduced in [47], while DMC has been first characterized and applied to a chemical process in [14]. The acceptance of these methods in the industry is backed by the use of impulse- and step response-based models that are very easy and convenient to identify. The drawback of FIR- and FSR-based models is, however that it is very difficult to generalize and apply them to more complex systems, moreover they cannot be formulated for unstable systems.

A second line of development is represented by generalized predictive control (GPC). GPC methods are based on single-input single-output (SISO) models such as the ones often utilized in adaptive control. Some of the control approaches falling under this category are the minimum variance (MV) [2] and the generalized minimum variance (GMV) [10] methods. Unfortunately, these methods have been sensitive to modeling errors and could not guarantee stability of non-minimum phase systems. These problems lead to further development of control theory and the introduction of GPC [12, 13], long-range predictive control [11], extended horizon adaptive control [53] and extended prediction self-adaptive control [15]. Later GPC has been formulated for a continuous time system [16] and MIMO models [27, 50] as well. A GPC predictive controller with guaranteed stability has been presented in [23].

The third and final line of development is called receding horizon control (RHC) where the research direction has been given on modifications and development of linear quadratic controllers. Initially the method did not assume the presence of system constraints. An RHC method minimizing a quadratic criterion with a terminal equality constraint ensuring stability has been introduced in [30, 31]. Reference tracking has been added to this formulation later in [29] while [33] has been dealing with state-space interpretation based closed-loop RHC control.

The different directions of research introduced previously have been evolving independently at first. Later the more general state-space representation of RHC allowed the investigation of the connections between the different predictive control approaches [38, 54]. We may state that receding horizon control is in fact the most general interpretation containing GPC or MPC as its special cases. In fact, the one-shot solution of GPC and recursive form of RHC is identical in the absence of constraints. Moreover, there is an analogy between the state observers used in RHC and the optimal predictors of GPC. The state-space representation of the predictive control problem allows the use of MIMO models and more intricate tools in ensuring stability. This book will assume the use of state-space models for representing vibration dynamics as well. Although the state-space representation allows more complex formulations and is now generally accepted in both theoretical and practical works, the simpler FIR- and FSR-based methods remain popular in the industry.

The difference among MPC, GPC and RHC research directions have been decreasing with time and predictive control has arrived at a merging point. Currently, the term *predictive control* or *model predictive control* is used in a general sense, and denotes the same concept. An overview of the connections between the different model predictive control interpretations and a unifying view is discussed for example in [4, 32, 51]. A review and discussion of the current predictive control methods and issues is presented for example in [34, 48].

6.1.2 Nonlinear Model Predictive Control

The plant models assumed in this work are linear or are linearized. Although in many cases this is only an approximation of real dynamics, control engineering practice has demonstrated that plants with complex dynamics can be often controlled using simplified linearized models. A version of MPC using nonlinear plant models is referred to as nonlinear MPC or NMPC. In linear MPC, the optimization task is convex and can be carried out relatively easily. NMPC however creates a non-convex optimization problem, which not only makes the online optimization task considerably difficult but also raises many questions associated with stability.

In practice the inherent mathematical properties of NMPC are exploited to speed up the online solution process, such as the fact that if NMPC problems are solved in sequence, they tend to be fairly similar to each other. This book does not deal with nonlinear models or the application of NMPC to vibration attenuation.

6.2 Prediction

Let us consider a linear system described by state-space equation:

$$\begin{aligned} \mathbf{x}_{k+1} &= \mathbf{A}\mathbf{x}_k + \mathbf{B}\mathbf{u}_k \\ \mathbf{y}_k &= \mathbf{C}\mathbf{x}_k + \mathbf{D}\mathbf{u}_k \end{aligned} \tag{6.1}$$

where $\mathbf{A}$ is the state matrix, $\mathbf{B}$ is the input matrix and $\mathbf{C}$ is the output matrix of dimensions $\mathbf{A} \in \mathbb{R}^{n_x \times n_x}$, $\mathbf{B} \in \mathbb{R}^{n_y \times n_x}$ and $\mathbf{C} \in \mathbb{R}^{n_y \times n_x}$. Since $\mathbf{D}$ represents the direct input–output feedthrough, it is omitted in most models based on real-life systems.

Vectors and matrices are marked with a bold upright font in this book, for example the state vector is denoted as $\mathbf{x}_k$, in the case of a general multi-input system inputs are $\mathbf{u}_k$ and the outputs with $\mathbf{y}_k$. To simplify our notation in the upcoming sections and chapters we will now replace these with italic fonts as in x_k, u_k and y_k. We will reserve the bold upright notation as in $\mathbf{x}_k$, $\mathbf{u}_k$ and $\mathbf{y}_k$ rather for *future predicted or computed sequences* of the quantities expressed by x_k, u_k and y_k.

In predictive control the state and successively the output of the system is predicted several time steps ahead of the current time. The idea is simply iterating the state-space model several times in succession, while always utilizing the new state update to get the next step. A discrete state-space system is in fact a state predictor, one step into the future. If our discrete time is marked by (k) and our state at that time is given by x_k we may iterate a one-step ahead prediction of the state using a discrete state-space model:

$$x_{k+1} = \mathbf{A}x_k + \mathbf{B}u_k \tag{6.2}$$

While the system output at the current step is defined by the second equation in (6.1), we can predict the output at the next step as well simply by substituting the actual predicted state x_{k+1}:

$$y_{k+1} = \mathbf{C}x_{k+1} \tag{6.3}$$

Let u suppose the current time is marked by the discrete time step (k). At time (k) we have the state x_k based on real readings from the available sensors.[2] We may calculate the predicted state at step $(k+1)$, that is x_{k+1}. If we take our predicted state x_{k+1} and perform the previous step once more, we get a prediction at time $(k+2)$, that is x_{k+2}. Therefore, the state is substituted into the basic state-space equation recursively. The process may be repeated arbitrary times: if we repeat it n_p times we have a n_p steps long *prediction horizon*:

$$\begin{array}{ll}
k & x_k = x_k \\
k+1 & x_{k+1} = \mathbf{A}x_k + \mathbf{B}u_k \\
k+2 & x_{k+2} = \mathbf{A}x_{k+1} + \mathbf{B}u_{k+1} = \mathbf{A}^2 x_k + \mathbf{AB}u_k + \mathbf{B}u_{k+1} \\
k+3 & x_{k+3} = \mathbf{A}x_{k+2} + \mathbf{B}u_{k+2} = \mathbf{A}^3 x_k + \mathbf{A}^2\mathbf{B}u_k + \mathbf{AB}u_{k+1} + \mathbf{B}u_{k+2} \\
\vdots & \vdots \\
k+n_p & x_{k+n_p} = \mathbf{A}x_{k+n_p-1} + \mathbf{B}u_{k+n_p-1} \\
 & \quad = \mathbf{A}^{n_p} x_k + \mathbf{A}^{n_p-1}\mathbf{B}u_k + \cdots + \mathbf{AB}u_{k+n_p} + \mathbf{B}u_{k+n_p-1}
\end{array} \tag{6.4}$$

An autonomous system does not assume an input to the system. It is possible to predict the dynamic behavior of a freely vibrating system subject to an initial disturbance by simply ignoring the terms featuring input u_i:

$$k+i \quad \longrightarrow x_{k+i} = \mathbf{A}^i x_k \tag{6.5}$$

with $i = 0, 1, 2, \ldots, n_p$. Figure 6.4 demonstrates that given a good model it is possible to predict the behavior of a vibrating system quite precisely. In this example, the state at time step 23 has been observed from the experimentally measured output—the deflection data of a vibrating cantilever beam. The state and successively the output have been iterated six steps forward, through using a state-space model as a basis for the predictions. Continuing this process tens or hundreds of steps onwards, the error would be likely to build up because of modeling errors and the recursive nature of the process.

A predictive controller iterates the state-space model several steps ahead in time to see how the system will behave in the future and adjusts the inputs u_k at the current time accordingly. Naturally, this must be done at each sampling instant, since the output measurements and the estimated states are always updated in real-time.

To automate the process of recursive iteration into the future the *prediction matrices* must be defined and constructed. Let us now denote the sequence of future predicted states at time (k) as an n_p elements long a row vector $\mathbf{x}_k$, the sequence of

[2] Or observed through the readings and reconstructive algorithms like the Kalman filter.

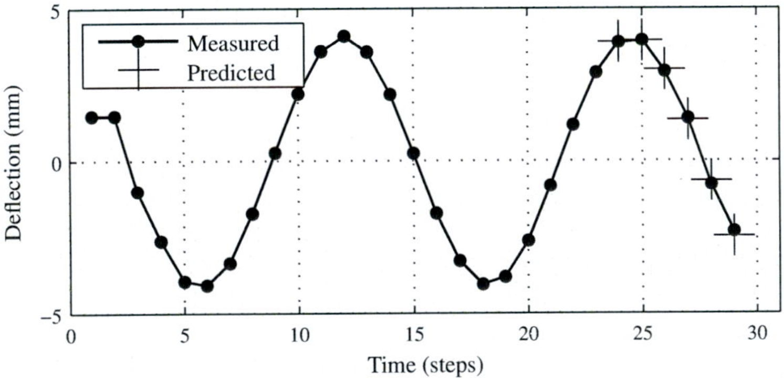

Fig. 6.4 Six (several) steps ahead model prediction of the state and original measured output

planned inputs as $\mathbf{u}_k$ and the sequence of predicted outputs as $\mathbf{y}_k$. The subscript k denotes that the vector has been last actualized at time (k) and contains the predictions starting from this discrete time point. According to this, these vectors can be described as follows:

$$\mathbf{x}_k = \begin{bmatrix} x_k \\ x_{k+1} \\ x_{k+2} \\ \vdots \\ x_{k+n_p-2} \\ x_{k+n_p-1} \\ x_{k+n_p} \end{bmatrix} \qquad \mathbf{y}_k = \begin{bmatrix} y_k \\ y_{k+1} \\ y_{k+2} \\ \vdots \\ y_{k+n_p-2} \\ y_{k+n_p-1} \\ y_{k+n_p} \end{bmatrix} \qquad \mathbf{u}_k = \begin{bmatrix} u_k \\ u_{k+1} \\ u_{k+2} \\ \vdots \\ u_{k+n_p-3} \\ u_{k+n_p-2} \\ u_{k+n_p-1} \end{bmatrix} \tag{6.6}$$

Note the terms in Eq. (6.4) which are multiplied by x_i. As we proceed from time (k) to the end of the prediction horizon, that is as $k = k, k+1, \ldots, k+n_p-1, k+n_p$ the term in front of x_k is simply A^i. Therefore, for an autonomous system the vector of predicted states can be calculated by:

$$\mathbf{x}_k = \mathbf{M}x_k = \begin{bmatrix} \mathbf{A}^0 \\ \mathbf{A}^1 \\ \mathbf{A}^2 \\ \vdots \\ \mathbf{A}^{n_p-2} \\ \mathbf{A}^{n_p-1} \\ \mathbf{A}^{n_p} \end{bmatrix} x_k \tag{6.7}$$

The prediction matrix[3] $\mathbf{M}$ for this autonomous system is also valid for systems with input (the general case), although here we must take care of the terms which are

3 Not to be confused with the identical notation of the mass matrix in Chap. 2.

multiplied by the input u_i. Again, if we look at Eq. (6.4) carefully, we may rearrange its terms so we can get the prediction matrices for a general case:

$$\mathbf{x}_k = \mathbf{M}x_k + \mathbf{N}\mathbf{u}_k \tag{6.8}$$

where $\mathbf{N}$, the second prediction matrix may be intuitively interpreted as the *impulse response matrix* which is an example of a *convolution matrix*. Note that instead of a direct input $\mathbf{u}$ most industrial controllers use a difference in input $\Delta\mathbf{u}$ as a degree of freedom. In that case, the prediction matrix can be interpreted as the *step response matrix*. The impulse response matrix $\mathbf{N}$ is calculated according to:

$$\mathbf{N} = \begin{bmatrix} \mathbf{0} & \mathbf{0} & \mathbf{0} & \dots & \mathbf{0} & \mathbf{0} & \mathbf{0} \\ \mathbf{B} & \mathbf{0} & \mathbf{0} & \dots & \mathbf{0} & \mathbf{0} & \mathbf{0} \\ \mathbf{AB} & \mathbf{B} & \mathbf{0} & \dots & \mathbf{0} & \mathbf{0} & \mathbf{0} \\ \mathbf{A}^2\mathbf{B} & \mathbf{AB} & \mathbf{B} & \dots & \mathbf{0} & \mathbf{0} & \mathbf{0} \\ \vdots & \vdots & \vdots & \ddots & \vdots & \vdots & \vdots \\ \mathbf{A}^{n_p-3}\mathbf{B} & \mathbf{A}^{n_p-4}\mathbf{B} & \mathbf{A}^{n_p-5}\mathbf{B} & \dots & \mathbf{B} & \mathbf{0} & \mathbf{0} \\ \mathbf{A}^{n_p-2}\mathbf{B} & \mathbf{A}^{n_p-3}\mathbf{B} & \mathbf{A}^{n_p-4}\mathbf{B} & \dots & \mathbf{AB} & \mathbf{B} & \mathbf{0} \\ \mathbf{A}^{n_p-1}\mathbf{B} & \mathbf{A}^{n_p-2}\mathbf{B} & \mathbf{A}^{n_p-3}\mathbf{B} & \dots & \mathbf{A}^2\mathbf{B} & \mathbf{AB} & \mathbf{B} \end{bmatrix} \tag{6.9}$$

Note that the first block row of matrix $\mathbf{N}$ is zero in order to get x_k as the first element of $\mathbf{x}_k$. For a time-invariant state-space model, we only have to construct the prediction matrices $\mathbf{M}$ and $\mathbf{N}$ once. For an adaptive system, these matrices have to be constructed online. To get the vector of predicted states, we simply substitute for the prediction matrices and obtain $\mathbf{x}_k$.

Let us denote the i-th block row[4] of matrix $\mathbf{M}$ with $\mathbf{M}_i$ according to:

$$\mathbf{M}_i = \mathbf{A}^i \tag{6.10}$$

with $\mathbf{M}_0 = \mathbf{I}$. Similarly, one may consider the i-th block row section of matrix $\mathbf{N}$ as defined by (6.9) and denote it with $\mathbf{N}_i$

$$\mathbf{N}_i = \left[\mathbf{A}^i\mathbf{B}\ \mathbf{A}^{i-1}\mathbf{B}\ \mathbf{A}^{i-2}\mathbf{B}\ \dots\ \mathbf{A}^2\mathbf{B}\ \mathbf{AB}\ \mathbf{B}\right] \tag{6.11}$$

$$\mathbf{N}_0 = \left[\mathbf{0}\ \mathbf{0}\ \mathbf{0}\ \dots\ \mathbf{0}\right] \tag{6.12}$$

and use it to get the predicted state at any time $(k+i)$, where $i = 0, 1, 2, \dots, n_p$ by using the following expression:

$$x_{k+i} = \mathbf{M}_i x_k + \mathbf{N}_i\mathbf{u}_k = \mathbf{A}^i x_k + \mathbf{N}_i\mathbf{u}_k \tag{6.13}$$

By computing the sequence of predicted states, we may also estimate the future system output by multiplying with $\mathbf{C}^{i+1}$ according to:

---

[4] Depending on the size of $\mathbf{B}$. Note that here in the interest of preserving notation customs, the index i of matrix block rows starts from zero. Thus the zeroth block row of $\mathbf{M}$ and $\mathbf{N}$ will generate the current state x_k without a change.

$$y_{k+i} = \mathbf{C}^{i+1} x_{k+i} = \mathbf{C}^{i+1} \mathbf{A}^i x_k + \mathbf{C}^{i+1} \mathbf{N}_i \mathbf{u}_k \tag{6.14}$$

Of course, it is not possible to calculate the vector of future states or outputs, unless the sequence of inputs $\mathbf{u_k}$ is known beforehand. Fortunately, this problem can be reversed, and instead of asking what the sequence of outputs or states will be, we can ask what sequence of inputs is necessary to achieve the desired sequence of outputs or states.

6.3 Cost Functions

A cost function is an important part of a predictive controller because it is an indicator of the degree of optimality of a dynamic response, resulting a sequence of control inputs $\mathbf{u}_k$ applied to the system. This degree of optimality may express how close we are to the desired output or state levels including how much effort is needed to get there, and in MPC this is referred to as the *cost*. In the controller itself the role of the cost function is reversed, and we are aiming to calculate the best series of control inputs $\mathbf{u}_k$ which results in a minimal cost.

In control engineering we want to keep output as close to the reference as possible. Reference is often located at zero; this is a common case in vibration attenuation, as we would like to keep the vibrating structure at equilibrium. We may designate the difference between the desired level with an error, which can be expressed at any moment by a numerical indicator. This numerical indicator called the *cost* does not necessarily have a physical meaning, and the mathematical way to calculate the cost is to set up a *cost function*.

Let us calculate a simple scalar indicator J, a cost describing how good our control will be in the future: from the next step up to the horizon defined by the prediction horizon n_p. This indicator only depends on the current measured or observed state x_k and the input sequence $\mathbf{u}_k$ we will implement at the next step and the time up to the end of the horizon:

$$J = f(x_k, \mathbf{u}_k) \tag{6.15}$$

Theoretically, this function can be arbitrary: for example it may contain numerical indicators expressing how close is the desired value at any given moment to the reference or how much input u_k is needed to keep it there, etc. Although a cost function can be arbitrary, it is better to prefer certain constructions for the cost function, as proper formulations may aid the evaluation and optimization procedures later. The most common form for the cost function in MPC is a *quadratic* cost function.

Having a cost, we can use it to calculate the sequence of future control inputs $\mathbf{u}_k$ by formulating the following problem:

Given a current state x_k what is the sequence of future control inputs, $\mathbf{u}_k$ which will keep the cost J at its minimum?

In other words, we have to minimize the cost function with the argument $\mathbf{u}_k$ to get an optimal sequence of inputs denoted with $\mathbf{u}_k^*$:

$$\mathbf{u}_k^* = arg \min_{\mathbf{u}} J(x_k, \mathbf{u}_k) \tag{6.16}$$

The above statement defines an optimal control problem.

6.3.1 Building a Quadratic Cost Function

As it has been previously stated, the aim of a control system is to keep the outputs y_k as close to the reference as possible. The difference between the reference value r_k and output y_k is called the control error and it is simply defined by:

$$e_k = r - y_k \tag{6.17}$$

where e_k is the error vector.[5] Let us consider a zero reference since defining a nonzero r is just a matter of shifting it to the desired tracking level or using a controller that produces input increments Δu_k. A zero reference is $r = 0$, in other words means that any type of output is an error:

$$e_k = r - y_k = 0 - y_k = -y_k \tag{6.18}$$

Instead of the relative value of the output it is better to consider the square of the error, this way negative values are eliminated and deviation from the equilibrium is penalized with an equal value. The square of the output is in this case a unit-less indicator of control quality j_k at the given sampling time (k). The less its value is, the better is our control:

$$j_k = y_k^T y_k \tag{6.19}$$

Note that due to a simplification in notation y_k is still a vector, therefore we are using $y_k^T y_k$ instead of y_k^2. Since the output in real systems is given by $y_k = \mathbf{C}x_k$ we may substitute that into (6.19) and get:

$$j_k = x_k^T \mathbf{C}^T \mathbf{C} x_k \tag{6.20}$$

For a state-space model this is a very good quality indicator, and expresses the *cost* of control at a given instant. In addition to removing negative outputs $-y$ from the cost the square representation has one more advantage. As the state is a column vector, taking a matrix square results in a scalar valued cost.

When formulating a cost function it is good to take into account the work performed by the actuators, or in other words the effect of control input u_k. In certain

5 Similarly as before e_k in general denotes a vector, while $\mathbf{e}_k$ would be a future sequence of errors.

situations, it is not necessary to limit actuation, nevertheless it may be necessary to preserve functionality and lengthen the lifetime of actuators. For example, if an actuator is adjusting the position of a mechanical structure continuously and aggressively, even if there is only a minimal disturbance—this may limit the lifetime of the actuators itself, the structure or possibly other system components. Aggressive controller action is another reason to include the effect of actuators into the cost function. For example, a vibration control system for aircraft or spacecraft shall not generate actuator inputs, which seriously affect the overall attitude and maneuverability of the system. Finally, the cost function may also express cost in the economic terms: as energy is needed to drive the actuators, there are financial aspects of every adjustment. A civil engineering structure requires actuators that may be expensive to drive: in a normal situation, their action should be minimal, however if the structure is subjected to an earthquake, the actuators should perform their task as well as possible. To summarize this paragraph, some of the reasons to include control input in the cost function are:

- lifetime prolongation
- design
- safety
- economic
- others...

In addition to the quadratic effect of the output defined by (6.20) we may therefore simply add the square of the control input. This will create a numerical indicator, a cost function for the time step (k):

$$j_k = x_k^T \mathbf{C}^T \mathbf{C} x_k + u_k^T u_k \tag{6.21}$$

6.4 State and Input Penalization

In a predictive controller, in addition to the prediction or control horizon there are two more important settings that can affect the overall type and performance of the controller and its resulting actions. These settings are the so-called *penalization matrices*. Let us introduce a matrix $\mathbf{R}$ that will affect the contribution of the second term into the overall cost:

$$j_k = x_k^T \mathbf{C}^T \mathbf{C} x_k + u_k^T \mathbf{R} u_k \tag{6.22}$$

where $\mathbf{R}$ is the so-called *input penalization matrix*. The value of $\mathbf{R}$ always depends on the application, and it is set by the control engineer. If it is not important to consider the effect of the work performed by the actuators, it is possible to use $\mathbf{R} = 0$ or a very low level and the second term will be practically eliminated from the cost function. On the other hand, the contribution of the second term can be fine tuned by raising the level of $\mathbf{R}$ to higher numbers.

With a multiple input system—that is having more actuators—by setting some elements of the matrix **R** to a higher value, we indicate the need to lower the input to certain actuators. Their effect will be *penalized*, as every action will be represented by a higher cost contribution. On the other hand, if one uses a very low number, the actuator will not be represented significantly in the overall cost and its actions are not penalized. **R** is therefore the input penalization matrix, a tuning parameter adjustable by the user.

The first part of the expression in (6.21) contains the state vector and its transpose, which encloses $\mathbf{C}^T\mathbf{C}$. In predictive control, we may replace this by the so-called *state penalization matrix* denoted by **Q**.

$$j_k = x_k^T \mathbf{Q} x_k + u_k^T \mathbf{R} u_k \tag{6.23}$$

The structure of this matrix depends on the given system and on the particular choice of the control engineer. If one chooses $\mathbf{Q} = \mathbf{C}^T\mathbf{C}$ then essentially the states are recalculated into outputs y_k. An arbitrary **Q** matrix may be chosen as well. This allows the control engineer to include or exclude effects of given states. For example consider the vibration of a system which is modeled by a second order state-space model, in which the first state describes the position of the structure while the second its velocity. If the state penalty matrix is constructed in the following way

$$\mathbf{Q} = \begin{bmatrix} 1 & 0 \\ 0 & 0 \end{bmatrix} \tag{6.24}$$

then one is penalizing the position, and the velocity does not play a role in the final cost. On the other hand, if one utilizes the penalization matrix

$$\mathbf{Q} = \begin{bmatrix} 1 & 0 \\ 0 & 10 \end{bmatrix} \tag{6.25}$$

the velocity is an order of magnitude more important in the final cost than position. The possible number of combinations to tune penalization matrices is endless and it is always up to the given application and the control engineer to choose a suitable **Q** and **R** matrix.

Based on the formerly introduced definition of the cost function it makes natural sense to use the output in the cost function, that is to choose $\mathbf{Q} = \mathbf{C}^T\mathbf{C}$. If there is no preference as to which state has to be penalized more (e.g. which is more important) then it is also common to choose **Q** to be equal with the identity matrix

$$\mathbf{Q} = \mathbf{I} = \begin{bmatrix} 1 & 0 \\ 0 & 1 \end{bmatrix} \tag{6.26}$$

for a second order system. The identity matrix **I** is a matrix of ones on the diagonal and zeros elsewhere. It is also a square matrix, and its size equals to the model order n_x.

In case of a single input system, if the input contribution of the actuators is not important in the cost, the input penalization $\mathbf{R} = r$ is chosen as a small number

$r =$ 1E-3. If the input contribution shall be included in the cost more dominantly, one may increase this value. Analogically, for a multiple input system we may use the identity matrix or its scalar multiples:

$$\mathbf{R} = \mathbf{I}r \tag{6.27}$$

where **I** is again the identity matrix, and r is a tunable scalar multiplier.

6.5 Cost of the Future States and Inputs

Instead of just considering the cost of the current step as in (6.23), a predictive controller needs to calculate the cost of future inputs, or in other words the *predicted cost*. A predicted cost at any time (k) is a sum of individual cost contributions according to (6.23) from the time (k) up to the end of the prediction horizon, that is $(k + n_p)$ [1, 20, 23, 36]:

$$J_k = \sum_{i=0}^{n_p-1} j_{k+i} = \sum_{i=0}^{n_p-1} \left(x_{k+i}^T \mathbf{Q} x_{k+i} + u_{k+i}^T \mathbf{R} u_{k+i} \right) \tag{6.28}$$

Note that while j_k is a cost just at the current time, J_k now denotes cost from (k) up to the end of the prediction (or control) horizon. This is called a finite horizon predicted cost calculated at the time (k). The cost in (6.28) expresses the cost of control inputs and its effects up to the prediction horizon.

Figure 6.5 illustrates a finite horizon MPC control law, where the effect of control inputs is only included in the predicted cost and thus the optimization problem for the length of the horizon. The effect of inputs is assumed zero beyond this horizon. The actual zero input level may or may not be ever reached, since the horizon is receding forward. However, input effects beyond the horizon are excluded from the optimization altogether.

For a finite horizon cost, there is no guarantee that the control law will achieve the optimal predicted performance. This situation may be solved by predicting the cost for an infinite horizon, that is [1, 42]:

$$J_k = \sum_{i=0}^{\infty} \left(x_{k+i}^T \mathbf{Q} x_{k+i} + u_{k+i}^T \mathbf{R} u_{k+i} \right) \tag{6.29}$$

Unfortunately, this would create an optimization problem with an infinite number of variables within $\mathbf{u}_k$. Luckily, it is possible to express the cost of the control inputs and their effects beyond the prediction horizon in such a way that the number of optimization parameters still remains finite.

For this, it is necessary to employ a method which approximates the cost for an infinite horizon but with a finite number of inputs: the so-called *dual-mode* control

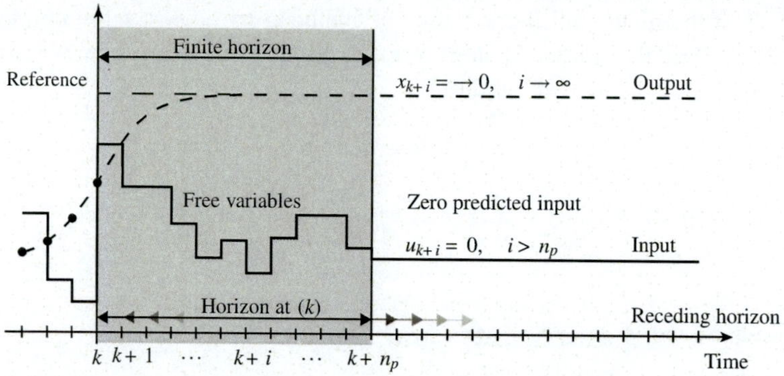

Fig. 6.5 Finite horizon model predictive control. The control law assumes free variables within the horizon, but predicts zero input afterward. The effect of inputs beyond the horizon is not included in the predicted cost

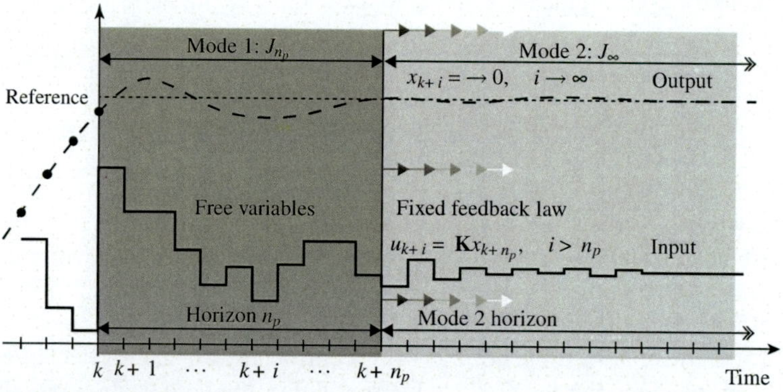

Fig. 6.6 Model predictive control demonstrating the receding horizon control idea. New state measurements (observations) are used to compute a new optimal sequence of inputs shifting the horizon forward at all times

paradigm. In the dual-mode paradigm, the predictive controller calculates the control explicitly up to a fixed horizon. The cost is calculated up to the very last step of the horizon, which is up to $(k+n_p-1)$. Instead of just considering the cost of the last state x_{k+n_p} as usual, this state is used to compute a so-called *terminal cost*. This terminal cost can be made equivalent to the cost J_∞ ranging from $(k+n_p-1)$ up to infinite time $k=\infty$. To express this cost, one has to use a special penalty matrix called the *terminal weighting matrix* denoted as $\mathbf{P}_f$ and express the new cost according to [34, 46, 48, 49]:

$$J_k = \sum_{i=0}^{n_p-1} \left(x_{k+i}^T \mathbf{Q} x_{k+i} + u_{k+i}^T \mathbf{R} u_{k+i} \right) + x_{k+n_p}^T \mathbf{P}_f x_{k+n_p} \tag{6.30}$$

In *dual-mode* control, not only will the cost be divided into two modes, but the control inputs as well. The first mode will contain free optimization variables, while the second mode (associated with the last state) will be steered into equilibrium by a stabilizing fixed feedback law, usually a linear quadratic control law. The two modes from the view of the inputs u_i are:

$$u_i = \begin{cases} \text{free variables} & i = k, k+1, \ldots, n_p - 1 \quad \text{Mode 1} \\ \mathbf{K}x_{k+i} & i = n_p, n_p + 1, \ldots, \infty \quad \text{Mode 2} \end{cases} \tag{6.31}$$

Since the cost in (6.30) needs to be evaluated explicitly only up to the end of the horizon, a proper choice for the terminal weighting matrix $\mathbf{P}_f$ is necessary. If we assume that the control moves in mode 2 will be computed by a fixed feedback law $u_{k+n_p+i} = \mathbf{K}x_{k+n_p}$, then the terminal weighting matrix $\mathbf{P}_f$ is the solution of the following Lyapunov equation [25, 39, 41, 48]:

$$\mathbf{P}_f - (\mathbf{A} + \mathbf{B}\mathbf{K})^T \mathbf{P}_f (\mathbf{A} + \mathbf{B}\mathbf{K}) = \mathbf{Q} + \mathbf{K}^T \mathbf{R}\mathbf{K} \tag{6.32}$$

where given the LTI system and the calculated fixed stabilizing feedback law $\mathbf{K}$, the terminal weighting matrix can be easily calculated.

The cost at a given time (k) and onwards up to the infinity is given by (6.30). Using the notation established earlier in (6.6), we can use a vector of a series of predicted inputs $\mathbf{u}_k$ and reformulate the infinite horizon cost to be more suitable for the optimization task. The cost function in the current sum based form is not appropriate for an MPC controller, where a compact matrix notation is preferred. If one substitutes for x_k at the current time and $\mathbf{u}_k$ for all future inputs up to the end of the horizon, obtains a transformed cost function after rearranging the terms. This transformed cost function does not use the sum operator anymore, matrix algebra is necessary to evaluate the cost of control actions $\mathbf{u}_k$ up to infinity. The transformed cost will be given by [8, 25, 39]:

$$J_k = \mathbf{u}_k^T \mathbf{H}\mathbf{u}_k + 2x_k^T \mathbf{G}^T \mathbf{u}_k + x_k^T \mathbf{F}x_k \tag{6.33}$$

where

$$\mathbf{H} = \sum_{i=0}^{n_p-1} \mathbf{N}_i^T \mathbf{Q}\mathbf{N}_i + \mathbf{N}_{n_p}^T \mathbf{P}_f \mathbf{N}_{n_p} + \mathscr{R} \tag{6.34}$$

$$\mathbf{G} = \sum_{i=0}^{n_p-1} \mathbf{N}_i^T \mathbf{Q}\mathbf{M}_i + \mathbf{N}_{n_p}^T \mathbf{P}_f \mathbf{M}_{n_p} \tag{6.35}$$

$$\mathbf{F} = \sum_{i=0}^{n_p-1} \mathbf{M}_i^T \mathbf{Q}\mathbf{M}_i + \mathbf{M}_{n_p}^T \mathbf{P}_f \mathbf{M}_{n_p} \tag{6.36}$$

where i denotes the i-th block row, respectively n_p denotes the last block row of $\mathbf{N}$ and $\mathbf{M}$. Matrix $\mathscr{R}$ is a block matrix having the input penalty $\mathbf{R}$ on its main diagonal, that is:

$$\mathscr{R} = \begin{bmatrix} \mathbf{R} & 0 & \dots & 0 \\ 0 & \mathbf{R} & \dots & 0 \\ \vdots & \vdots & \ddots & \vdots \\ 0 & 0 & \dots & \mathbf{R} \end{bmatrix} \tag{6.37}$$

Given a linear time-invariant model matrices **H**, **G** and **F** can be computed offline.

If one inspects the cost function in (6.33) closely, finds a vector of future predicted control inputs $\mathbf{u}_k$ in the first two terms. In the MPC controller this is the unknown or free variable, in other words the aim of the controller is to find the optimal $\mathbf{u}_k$ which minimizes the cost function (6.33). There is one more variable in the cost function x_k but that is dependent on the actual measured or observed state, and is updated accordingly at every (k). If everything is constant in (6.33) except the input vector, then the cost function is in fact a matrix analogy of the scalar quadratic function:

$$j = au^2 + bu + c \tag{6.38}$$

The cost function is therefore quadratic. From the pure mathematical optimization point of view, such a quadratic function has beneficial properties making its evaluation easier. There is one more fact to note about the formulation in (6.33). The last part of the expression containing $x_k^T \mathbf{F} x_k$ does not depend on the inputs $\mathbf{u}_k$. The last part contributes to the final cost a steady amount regardless of the sequence of planned inputs. Since in MPC we are interested in minimizing the cost, we may just simply leave out this static part since it does not carry any useful information:

$$J_k = \mathbf{u}_k^T \mathbf{H} \mathbf{u}_k + 2x_k^T \mathbf{G}^T \mathbf{u}_k \tag{6.39}$$

6.6 Unconstrained Model Predictive Control

The simplest possible formulation of a model predictive controller is a controller without constraints: in other words without limits on the input, output or the states. This is actually an exception in the field of MPC, as it can be expressed in a *closed form*. If one minimizes the cost function in (6.33) without a regard to constraints, a closed form expressed by a fixed matrix feedback law is obtained. This feedback law is explicit and it does not have to be recalculated at each iteration in the case of an LTI system. Unlike in the absence of constraints, it is not possible to obtain constrained MPC in a closed form. If constraints are assumed the optimization has to be performed by a numerical optimization algorithm, online and at every sampling instant (k).

The aim of an MPC controller can be defined stating that we want to find the sequence of input values $\mathbf{u}_k$ which produces a minimal cost function J_k given a measured or observed state x_k at time (k). In this way we get a sequence of optimal inputs $\mathbf{u}_k^*$ or mathematically [34, 48]:

$$\mathbf{u}_k^* = arg \min_{\mathbf{u}_k} J_k(x_k, \mathbf{u}_k) \tag{6.40}$$

$$\mathbf{u}_k = arg \min_{\mathbf{u}_k} \left\{ \sum_{i=0}^{n_p-1} \left(x_{k+i}^T \mathbf{Q} x_{k+i} + u_{k+i}^T \mathbf{R} u_{k+i} \right) + x_{k+n_p}^T \mathbf{P}_f x_{k+n_p} \right\} \tag{6.41}$$

The quadratic infinite time cost in a simplified collected way has been defined previously by (6.39). The matrices **H**, **G** can be calculated in advance and are not variable given an LTI system while the optimization task can be computed beforehand as well, resulting a closed form control law. The optimization task is to minimize:

$$\mathbf{u}_k^* = arg \min_{\mathbf{u}_k} \left(\mathbf{u}_k^T \mathbf{H} \mathbf{u}_k + 2x_k^T \mathbf{G}^T \mathbf{u}_k \right) \tag{6.42}$$

The gradient of the cost function with respect to **u** will be [6, 8]:

$$\nabla J_k = 2\mathbf{H}\mathbf{u}_k + 2\mathbf{G}x_k \tag{6.43}$$

The minimum of J_k is satisfied at ∇J_k. If **H** is nonsingular, then the optimization result is unique and is given by [6, 25]

$$\mathbf{u}_k^* = -\mathbf{H}^{-1}\mathbf{G}x_k \tag{6.44}$$

where **H** and **G** are according to (6.34) and the actual measured or observed state is x_k. The result is a sequence of optimal inputs $\mathbf{u}_k$ of which only the first element or in the case of a multiple input system the first block element is required. An actual controller uses the first element of $\mathbf{u}_k$, then the inputs are re-evaluated based on new observed states. For an LTI system this reduces to a static control law, which is the first n_u wide block row of $-\mathbf{H}^{-1}\mathbf{G}$ used in the following fashion:

$$u_k = -\mathscr{K} x_k = -[\mathbf{I}_{n_u} \mathbf{0}\ \mathbf{0} \dots \mathbf{0}]\mathbf{H}^{-1}\mathbf{G}x_k \tag{6.45}$$

where $\mathscr{K}$ is the resulting fixed unconstrained MPC law and $\mathbf{I}_{n_u}$ is a square matrix of the size equivalent to the number of inputs n_u.

The usual and obvious choice for the mode 2 fixed stabilizing feedback **K** is the LQ optimal gain. Because the predictions in the unconstrained MPC law will be optimal in both modes and the previously formulated cost is equivalent to the cost used in LQ controllers, the unconstrained MPC gain $\mathscr{K}$ will be identical to the LQ gain **K**. The future optimal sequence of the outputs $\mathbf{u}_k^*$ will be related to the LQ gain as well.

6.7 Constraint Formulation

The real power of MPC lies in computing optimal control actions for systems, which incorporate constraints. In this case, however, the feedback law cannot be computed beforehand, given a knowledge of the plant model. Instead, the optimization procedure must be performed online, in between samples (k) and $(k+1)$.

The quadratic cost determined by (6.30) and then subsequently simplified by (6.33) still holds. What we have in addition is a set of constraints, or limits on the input variables, output or possibly states. These inputs are defined as follows [1, 8, 39]:

$$\underline{u} \leq u_k \leq \overline{u} \tag{6.46}$$

$$\underline{x} \leq x_k \leq \overline{x} \tag{6.47}$$

These constraints have to be rewritten in a form suitable for a predictive controller. This form is more compact than the above notation and collects the constraints in such a way that they are expressed in terms of the argument, that is u_k. The constrained MPC control law has to be evaluated online using a *quadratic programming* (QP) algorithm and most QP solvers process optimization constraints in the following generic form [5]:

$$\mathbf{A_c} u_k \leq \mathbf{b_0} \tag{6.48}$$

where $\mathbf{A_c}$ and $\mathbf{b_0}$ define the constraints. The input constraints 6.47 can be divided into the following two equivalent constraints:

$$\begin{aligned} \underline{u} &\leq u_k \\ -\overline{u} &\leq -u_k \end{aligned} \tag{6.49}$$

If this holds for each input u_k then it is necessary to redefine it until the end of the mode 1 predictions,[6] that is for all free variables. The constraints shall cover $u_k, u_{k+1}, \ldots, u_{k+n_p-1}$. This can be written in the following equivalent matrix form:

$$\left[\begin{array}{c} \mathbf{I} \\ \hline -\mathbf{I} \end{array}\right] \mathbf{u}_k \leq \left[\begin{array}{c} \mathbf{1}\overline{u} \\ \hline -\mathbf{1}\underline{u} \end{array}\right] \tag{6.50}$$

where $\mathbf{I}$ is an identity matrix, while $\mathbf{1}$ is a vector of ones for a single input system and $\left[\mathbf{1} = \mathbf{I}\,\mathbf{I}\ldots\mathbf{I}\right]$ with n_u sized identity matrices if the system has n_u inputs.

State constraints can be similarly rewritten. State constraints from (6.47) have to be applied not for the current step x_k but similarly to the input for the future free variables $x_{k+1}, x_{k+2}, \ldots, x_{k+n_p}$ as well. We can divide the state constraints (6.47) in two parts and we get

$$\begin{aligned} \underline{x} &\leq x_k \\ -\overline{x} &\leq -x_k \end{aligned} \tag{6.51}$$

[6] In reality, enforcing the constraints only for the free variables creates stability issues. These problems and a solution to guarantee stability are discussed in the following chapter.

One must change this direct state formulation, so that the states are expressed in the terms of the inputs. From the prediction equations it is possible to calculate the next state $x_{k+1} = \mathbf{M}_i x_k + \mathbf{N}_i \mathbf{u}_k$ at each $i = 1, 2, 3, \ldots, n_p$, we can rewrite the state constraints to obtain the following simplified form [8]:

$$\begin{bmatrix} \mathbf{N}_i \\ -\mathbf{N}_i \end{bmatrix} \mathbf{u}_k \leq \begin{bmatrix} \overline{x} \\ -\underline{x} \end{bmatrix} + \begin{bmatrix} -\mathbf{M}_i \\ \mathbf{M}_i \end{bmatrix} x_k \tag{6.52}$$

substituting for each block row of $\mathbf{M}_i$ and $\mathbf{N}_i$ $i = 1, 2, \ldots, n_p$. Note that $i = 0$, or first block row of $\mathbf{M}$ and $\mathbf{N}$ is missing, since we cannot take into account the currently measured or observed state x_k at the time $(k + i)$, $i = 0$.

The input and state constraints from (6.50) and (6.52) may be combined, since the usual QP algorithm accepts constraints on the optimized variables in the following form:

$$\mathbf{A_c} u_k \leq \mathbf{b}_0 + \mathbf{B}_0 x_k \tag{6.53}$$

where the matrices $\mathbf{A_c}$, $\mathbf{b}_0$ and $\mathbf{B}_0$ are constant and can be determined offline.

6.7.1 Hard Saturation Versus Constraint Handling

One might wonder why we need constrained MPC if natural actuator boundaries can be effectively incorporated into a control law by using simple saturation limits. *Saturation* is a commonly used technique, where the real inputs computed by an arbitrary control strategy are clipped to the allowable level according to:

$$\begin{aligned} u_k &= u_k && \text{if} \quad \underline{u} < u_k < \overline{u} \\ u_k &= \overline{u} && \text{if} \quad u_k \geq \overline{u} \\ u_k &= \underline{u} && \text{if} \quad u_k \leq \underline{u} \end{aligned} \tag{6.54}$$

In case the controller computes an input in between the lower and upper bounds, the input is directly used in the closed-loop system as intended. As soon as the input exceeds the lower or upper limits, it is clipped to the allowable level by hard saturation limits. This is the essential technique used by most traditional controllers to incorporate constraints imposed by physical or technological limits of the actuators.

The use of saturation limits creates a discrepancy between the inputs computed by the controller (assumed to be ideal or even optimal), and the real ones which are simply clipped to the allowable maximal or minimal level. The closed-loop system acquires a completely different set of inputs than it was originally intended by the controller. By clipping the inputs using saturation, we also introduce a level of nonlinearity into the control law. All of this essentially creates two very serious problems with:

- optimality
- stability.

Let us use the linear quadratic controller introduced earlier in Sect. 4.3 to illustrate the problems with saturation limits. LQ is an ideal controller to compare with MPC, since a properly formulated infinite horizon unconstrained MPC law is essentially equivalent to its LQ analogy. If one designs an LQ controller without the constraints in mind, but then imposes saturation limits on the inputs, eventually ends up with an completely different control law than originally intended. This different and nonlinear control law may not work as planned and may perform much less efficiently. Moreover,

traditional closed-loop stability guarantees will not be valid anymore, as a fundamentally different nonlinear strategy is taking over the plant instead of the initial design. It is also possible to take into account the saturation limits right at the stage of control system design. If the LQ controller is penalized enough in order not to invoke constraints too often, there is a greater chance that the closed-loop system will remain stable, although stability still cannot be guaranteed beforehand. Nevertheless, this conservativeness implies a great loss of efficiency and performance.

A constrained MPC control law with stability guarantees works in a completely different manner. Inputs are not simply clipped to the level of constraints, but actively considered at the online optimization task. Thanks to the unique formulation of the MPC law, at each sampling interval inputs are computed in a manner that they are as optimal as possible while still guaranteeing closed-loop stability. As one may clearly see, not even an input saturated LQ (or other) control law shall be directly compared to constrained MPC, since the essential methodology of manipulating with input limits is entirely different. Saturated LQ is simply not equivalent to constrained MPC, imposing hard saturation limits is a separate concept to constraint handling through online optimization.

Note that the performance difference between saturated LQ (or any other saturated control law) and constrained MPC with stability guarantees is likely to manifest with increasing problem dimensionality and plant complexity. Although the performance loss of a saturated LQ law in comparison with MPC is only barely detectable in the simple SISO case introduced in the upcoming chapters, a MIMO system with several inputs and outputs is much more likely to demonstrate the advantages of constrained MPC. We are not interested in making a point that constrained MPC is better than a simple saturated law, as this point has been made by numerous authors before us [34, 48]. What is more important, industrial users have clearly taken a stance with implementing numerous MPC controllers in MIMO plants with slow dynamics [43–45], ultimately gaining a performance advantage over saturated control strategies. After all, what is advantageous for a slow plant shall also be favorable for active vibration control.

6.8 Constrained Quadratic Programming-Based MPC

The unconstrained model predictive control law can be expressed in a closed form as demonstrated previously in Sect. 6.6. For this, it is enough to perform differentiation on the matrix cost function and we obtain a fixed feedback matrix as a result, which in the case of a linear time-invariant system can be used at each sampling interval. With the addition of constraints, the problem cannot be expressed explicitly in closed form as before, and the optimization procedure has to be performed at every sampling interval repeatedly.

Once again, our aim is to minimize the quadratic cost function, only this time with constraints. The minimization of a quadratic function with constraints is known in mathematics as a *quadratic programming problem*. The general logic of the MPC algorithm will stay the same, only the means and the method of the optimization task will be changed.

Let us now assume that we would like to steer the system state of (6.1) into the origin so that we drive our vibrating system into equilibrium. Furthermore, let us assume that the problem requires the implementation of constraints as well. We may define the following MPC algorithm [8, 26]:

Algorithm 6.1 To find the solution of the constrained model predictive control problem, perform the following set of operations at each sampling instant:

- Observe or measure actual system state at sample x_k.
- Minimize the following cost function with respect to constraints:

$$\min_{\mathbf{u}_k} J(\mathbf{u}_k, x_k) = \sum_{i=0}^{n_c-1} \left(x_{k+i}^T \mathbf{Q} x_{k+i} + \mathbf{u}_{k+i}^T \mathbf{R} \mathbf{u}_{k+i} \right) + x_{k+n_c}^T \mathbf{P}_f x_{k+n_c}$$

where $\mathbf{u}_k = \left[u_k, u_{k+1}, u_{k+2}, \ldots, u_{k+n_c-1} \right]$, $\mathbf{Q} = \mathbf{Q}^T \geq 0$ is a state penalization matrix, $\mathbf{R} = \mathbf{R}^T \geq 0$ is an input penalization matrix, n_c is both the prediction and control horizon[7] and $\mathbf{P}_f$ is the solution of the unconstrained, infinite horizon quadratic regulation problem. The typical MPC cost function must be subject to the following constraints:

$$\underline{u} \leq u_{k+i} \leq \overline{u}, \qquad i = 0, 1, 2, \ldots, n_c - 1 \tag{6.55}$$

$$\underline{x} \leq x_{k+i} \leq \overline{x}, \qquad i = 1, 2, \ldots, n_c \tag{6.56}$$

$$x_{k+0} = x_k \tag{6.57}$$

$$x_{k+i+1} = \mathbf{A} x_{k+i} + \mathbf{B} u_{k+i}, \qquad i \geq 0 \tag{6.58}$$

7 In this book, we will use a prediction horizon with a length equal to the number of free optimization variables. In other words, the length of the prediction horizon will be identical to the length of the control horizon. Generally, the prediction and control horizons do not necessarily need to have equal lengths.

$$y_{k+i} = \mathbf{C}x_{k+i}, \qquad i \geq 0 \tag{6.59}$$

$$u_{k+i} = \mathbf{K}x_{k+i}, \qquad i \geq n_c \tag{6.60}$$

where **K** is a stabilizing feedback gain.

- Apply the first element of the vector of optimal control moves $\mathbf{u}_k$ to the controlled system, and re-start the procedure.

6.8.1 Quadratic Programming

The core of algorithm 6.1 defined above is the minimization of the cost function with subject to constraints. The quadratic cost function can be rewritten in a compact matrix form as introduced by (6.33), while the constraints can be collected by (6.53). This minimization task is referred to as a *quadratic programming* (QP) optimization in mathematics.

In general, a quadratic programming optimization problem minimizes a multi-variable quadratic function, which is subject to linear constraints on the variables. Let us assume **u** is in general a vector containing the optimization variables, while **H** is a symmetric matrix and **G** is a vector. A quadratic programming problem is then defined by [5, 28]:

$$\begin{aligned} \text{minimize} \quad & f(\mathbf{u}) = \tfrac{1}{2}\mathbf{u}^T\mathbf{H}\mathbf{u} + \mathbf{G}^T\mathbf{u} \\ \text{subject to} \quad & \mathbf{A_c}\mathbf{u} \leq \mathbf{b}_0 \\ & \mathbf{A_e}\mathbf{u} = \mathbf{b}_e \end{aligned} \tag{6.61}$$

where the first constraint is a so-called *inequality constraint* and the second is an *equality constraint*. If **H** is a positive semidefinite matrix, then the function $f(\mathbf{u})$ is convex and it has a global minimizer, if there exists a feasible vector **u**. Feasibility means that the variable **u** satisfies all constraints. Given a positive definite **H** and a feasible **u** the global minimizer of the QP is unique.

6.8.1.1 Geometric Representation of QP

One may interpret the definition of the quadratic programming problem in an intuitive geometrical representation as well. Instead of presenting a generic treatment of the geometrical representation of the QP problem, we will use an example to illustrate the idea of quadratic programming. In order to preserve the clarity of the presentation, let us imagine an optimization variable **u** which is defined in a two-dimensional space, having only two elements. To illustrate the various aspects of quadratic programming, let us choose a simple quadratic cost function: $f(\mathbf{u}) = f(u_1, u_2)$. This function depends on variables u_1 and u_2 for which we will try to find the minimum:

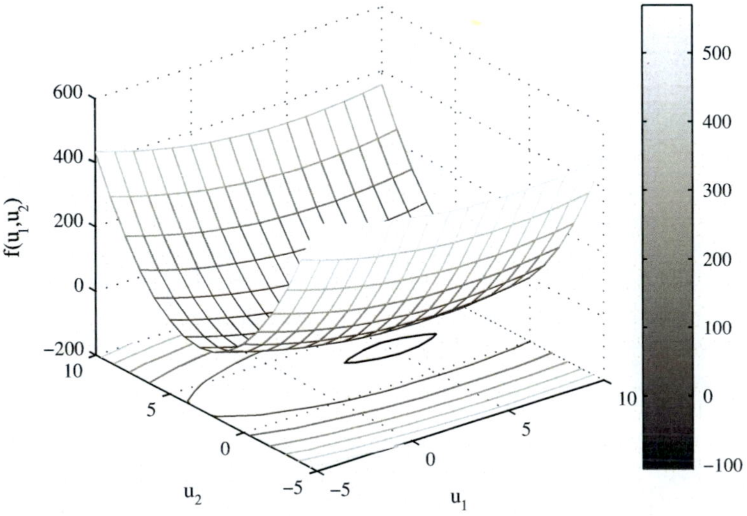

Fig. 6.7 Surface and contour plot of the cost function in quadratic programming

$$f(\mathbf{u}) = f(u_1, u_2) = u_1^2 + 9u_2^2 - 10u_1 - 54u_2 \tag{6.62}$$

This cost function has different values for different combinations of u_1 and u_2 and it can be plotted in a three-dimensional space. The surface created by this example is illustrated in Fig. 6.7 where the different contours of the function are shown as well. Without constraints, the minimum of our function is the "bottom" of the three-dimensional surface. In mathematical terms, the minimum of the unconstrained cost function is found if we find where the partial derivative equals zero with respect to all of the variables:

$$\nabla_{\mathbf{u}} f(\mathbf{u}) = 0 \tag{6.63}$$

$$\frac{\partial f(u_1, u_2)}{\partial u_1} = 0 \tag{6.64}$$

$$\frac{\partial f(u_1, u_2)}{\partial u_2} = 0 \tag{6.65}$$

Evaluating the partial differentiation with respect to u_1 yields 5, while with respect to u_2 yields 3, meaning that the unconstrained minimum of this cost function is to be found at the coordinates $\mathbf{u}^{\blacktriangle} = [5 \quad 3]^T$. In absence of constraints it is trivial to find the minimum of a quadratic function, but the addition of constraints creates a quadratic programming problem.

Let us confine the pool of valid solutions for our minimization problem in (6.62) in the $\mathbf{u}$-space by a set of linear constraints:

$$\begin{aligned} u_1 &\geq 0 \\ u_2 &\geq 0 \\ u_2 &\leq 4 - u_1 \end{aligned} \tag{6.66}$$

and let us not consider equality constraints for our example. In the two-dimensional space these constraints form lines, while in a general multidimensional space constraints form hyperplanes. The first constraint here is coincident with the horizontal axis, the second with the vertical axis and the third is simply a diagonal line. The constraints and the *feasible* space from which it is possible to select the optimization variables u_1 and u_2 are shown in Fig. 6.8. The shaded area represents the feasible subspace of $\mathbf{u}$, while the thick black lines illustrate the linear constraints given by (6.66).

To evaluate this constrained minimization problem, we will utilize the *quadprog* quadratic programming solver, which is a part of the Matlab Optimization Toolbox [52]. First, it is necessary to convert the problem formulation so that it resembles the general formulation of (6.61). We may write the cost function (6.62) in a matrix form by expressing it as:

$$\begin{aligned} f(\mathbf{u}) = f(u_1, u_2) &= \frac{1}{2}\begin{bmatrix} u_1 & u_2 \end{bmatrix}\begin{bmatrix} 2 & 0 \\ 0 & 18 \end{bmatrix}\begin{bmatrix} u_1 \\ u_2 \end{bmatrix} + \begin{bmatrix} -10 & -54 \end{bmatrix}\begin{bmatrix} u_1 \\ u_2 \end{bmatrix} \\ &= \frac{1}{2}\mathbf{u}^T\mathbf{H}\mathbf{u} + \mathbf{G}^T\mathbf{u} \end{aligned} \tag{6.67}$$

The constraints may also be rearranged to the form given by (6.61) by changing them to:

$$\begin{aligned} -u_1 &\leq 0 \\ -u_2 &\leq 0 \\ u_1 + u_2 &\leq 4 \end{aligned} \tag{6.68}$$

which in a matrix form will be rendered to

$$\begin{gathered} \begin{bmatrix} -1 & 0 \\ 0 & -1 \\ 1 & 1 \end{bmatrix}\begin{bmatrix} u_1 \\ u_2 \end{bmatrix} \leq \begin{bmatrix} 0 \\ 0 \\ 4 \end{bmatrix} \\ \mathbf{A_c}\mathbf{u} \leq \mathbf{b}_0 \end{gathered} \tag{6.69}$$

Passing the matrices and vectors defined by $\mathbf{H}$, $\mathbf{G}$, $\mathbf{A_c}$ and $\mathbf{b}_0$ to the *quadprog* solver in Matlab using the syntax

[u_ast, f] = `quadprog`(H, G, Ac, b0)

will yield in the optimal solution $\mathbf{u}^* = \begin{bmatrix} 1.4 & 2.6 \end{bmatrix}^T$ in merely two iterations. Figure 6.9 illustrates this process in the space of the variables $\mathbf{u}$. The optimal solution for the unconstrained case $\mathbf{u}^\blacktriangle$ is visible as the center point of the contour lines of the

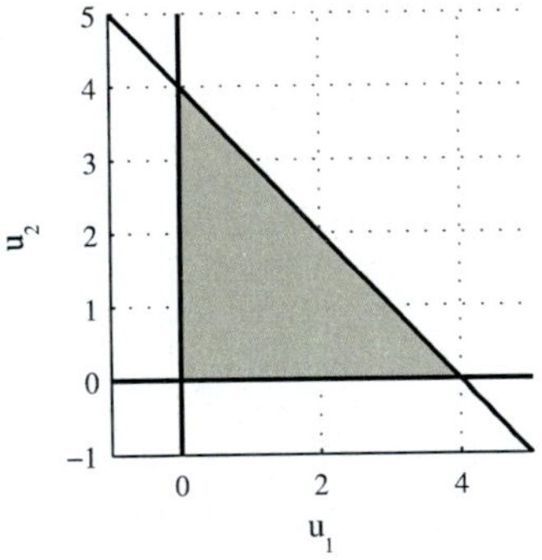

(a) Constraints in **u**-space

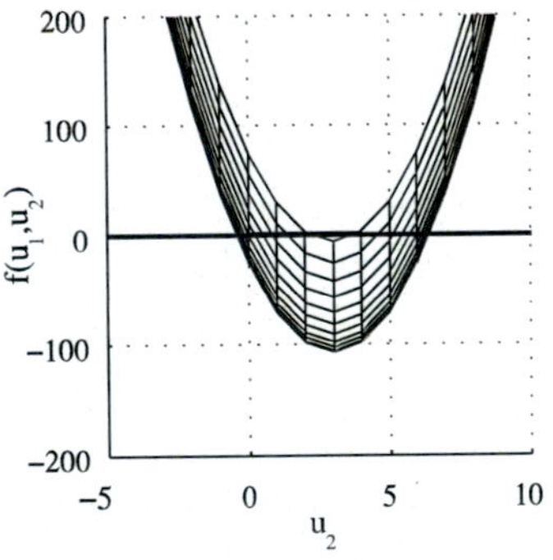

(b) Constraints in $u_2, f(\mathbf{u})$-space

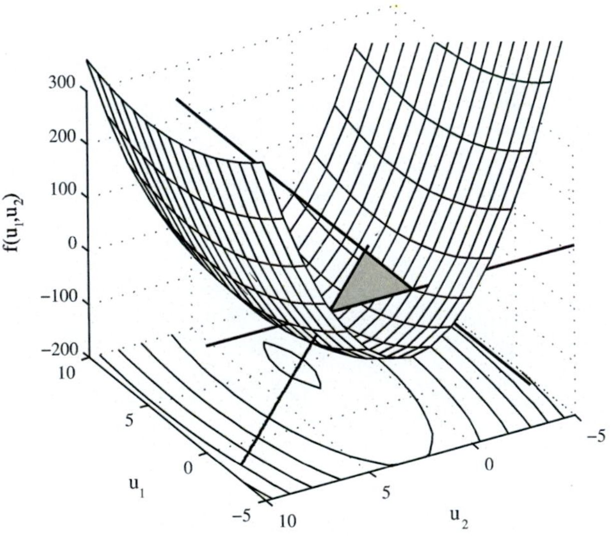

(c) Constraints in $\mathbf{u}, f(\mathbf{u})$-space

Fig. 6.8 The *shaded triangular area* illustrates the restricted space in which the choice of variables **u** are limited by the linear constraints represented by the *black lines*. The constraints are presented in (**a**) in the **u** space, a view of the quadratic cost surface in the space of u_2 is depicted in (**b**), while the constraints are shown with the cost function $f(\mathbf{u})$ in (**c**) in three dimensions

cost function $f(\mathbf{u})$. Because the problem is constrained, this solution is not feasible anymore and the algorithm has to choose from the inside of (including the boundaries) the triangular space enclosed by the constraints. The optimal solution $\mathbf{u}^*$ lies on the third constraint and the contour-line encompassing it is also illustrated. Different methods and algorithms may be used to solve a quadratic programming problem, including [5, 8, 34]:

- active set

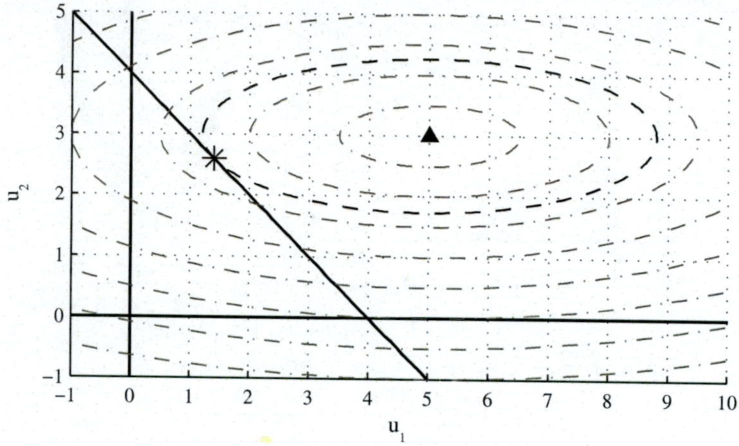

Fig. 6.9 Quadratic programming example showing the contour lines of the cost function $f(\mathbf{u})$ (*dash-dot*), the constraints (*continuous*), the unconstrained solution $\mathbf{u}^\blacktriangle$ (*triangle*) and the constrained QP solution $\mathbf{u}^*$ (*asterisk*) and the contour line containing the solution (*dashed*)

- interior point
- conjugate gradient
- gradient projection
- variations of the simplex algorithm
- others.

6.8.1.2 Active Set Quadratic Programming Method

One of the best-known quadratic programming algorithms is referred to as the *active set* (AS) algorithm [20]. In essence, the active set algorithm finds the optimal solution of the constrained QP optimization by evaluating problems involving only equality constraints. The active set QP algorithm is used with medium-scale optimization problems, which are typical for model predictive control. The big advantage of the active set algorithm is that it can be *warm started*. Warm starting means that, due to the nature of the MPC problem, it is possible to find the solution much faster at time step $(k+1)$ if the algorithm can use knowledge gained from the solution evaluated at previous time (k).

Let us have a look at the general inequality constraint formulation in (6.61) again. Each row of the matrix inequality formulation is an individual linear constraint, which we can denote with a simple sequential number i. Therefore, we may separate the individual constraints if we denote the rows of $\mathbf{A}_c$ with A_c^i and the rows of $\mathbf{b}_0$ with b_0^i to get $A_c^i\mathbf{u} \leq b_0^i$ for each constraint. A certain constraint is said to be *inactive* if the term $A_c^i\mathbf{u}^* < b_0^i$ holds. However, in case the term $A_c^i\mathbf{u}^* = b_0^i$ holds, then the

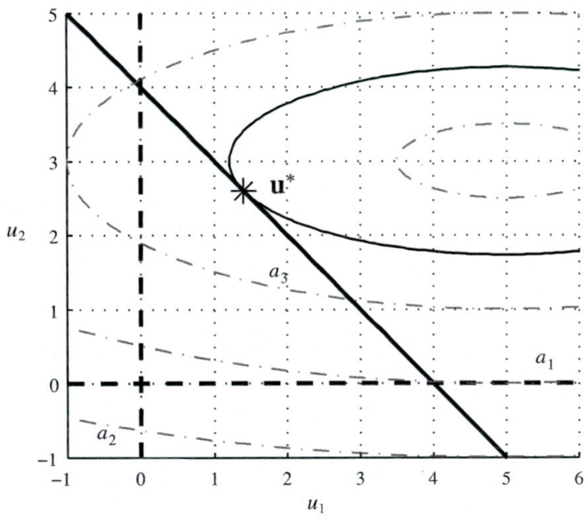

Fig. 6.10 Illustration of the active set in the active set quadratic programming method. Two constraints are inactive, while one is active. The inactive constraints can be removed from the problem, leaving only a constrained quadratic minimization problem to be solved

i-th constraint is said to be **active**. It is also possible for the active set to be empty if none of the inequality constraints are in effect at the solution [34].

If we number the constraints (6.69) in our previous example with the sequential number i, we will get three constraints, which can be either active or inactive. Figure 6.10 illustrates the solution for our example, from which it is clear that two constraints are inactive (1, 2) while number 3 is active for $\mathbf{u}^*$. It is possible to remove the inactive constraints without affecting the solution [8]. The optimal solution of the problem is the solution of an equality constrained problem where only the active equality constraints are considered while the rest is discarded. We can denote this by:

$$\begin{aligned} &\text{minimize} \quad f(\mathbf{u}) = \tfrac{1}{2}\mathbf{u}^T\mathbf{H}\mathbf{u} + \mathbf{G}^T\mathbf{u} \\ &\text{subject to} \quad A_c^i\mathbf{u} = b_0^i, \quad i \in \mathbf{a}^* \end{aligned} \tag{6.70}$$

where $\mathbf{a}^* = \{i : A_c^i\mathbf{u}^* = b_0^i\}$ is the configuration of the active constraints at the solution of the quadratic programming problem [8].

Naturally, one does not know in advance the ideal combination of the active and inactive constraints $\mathbf{a}^*$ which eventually leads to an optimal solution $\mathbf{u}^*$. However, it is possible to devise a strategy in which the equality problem associated with the current active set yields an improved solution at each iteration, while determining optimality and managing the active set at the same time. In a well-designed active set QP algorithm the active sets are not selected randomly, various strategies exist which help to avoid the necessity to test a large number of active/inactive constraint permutations. For example, it is possible to select successive sets in a way that the value of the cost function $f(\mathbf{u})$ will be decreased at each successive iteration.

The rough outline of an active set quadratic programming algorithm may be given by [5, 8, 19, 34]:

Algorithm 6.2 To solve a quadratic programming problem (given a feasible initial solution $\mathbf{u}_0$ and an active set $\mathbf{a}_0$), at each iteration p perform the following procedure:

- solve the equality problem corresponding to the active set $\mathbf{a}_p$ given by (6.70)
- test the optimality of the solution
  - if the solution is optimal ($\mathbf{u}_p = \mathbf{u}^*$), terminate
  - if the solution is not optimal, continue
- find which constraint i violated feasibility and add it to the active set to create a new active set $\mathbf{a}_{p+1}$. Repeat the procedure to find an improved solution $\mathbf{u}_{p+1}$.

Let us now focus on the subproblem described by the second item in Algorithm 6.2, namely determining whether the solution of the equality problem is really the global solution of the QP. The optimality of the partial solution with a certain active set can be determined from the *Lagrange multipliers* of the associated equality problem. To verify the optimality in an unconstrained optimization problem defined by:

$$\text{minimize} \quad f(\mathbf{u}) \tag{6.71}$$

we have to test whether its gradient equals zero, that is whether

$$\nabla_{\mathbf{u}} f(\mathbf{u}^*) = 0 \tag{6.72}$$

holds. In general, a constrained optimization problem with m constraints can be expressed by the following formulation:

$$\begin{aligned} &\text{minimize} && f(\mathbf{u}) \\ &\text{subject to} && g_i(\mathbf{u}) = 0, \quad i = 1, \ldots, m \end{aligned} \tag{6.73}$$

which is also true for a given active set $\mathbf{a}$ and the corresponding equality problem of (6.70), where m constraints are given by $g_i(\mathbf{u}) = A_c^i \mathbf{u} - b_0^i$. To test the optimality of the solution after evaluating the equality problem it is not sufficient to fulfill the gradient condition (6.72) anymore. Instead, this gradient condition is augmented by the effect of the constraints. For a single constraint (m = 1) the optimality condition would be transformed to:

$$\nabla_{\mathbf{u}} f(\mathbf{u}^*) = -\lambda \nabla_{\mathbf{u}} g(\mathbf{u}^*) \tag{6.74}$$

meaning that the direction of the gradient of the minimized function at $\mathbf{u}^*$ must be exactly opposite to the direction of the gradient of the constraint. The scalar λ is a Lagrange multiplier and it is necessary in the formulation because, although the directions of the gradients are opposite, their magnitude is not necessarily the same.

This concept is illustrated in Fig. 6.11 on the example discussed before. Here constraints number 1 and 2 can be removed from the active set without affecting the solution, only constraint 3 is considered. After the equality problem (6.70) is solved, we have to make sure whether the trial solution is truly the optimal solution. As we are following the contour line of the constraint $g(\mathbf{u})$ the value of the cost function

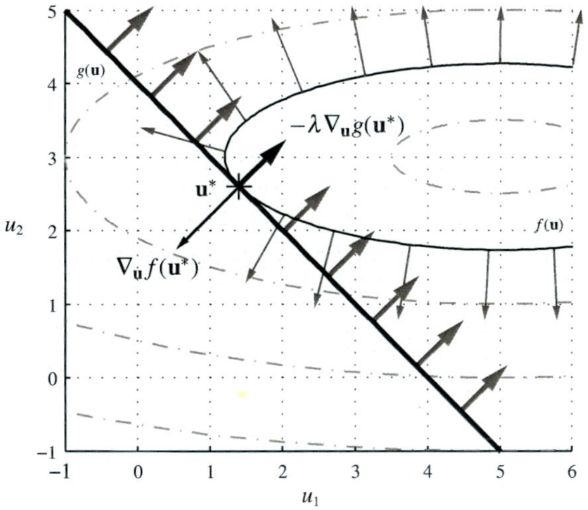

Fig. 6.11 Illustration of the gradients of the minimized function $f(\mathbf{u})$ and the active equality constraint $g(\mathbf{u}) = 0$. At the optimal solution, the direction of the gradients is opposite, their magnitude is scaled by the Lagrange multiplier λ

may vary, the contour lines of $g(\mathbf{u})$ cross the constraint at several points. The value of the cost function is neither increased, nor decreased only when the contour lines touch but do not cross. In other words, when the constraint touches the contour lines of the cost function tangentially we have found our solution $\mathbf{u}^*$. Mathematically expressed, the two function contours touch when the tangent vectors of the contours are parallel. The gradients of a function are perpendicular to the contour line, thus we may equivalently say that the gradients of functions $f(\mathbf{u})$ and $g(\mathbf{u})$ are parallel but with magnitudes different by a scalar value λ.

This constrained optimality condition can be easily converted into a multi constraint formulation as well. If for the actual active set $\mathbf{a}_p$ at iteration p the variable $\mathbf{u}_p = \mathbf{u}^*$ is really the solution of the constrained equality problem given by (6.70), then we can find Lagrange multipliers $\lambda_i^*, i = 1, 2, \ldots, m$ which will satisfy [9]:

$$\begin{aligned} \nabla_{\mathbf{u}} f(\mathbf{u}^*) + \sum_{i=1}^{m} \lambda_i^* \nabla_{\mathbf{u}} g_i(\mathbf{u}^*) = 0, \quad i = 1, 2, \ldots, m \\ g_i(\mathbf{u}^*) = 0, \quad i = 1, 2, \ldots, m \end{aligned} \tag{6.75}$$

In other words, if we can find a set of λ_i^* for the trial solution which fulfills the condition above, the trial solution is in fact the solution of the QP problem.

The optimality condition of (6.75) can be extended to a case of n additional equality constraints given by $h(\mathbf{u}) = \mathbf{A}_e\mathbf{u} - \mathbf{b}_e$ as well. The necessary and sufficient conditions for $\mathbf{u}^*$ to be the global optimum are given by the so-called *Karush–Kuhn–Tucker* (KKT) conditions [5, 19]:

$$
\begin{aligned}
\nabla_{\mathbf{u}} f(\mathbf{u}^*) + \textstyle\sum_{i=1}^{m} \lambda_i^* \nabla_{\mathbf{u}} g_i(\mathbf{u}^*) + \sum_{i=1}^{n} v_i^* \nabla_{\mathbf{u}} h_i(\mathbf{u}^*) = 0, \quad & i = 1, 2, \ldots, m \\
& i = 1, 2, \ldots, n \\
g_i(\mathbf{u}^*) = 0, \quad & i = 1, 2, \ldots, m \\
h_i(\mathbf{u}^*) = 0, \quad & i = 1, 2, \ldots, n
\end{aligned}
\tag{6.76}
$$

According to the KKT conditions if vectors of Lagrange multipliers $\lambda \geq 0$ and v exist, then $\mathbf{u}^*$ is the global optimizer. After substitution and differentiation, the KKT conditions for the quadratic cost function $f(\mathbf{u})$, inequality and equality constraints defined by (6.61) will be reduced to [34]:

$$\mathbf{Hu} + \mathbf{A}_c^T \lambda + \mathbf{A}_e^T v = -\mathbf{G} \tag{6.77}$$

$$\mathbf{A}_c \mathbf{u} = \mathbf{b}_0 \tag{6.78}$$

$$\mathbf{A}_e \mathbf{u} = \mathbf{b}_e \tag{6.79}$$

where the elements of λ corresponding to the inactive inequality constraints must remain zero (this can be ensured by an additional complementary condition). To put it differently, only elements corresponding to the active set need to be evaluated and, if all of them are nonnegative, then the solution $\mathbf{u}$ is the global optimum of the QP problem.

A valid strategy to implement the rough algorithm outlined in Algorithm 6.2 [34] can be created by modifying our original cost function in (6.70) by replacing absolute values of the solution $\mathbf{u}_p$ at iteration p with increments $\mathbf{u}_p + \Delta\mathbf{u}$. Substituting this into Algorithm 6.2 in $f(\mathbf{u})$ yields a new incremented cost:

$$f(\mathbf{u}_p + \Delta\mathbf{u}) = \frac{1}{2}(\mathbf{u}_p + \Delta\mathbf{u})^T \mathbf{H}(\mathbf{u}_p + \Delta\mathbf{u}) + \mathbf{G}^T(\mathbf{u}_p + \Delta\mathbf{u}) \tag{6.80}$$

which after factoring out and simplification can be reduced to

$$f(\mathbf{u}_p + \Delta\mathbf{u}) = f(\mathbf{u}_p) + \frac{1}{2}\Delta\mathbf{u}^T \mathbf{H} \Delta\mathbf{u} + (\mathbf{G}^T + \mathbf{u}_p^T \mathbf{H})\Delta\mathbf{u} \tag{6.81}$$

and the equality constrained minimization problem (6.70) can be equivalently stated by the new cost function $f(\Delta\mathbf{u})$:

$$
\begin{aligned}
\text{minimize} \quad & f(\Delta\mathbf{u}) = \tfrac{1}{2}\Delta\mathbf{u}^T \mathbf{H} \Delta\mathbf{u} + \mathbf{G}_p^T \Delta\mathbf{u} \\
\text{subject to} \quad & A_c^i \Delta\mathbf{u} = 0, \quad i \in \mathbf{a}^*
\end{aligned}
\tag{6.82}
$$

where $\mathbf{G}_p^T = (\mathbf{G}^T + \mathbf{u}_p^T \mathbf{H})$. Note that the inequality constraints in this new formulation have also changed, b_0^i has been removed. This can be explained by substituting the new incremental optimization variable $\mathbf{u}_p + \Delta\mathbf{u}$ into the equality constraint in (6.70) to get:

$$A_c^i \mathbf{u}_p + A_c^i \Delta\mathbf{u} = b_0^i \tag{6.83}$$

from which we only have to ensure that $g_i(\Delta\mathbf{u}) = A_c^i \Delta\mathbf{u} = 0$ in order to maintain the validity of the original equality constraint.

Let us now use this new equality constraint $g(\Delta\mathbf{u})$ with the cost function $f(\Delta\mathbf{u})$ and substitute it back to the conditions of optimality in (6.75), with using a new vector of Lagrange multipliers $\Delta\lambda$:

$$\begin{aligned} \nabla f(\Delta\mathbf{u}) + \Delta\lambda\nabla g_{\mathbf{a}}(\Delta\mathbf{u}) &= 0 \\ g_{\mathbf{a}}(\Delta\mathbf{u}) &= 0 \end{aligned} \tag{6.84}$$

where subscript $\mathbf{a}$ marks that only equality constraints from the active set $\mathbf{a}_p$ at iteration p is utilized. After substitution and differentiation with respect to $\Delta\mathbf{u}$ this will be reduced to [5, 34]:

$$\mathbf{H}\Delta\mathbf{u} + \mathbf{A}_c^{\mathbf{a}T}\Delta\lambda = -\,\mathbf{G}_p \tag{6.85}$$

$$\mathbf{A}_c^{\mathbf{a}}\Delta\mathbf{u} = 0 \tag{6.86}$$

which can be easily expressed in a matrix form:

$$\begin{bmatrix} \mathbf{H} & \mathbf{A}_c^{\mathbf{a}T} \\ \mathbf{A}_c^{\mathbf{a}} & 0 \end{bmatrix} \begin{bmatrix} \Delta\mathbf{u} \\ \Delta\lambda \end{bmatrix} = \begin{bmatrix} -\mathbf{G}_p \\ 0 \end{bmatrix} \tag{6.87}$$

The equality optimization problem expressed by (6.82) thus can be conveniently expressed with the above matrix equation. The problem may be expanded for a general quadratic programming case with both equality and inequality constraints by using the Karush-Kuhn-Tucker conditions in (6.76) to get [5, 20, 34]:

$$\mathbf{H}\Delta\mathbf{u} + \mathbf{A}_c^{\mathbf{a}T}\Delta\lambda + \mathbf{A}_e^{\mathbf{a}T}\Delta\nu = -\,\mathbf{G}_p \tag{6.88}$$

$$\mathbf{A}_c^{\mathbf{a}}\Delta\mathbf{u} = 0 \tag{6.89}$$

$$\mathbf{A}_e^{\mathbf{a}}\Delta\mathbf{u} = 0 \tag{6.90}$$

which can be expressed in a matrix form as well:

$$\begin{bmatrix} \mathbf{H} & \mathbf{A}_c^{\mathbf{a}T} & \mathbf{A}_e^{\mathbf{a}T} \\ \mathbf{A}_c^{\mathbf{a}} & 0 & 0 \\ \mathbf{A}_e^{\mathbf{a}} & 0 & 0 \end{bmatrix} \begin{bmatrix} \Delta\mathbf{u} \\ \Delta\lambda \\ \Delta\nu \end{bmatrix} = \begin{bmatrix} -\mathbf{G}_p \\ 0 \\ 0 \end{bmatrix} \tag{6.91}$$

A new active set strategy using these ideas can be outlined [34]:

Algorithm 6.3 To solve a quadratic programming problem (given a feasible initial solution $\mathbf{u}_0$ and an initial active set $\mathbf{a}_0$), at each iteration p perform the following procedure [34]:

- given $\mathbf{u}_p$ solve the modified equality problem (6.70) corresponding to the active set $\mathbf{a}_p$ by evaluating (6.87) to get a solution improvement $\mathbf{u}_p + \Delta\mathbf{u}$

- test the feasibility of this improved solution $\mathbf{u}_p + \Delta\mathbf{u}$ by evaluating $\mathbf{A}_c\mathbf{u}_p \leq \mathbf{b}_0$ and
  - if the solution is feasible, accept as an improved optimization variable at the next iteration $\mathbf{u_{p+1}} = \mathbf{u}_p + \Delta\mathbf{u}$ and maintain the current active set $\mathbf{a}_p = \mathbf{a}_{p+1}$
  - if the solution is infeasible perform a line search pointing in the direction of $\Delta\mathbf{u}$ and locate the point $\mathbf{u_{p+1}} = \mathbf{u}_p + \alpha_l\Delta\mathbf{u}$ at which the feasibility is lost (where $0 < \alpha_l < 1$). Add this new active constraint to the active set, creating $\mathbf{a}_{p+1}$
- determine the optimality of the improved solution $\mathbf{u}_{p+1}$ by checking the Lagrange multipliers according to (6.75) and then
  - if it is not the constrained optimum of the QP, restart the procedure
  - if it is the constrained optimum of the QP, terminate the procedure

Let us review the procedure of Algorithm 6.3 once more on the simple example familiar from our previous discussion and illustrate the steps in Fig. 6.12. Without going into details, suppose that we have a feasible initial solution $\mathbf{u}_0$ at our disposition and an empty active set $\mathbf{a}_0 = \begin{bmatrix} 0 & 0 & 0 \end{bmatrix}^T$. In the first step $p = 1$(a) the matrix expression (6.87) is evaluated, obtaining an improvement $\Delta\mathbf{u}_1$ in the solution. As the active set is empty, the new solution is equivalent to the unconstrained optimum of the cost function. The feasibility of this new solution $\mathbf{u}_1 = \mathbf{u}^{\blacktriangle} = \mathbf{u}_0 + \Delta\mathbf{u}_0$ must be then determined. By evaluating $\mathbf{A}_c\mathbf{u}_1 \leq \mathbf{b}_0$ the solution is clearly infeasible, thus $\mathbf{u}_1 \neq \mathbf{u}_0 + \Delta\mathbf{u}_0$. A line search is made in the direction of $\mathbf{u}_0$ to determine where feasibility has been lost. The line search coefficient α_l determines the new partial solution at iteration $p = 1$ (b) which is given by $\mathbf{u}_1 = \mathbf{u}_0 + \alpha_l\Delta\mathbf{u}_0$. A new active set is created by adding constraint number 3, resulting in $\mathbf{a}_1 = \begin{bmatrix} 0 & 0 & 1 \end{bmatrix}^T$. We have to see whether this solution is the global optimum of the quadratic programming problem, by computing the Lagrange coefficients of the problem at $\mathbf{u}_1$ according to (6.75). As it turns out this is not the global optimum of the QP, so one more iteration $p = 2$ is initiated.

By evaluating (6.87) utilizing the new solution improvement $\mathbf{u}_1$ and the next active set $\mathbf{a}_1$ we arrive at the solution $\mathbf{u}_2 = \mathbf{u}_1 + \Delta\mathbf{u}_1$. This solution is feasible, as $\mathbf{A}_c\mathbf{u}_2 \leq \mathbf{b}_0$ holds. Testing the optimality of the solution by computing the Lagrange multipliers according to (6.75) shows that this new solution is the global optimum. As no further improvement is possible $\mathbf{u}^* = \mathbf{u}_2$ and the sequence is terminated.[8]

6.8.1.3 Interior Point Quadratic Programming Method

The interior point (IP) quadratic programming algorithm [37, 40] is mostly used for large-scale problems in the range of hundreds of optimization variables. Although it is a more recent algorithm, it is not necessarily better than active set methods [34]. Unlike the active set algorithm the interior point algorithm cannot be warm started,

[8] The above example evaluated with the active set solver in *quadprog* terminates in two iterations as well.

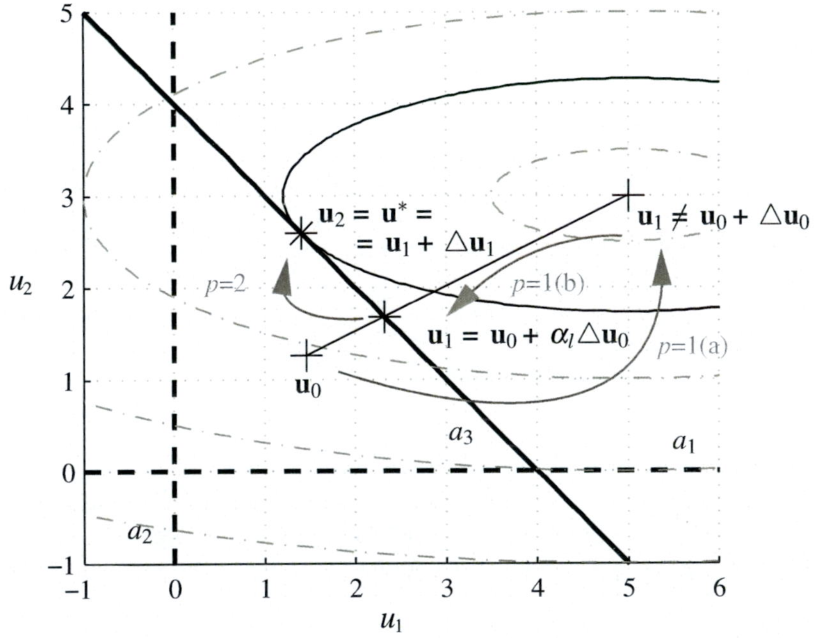

Fig. 6.12 Illustration of the iterations in an active set quadratic programming method. The solution for this simple problem is found in merely two iterations of the algorithm

in other words, we cannot reduce computational load by applying prior knowledge on the previous solution. The advantage of interior point methods is that their computational complexity is generally a polynomial function of parameters (for example $3n_x + n_u$) whereas active set and most other algorithms require a computational time which grows exponentially depending on the problem parameters (for example $n_x + {n_u}^3$) [34]. The use of interior point algorithms in model predictive control is less typical for this reason.

The idea behind the interior point algorithm is to convert the QP problem into an unconstrained optimization problem by augmenting the cost function by a so-called *barrier function*. The role of the barrier function is to supply a finite value to the minimized function, when the solution satisfies the constraints. The value of the boundary function tends to infinity, whenever solution approaches the constraint boundary. This can be expressed mathematically as [9, 34]:

$$\text{minimize} \quad \mu[f(\mathbf{u}) = \frac{1}{2}\mathbf{u}^T\mathbf{H}\mathbf{u} + \mathbf{G}^T\mathbf{u})] + \psi(\mathbf{u}) \tag{6.92}$$

where $f(\mathbf{u})$ is the quadratic cost, $\psi(\mathbf{u})$ is the boundary function based on the constraints and μ is a scalar value. In case we have constraints formed as $A_c^i\mathbf{u} = b_0^i$ a typical choice for a barrier function could be:

$$\psi(\mathbf{u}) = \sum_{i=1}^{m} -\log(b_0^i - A_c^{iT}\mathbf{u}) \tag{6.93}$$

although other formulations are also possible. As it has been previously noted, the value of $\psi(\mathbf{u})$ approaches infinity as the solution $\mathbf{u}$ approaches the constraint boundaries. The solution of the unconstrained interior point minimization problem (6.93) will satisfy constraints for any given scalar μ. In other words, by minimizing $f(\mathbf{u})$ the solution will always remain within the feasible region, because the barrier function prevents the search from leaving it [34].

As we increase μ the solution will tend to get closer and closer to the optimal solution $\mathbf{u}^*$:

$$\mathbf{u} \longrightarrow \mathbf{u}^* \quad \text{as} \quad \mu \longrightarrow \infty \tag{6.94}$$

If an initial feasible solution can be found in the vicinity of $\mathbf{u}_0$, then this solution can be continually improved by increasing μ until the difference between the partial solution $\mathbf{u}$ and the real optimal solution $\mathbf{u}^*$. An interior point algorithm therefore successively increases the value of μ until the constraints are met within a given tolerance [8]:

Algorithm 6.4

- increase the value of μ
- minimize the unconstrained problem (6.93)
- if solution meets constraints within tolerance terminate, otherwise restart the procedure

The initial point of the algorithm $\mathbf{u}_0$ is known as the *analytic center* of the constraints, while the path traced out by the successively improving partial solutions $\mathbf{u}$ is known as the *central path* [34].

While iterations of the active set algorithm search along the points on the boundary of the feasible region, as the name implies interior point algorithms search in the interior of this region. This approach requires an initial feasible solution but will remain feasible through all iterates. This strategy is not the most efficient and the iterates of modern IP algorithms are not always feasible, that is until the end of the search [34].

A central problem to active set algorithms is the possible numerical ill-conditioning caused by the logarithmic nature of the barrier function. As we are successively increasing the value of μ to get better and better solutions, the value of the barrier function starts to rise very rapidly, thus rendering the optimization problem (6.92) numerically ill-conditioned. This may cause that the algorithm results infeasible or sub-optimal solutions. A well-designed interior point algorithm is relatively complicated and amongst others contains algorithmic tools to prevent the problem described above. Currently the most effective interior point algorithms referred to as *primal-dual* are based on solving the convex optimization problems and their dual problems simultaneously [5].

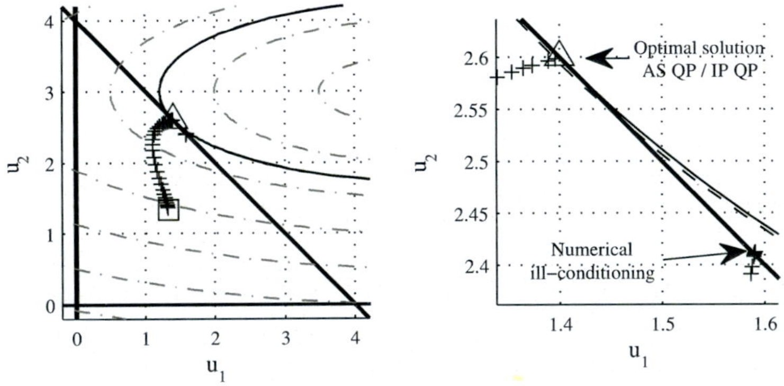

(a) Interior point QP example **(b)** Demonstration of numerical issues

Fig. 6.13 The iterations of an interior point quadratic programming algorithm are shown in (**a**) while sub optimal and infeasible solutions due to numerical ill-conditioning are demonstrated in (**b**). The figures show the analytical center (*square*), the solution of an active set algorithm (*triangle*) the iterations of the interior point QP algorithm (*cross*), constraints (*thick black line*), contours of the original cost (*dashed gray line*), contour of the active set solution (*thin black line*), contour of the interior point solution (*thin dashed black line*)

Figure 6.13 illustrates a very simple interior point algorithm solving the demonstration example familiar from our previous discussion on quadratic programming. The algorithm starts from the analytic center of the feasible region, located at $\mathbf{u}_0 = [4/3 \quad 4/3]$. The original cost function is augmented by the barrier function, and the new unconstrained cost function (6.93) is minimized. To solve our simple example in Matlab the new cost function (6.93) has been programmed as a function *objfun*, which is then minimized by a derivative-free method through the built-in *fminsearch* algorithm. The syntax to minimize the objective function is:

```
[u, fval] = fminsearch(f, u0)
```

where $\mathbf{u}$ is the solution of the interior point algorithm and $\mathbf{u}_0$ is the analytical center of the feasible region. The central path is generated by supplying this problem with different values of μ, spanning a logarithmic space from 0.001 to 100 in 40 steps. The algorithm starts from the analytic center and the solutions marked with a cross are improving with increasing μ along the central path. The solutions get very close to the optimal solution marked with a triangle, as computed previously by the active set algorithm implemented in *quadprog*. The two solutions match very closely, the contour lines of $f(\mathbf{u})$ corresponding to the different algorithms are practically indistinguishable.

The possible numerical issues with simple interior point algorithms are demonstrated in Fig. 6.13b where the detail in the neighborhood of the solution is shown. As it is clear from the figure, even this very basic interior point algorithm yields good approximations of the optimal $\mathbf{u}^*$. Note, however, that certain solutions for increasing μ tend to be very suboptimal, while some are even infeasible. This is caused by the previously mentioned numerical ill-conditioning of the unconstrained optimization problem. There is for example a point $\mathbf{u}_0 = [1.59 \quad 2.41]$ which is not only suboptimal but also outside the feasible region.

6.8.2 MPC and Quadratic Programming

Now, it is clear that the cost function in (6.33) and the constraints (6.53) in fact formulate a quadratic programming problem, which is given by [48]:

$$\begin{array}{ll} \text{minimize} & J_k(\mathbf{u}_k) = \mathbf{u}_k^T \mathbf{H} \mathbf{u}_k + 2x_k^T \mathbf{G}^T \mathbf{u}_k \\ \text{subject to} & \mathbf{A_c} \mathbf{u}_k \leq \mathbf{b}_0 + \mathbf{B}_0 x_k \end{array} \tag{6.95}$$

where $\mathbf{H}$ and $\mathbf{G}$ are the predetermined prediction matrices, $\mathbf{A_c}$, $\mathbf{b}_0$, and $\mathbf{B}_0$ define the constraints, x_k is the actually measured or observed state and $\mathbf{u}_k$ is the optimization variable—that is the unknown sequence of future optimal control inputs.

To solve the quadratic programming problem given by the MPC formulation (6.95) in Matlab, we can use the built-in *quadprog* function, which at each time step will solve the QP given by the following syntax[9]:

```
u=quadprog(H, G*x, Ac, b0+B0*x)
```

where x is the actual measured or observed state, $\mathbf{H}$ and $\mathbf{G}$ are prediction matrices, $\mathbf{A}_c$, $\mathbf{b}_0$ and $\mathbf{B}_0$ are given by the constraints and $\mathbf{u}$ is the optimal input trajectory of which only the first element is utilized. The function *quadprog* is only usable in the Matlab environment, and cannot be employed in a real-time environment. For the online optimization task in traditional QP based MPC (QPMPC) a dedicated solver is required, preferably a C language version of either generic quadratic programming software or one which is optimized for MPC usage. Such a solver is for example qpOASES developed by Ferreau et al. [17–19] which is utilized as a benchmark for traditional optimal MPC in the upcoming chapters.

Quadratic programming solvers specifically designed to solve MPC problems utilize some of the unique properties of the QP problem arising due to the predictive control formulation. These properties are [34]:

- the QP problem is *sparse*
- it is possible to *warm start*

After reordering the variables in the QP problem arising from the MPC formulation the problem becomes *sparse*, meaning that the matrices are mainly populated with zeros. The second condition comes from the fact that unless there are excessive

[9] Note that unlike the cost function in the MPC formulation given by (6.95), the solver *quadprog* minimizes the function $J_k(\mathbf{u}_k) = \frac{1}{2}\mathbf{u}_k^T \mathbf{H} \mathbf{u}_k + x_k^T \mathbf{G}^T \mathbf{u}_k$. Because only half of $\mathbf{H}$ is assumed by *quadprog*, the number "2" in front of the second part of the original cost function is omitted in the syntax.

disturbances acting on the controlled plant, the successive solutions at times (k) and $(k+1)$ are very much alike. For this reason, it is possible to warm start the algorithm based on the previous solution to save on computational time. The MPC tuned implementations of active set algorithms are more common; however, examples of interior point algorithms utilizing the structure and properties of MPC exist as well.

6.9 Prediction and Control Horizon

In the previous sections, the prediction horizon—that is the point in the future up to which predictions are made—was identical to the control horizon. The control horizon can be understood as a point in the future up to which control moves are calculated.

Let us denote the control horizon by n_c and the prediction horizon as n_p as illustrated in Fig. 6.14. The control and prediction horizon can be equivalent, though they are often different: the cost is predicted further into the future, but in order to reduce the number of free variables the control horizon n_c is considered to be shorter than the prediction horizon n_p. A common approach is when the controller computes the control moves in such a manner that it assumes that the control inputs will remain on the same level after the control horizon ends. The control input may vary up to the end of n_c. The control input vector will have two different domains:

$$\mathbf{u}_k = [\, u_k \; u_{k+1} \; \ldots \; u_{k+n_c} \; u_{ss} \; u_{ss} \; u_{ss} \,]^T \tag{6.96}$$

where u_{ss} is the steady-state control input. If we take a simple regulation example where the output needs to be driven to zero, we can divide the prediction matrix $\mathbf{N}$ to two parts: one for the control moves which may vary, and one which is assumed to be steady-state:

$$\begin{aligned}
\mathbf{x}_k &= \mathbf{M}x_k + \mathbf{N}\mathbf{u}_k \\
&= \mathbf{M}x_k + \mathbf{N}_1[u_0 \ldots u_{n_c}]^T + \mathbf{N}_2[1 \ldots 1]^T u_{ss} \\
&= \mathbf{M}x_k + \tilde{\mathbf{N}}[u_0 \ldots u_{n_c} u_{ss}]^T
\end{aligned} \tag{6.97}$$

In most predictive control applications the control horizon n_c actually equals the prediction horizon n_p. Similarly, the ongoing discussion will not differentiate these two concepts. From now on it will be assumed that $n_c = n_p$ and the control horizon will be simply marked by n_c, expecting predictions to be computed up to this point in time as well and vice versa.

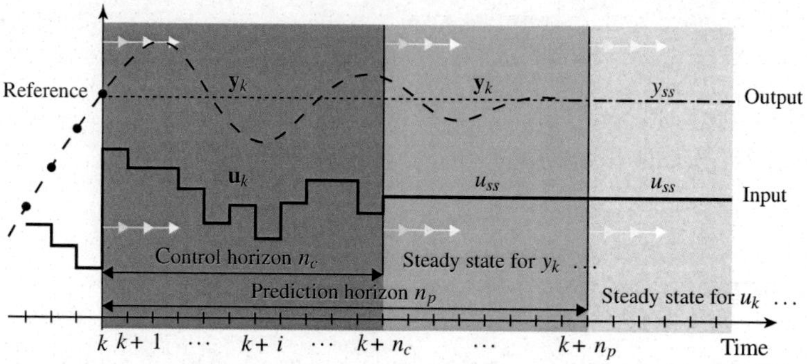

Fig. 6.14 The control horizon n_c may be shorter than the prediction horizon n_p, however it is very common to choose these two horizons to have equal lengths

6.10 Fixed Target Tracking

The problem statement has to be slightly changed if the goal of the control application is to keep the output at a predetermined level, instead of keeping it at zero. The formulation is similar, but the zero is shifted toward the new required value. The desired output will be designated as y_d and the corresponding control input as u_d. When the system reaches y_d it will be true:

$$\begin{aligned} x_d &= \mathbf{A}x_d + \mathbf{B}u_d \\ y_d &= r = \mathbf{C}x_d \end{aligned} \tag{6.98}$$

From this, the reference state and the required steady-state input can be calculated:

$$\begin{aligned} x_d &= (\mathbf{I} - \mathbf{A})^{-1}\mathbf{B}u_d \\ r &= \mathbf{C}(\mathbf{I} - \mathbf{A})^{-1}\mathbf{B}u_d \end{aligned} \tag{6.99}$$

The term $\mathbf{C}(\mathbf{I} - \mathbf{A})^{-1}\mathbf{B}$ is often referred to as the DC gain. If we denote the difference between the actual and the desired state as $x_e = x - x_d$, and also $u_e = u - u_d$ the model may be modified to:

$$x_{e(k+1)} = \mathbf{A}x_{e(k)} + \mathbf{B}(u - u_d) \tag{6.100}$$

Using this notation to create and evaluate a cost function results a calculated set of future control differences u_e. One must use u_e instead of u, and similarly x_e instead of x. The control move must be recalculated properly as well, before it can be applied to the system. For example a controller may take a form of $u = \mathbf{K}x_e$ and the required output shall be $u = \mathbf{I}u_d + \mathbf{K}x_e$.

Problems arise with calculating the required steady-state values. Since no model can be perfectly correct, there will always be some challenges. If the desired steady-state values are incorrect, offset-free tracking is not possible. There are several techniques to overcome this problem and ensure offset-free regulation.

The most common control aim in active vibration control is to steer the structure into equilibrium and to keep it there. This equilibrium point is the origin of the state-space system. Alternatively, the system model can be transformed and shifted to have the origin as an equilibrium point. Further discussion in this book will not assume the use of tracking (fixed target or any other sort), the systems will rather be steered into the origin of the state-space formulation, which is coincident with the physical equilibrium of the mechanical system.

6.11 Integral Action

In this work, we will assume that the control input u_k will be computed by the predictive controller in its absolute form. However, this is not the only way to produce inputs to the plant. Just as PID controllers often use an *incremental* or *velocity*[10] input [3], we may also define an MPC controller in a similar way. Instead of computing the absolute level of the input u_k, only its changes will be computed which are given by Δu_k. This formulation is also referred to as *discrete-time integration* formulation [34]. The reason to use a controller formulation with integral action is to get offset-free tracking. To achieve offset-free tracking, the minimum of the cost function J must be consistent with zero tracking errors in steady-state and the predicted control move must also be zero to maintain zero tracking [48].

There are several valid ways to express the controller input in its incremental, integrating form. All of the methods involve augmenting the state vector with an additional block of elements. Let us therefore consider a new augmented state, which is denoted by:

$$\tilde{x}_k = \begin{bmatrix} x_k \\ u_{k-1} \end{bmatrix} \tag{6.101}$$

where our new augmented vector[11] will be denoted by x_k and it will contain the original state vector x_k and the previous value of the input u_{k-1} at time $(k-1)$. The next iteration of this vector at time $k+1$ would be [34]:

$$\begin{bmatrix} x_{k+1} \\ u_k \end{bmatrix} = \begin{bmatrix} \mathbf{A} & \mathbf{B} \\ \mathbf{0} & \mathbf{I} \end{bmatrix} \begin{bmatrix} x_k \\ u_{k-1} \end{bmatrix} + \begin{bmatrix} \mathbf{B} \\ \mathbf{I} \end{bmatrix} \Delta u_k \tag{6.102}$$

where the dimensionality of the zero matrix $\mathbf{0}$ and $\mathbf{I}$ depend on the number of inputs. Similarly, the outputs can be defined by:

[10] The term velocity in the velocity formulation does not refer to the physical interpretation of the concept, it merely denotes the differentiation (difference) of an absolute variable.

[11] Note that in this book we have used a scalar notation for x_k, u_k, y_k, despite their general vector nature in order to reserve the bold notation for vectors of *predicted* sequences.

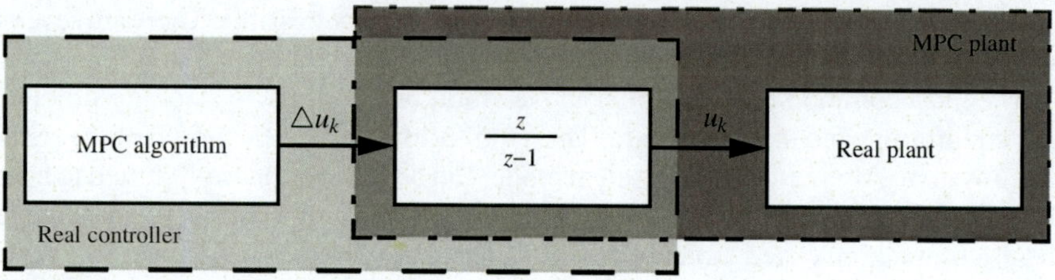

Fig. 6.15 The boundary between the actual controller and plant depends on our view. A real plant augmented by an integration operation may be controlled by an algorithm computing input increments Δu_k

$$y_k = \left[\mathbf{C}\ \mathbf{0}\right]\begin{bmatrix} x_k \\ u_{k-1} \end{bmatrix} \tag{6.103}$$

Figure 6.15 illustrates the boundary between the actual MPC controller and the controlled plant. It is often convenient to consider the discrete-time integration operation to be a part of the plant dynamics. This integration operator computes the absolute value of the inputs u_k from the increments Δu_k, thus creating an augmented MPC plant. In this way we may utilize an MPC controller producing an incremental input Δu_k, which may be used to control the plant augmented by the discrete-time integration operation.

A slightly less intuitive way to create a discrete-time integration formulation of the predictive controller is to augment the state-space equation in the following way [34]:

$$\tilde{x}_k = \begin{bmatrix} \Delta x_k \\ y_k \end{bmatrix} \tag{6.104}$$

where Δx_k is simply the first difference of the state $\Delta x_k = x_k - x_{k-1}$ and $\tilde{x}_k$ is the augmented state. Considering a simple state-space system given by 1 for time (k) and $(k-1)$ we will get

$$x_{k+1} = \mathbf{A}x_k + \mathbf{B}u_k \tag{6.105}$$

$$x_k = \mathbf{A}x_{k-1} + \mathbf{B}u_{k-1} \tag{6.106}$$

and subtracting these will yield the state difference Δx_k according to

$$\Delta x_{k+1} = \mathbf{A}\Delta x_k + \mathbf{B}\Delta u_k \tag{6.107}$$

From the second output equation in (6.1) we can also deduct:

$$y_{k+1} = \mathbf{C}x_{k+1} \tag{6.108}$$

$$= \mathbf{C}[\Delta x_{k+1} + x_k] \tag{6.109}$$

$$= \mathbf{C}[\mathbf{A}\Delta x_k + \mathbf{B}\Delta u_k] + y_k \tag{6.110}$$

so the new state-space representation will be changed to

$$\begin{bmatrix} \Delta x_{k+1} \\ y_{k+1} \end{bmatrix} = \begin{bmatrix} \mathbf{A} & \mathbf{0} \\ \mathbf{CA} & \mathbf{I} \end{bmatrix} \begin{bmatrix} \Delta x_k \\ y_k \end{bmatrix} + \begin{bmatrix} \mathbf{B} \\ \mathbf{CB} \end{bmatrix} \Delta u_k \tag{6.111}$$

$$y_k = \begin{bmatrix} \mathbf{0} \ \mathbf{I} \end{bmatrix} \begin{bmatrix} \Delta x_k \ y_k \end{bmatrix} \tag{6.112}$$

References

1. Agachi PS, Nagy Z, Cristea MV, Imre-Lucaci A (2006) Model based control: case studies in process engineering, 1st edn. Wiley-VCH, Weinheim
2. Åström KJ, Wittenmark B (1973) On self tuning regulators. Automatica 9:185–199
3. Belavý C (2009) Teória Automatického Riadenia II: Návody na cvičenia, 1st edn. Slovenská vysoká škola technická v Bratislave: Strojnícka Fakulta, Bratislava (Theory of automatic control II: seminar guide) In Slovak language
4. Bitmead RR, Gevers M, Wertz V (1990) Adaptive optimal control: the thinking man's GPC. Prentice Hall, San Francisco
5. Boyd S, Vandenberghe L (2004) Convex optimization. Cambridge University Press, Cambridge
6. Camacho EF, Bordons C (1995) Model predictive control in the process industry, advances in industrial control, vol 104. Springer, London
7. Camacho EF, Bordons C (2007) Model predictive control, 2nd edn. Springer, London
8. Cannon M (2005) Model predictive control, Lecture notes, Michaelmas Term 2005 (4 Lectures). Course code 4ME44. University of Oxford, Oxford
9. Cannon M, Kouvaritakis B (2005) Optimizing prediction dynamics for robust MPC. IEEE Trans Autom Control 50(11):1892–1897. doi: 10.1109/TAC.2005.858679
10. Clarke DW, Gawthrop PJ (1979) A self-tuning controller. In: IEE proceedings Part D 123:633–640
11. Clarke DW, Zhang L (1987) Long-range predictive control using weighting-sequence models. In: IEE Proceedings Part D 134(3):187–195
12. Clarke DW, Mohtadi C, Tuffs PS (1987) Generalized predictive control, Part I: The basic algorithm. Automatica 23(2):137–148
13. Clarke DW, Mohtadi C, Tuffs PS (1987) Generalized predictive control, Part II: extensions and interpretations. Automatica 23(2):149–160
14. Cutler CR, Ramaker BC (1980) Dynamic matrix control—a computer control algorithm. In: Automatic control conference, San Francisco
15. de Keyser RMC, Cauwenberghe ARV (1985) Extended prediction self-adaptive control. In: IFAC symposium on identification and system parameter estimation, Yorkshire pp 1317–1322
16. Demircioğlu H, Gawthrop PJ (1991) Continuous-time generalised predictive control. Automatica 27(1):55–74
17. Ferreau HJ (2006) An online active set strategy for fast solution of parametric quadratic programs with applications to predictive engine control. Master's thesis, University of Heidelberg
18. Ferreau HJ (2011) qpOASES—Online active set strategy, Leuven. Available: http://www.qpoases.org
19. Ferreau HJ, Bock HG, Diehl M (2008) An online active set strategy to overcome the limitations of explicit MPC. Int J Robust Nonlinear Control 18(8):816–830

20. Fletcher R (2000) Practical methods of optimization. Wiley, New York
21. Fuller CR, Elliott SJ, Nelson PA (1996) Active Control of Vibration, 1st edn. Academic Press, San Francisco
22. Grafixar / morgueFile (2008) Oil refinery in Texas City, Texas. In agreement with the morgueFile free license: free to use for commercial work without attribution. Online, http://www.morguefile.com/archive/display/212157
23. Grossner J, Kouvaritakis B, Rossiter J (1997) Cautious stable predictive control: a guaranteed stable predictive control algorithm with low input activity and good robustness. Int J Control 67(5):675–697
24. Inman DJ (2006) Vibration with control. Wiley, Chichester
25. Karas A (2002) Stabilizujúce prediktívne riadenie systémov s obmedzeniami. PhD thesis, Slovak University of Technology in Bratislava, Bratislava (Stabilizing predictive control of systems with constraints.) In Slovak language
26. Karas A, Rohal'-Ilkiv B, Belavý C (2007) Praktické aspekty prediktívneho riadenia, 1st edn. Slovak University of Technology in Bratislava / Slovenská E-Akadémia, Bratislava (Practical aspects of predictive control.) In Slovak language.
27. Kinnaert M (1989) Adaptive generalized predictive controller for MIMO systems. J Process Control 50(1):161–172
28. Kvasnica M (2009) Real-time model predictive control via multi-parametric programming: theory and tools, 1st edn. VDM Verlag, Saarbrücken
29. Kwon WH, Byun DG (1989) Receding horizon tracking control as a predictive control and its stability properties. Int J Control 50(5):1807–1824
30. Kwon WH, Pearson AE (1975) On the stabilisation of a discrete constant linear system. IEEE Trans Autom Control 20(6):800–801
31. Kwon WH, Pearson AE (1978) On feedback stabilization of time-varying discrete linear systems. IEEE Trans Autom Control 23:479–481
32. Kwon WH, Choi HH, Byun DG, Noh SB (1992) Recursive solution of generalized predictive control and its equivalence to receding horizon tracking control. Automatica 28(6):1235–1238
33. Lee JH, Morari M, Garcia CE (1994) State space interpretation of model predictive control. Automatica 30(4):707–717
34. Maciejowski JM (2000) Predictive control with constraints, 1st edn. Prentice Hall, New Jersey
35. Morari M, Lee JH (1999) Model predictive control: past, present and future. Comput Chem Eng 23(4):667–682
36. Mosca E (1994) Optimal, predictive and adaptive control, 1st edn. Prentice Hall, New Jersey
37. Nesterov Y, Nemirovskii A (1994) Interior point polynomial methods in convex programming, studies in applied mathematics, vol 13. SIAM, Pennsylvania
38. Ordys AW, Clarke DW (1993) A state-space description for GPC controllers. Int J Syst Sci 24(9):1727–1744
39. Pistikopoulos EN, Georgiadis MC, Dua V (eds) (2007) Multi-parametric model-based control, vol 2., 1st edn. Wiley-VCH, Weinheim
40. Potra FA, Wright SJ (1997) Primal-dual interior-point methods. SIAM, Pennsylvania
41. Preumont A (2002) Vibration control of active structures. 2nd edn. Kluwer Academic Publishers, Dordrecht
42. Preumont A, Seto K (2008) Active control of structures. 3rd edn. Wiley, Chichester
43. Qin SJ, Badgwell TA (1997) An overview of industrial model predictive control technology. In: Proceedings of chemical process control—V. Tahoe City. In: AIChE symposium series, vol 93, pp 232–256
44. Qin SJ, Badgwell TA (1999) An overview of nonlinear model predictive control applications. In: Zheng FAA (ed) Nonlinear model predictive control, Birkhauser Verlag, pp 369–392
45. Qin SJ, Badgwell TA (2003) A survey of industrial model predictive control technology. Control Eng Pract 11(7):733–764. doi: 10.1016/S0967-0661(02)00186-7, http://www.sciencedirect.com/science/article/B6V2H-47BX35T-1/2/1e355f78abeb6d9ee76d726330e7ca54

46. Rawlings JB, Muske KR (1993) The stability of constrained receding horizon control. IEEE Trans Autom Cont 38(10):1512–1516
47. Richalet J, Rault A, Testud JL, Papon J (1978) Model predictive heuristic control: application to industrial processes. Automatica 14(2):413–428
48. Rossiter JA (2003) Model-based predictive control: a practical approach. 1st edn. CRC Press, Florida
49. Scokaert POM, Rawlings JB (1996) Infinite horizon linear quadratic control with constraints. In: Proceedings of IFAC'96 World Congress, San Francisco, vol 7a-04 1, pp 109–114
50. Shah SL, Mohtadi C, Clarke DW (1987) Multivariable adaptive zrol without a prior knowledge of the delay matrix. Syst Control Lett 9:295–306
51. Soeterboek R (1992) Predictive control—a unified approach. Prentice Hall, New York
52. The Mathworks (2011) Matlab optimization toolbox v6.0 (R2011a). Software. The Mathworks Inc., Natick. Available at http://www.mathworks.com/products/optimization/
53. Ydstie BE (1984) Extended horizon adaptive control. In: Proceedings of 9th IFAC World Congress, Budapest
54. Yoon TW (1994) Robust adaptive predictive control. PhD thesis, Department of Engineering Science, Oxford University, Oxford

Chapter 7
Stability and Feasibility of MPC

Stability is not merely a distant abstract theoretical feature of feedback controller systems. An unstable controller may supply inputs to the closed-loop plant to a point where the actuators fail or even functional damage occurs to the plant. In the application field of active vibration control, functional damage is easily interpreted in practical terms: it is structural failure due to excessive stress and strain. In case instability results in oscillations, the cause of failure may be material fatigue as well. These dramatic effects imply not only the loss of material goods and damage of equipment but also potential injury or loss of life.

One of the classic educational toys for the control engineering community is the so-called Furuta pendulum [1, 19, 24, 63], or inverted rotational pendulum which is illustrated in Fig. 7.1. This pendulum consists of an electric motor, which drives an arm rotating in the horizontal plane. The end of this arm is equipped with a pendulum freely rotating in the vertical plane and is connected to the arm with a simple joint. The aim is to swing the pendulum to the upright position and keep it there afterward by the rotational movement of the arm. This system is underactuated, nonlinear and it is easy to see that the upper equilibrium position of the pendulum is open-loop unstable. The stability of the closed-loop control scheme is an essential feature in this demonstration application, while the loss of stable control presents itself with striking clarity: the pendulum falls down to its stable equilibrium located at the bottom of its path. Although practical issues with control stability are not so tangible and evident in all control systems, it is nevertheless important to make sure that a structure manipulated by model predictive vibration control is guaranteed to remain stable at all times.

Probably the simplest definition of stability is bounded input–bounded output (BIBO) stability. As the name implies, BIBO stability means that if a finite (bounded) input is supplied to the system a finite output response must result. On the contrary, if a finite response excites the system in a way that an infinite response would theoretically result, the system is said to be unstable. The stability of traditional control systems can be guaranteed very easily. In a case of a discrete controller in a closed form, all

G. Takács and B. Rohal'-Ilkiv, *Model Predictive Vibration Control*,
DOI: 10.1007/978-1-4471-2333-0_7, © Springer-Verlag London Limited 2012

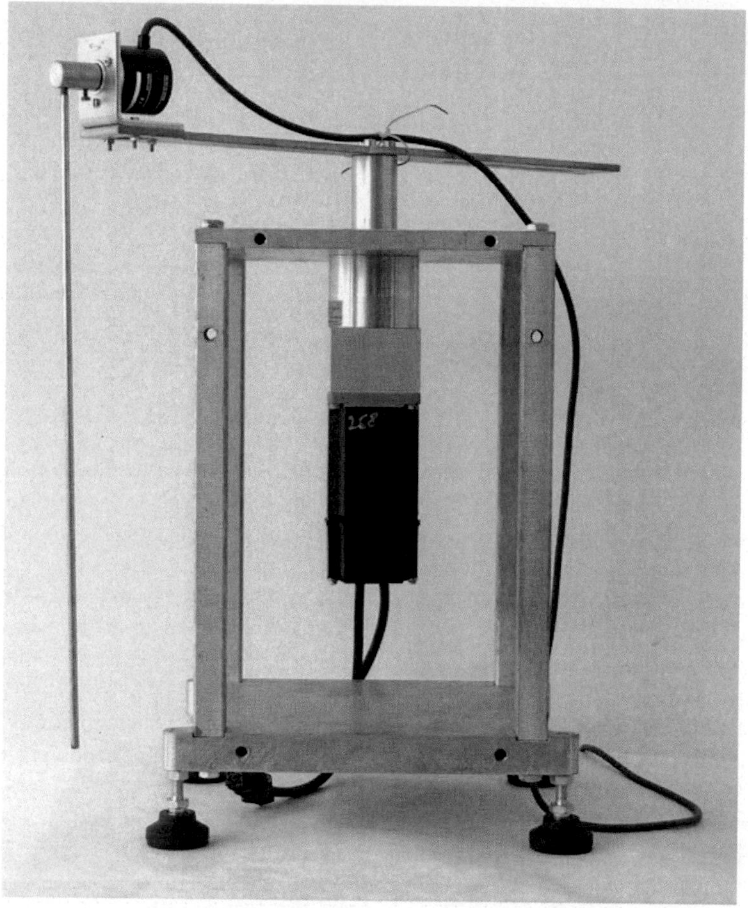

Fig. 7.1 The rotational inverted pendulum, or the Furuta pendulum, is a classic example of an open-loop unstable system. Stability of the MPC (or any other) controller is an essential feature here as well, since an insufficiently designed control strategy could easily drive the system away from its unstable equilibrium

poles of the transfer function must lie within the unit circle. However, in the case of a constrained MPC law this is not so simple anymore.

Given that one assumes no constraints for the model predictive control (MPC) problem with an infinite horizon cost, the predicted and actual input and state trajectories are identical, [1] and the stability of the process is guaranteed. This is a tractable problem, as we can readily calculate a fixed feedback matrix based on an infinite horizon cost. The predicted and actual responses would be theoretically identical in a constrained MPC problem as well, if one would use an infinitely long prediction

[1] Of course assuming no model errors or external disturbances are present.

horizon and the corresponding infinite cost. Although this would also imply a process with guaranteed stability, an infinitely long horizon would require an infinite number of optimization variables. This is clearly not possible. Instead of using an infinitely long horizon, one may utilize a finite horizon with a finite number of optimization variables in combination with terminal cost replacing the cost contribution up to infinite time—in other words the dual mode control paradigm introduced in the previous chapter. Unfortunately, this complicates the question of stability quite a bit as the constraints are only considered on a finite horizon, while beyond the horizon the feasibility of the constraints—the match of the predicted and actual trajectories and thus ultimately stability—is not guaranteed. For certain disturbances, the controller may produce an input sequence which results in an unstable response, leading to potentially catastrophic results.

The issue of MPC controller stability is reviewed in this chapter. The ultimate aim here is to provide the reader with the fundamentals to formulate a modern MPC control law, which is able to guarantee both system stability and constraint feasibility while providing maximum performance. Alternative formulations allowing for stability guarantees are also discussed, while the details of how those may help to increase computation speeds are left to Chap. 8. To the reader seeking more information on stability and alternative formulations; we may recommend the well-known books of Maciejowski and Rossiter [36, 51] and particularly the seminal work [38].

A brief review of the development of stabilized model predictive control starts our discussion on the stability of the MPC strategy. After this Sect. 7.2 inspects what the conditions of stability are for an MPC control law. Calling the well-known Lyapunov stability analysis to assistance, a very powerful conclusion can be deducted. Considering the cost as Lyapunov function will imply that if the succession of costs J_k is monotonically decreasing for the time $k \rightarrow \infty$, the controller will remain stable. The cost will be nonincreasing as long as the elements of the predicted optimal sequence $\mathbf{u}_k$ ranging from the discrete time iteration $(k+1)$ up to infinity are feasible. Next the concept of terminal constraints, and a special region within the state-space called the target set is introduced. The target set is actually bounded by the process constraints which are enforced for states ranging from the control horizon n_c up to infinity. If the states are forced to remain in this set, the system will remain feasible and thus stable as well. Fortunately, one does not have to inspect and force the inputs to conform to the constraints up to infinite time, instead it can be proved that it is enough to enforce feasibility of the constraints for an additional constraint checking horizon. The finite number of additional constraints creates a target set, which is actually identical to the maximal possible target set thus ensuring an MPC control with optimal performance while still guaranteeing stability. Based on this discussion, Sect. 7.4 will introduce the modified dual-mode quadratic programming-based MPC algorithm, which by utilizing the extra process constraints exceeding the control horizon by the new constraint checking horizon guarantees stability at all times. The following two sections of this chapter discuss alternative formulations, which can ensure the stability of MPC as well. The maximal invariant target set created by the additional constraints is a polyhedron of high complexity that further increases the necessary computational power of the MPC law. To relieve this situation partly, simplifications may be introduced. One such formulation replaces the high complexity polyhedron with a low complexity one, which is actually just a multidimensional cube in hyperspace. The other possibility is to replace the maximal invariant target set with a hyper-ellipsoid, as introduced in Sect. 7.6. A similar ellipsoidal constraint formulation is used in the upcoming chapter to create a computationally efficient MPC controller. The chapter is finished by Sect. 7.7, briefly reviewing the issues caused by mutually incompatible constraints and some strategies to avoid the infeasibility of the MPC optimization problem.

7.1 Development of MPC with Stability Guarantees

Early MPC formulations could not guarantee the stability of the closed-loop system. This however did not prevent industrial practitioners from using MPC in practical applications. The use of open-loop stable plants and long horizons in the absence of constraints prevented most stability issues; nevertheless, stability guarantees in the strict theoretical sense could not have been given.

In the beginning, stability has been investigated for finite horizon predictive controllers with a quadratic cost and in the absence of system constraints [20]. Essentially the effects of the horizon length and parameter choices were evaluated for a given controller, determining whether it is stabilizing or not. Later this rather basic approach has been deemed inappropriate [6] as several examples have shown the need for an a priori method of guaranteeing stability [58].

7.1.1 Equality Terminal Constraints

Stability guarantees for linear plants with constraints have been given later using the so-called *terminal constraints*. These terminal constraints posed a requirement on the controller, namely that the system state must equal to zero in a given number of steps beyond the horizon. By the end of the horizon n_c and beyond, the states are assumed to be zero ($x_{n_c} = 0$) while inputs assume a zero level as well [33]. Mathematically this can be translated as:

$$x_i = 0 \quad \text{for} \quad i \geq n_c \quad \text{and} \quad u_i = 0 \quad \text{for} \quad i \geq n_c \tag{7.1}$$

We may consider using an alternative interpretation to define terminal constraints. Let us require the terminal state x_{n_c} and all following states to be a part of a terminal set $x_{n_c} \in \Omega$. This terminal set is actually just a zero set $\Omega = \{0\}$.

It is possible to guarantee stability analytically for linear, unconstrained systems with an equality terminal constraint and a quadratic cost by proving that the cost function is nonincreasing. The work of Keerthi and Gilbert [28] has become a de facto basis for further stable MPC approaches. Their work proposed a constrained MPC controller based on this idea for nonlinear time varying systems, while pointing out that the cost function of the finite horizon MPC controller approaches the infinite horizon cost if the horizon is increased. Several works have proposed variations and novel algorithms which essentially make use of the fundamental idea of equality terminal constraints, such as the works of Clarke and Scattolini [17] and others [30, 40, 41]. The cost function in these early methods was finite and up to the end of the horizon J_{n_c}. A terminal cost has not been considered here.

7.1.2 Penalty on the Terminal State

One of the first attempts to ensure the stability of a controller is the use of a terminal penalty $\mathbf{P}_f$. The penalty of the terminal state which is added to the cost function is $J_T = x_{n_c}^T \mathbf{P}_f x_{n_c}$. As it has been previously introduced in Sect. 6.5, the most logical choice for this matrix $\mathbf{P}_f$ is the solution of the Lyapunov equation [48], which will then ensure stability through a nonincreasing cost J. This choice for $\mathbf{P}_f$ actually ensures that the addition of $J_T = x_{n_c}^T \mathbf{P}_f x_{n_c}$ to the finite horizon cost has the value of the cost from $n_c \to \infty$, thus overall giving an infinite horizon cost $J_\infty = J_{n_c} + J_T$. Mathematically express this requirement as [27]:

$$x_i \to 0 \quad \text{for} \quad i \to \infty \quad \text{and} \quad u_i \to 0 \quad \text{for} \quad i \geq n_c \tag{7.2}$$

meaning that the state and the inputs must approach zero as the time progresses toward infinity. The terminal penalty by itself would require the state to remain in a terminal set Ω which is infinitely large, that is, equal to the whole state-space $\Omega = \mathbb{R}^{n_x \times n_x}$. If the controlled system is unstable, the unstable poles must be equal to zero by the end of the horizon, while the method is utilized for the rest of the stable poles.

7.1.3 Target Sets

A terminal constraint in the form of an equality is too strict for most applications, as it severely limits the pool of possible initial conditions from which it is possible to steer the system into equilibrium within a finite number of steps. An equality constraint also causes a controller course with overly aggressive inputs. This strict equality requirement has been later replaced by the use of so-called terminal sets, which formulate an additional constraint in the form of an inequality. In this case, the terminal set Ω is chosen as a finite subset of the state-space $\Omega = \mathbb{R}^{n_x \times n_x}$. The aim of the controller was to steer the state into this set in a finite number of steps. The value of the cost function after the finite part J_{n_c} has been however considered to be equal to zero for this approach $J_T = 0$.

Inside the set Ω instead of requiring the states and inputs to be zero as in the case of equality constraints, a local stabilizing control law took over. The course of inputs

u_i with $i \geq n_c$ was no more equal to zero instead the fixed feedback control law $\mathbf{K}$ then steers the state to the origin as time progresses toward infinity. Mathematically this can be given as:

$$x_i \to 0 \quad \text{for} \quad i \to \infty \quad \text{and} \quad u_i = \mathbf{K}x_i \quad \text{for} \quad i \geq n_c \tag{7.3}$$

This concept has been introduced previously in Sect. 6.5 as dual-mode predictive control, where mode 1 assumes free variables and mode 2 the fixed feedback law. This approach has been introduced in [60] and later in [39] for continuous nonlinear systems with constraints. The local stabilizing feedback law $\mathbf{K}$ is most frequently chosen as the linear quadratic gain [55, 56].

7.1.4 Combination of Target Sets and Terminal Penalties

The stable MPC formulation used nowadays is the combination of the concepts presented in the previous two subsections: that is, terminal penalty $\mathbf{P}_f$ and target sets. Here the terminal cost is nonzero, rendering the total predicted cost is equal to the infinite horizon cost $J_\infty = J_{n_c} + J_T$. Moreover, the state is forced to a target set, where a stabilized fixed feedback law is taking over according to (7.3). This approach will be expanded in more detail in the rest of the chapter, while the dual-mode approach has been already considered in Chap. 6. To recapitulate the basics of modern, constrained MPC approaches with stability guarantees, we may state that they utilize the concepts of:

- Nominal stabilizing control law $u_i = \mathbf{K}x_i$ for $i \geq n_c$ also referred to as dual-mode control; which assumes the use of nonzero inputs beyond the control horizon n_c
- Target set Ω defining a state constraint at the end of the control horizon n_c, with the property that this set is contracting, e.g. once the state enters it cannot leave anymore
- Terminal penalty, $J_T = x_{n_c}^T \mathbf{P}_f x_{n_c}$ which allows to take into account a total cost $J_\infty = J_{n_c} + J_T$ equivalent to an infinitely long horizon.

In the absence of constraints, choosing a terminal penalty creates an infinite horizon cost which results in an ideal situation, where stability is ensured and online optimization is not needed. This in fact is the basis of the unconstrained controller presented previously in Sect. 6.6. However, by introducing input, state, and output constraints to the system, the predicted cost J_∞ diverges from the real cost. To relieve this situation the idea was to use a finite set Ω around the state-space equilibrium, in which a local stabilizing law took over, thus the sum of the finite horizon and terminal cost was in fact equal to the real cost.

The use of a Lyapunov function as a cost to ensure stability for systems without system constraints has been considered relatively early in [16] by Chen and Shaw. This approach turned out to be valid and current stable MPC formulations are based on this idea. Subsequently a Lyapunov function-based stability guarantee has been worked out for continuous systems [37], while the first application of this concept for a discrete constrained system has been presented in [28] and later in [4]. These works present a general treatment of the stability issue in MPC which is based on the monotonicity of the decreasing cost function.

7.1.5 State Contractility and Others

Instead of using a Lyapunov cost function, an alternative way to ensure stability is to require that the two-norm of the state is contracting. Mathematically this is given as:

$$\|x_{k+1}\| \leq \alpha \|x_k\| \quad \text{where} \quad \alpha < 1 \tag{7.4}$$

Stability is achieved independently of the parameters of the cost function. This method has been characterized in [44, 45, 64] while it is further expanded for nonlinear systems by de Oliveira [18]. Later Bemporad has proposed an MPC method with stability guarantees utilizing a quadratic Lyapunov function similarly to the methods employing target sets [3]. As it turns out, contractility-based methods have been useful to earn guaranteed stability, albeit with considerable performance loss. Because the norm of the system state is required to be contracting at all times, its course is constantly forced to be outside the ideal trajectory. In essence, the contractility condition introduced above is a Lyapunov function, thus suitable for stability guarantees.

Yet other approaches are based on confining the final state to a terminal set while requiring this set to be stabilizing [38, 43, 47].

7.2 Closed-Loop Stability of the Infinite Horizon MPC Law

The possible issues with stability of a constrained MPC system are demonstrated if one plots the evolution of the cost J_k overtime for a system without disturbance and a perfect model match. The value of the optimal cost function should be decreasing steadily; however due to nonlinear nature of the closed-loop MPC law the cost can be increasing, even if the overall response remains stable. This variation indicates that the closed-loop input trajectory does not follow an optimal predicted trajectory since the predicted cost does not steadily decrease overtime. Given an otherwise stable linear time-invariant system, it is possible to choose an initial condition, which will render the MPC controlled constrained system unstable.

Formally, the stability of the constrained MPC law can be evaluated by considering the cost function J_k as a Lyapunov function. The aim of the Lyapunov stability analysis is not to assess stability for each individual controller; on the contrary, the aim is to establish the conditions under which the MPC control law will be stable in general. Given the knowledge of these conditions, one may create a modified MPC controller that does ensure guaranteed stability. As it will be demonstrated, the stability of the closed-loop system is closely related to the feasibility of the process. The generic stability analysis for a constrained MPC strategy has been first introduced by Mayne et al. in [38].

According to the Lyapunov stability theorem [35] which has been applied to discrete systems by Bertram and Kalman [5], we may define x_0 as an *equilibrium* point of a system $x_{k+1} = f(x)$ if and only if $f(x_0) = x_0$. The function $f(x)$ is actually the state equation as defined by (6.2). The natural equilibrium point of our interest is then $x = 0$ which is located at $f(0) = 0$. It is possible to define a *stable equilibrium* similarly. The state $x = 0$ is a stable equilibrium of a system if for all $k > 0$ the state remains within an arbitrarily small region of the state-space containing $x = 0$ whenever the initial condition $x(0)$ lies sufficiently close to $x = 0$. Mathematically this can be given as for all $R > 0$ exists $r > 0$ such that for all $k > 0$ [2, 14, 35]:

$$\|x(0)\| < r \Rightarrow \|x(k)\| < R \tag{7.5}$$

According to Lyapunov's second method for stability whenever $\|x\|$ is sufficiently small, $x = 0$ will be a stable equilibrium point if there exists a continuously differentiable scalar function $V(x)$ which [14, 35]:

1. positive definite
2. $V(f(x)) - V(x) \leq 0$ holds

and where $x_{k+1} = f(x)$.

It is possible to define asymptotic convergence in a similar fashion. According to this $l(x)$ will converge to zero, meaning that as $k \to \infty$ the series $l(x) \to 0$, if $V(x)$ is a continuously differentiable scalar function and is

1. positive definite
2. $V(f(x)) - V(x) \leq -l(x) \leq 0$ holds

After rearranging the second condition we get the following statement

$$l(x) \leq V(x_k) - V(x_{k+1}) \tag{7.6}$$

which after summing both sides $k = 0, 1, 2, \ldots, \infty$ will render to

$$\sum_{k=0}^{\infty} l(x) \leq V(x_0) - \lim_{k\to\infty} V(x_k) \tag{7.7}$$

The right hand side of this equation is finite, which has an important consequence on the convergence of $l(x)$. The finite nature of the right hand side of the above equation comes from the fact that the $V(x_k) \geq 0$ or in other words the function is positive definite. The second Lyapunov stability gives $V(x_k) \geq V(x_{k+1})$, meaning that the function value at the next step is smaller than the previous. This implies that $V(x_k)$

approaches a finite limit as $k \to \infty$. This is because if the first term on the right side is finite, the second term must be smaller than the first one, and their difference is also finite. *If the right hand side of the equation is a finite number the value of $l(x)$ must approach zero with $k \to \infty$*. This comes from the fact that if one sums an infinite series and the sum is a finite value, then the series must converge to zero.

In this light, it is possible to perform a stability analysis for the MPC control law. Assuming that the MPC problem is feasible, the optimal predicted cost (6.30) is simply a function of the current state x_k, considering the cost function J_k as a Lyapunov function:

$$J_k(x_k) = V(x_k) \tag{7.8}$$

To fulfill the asymptotic convergence property defined previously, the cost function has to be positive definite. The optimal predicted cost J_k is a positive definite function of x_k if either of the following two conditions hold [14, 36]:

1. $\mathbf{Q}$ is positive definite
2. The pair $(\mathbf{A}, \mathbf{Q}^{1/2})$ is observable

where $\mathbf{Q} = \mathbf{Q}^{1/2T}\mathbf{Q}^{1/2}$. The first condition ensures that the first term and hence the entire sum in (6.30) is positive definite.

If the terminal weight $\mathbf{P}_f$ is chosen in a way that J_k is an infinite horizon cost and the optimal predicted input sequence $\mathbf{u}_k$ computed at time (k) is feasible for the optimization problem at $(k+1)$ then the optimal predicted cost is nonincreasing and satisfies [38]:

$$J_{k+1} - J_k \leq -\left(x_k^T \mathbf{Q} x_k + u_k^T \mathbf{R} u_k\right) \tag{7.9}$$

Given that the optimal predicted input sequence $\mathbf{u}_k$ is feasible for the problem at $(k+1)$, (7.9) holds because the optimal cost at current time (k) must be at least as small (or smaller) as the cost for the *tail* of the optimal sequence predicted for the previous sample. The condition of the nonincreasing cost in (7.9) is also referred to as the *direct stability* method and it is utilized to prove the stability of the constrained MPC law in several works [15, 28, 37, 39, 56].

To demonstrate the previous statement, let us take the optimal predicted sequence $\mathbf{u}_k$ at (k) and take it as a basis for the prediction at the next time step, that is, at $(k+1)$:

$$\tilde{\mathbf{u}}_{k+1} = [u_{k+1} \; u_{k+2} \; \ldots \; u_{k+n_c-1} \; \mathbf{K}x_{k+n_c}] \tag{7.10}$$

where $\tilde{\mathbf{u}}_{k+1}$ is referred to as the *tail* of $\mathbf{u}_k$ and it is illustrated in Fig. 7.2. The last element of the tail is given by the fixed feedback law in mode 2, therefore it is according to $u_{k+n_c} = \mathbf{K}x_{k+n_c}$.

The cost function J_k in (6.30) expresses an infinite horizon cost. The cost $\tilde{J}_k$ associated with the tail $\tilde{\mathbf{u}}_{k+1}$ is the cost J_k at (k) minus the term which remains to that time:

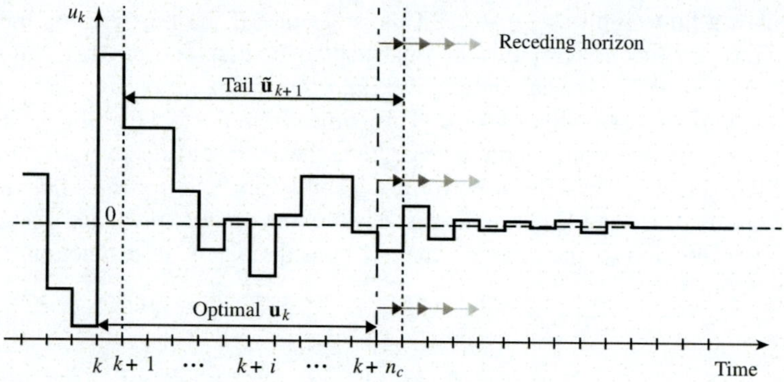

Fig. 7.2 The optimal prediction $\mathbf{u}_k$ at time (k) and its tail $\tilde{\mathbf{u}}_{k+1}$

$$
\begin{aligned}
\tilde{J}_{k+1} &= \sum_{i=1}^{n_c}\left(x_{k+i}^T\mathbf{Q}x_{k+i} + u_{k+i}^T\mathbf{R}u_{k+i}\right) + x_{k+n_c+1}^T\mathbf{P}_f x_{k+n_c+1} \\
&= \sum_{i=1}^{\infty}\left(x_{k+i}^T\mathbf{Q}x_{k+i} + u_{k+i}^T\mathbf{R}u_{k+i}\right) \\
&= \sum_{i=0}^{\infty}\left(x_{k+i}^T\mathbf{Q}x_{k+i} + u_{k+i}^T\mathbf{R}u_{k+i}\right) - \left(x_k^T\mathbf{Q}x_k + u_k^T\mathbf{R}u_k\right) \\
&= J_k - \left(x_k^T\mathbf{Q}x_k + u_k^T\mathbf{R}u_k\right)
\end{aligned}
$$

In reality, the tail $\tilde{\mathbf{u}}_{k+1}$ will be suboptimal at that time because it is based on the optimal predictions at the previous step (k). The optimal value at $(k+1)$ will satisfy [14]

$$
J_{k+1} \leq \tilde{J}_{k+1} = J_k - \left(x_k^T\mathbf{Q}x_k + u_k^T\mathbf{R}u_k\right) \tag{7.12}
$$

implying the condition given by (7.9).

The important aspect of this stability analysis helps to formulate algorithms that do ensure the stability of the closed-loop system. To summarize, the previously introduced conditions [14]:

If J_k is a positive definite infinite horizon cost, then $x_k = 0$ is a stable equilibrium for the closed-loop system and x_k converges asymptotically to zero, if the tail $\tilde{\mathbf{u}}_{k+1}$ is feasible for all $k > 0$.

The second method to prove the stability of a constrained MPC law also referred to as the *indirect stability* method originates from [16]:

$$
J_k - J_{k+1} > 0 \quad \text{for} \quad x \neq 0 \tag{7.13}
$$

If it is possible to prove that the left side of Eq. (7.13) is positive, the stability of the control course is proven as well. This approach is discussed in more detail in the works of Chen and Shaw [16] and others [6, 46].

7.3 Stability Through Terminal Constraints

It has been demonstrated in the Sect. 7.2 that a closed-loop MPC control law will remain asymptotically stable, given that the *tail* of the input predictions generated at time (k) will satisfy constraints at times $k = 1, 2, 3, \ldots$ and onwards. Fortunately, it is not necessary to force the system to comply with the constraints from the present time all to infinity, since this problem would not be possible to formulate with a finite number of constraints.

This requirement on the feasibility of the tail $\tilde{\mathbf{u}}_{k+1}$ is at least partly satisfied by the problem formulation itself. The constraints will be satisfied for $\tilde{\mathbf{u}}_{k+1}$ in the first $n_c - 1$ sampling intervals of the n_c steps long prediction horizon. This is true because the optimal predictions at time (k) must satisfy the constraints. However, $\tilde{\mathbf{u}}_{k+1}$ has one more element, that is, the n_c-th element $u_{k+n_c} = \mathbf{K}x_{k+n_c}$ as defined by the mode 2 fixed feedback control law.

To guarantee the feasibility of the last element and therefore the whole tail at $(k+1)$, we must define an additional constraint at time (k). Constraints additional to the constraints arising from the problem definition are referred to as *terminal constraints*. Terminal constraints are defined in mode 2, where the fixed feedback law is active and therefore they are defined in terms of the terminal state prediction that is, x_{k+n_c}.

We can define a region of the state-space Ω in which the terminal state prediction x_{k+n_c} must lie in order to satisfy the terminal constraints. If we have a system with input constraints $\underline{u} \leq u_k \leq \overline{u}$ and state constraints $\underline{x} \leq x_k \leq \overline{x}$, to ensure that the tail will satisfy constraints over the whole prediction horizon n_c we must include a terminal constraint in the following form

$$x_{k+n_c} \in \Omega \quad \Rightarrow \quad \begin{array}{c} \underline{u} \leq \mathbf{K}x_{k+n_c} \leq \overline{u} \\ \underline{x} \leq x_{k+n_c} \leq \overline{x} \end{array} \tag{7.14}$$

defining a region in the state-space Ω. The set Ω is a region where we want to steer the state by the end of the horizon, and it is referred to as the *target set*. The terminal constraints must be computed in a way that they ensure the feasibility of the MPC optimization recursively, that is the tail predictions ensure constraints including the terminal constraints themselves.

This is however not enough, as the predictions generated by the tail $\tilde{\mathbf{u}}_{k+1}$ at $(k+1)$ must also satisfy the next terminal constraint, or in other words the terminal state x_{k+n_c+1} at $(k+1)$ must also be a part of the region Ω:

$$x_{k+n_c+1} \in \Omega \tag{7.15}$$

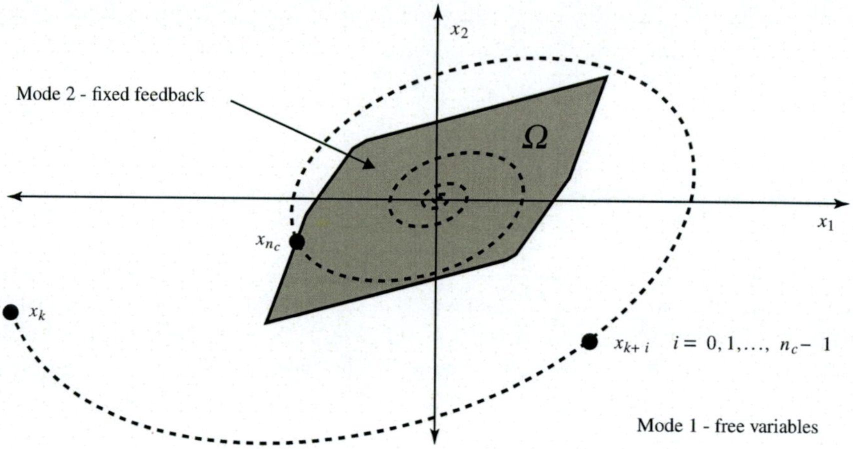

Fig. 7.3 Illustration of the invariant target set Ω with the state trajectory. The area outside the target set represents mode 1 predictions, where inputs are free optimization variables. The target set is mode 2, where the inputs are assumed to be calculated by a fixed feedback law

A target set is illustrated for a two-dimensional state-space in Fig. 7.3. The area outside the target set is where the inputs are assumed to be free optimization variables, while inside the target set the inputs are calculated by the fixed feedback law. Once the state trajectory enters target set Ω it cannot exit it. This is in fact the invariance property of the set Ω, which can be interpreted in a way that the set is contracting. If the state is the part of the set at (k) so must be at the next and subsequent times as well.

To ensure the stability of the closed-loop MPC system, we must ensure the feasibility of the tail $\tilde{\mathbf{u}}_{k+1}$ at time $(k + 1)$ whenever the MPC optimization is feasible at time (k). To achieve this we must ensure recursive feasibility. The conditions to ensure this are

1. system constraints are in Ω
2. Ω is invariant in mode 2

The first condition is simply a restatement of (7.14). Invariance of the region Ω then means that if the terminal state is part of the region, then so must be its closed-loop iteration according to the fixed feedback law [8, 9]:

$$x_{k+n_c} \in \Omega \Longrightarrow (\mathbf{A} + \mathbf{BK})x_{k+n_c} \in \Omega \tag{7.16}$$

The constrained MPC optimization has been defined by Algorithm 6.1 without stability guarantees. Now we will add the terminal constraints, so that the stability of the constrained MPC optimization will be now guaranteed. If we would like to steer system (6.1) into the origin, we may define the following algorithm:

Algorithm 7.1 To find the solution of the constrained model predictive control problem with guaranteed stability, perform the following set of operations at each sampling instant:

- Observe or measure actual system state at sample x_k.
- Minimize the following cost function with respect to constraints:

$$\min_{\mathbf{u}_k} J(\mathbf{u}_k, x_k) = \sum_{i=0}^{n_c-1} \left(x_{k+i}^T \mathbf{Q} x_{k+i} + \mathbf{u}_{k+i}^T \mathbf{R} \mathbf{u}_{k+i} \right) + x_{k+n_c}^T \mathbf{P}_f x_{k+n_c}$$

where $\mathbf{u}_k = [u_k, u_{k+1}, u_{k+2}, \ldots, u_{k+n_c-1}]$, $\mathbf{Q} = \mathbf{Q}^T \geq 0$ is a state penalization matrix, $\mathbf{R} = \mathbf{R}^T \geq 0$ is an input penalization matrix, n_c is the prediction horizon and $\mathbf{P}_f$ is the solution of the unconstrained, infinite horizon quadratic regulation problem. The typical MPC cost function must be subject to the following system and terminal constraints:

$$\underline{u} \leq u_{k+i} \leq \overline{u}, \qquad i = 0, 1, 2, \ldots, n_c - 1 \tag{7.17}$$

$$\underline{x} \leq x_{k+i} \leq \overline{x}, \qquad i = 1, 2, \ldots, n_c \tag{7.18}$$

$$x_{k+n_c} \in \Omega \tag{7.19}$$

$$x_{k+0} = x_k \tag{7.20}$$

$$x_{k+i+1} = \mathbf{A} x_{k+i} + \mathbf{B} u_{k+i}, \qquad i \geq 0 \tag{7.21}$$

$$y_{k+i} = \mathbf{C} x_{k+i}, \qquad i \geq 0 \tag{7.22}$$

$$u_{k+i} = \mathbf{K} x_{k+i}, \qquad i \geq n_c \tag{7.23}$$

where $\mathbf{K}$ is a stabilizing feedback gain.
- Apply the first element of the vector of optimal control moves $\mathbf{u}_k$ to the controlled system, and restart the procedure.

If Ω satisfies system constraints and is invariant in mode 2, then a system which is feasible at k will also remain feasible for $k = 1, 2, 3, \ldots, \infty$.

We may define the *region of attraction*, which is a subset of the state-space defining the set of all initial conditions from which is possible to drive the state predictions inside Ω over the n_c steps long mode 1 horizon [29]. Let us denote this set with S_Ω, then formally the above statement means that S_Ω is a collection of all initial conditions x_0 for which exists a vector of inputs $\mathbf{u}$ such that by the end of the horizon x_{n_c} will be a part of the set Ω:

$$S_\Omega = \left\{ x_0 : \text{exists } \mathbf{u}_k \text{ such that } \begin{array}{c} x_{n_c} \in \Omega \\ \underline{u} \leq u_{k+i} \leq \overline{u} \\ \underline{x} \leq x_{k+i+1} \leq \overline{x} \end{array} \right\} \tag{7.24}$$

where $i = 0, 1, \ldots, n_c - 1$. The region of attraction is the operating region of the MPC law. No state outside the region of attraction is feasible; therefore, it is in our

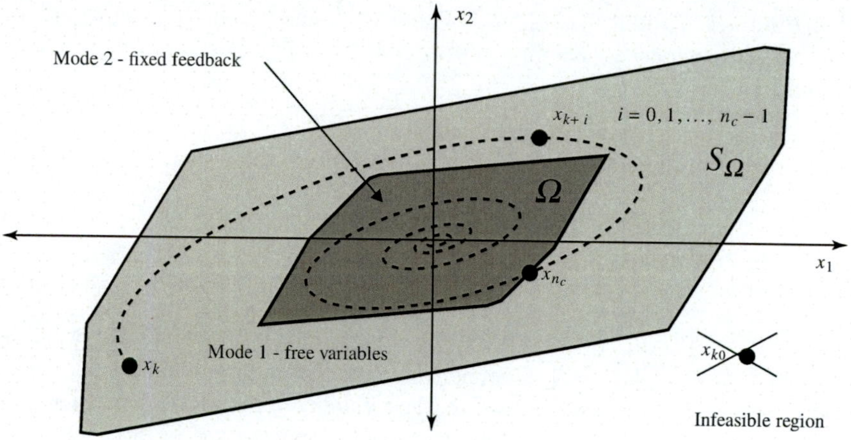

Fig. 7.4 Illustration of the region of attraction S_Ω, which is a set of all feasible initial states x_k. The target set Ω is illustrated as the smaller *dark* area and it is a subset of the region of attraction

interest to make it as large as possible. If one chooses to increase the horizon n_c the size of the region of attraction S_Ω will also increase as well, since the number of steps in which is possible to reach the target set is larger. Another approach is to increase the size of the target set Ω.

The region of attraction S_Ω is illustrated in Fig. 7.4 as the larger *shaded* area. All states within the region of attraction are feasible, which means that it is possible to steer them into the target set within n_c steps. The target set Ω is a subset of the region of attraction. States outside the region of attraction are infeasible and would violate constraints sometime in the future. A state trajectory is denoted with the *dotted* spiral shaped line. Mode 1 control is effective within the region of attraction, however the fixed feedback law in mode 2 is assumed active within the target set.

7.4 Maximal Invariant Terminal Set

For a given horizon length it is necessary to ensure the largest possible region of attraction, thus enlarging the operating region of the MPC controller. To do this, one must formulate the terminal constraints in a way that the largest possible target set Ω is created. The question of maximal admissible sets has been discussed by Gilbert and Tan in [23], while its ellipsoidal approximation has been given by Mayne et al. [38].

The largest possible target set is ensured if one creates a set of terminal constraints, which enforce system constraints over a horizon n_a called the *constraint checking horizon*. For an LTI system, the length of this constraint checking horizon is constant and determined offline.

Recursive feasibility requires that the system constraints are satisfied over the entire mode 2 horizon. This means that state and input constraints must be enforced for:

$$\underline{u} \leq u_{k+i} \leq \overline{u}, \quad i = n_c, n_c+1, n_c+2, \ldots, \infty \tag{7.25}$$

$$\underline{x} \leq x_{k+i} \leq \overline{x}, \quad i = n_c, n_c+1, n_c+2, \ldots, \infty \tag{7.26}$$

from which it is logical to assume that the largest possible target set is defined, if we enforce system constraints over the mode 2 horizon by assuming that Ω is formulated as follows:

$$\Omega = \left\{ x : \underline{u} \leq \mathbf{K}(\mathbf{A}+\mathbf{B}\mathbf{K})^i x \leq \overline{u}, \underline{x} \leq (\mathbf{A}+\mathbf{B}\mathbf{K})^i x \leq \overline{x} \right\} \tag{7.27}$$

for $i = n_c, n_c+1, n_c+2, \ldots, \infty$. Fortunately, instead of having an infinite set of constraints one may satisfy the constraints over the whole infinitely long mode 2 horizon by an n_a steps long finite horizon. Therefore, (7.27) will be defined instead of an infinitely long constraint checking horizon over $i = n_c, n_c+1, \ldots, n_a-1, n_a$.

To prove that a finite steps long constraint checking horizon can ensure system constraints over the infinite mode 2 horizon, let us assume a case where only input constraints $\overline{u}$ and $\underline{u}$ are considered. Moreover, let us denote Π_i as the set of initial conditions for which the input constraints are satisfied over a certain horizon length of n steps under a mode 2 fixed feedback law given by $u = \mathbf{K}x$ [14, 36]:

$$\Pi_i = \left\{ x : \underline{u} \leq \mathbf{K}(\mathbf{A}+\mathbf{B}\mathbf{K})^i x \leq \overline{u} \right\} \tag{7.28}$$

For this case, we may also assume that the set of initial conditions satisfying system constraints for the infinitely long mode 2 horizon can be replaced by a set of initial conditions determined by a finite steps long constraint checking horizon:

$$\Pi_\infty = \Pi_{n_a} \tag{7.29}$$

The closed-loop system $(\mathbf{A}+\mathbf{B}\mathbf{K})$ defines a stable matrix with eigenvalues smaller than the one according to $|\kappa\{\mathbf{A}+\mathbf{B}\mathbf{K}\}| < 1$. As we increase i into infinity, this matrix term will approach zero:

$$(\mathbf{A}+\mathbf{B}\mathbf{K})^i \to 0 \quad \text{as} \quad i \to \infty \tag{7.30}$$

Let us denote the perpendicular distance of the hyperplane defined by $(\mathbf{A}+\mathbf{B}\mathbf{K})^i x = \overline{u}$ from $x = 0$ by o_i. For any given x this distance o_i will tend to infinity as $i \to \infty$ [14]:

$$o_i = \frac{\overline{u}}{\left\| \mathbf{K}(\mathbf{A}+\mathbf{B}\mathbf{K})^i \right\|_2} \to \infty \quad \text{as} \quad i \to \infty \tag{7.31}$$

since the term $(\mathbf{A}+\mathbf{B}\mathbf{K})^i$ will tend to zero with increasing i because the closed-loop matrix is stable. This geometrically means that the upper constraints $\overline{u}$ and lower constraints $\underline{u}$ define hyperplanes associated with the increasing horizon. The strip between these hyperplanes will increase in size, and will be wider and wider as i approaches infinity.

For a constrained system, it is always possible to conceive an initial state that will go over system constraints sooner or later in the future, indicating that mode 2 constraints create a finite set in the overall state-space. This means that constraints must be violated sometimes in the future if the initial state x_0 is large enough, therefore Π_∞ must be actually finite assuming an observable pair $(\mathbf{A}+\mathbf{BK}), \mathbf{K}$. The strip of state-space defined by the upper and lower input constraint

$$\left\{x : \underline{u} \leq \mathbf{K}(\mathbf{A}+\mathbf{BK})^i x \leq \overline{u}\right\} \tag{7.32}$$

contains Π_∞ for all $i > n_a$ for a finite n_a. This implies that a finite horizon n_a must exist which ensures system constraint feasibility for the whole infinitely long mode two horizon. Then according to this, all points located in the finite Π_{n_a} are also located in Π_∞ .

Let us search for $\Pi_\infty = \Pi_{n_a}$ by assuming that increasing the horizon n_a will eventually cause that the next set will be equivalent to the previous and no improvements can be made in the size or $\Pi_{n_a+1} = \Pi_{n_a}$. Formally we can state that [14]:

Theorem 7.1 $\Pi_\infty = \Pi_{n_a}$ *if and only if* $\Pi_{n_a+1} = \Pi_{n_a}$ To see that Theorem 7.1 is true, let us follow the next line of thought:

Proof Assume that if $x \in \Pi_{n_a}$ then x is also in the next set $x \in \Pi_{n_a+1}$ for some n_a:

$$x \in \Pi_{n_a} \Longrightarrow x \in \Pi_{n_a+1} \Longrightarrow \underline{u} \leq \mathbf{K}(\mathbf{A}+\mathbf{BK})^{n_a+1} x \leq \overline{u} \tag{7.33}$$

Let us replace x with $x = (\mathbf{A}+\mathbf{BK})\check{x}$ and substitute it to the previous equation to get

$$\underline{u} \leq \mathbf{K}(\mathbf{A}+\mathbf{BK})^{n_a+1}\check{x} \leq \overline{u} \Longrightarrow \underline{u} \leq \mathbf{K}(\mathbf{A}+\mathbf{BK})^{n_a+2}\check{x} \leq \overline{u} \tag{7.34}$$

This means that whenever the state $x \in \Pi_{n_a+1}$ it is also $x \in \Pi_{n_a+2}$. If we apply this repeatedly, we can get to the conclusion that Eq. (7.33) also implies that for some n_a the state will be in the infinite mode 2 horizon $x \in \Pi_{n_a+1}$.

The practical meaning of Theorem 7.1 and Eq. (7.33) is that, in the case of input constraints it is sufficient to check whether the mode 2 constraints enforced over n_a also satisfy constraints over $n_a + 1$ steps. If this is true, then n_a is the constraint checking horizon. At each iteration, one must check whether the following statement is true:

$$\underline{u} \leq \mathbf{K}(\mathbf{A}+\mathbf{BK})^i x \leq \overline{u}, i = 0, 1, 2, \ldots, n_a \Longrightarrow \underline{u} \leq \mathbf{K}(\mathbf{A}+\mathbf{BK})^{n_a+1} x \leq \overline{u} \tag{7.35}$$

For a case with input constraints, the constraint checking horizon algorithm is [13]:

Algorithm 7.2 To compute the constraint checking horizon, initialize with an $n_a = 0$ long horizon in mode 2 and

1. Evaluate two linear programs,[2] one for the lower constraint $\underline{u}$ and one for the upper constraint $\overline{u}$ subject to constraints:

$$u_{\max,j} = \max_x \mathbf{K}_j(\mathbf{A}+\mathbf{BK})^{n_a+1}x \quad s.t. \quad \underline{u} \leq \mathbf{K}(\mathbf{A}+\mathbf{BK})^i x \leq \overline{u} \tag{7.36}$$

$$u_{\min,j} = \min_x \mathbf{K}_j(\mathbf{A}+\mathbf{BK})^{n_a+1}x \quad s.t. \quad \underline{u} \leq \mathbf{K}(\mathbf{A}+\mathbf{BK})^i x \leq \overline{u} \tag{7.37}$$

   where the constraints are formulated for $i = 0, 1, 2, \ldots, n_a$ and for input constraints[3] $j = 1, 2, \ldots, n_u$.
2. Check whether $u_{\max,j} \leq \overline{u}_j$ and $u_{\min,j} \geq \overline{u}_j$ for each $j = 1, 2, \ldots, n_u$. If this is satisfied, terminate the program and note n_a as the constraint checking horizon, otherwise continue.
3. Increase $n_a = n_a + 1$ horizon and resume at 2...

In other words, at each iteration (for each constrained input) we check whether the constraints over the first n_a steps will also satisfy constraints for the $n_a + 1$-th step. To do this we simply compute the maximal (minimal) input value which would be possible to compute with the fixed feedback law $\mathbf{K}(\mathbf{A} + \mathbf{BK})^{n_a+1}x$ at the next step at $n_a + 1$, given that the computation is constrained from steps $i = 0, \ldots, n_a$ with $\underline{u} \leq \mathbf{K}(\mathbf{A} + \mathbf{BK})^i x \leq \overline{u}$. If the computed maximal (minimal) value is larger (smaller) than the constraints, the process is repeated and the horizon is increased. On the other hand, if the constraints are satisfied, the program is terminated and the current n_a is the constraint checking horizon. To satisfy constraints over the infinite mode 2 long horizon, it is enough to solve a finite set of linear programs at each iteration to get to a finite constraint checking horizon n_a.

A maximal invariant target set is illustrated for the two-dimensional state-space ($n_x = 2$) for a system with symmetric input constraints in Fig. 7.5. The terminal constraints iterated through the constraint checking horizon enclose strips in the hyperspace. The width of these hyperplanes grows and they will be rotated around the origin as the constraint checking horizon grows. Finally there will be a set of strips, which encloses the maximal possible invariant target set $\Pi_{n_a} = \Pi_\infty$.

7.4.1 Implementing the Terminal Constraints

With the explicit knowledge of the constraint checking horizon n_a after performing Algorithm 7.2 we can change Algorithm 7.1 defining the dual-mode infinite horizon MPC problem with guaranteed stability to:

Algorithm 7.3 To find the solution of the constrained infinite horizon dual-mode model predictive control problem with guaranteed stability, perform the following set of operations at each sampling instant:

---

[2] Each with n_u complexity, if the input has n_u dimensions.

[3] Note that although x is the optimization variable, we are not searching which x maximizes (minimizes) this function but on the contrary, the value of the function which is not a state but an input value.

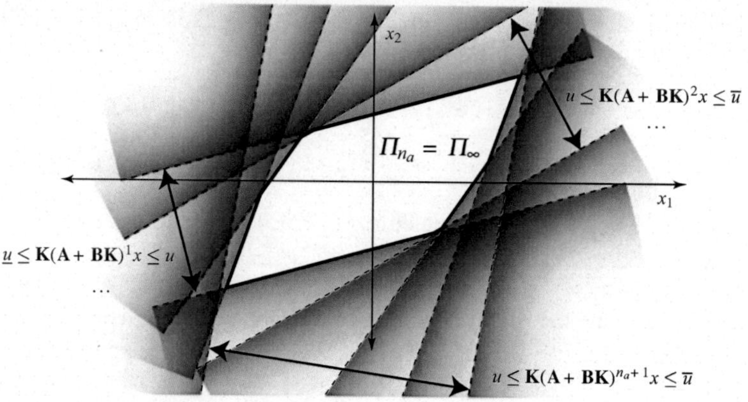

Fig. 7.5 Illustration of a maximal invariant target set in two-dimensional state-space. The strips defined by the terminal constraints create a maximal invariant target set $\Pi_{n_a} = \Pi_\infty$

- Observe or measure actual system state at sample x_k.
- Minimize the following cost function with respect to constraints:

$$\min_{\mathbf{u}_k} J(\mathbf{u}_k, x_k) = \sum_{i=0}^{n_c-1} \left(x_{k+i}^T \mathbf{Q} x_{k+i} + \mathbf{u}_{k+i}^T \mathbf{R} \mathbf{u}_{k+i} \right) + x_{k+n_c}^T \mathbf{P}_f x_{k+n_c}$$

where $\mathbf{u}_k = \left[u_k, u_{k+1}, u_{k+2}, \ldots, u_{k+n_c-1} \right]$, $\mathbf{Q} = \mathbf{Q}^T \geq 0$ is a state penalization matrix, $\mathbf{R} = \mathbf{R}^T \geq 0$ is an input penalization matrix, n_c is the prediction horizon and $\mathbf{P}_f$ is the solution of the unconstrained, infinite horizon quadratic regulation problem, The MPC cost function must be subject to the following system and terminal constraints:

$$\underline{u} \leq u_{k+i} \leq \overline{u}, \qquad i = 0, 1, 2, \ldots, n_c - 1 \tag{7.38}$$

$$\underline{x} \leq x_{k+i} \leq \overline{x}, \qquad i = 1, 2, \ldots, n_c \tag{7.39}$$

$$\underline{u} \leq \mathbf{K}(\mathbf{A} + \mathbf{BK})^i x_{k+n_c} \leq \overline{u}, \qquad i = 0, 1, 2, \ldots, n_c, \ldots, n_c + n_a \tag{7.40}$$

$$\underline{x} \leq (\mathbf{A} + \mathbf{BK})^i x_{k+n_c} \leq \overline{x}, \qquad i = 0, 1, 2, \ldots, n_c, \ldots, n_c + n_a \tag{7.41}$$

$$x_{k+0} = x_k \tag{7.42}$$

$$x_{k+i+1} = \mathbf{A} x_{k+i} + \mathbf{B} u_{k+i}, \qquad i \geq 0 \tag{7.43}$$

$$y_{k+i} = \mathbf{C} x_{k+i}, \qquad i \geq 0 \tag{7.44}$$

$$u_{k+i} = \mathbf{K} x_{k+i}, \qquad i \geq n_c \tag{7.45}$$

where $\mathbf{K}$ is a stabilizing feedback gain, n_c is the control and prediction horizon and n_a is the control checking horizon.

- Apply the first element of the vector of optimal control moves $\mathbf{u}_k$ to the controlled system, and restart the procedure.

7.4.2 Horizon Length

The inclusion of the terminal constraints in this quadratic programming problem does not increase the computational cost significantly, since most of the online computation effort is spent on the QP itself. The additional terminal constraints over the n_a steps long constraint checking horizon are linear. As it will be demonstrated later, computational time may be saved by using other types of terminal constraints. This is not resulting because of the direct online computational savings on the additional constraints, but rather because alternative constraint formulations may also allow for alternative formulations of the minimization of the cost itself.

The size of the feasible initial conditions S_Ω will increase with an increased horizon n_c. This proves to be essential with lightly damped vibrating systems, as the large discrepancy between actuator capabilities and expected deformations calls for a large region of attraction. It is likely that in a vibration attenuation application the horizon n_c will be kept at high values anyways.

In addition to enlarging the region of attraction S_Ω, increasing the horizon has other effects as well. That is the increase of the performance of the closed-loop system, or the decrease of the closed-loop cost function

$$J = \sum_{k=0}^{\infty} \left(x_k^T \mathbf{Q} x_k + u_k^T \mathbf{R} u_k \right) \tag{7.46}$$

The performance of the MPC law will increase because of the reduction of the predicted cost.

However, this performance increase has its limits, and there is a certain horizon n_c over which the optimality of the closed-loop system will not improve. This limit is known as the *constrained LQ-optimal performance* and it ensures the performance equivalent to an infinite number of degrees of freedom MPC problem. The reason why the performance cannot be increased beyond a further limit is given by the fact that terminal constraints must be inactive for a sufficiently large n_c, so there cannot be any further reduction to the cost. In the problem of active vibration damping the requirement for a large region of attraction will dominate when the horizon n_c is designed; therefore it is likely that a constrained LQ optimal performance will be reached anyway. It is possible to perform a simulation analysis, where one calculates closed-loop performance J for different horizons in order to assess whether an increased horizon brings an optimality improvement. Alternatively, one may compare the closed-loop cost J to the cost predicted at the initial time J_0 and see if they are identical. If the closed-loop cost is smaller, improvement can be made, but if the two are identical, the constrained LQ-optimal performance has been already reached.

7.5 Simplified Polyhedral Target Sets

As demonstrated before, the inclusion of a constraint checking horizon n_a in dual-mode MPC is the most straightforward way to ensure stability a priori while also reaching maximum performance. However, polyhedral target sets created by process constraints may be very complex in certain cases, especially with higher order prediction models. This complexity directly translates to the computational effort necessary to acquire the evolution of future process inputs at each sampling instant. It is therefore sometimes desirable to approximate the complex polytope $\Pi_{n_a} = \Pi_\infty$ created by constraints with a simplified shape. These polyhedral target sets are in fact created by an assembly of simpler elements: hyperspaces[4] and the half spaces bounded by them [25]. According to this, it is also possible to create an invariant target set that is bounded by a smaller number of hyperspaces but still ensures the stability of the MPC law. Several alternative stabilized MPC approaches rely on such simplifications. The upside is the reduction of the computational effort; however, as the ideal target set is only approximated with a simplified equivalent, the performance of the control law will suffer as well.

Let us consider a simple regulation problem where we would like to steer our system state into zero. For this regulation problem, we are aiming to minimize the cost function such as in (6.30) with respect to constraints. But, instead of using the high complexity target set Π_{n_a} to ensure stability as discussed before, let us imagine a simplified polyhedral target set Π_s instead. The state x shall remain within this set which shall be defined by:

$$\{x : \mathbf{V}_s x \leq 1\} \quad \text{where} \quad \mathbf{V}_s \in \mathbb{R}^{n \times n} \tag{7.47}$$

where $\mathbf{V}_s$ is a matrix defining a simplified polyhedral target set. Essentially, this defines a hypercube and our aim is to determine what matrix $\mathbf{V}_s$ will be. The conditions for invariance are defined by the properties of the set, see for example the paper by Bitsoris [7].

A low complexity polyhedral invariant set is illustrated for a second order system in Fig. 7.6. If at time (k) the set is defined by $|\mathbf{V}_s x_k| \leq 1$, then at the next time step $(k+1)$ should be smaller and described by the following relation:

$$|\mathbf{V}_s x_k| \leq 1 \longrightarrow |\mathbf{V}_s \boldsymbol{\Phi} x_k| \leq 1 \quad \text{where} \quad \boldsymbol{\Phi} = (\mathbf{A} + \mathbf{BK}) \tag{7.48}$$

It is possible to rewrite this by inserting $\mathbf{V}_s^{-1}\mathbf{V}_s = 1$ and obtain $\left|\mathbf{V}_s \boldsymbol{\Phi} \mathbf{V}_s^{-1} \mathbf{V}_s x_k\right| \leq 1$. Let us denote elements of the product $\mathbf{V}_s \boldsymbol{\Phi} \mathbf{V}_s^{-1}$ with k_{ij} and elements of the product $\mathbf{V}_s x_k$ with l_j. The definition of the invariant set (7.47) actually states that in the worst-case scenario the elements of $\mathbf{V}_s x_k$ will equal to 1: $|l_1| = \cdots = |l_n| = 1$. We can utilize a second order system to illustrate the situation:

$$\max_{x \in \Pi_s} |k_{11} l_1 + k_{12} l_2| = |k_{11}| + |k_{12}| \tag{7.49}$$

$$\max_{x \in \Pi_s} |k_{21} l_1 + k_{22} l_2| = |k_{21}| + |k_{22}| \tag{7.50}$$

---

[4] It is also possible to represent polyhedra in a vertex-based representation instead of hyperspaces, see the book by Ziegler [66].

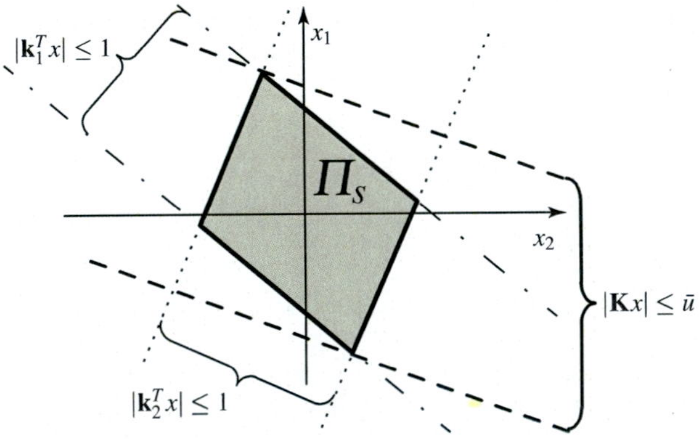

Fig. 7.6 Low complexity polyhedral invariant set, where $\mathbf{k}_1 = [k_{11} \quad k_{12}]^T$ and $\mathbf{k}_1 = [k_{21} \quad k_{22}]^T$

then

$$\begin{vmatrix} |k_{11}| & |k_{12}| \\ |k_{21}| & |k_{22}| \end{vmatrix} \begin{vmatrix} 1 \\ 1 \end{vmatrix} = \begin{vmatrix} 1 \\ 1 \end{vmatrix} \tag{7.51}$$

From this, it is clear that the condition of invariance may be transformed to a very convenient form:

$$\left| \mathbf{V}_s \boldsymbol{\Phi} \mathbf{V}_s^{-1} \right| 1 = 1 \tag{7.52}$$

In addition to (7.52) which sufficiently defines invariance, feasibility conditions need to be defined as well. The simplest case is to have symmetric constraints only on the control input. In this case we have to ensure:

$$|\mathbf{K}x_k| \leq \bar{u} \longrightarrow |\mathbf{K}\mathbf{V}_s^{-1}\mathbf{V}_s x_k| \leq \bar{u} \tag{7.53}$$

where $\mathbf{V}_s^{-1}\mathbf{V}_s$ was inserted to the second equation. Similarly to the invariance condition, the definition of the invariant set ensures that $|\mathbf{V}_s x_k| \leq 1$, therefore in the worst case $|l_1| = \cdots = |l_n| = 1$. This simplifies the problem of feasibility to:

$$|\mathbf{K}\mathbf{V}_s^{-1}| 1 \leq \bar{u} \tag{7.54}$$

To find our invariant set defined by $\mathbf{V}_s$ let us state the eigenvalue problem for the matrix $\boldsymbol{\Phi}$:

$$\begin{aligned} \boldsymbol{\Phi}\Delta &= \Delta\Lambda \\ \Delta^{-1}\boldsymbol{\Phi}\Delta &= \Lambda \\ \Delta &= [\delta_1\ \delta_2\ \ldots \delta_{n_c}] \\ \Lambda &= \begin{bmatrix} \kappa_1 & & \cdots 0 \\ \kappa_2 & & \vdots \\ \vdots & & \ddots \\ 0 & \cdots & \kappa_n \end{bmatrix} \end{aligned} \tag{7.55}$$

where Δ is a matrix of eigenvectors δ_i, Λ is a diagonal matrix containing eigenvalues κ_i. We can utilize the inverse of the eigenvector matrix scaled with α_s to choose a suitable $\mathbf{V}_s$:

$$\mathbf{V}_s = \alpha_s \Delta^{-1} \tag{7.56}$$

the conditions for invariance will transform to

$$|\alpha_s \Delta^{-1} \boldsymbol{\Phi} \alpha_s^{-1} \Delta| 1 \leq 1 \longrightarrow |\Lambda| 1 \leq 1 \tag{7.57}$$

Equation (7.57) tells us that the eigenvalues of the closed-loop system need to be real and from within the unit disk, more formally: $|\kappa_i| \leq 1$. It is possible to manipulate the eigenvalues of the closed-loop system by pole placement, although this fundamental problem may have a surprisingly large computational complexity [10, 11]. By conforming to the former requirement, it is possible now to rewrite the conditions for feasibility:

$$|\mathbf{K}\alpha_s^{-1} \Delta| \leq \bar{u} \longrightarrow \alpha_s^{-1} |\mathbf{K}\Delta| 1 \leq \bar{u} \tag{7.58}$$

Finally, the conditions for feasibility will transform into a convenient form:

$$\alpha_s \geq |\mathbf{K}\Delta| 1 / \bar{u} \tag{7.59}$$

We can summarize the algorithm for model predictive control with guaranteed stability, utilizing simplified polyhedral invariant target sets as follows:

Algorithm 7.4

- Find the multiplier α_s using relation (7.59)
- Find $\mathbf{V}_s$ defining the simplified polyhedral target set using (7.56)
- Perform the minimization of J_k subject to constraints $|\mathbf{u}_k| \leq \bar{u}$ and $|\mathbf{V}_s x_{n_c}| \leq 1$

The constraints need to be fed to the quadratic programming solver, mostly in the form $\mathbf{A}_c \mathbf{u}_k \leq \mathbf{b}_0 + \mathbf{B}_c x_k$. We have an additional constraint, defining the terminal state $-1 \leq \mathbf{V}_s x_{n_c} \leq 1$, where $x_{k+n_c} = \mathbf{M}_{n_c} x_k + \mathbf{N}_{n_c} \mathbf{u}_k$. Generally, we can define these constraints according to the following relation:

$$\begin{bmatrix} 1 & \cdots & 0 \\ \vdots & 1 & \vdots \\ 0 & \cdots & 1 \\ & \mathbf{V}_s \mathbf{N}_{n_c} & \end{bmatrix} \mathbf{u}_k \leq \begin{bmatrix} \bar{u} \\ \vdots \\ \bar{u} \\ \mathbf{I} \end{bmatrix} + \begin{bmatrix} 0 & \cdots & 0 \\ \vdots & \ddots & \vdots \\ 0 & \cdots & 0 \\ & \mathbf{V}_s \mathbf{M}_{n_c} & \end{bmatrix} x_k \tag{7.60}$$

State and output constraints may be defined similarly.

A practical issue with the construction of low complexity invariant target sets is the occurrence of complex eigenvalues κ_i of the closed-loop matrix $\boldsymbol{\Phi}$. In this case, the definition matrix of the polyhedral set Γ will contain a pair of complex conjugate eigenvectors and the set will remain open [27]. To solve this situation and close the target set elementary rotation matrices may be used [50], which essentially break down eigenvectors to their real and imaginary components. Such and similar operations may cause that the condition of invariance is not met for certain complex conjugate eigenvalues. Kouvaritakis et al. in their conference article in [31] review for which eigenvalues is the invariance condition still valid. A formulation for continuous systems is also possible [49], while its application to the pole-placement of gain matrices for fixed feedback systems is described by Rusko in [52].

The introduction of non-symmetric constraints $\overline{u} \neq \underline{u}$ or rate of change constraints $\Delta\overline{u}, \Delta\underline{u}, \Delta\overline{x}, \Delta\underline{x},$ and $\Delta\overline{y}, \Delta\underline{y},$ may require more complex formulations [26, 27, 31].

7.6 Elliptic Invariant Target Sets

The previous section introduced a formulation where the complex polyhedral set $\Pi_{n_a} = \Pi_{\infty}$ created by the process constraints in the constraint checking horizon of optimal dual-mode QPMPC were replaced by a simplified invariant target set Π_s. An additional possibility to replace the maximal target set Π_{∞} with a simpler approximation is the use of elliptic invariant target sets [38]. Geometric stability guarantees based on the elliptic invariant set formulation are the cornerstone of the efficient algorithm considered in Sect. 8.1 of the upcoming chapter. In general, the construction of ellipsoidal target sets is based on the Lyapunov or Ricatti equation [65] and linear matrix inequalities (LMI).

The shape of the invariant ellipsoidal target set will be an ellipse in the case of a second, an ellipsoid for a third order system. For a second order system, an ellipsoidal set is illustrated in Fig. 7.7. In case the system order is larger than three, we talk about a hyperellipsoid. Unfortunately, it is difficult to illustrate hyperellipsoids graphically without creating confusion; therefore, the illustrations will assume a second order system. Generally, we may describe the target set by the following expression:

$$E_x = \left\{ x | x^T \Gamma x \leq 1 \right\} \tag{7.61}$$

The aim is to find the matrix Γ such that the ellipsoid will enclose the largest invariant target set within the bounds and conforming the constraints. Naturally, it is desirable to make the target set—this case an ellipsoid—as large as possible. The conditions for invariance relate to the basic idea of invariant sets. If a system state in a certain point of time is contained within the ellipsoid, so must it be at the next time step:

$$x_k \in E_x \rightarrow x_{k+1} \in E_x \rightarrow x_{k+2} \in E_x \rightarrow \ldots \tag{7.62}$$

This condition may be expressed by stating that the ellipsoid in the next time step must be smaller or at least equal to the one in the current time steps:

$$x_{k+1}^T \Gamma x_{k+1} \leq x_k^T \Gamma x_k \tag{7.63}$$

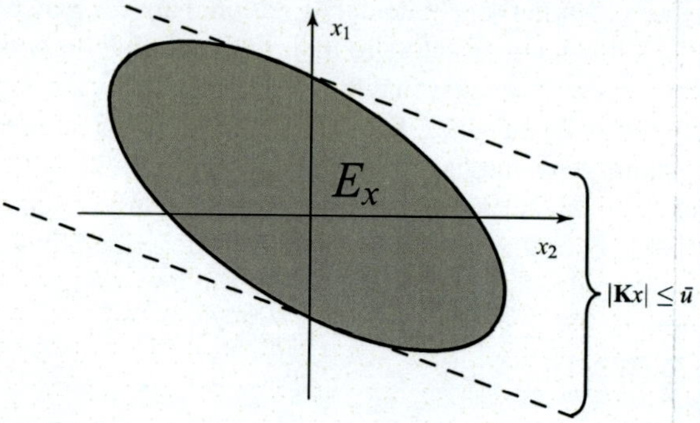

Fig. 7.7 Illustration of an elliptic invariant target set

We can take advantage of the fact that $x_{k+1} = (\mathbf{A}+\mathbf{BK})x_k$, where $\mathbf{K}$ is actually $\mathbf{K}_{LQ}$ in dual-mode. Therefore, if $(\mathbf{A} + \mathbf{BK})$ is substituted by the closed-loop matrix $\boldsymbol{\Phi}$, we obtain

$$x_k^T \boldsymbol{\Phi}^T \boldsymbol{\Gamma} \boldsymbol{\Phi} x_k \leq x_k^T \boldsymbol{\Gamma} x_k \tag{7.64}$$

From this, by rearranging we get

$$-x_k^T \boldsymbol{\Gamma} x_k + x_k^T \boldsymbol{\Phi}^T \boldsymbol{\Gamma} \boldsymbol{\Phi} x_k \leq 0 \tag{7.65}$$

The final condition for invariance of the hyperellipsoid is:

$$\boldsymbol{\Gamma} - \boldsymbol{\Phi}^T \boldsymbol{\Gamma} \boldsymbol{\Phi} \geq 0 \tag{7.66}$$

In addition to the invariance condition, there are input and possibly state constraints present. Let us consider the case of the simple symmetric input constraints, defined by:

$$|\mathbf{K}x| \leq \overline{u} \tag{7.67}$$

Utilizing the identity $\boldsymbol{\Gamma}^{-\frac{1}{2}} \boldsymbol{\Gamma}^{\frac{1}{2}} = \mathbf{I}$, we may transform (7.67) to:

$$|\mathbf{K}\boldsymbol{\Gamma}^{-\frac{1}{2}} \boldsymbol{\Gamma}^{\frac{1}{2}} x|^2 \leq \overline{u}^2 \tag{7.68}$$

which may also be equivalently denoted as:

$$||\mathbf{K}\boldsymbol{\Gamma}^{-\frac{1}{2}}||^2 ||\boldsymbol{\Gamma}^{\frac{1}{2}} x||^2 \leq \overline{u}^2 \tag{7.69}$$

The second term on the left side of (7.69) is simply $x^T \boldsymbol{\Gamma}^{\frac{1}{2}^T} \boldsymbol{\Gamma}^{\frac{1}{2}} x = x^T \boldsymbol{\Gamma} x$. According to Eq. (7.61) $x^T \boldsymbol{\Gamma} x \leq 1$ which implies that the second term can have a value of one in the worst case. Therefore, we may rewrite the conditions of feasibility:

$$||\mathbf{K}\boldsymbol{\Gamma}^{-1} \mathbf{K}^T|| \leq \overline{u}^2 \tag{7.70}$$

To calculate Γ from the conditions of invariance, one needs to employ semidefinite programming or SDP as it is often referred to. First, it is necessary to transform the invariance conditions to a more convenient form using Schur complements. According to Schur complements, the following is valid [13]:

$$\begin{vmatrix} \mathbf{A} & \mathbf{B} \\ \mathbf{B}^T & \mathbf{C} \end{vmatrix} \geq 0 \Longleftrightarrow \begin{matrix} \mathbf{A} - \mathbf{B}\mathbf{C}^{-1}\mathbf{B}^T \geq 0, \ \mathbf{C} > 0 \\ \mathbf{C} - \mathbf{B}^T\mathbf{A}\mathbf{B} \geq 0, \ \mathbf{A} > 0 \end{matrix} \tag{7.71}$$

A function in the form $\mathbf{F}(p) > 0$ where $\mathbf{F}(p) = \mathbf{p}_{11}\mathbf{I}_{11} + \mathbf{p}_{12}\mathbf{I}_{12} + \cdots$ is called a linear matrix inequality[5] [12]. The invariance condition is (7.66) and in addition to that it is necessary for the eigenvalues of Γ to be positive, that is, $\Gamma > 0$. If we multiply both sides of the invariance condition by Γ^{-1} we obtain:

$$\begin{matrix} \Gamma^{-1}(\Gamma - \boldsymbol{\Phi}^T\Gamma\boldsymbol{\Phi})\Gamma^{-1} \geq 0 \\ \Gamma > 0 \end{matrix} \Longleftrightarrow \begin{matrix} \Gamma^{-1} - \Gamma^{-1}\boldsymbol{\Phi}^T\Gamma\boldsymbol{\Phi}\Gamma^{-1} \geq 0 \\ \Gamma > 0 \end{matrix} \tag{7.72}$$

According to Schur complements, it is possible to rewrite this relation to

$$\begin{vmatrix} \Gamma^{-1} & \Gamma^{-1}\boldsymbol{\Phi} \\ \boldsymbol{\Phi}\Gamma^{-1} & \Gamma^{-1} \end{vmatrix} \geq 0 \tag{7.73}$$

Although we have Γ^{-1} instead of Γ, it is possible to calculate Γ^{-1} and invert it afterward. For feasibility, we have

$$\mathbf{K}_i\Gamma^{-1}\mathbf{K}_i^T < \overline{u}_i^2 \tag{7.74}$$

$$\mathbf{K} = \begin{bmatrix} \mathbf{K}_1 \\ \mathbf{K}_2 \\ \vdots \\ \mathbf{K}_{nu} \end{bmatrix}, \quad \bar{\mathbf{u}} = \begin{bmatrix} \overline{u}_1 \\ \overline{u}_2 \\ \vdots \\ \overline{u}_{nu} \end{bmatrix}$$

where the index i represents the i-th row of the LQ optimal gain matrix $\mathbf{K}$ and the i-th element of the vector $\bar{\mathbf{u}}$ for a general multiple-input system. As previously mentioned, it is necessary to maximize the volume of the hyperellipsoid, or in the case of a second order system, the area of an ellipse-subject to feasibility and invariance constraints. If an ellipse is described by (7.61), then its major and minor axes are defined by the reciprocals of the eigenvalues of the matrix Γ (Fig. 7.8). The area of an ellipse (valid for a second order system) may be calculated by:

$$\mathrm{Vol} E_x = \frac{\pi}{\kappa_1\kappa_2} = \pi \det \Gamma^{-1} \tag{7.75}$$

[5] Most mathematics and optimization-related publications use the notation $\mathbf{F}(p) \succ 0$ to denote the concept of positive definiteness. This book will denote this concept with simple relation signs.

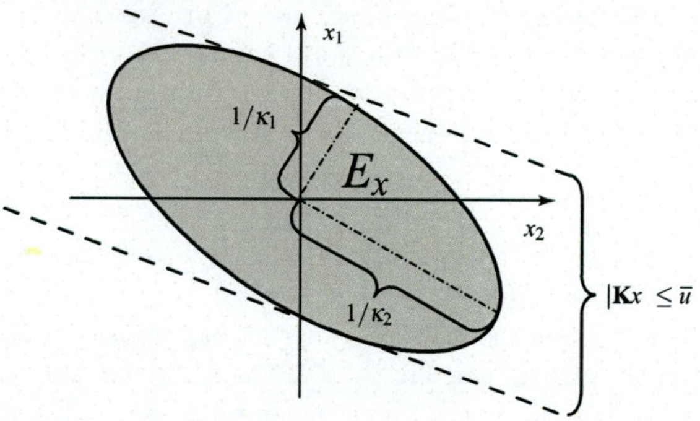

Fig. 7.8 Major and minor axes of an ellipsoid E_x expressed by the reciprocals of the eigenvalues of the definition matrix Γ

and the volume of a generic hyperellipsoid is calculated by evaluating:

$$\mathrm{Vol} E_x = h \det \Gamma^{-1} \tag{7.76}$$

where h is an unknown number. Finding Γ is an offline optimization problem:

Algorithm 7.5 Perform the maximization of the determinant according to [62] in offline mode:

$$\max \mathrm{Vol} E_x = h \det \Gamma^{-1} \tag{7.77}$$

Subject to the invariance condition (7.73), feasibility condition (7.74) and possibly a shape conditioning constraint, for example:

$$h\mathbf{I} < \Gamma^{-1} \text{ or trace } (\Gamma) < 1 \text{ etc} \ldots \tag{7.78}$$

The so-called shape conditioning ensures that the hyperellipsoid axes will not be distorted in a particular direction. That said, it avoids an infinitely thin and long ellipsoid-which otherwise would have the maximal volume and conform to all the conditions. To realize this in practice, one needs to deploy a solver for semidefinite programming (SDP) [42]. For this purpose, a rational choice is the freely available *SeDuMi* solver [46, 59]; with an LMI parser called *YALMIP* [34]. The next problem to solve is the actual online computation. The online algorithm may be described by:

Algorithm 7.6 Solve the following optimization problem at each time instant:

$$\min_u J_k = \mathbf{u}_k^T \mathbf{H} \mathbf{u}_k + 2x_k^T \mathbf{G} \mathbf{u}_k + x_k^T \mathbf{F} x_k \tag{7.79}$$

Feasibility: $\mathbf{A}_c \mathbf{u}_k \leq \mathbf{b}_0 + \mathbf{B}_c x_k$

Invariance: $x_{n_c}^T \Gamma x_{n_c} \leq 1$

This algorithm presents a quadratic optimization problem with quadratic constraints. The evaluation of this task in the original form may be formidable. Fortunately, this formulation may be changed to a second order cone programming (SOCP) problem. Constraints for a second order cone programming problem are given in the following form:

$$||\mathbf{A}x + \mathbf{b}||_2 \leq \mathbf{C}x + \mathbf{d} \tag{7.80}$$

where x is the variable to be optimized, and $\mathbf{A}$, $\mathbf{b}$, $\mathbf{C}$ and $\mathbf{d}$ are given optimization parameters. The expression on the left side of the equation is in the two norm, also referred to as Euclidean norm, where

$$||\mathbf{A}x + \mathbf{b}||_2 = \left[(\mathbf{A}x + \mathbf{b})^T (\mathbf{A}x + \mathbf{b})\right]^{\frac{1}{2}} \tag{7.81}$$

In the light of this information, the original definition of ellipse $x_{n_c}^T \Gamma x_{n_c} \leq 1$ can be described equivalently as

$$||\Gamma^{\frac{1}{2}} x_{n_c}||_2 \leq 1 \tag{7.82}$$

where x_{n_c} denotes the state at the end of the prediction horizon: $x_{n_c} = \mathbf{M}_{n_c} x_k + \mathbf{N}_{n_c} \mathbf{u}_k$. The new invariance condition in (7.79) will be

$$||\Gamma^{\frac{1}{2}} \mathbf{M}_{n_c} x_k + \Gamma^{\frac{1}{2}} \mathbf{N}_{n_c} \mathbf{u}_k||_2 \leq 1 \tag{7.83}$$

The quadratic optimization problem can be transformed into a second order cone programming form as well:

$$\begin{aligned} J_k &= (\mathbf{H}^{\frac{1}{2}} \mathbf{u}_k + \mathbf{H}^{-\frac{1}{2}} \mathbf{G}^T x_k)^T (\mathbf{H}^{\frac{1}{2}} \mathbf{u}_k + \mathbf{H}^{-\frac{1}{2}} \mathbf{G}^T x_k) \\ &\quad + x_k^T \mathbf{F} x_k - x_k^T \mathbf{G} \mathbf{H}^{-1} \mathbf{G}^T x_k \end{aligned} \tag{7.84}$$

The last two terms of the equation are negligible, since we have a minimization problem to solve. Therefore, we have a new expression in the following form:

$$||\mathbf{H}^{\frac{1}{2}} \mathbf{u}_k + \mathbf{H}^{-\frac{1}{2}} \mathbf{G}^T x_k||_2 \leq h \tag{7.85}$$

where h is a new scalar optimization variable. The optimization algorithm is transformed to:

Algorithm 7.7 Evaluate the following second order cone programming problem at each sampling instant:

$$\min_{\mathbf{u}_k, h}(h) \tag{7.86}$$

Subject to the transformed invariance condition (7.83) and the feasibility condition. The new transformed optimization problem is now expressed as an additional constraint:

$$||\mathbf{H}^{\frac{1}{2}} \mathbf{u}_k + \mathbf{H}^{-\frac{1}{2}} \mathbf{G}^T x_k||_2 \leq h \tag{7.87}$$

7.7 Infeasibility Handling

There are many kinds of constraints in the MPC formulation that must be taken into account in practical control applications: safety limitations, physical restrictions, technological requirements, product quality specifications, etc. The importance of constraints is also reinforced by the fact that, in practice, the optimal operating point very often lies on one or more of the boundaries defined by the applied constraints.

The relatively simple implementation of equality or inequality constraints into the task of minimizing the criterion function in predictive control [see e.g. the generic QP problem in Eq. (6.61)] may introduce very significant problems related to their compatibility. Incompatible constraints can cause that the optimization problem may be insolvable, respectively the given optimization problem is *infeasible*. The term infeasibility can be defined as an inability to satisfy all the constraints simultaneously. These problems arise when the restrictions on the relevant variables define an empty area and the optimization problem of minimizing the criterion function does not provide an adequate solution. An infeasible set of constraints may occur as a result of disturbances, operator failure, actuator or control system failure, bad design; ultimately causing the optimization problem to become incompatible in certain steps. It may also happen that the numerical algorithm minimizing the criterion leads the system outside the feasible region. The problem of feasibility is often referred to as a compatibility problem or the realization problem of the constraints. A general solution to the feasibility problem does not exist and therefore the issue needs attention in a real control application. In general, the constraints in MPC algorithms can be interpreted as [51]:

- *Hard constraints* are constraints that must always be satisfied. For example, hard constraints may be physical limits on actuators or safety limits. A control scheme ideally shall not use tactics to violate hard constraints, as this is either physically impossible or would lead to catastrophic results. If hard constraints would be violated, a mismatch between the predicted and actual closed-loop plant would occur leading to serious consequences and even loss of stability.
- *Soft constraints* are those, which should be satisfied only if possible. It is assumed that if necessary, soft constraints can be violated (ignored). Soft constraints are usually enforced on output or state variables, although they could also be applied to inputs. In a practical sense, the constraints are nonessential, only preferred.
- *Terminal constraints* are somewhat artificial in a sense that they arise from the stability guarantee conditions of the control algorithm, which have been discussed in detail previously. They can be defined in the form of equality or inequality conditions given on terminal state and terminal region. In fact, they represent a mixture of hard and soft constraints.

We may distinguish two types of infeasibility problems for various incompatible constraint configurations [53]:

- *Type I infeasibility* is caused by incompatibility between equality and inequality constraints, e.g. the inequality constraints define a nonempty region $\Phi_{\text{in}} \neq 0$ and $\Phi_{\text{in}} \cap \Phi_{\text{eq}} \equiv 0$, where Φ_{eq} is the region created by the equality constraints
- *Type II infeasibilities* are caused by incompatibility between the inequality constraints because they define an empty region $\Phi_{\text{in}} \equiv 0$, e.g. $u_k \leq 1$ and $y_k \geq 2$ when the process has unity DC gain

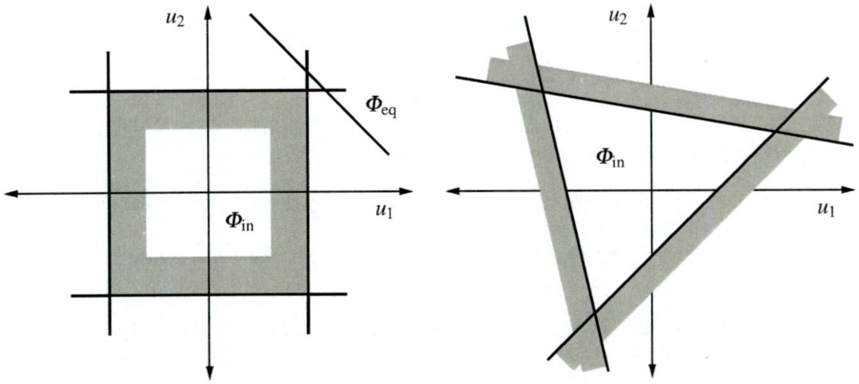

(a) Type I infeasibility **(b)** Type II infeasibility

Fig. 7.9 Illustration of the two main types of infeasible constraint configurations for an input constrained problem with two elements. **a** Type I infeasibility. **b** Type II infeasibility

The issue of Type I infeasibilities is very important since some earlier MPC stabilizing strategies (such as CRHPC, SGPC or SIORHC) rely on the end point equality constraints in Eq. (7.1) to ensure stability. It can be shown that, if at least one inequality constraint is imposed, it is always possible to find a set-point sequence causing this type of incompatibility. Careful design of constraints cannot guarantee feasibility; hence, there are situations when such a stabilizing control is infeasible. Type II infeasibilities usually arise because of either poor design, or the nature of the plant and the presence of disturbances. The resulting mismatch between the predicted and actual plant behavior can than lead to serious consequences in online control. A simple example of the both types of infeasibilities is illustrated in Fig. 7.9, where incompatible constraint configurations of Type I and II are given for a two-dimensional input vector $u = [u_1 \quad u_2]^T$.

One may see that there is a possibility that the minimization problem of the MPC cost function subject to design constraints may not have a solution at all. This is why it is necessary to devise procedures for the effective handling of infeasibility. All practical MPC implementations should have means to recover from infeasibility, shall that occur during the real-time control process.

Let us briefly discuss some typical techniques for avoiding infeasibility. One way of handling (Type I) infeasibilities is the *set point management* technique [21, 22]. An obvious case of infeasibility is due to rapid set point changes. This implies a large change in the terminal constraint set (due to a shift in steady-state values) and hence these may become inconsistent. The key philosophy of set point management algorithms is to establish a set point different from the true one, when changes in true set point would otherwise cause infeasibility. The controller set point therefore implements slower changes in the value than the true set point would necessitate. Simple algorithms implementing this set point change strategy can be found in [51].

The next two techniques referred to as *constraint removal* and *constraint softening* can be applied for handling infeasibilities of both types. The idea is, when an infeasibility arises, control is continued as unconstrained. Feasibility is checked at every sample and a full set of constraints are reintroduced back as soon as they become feasible. If the set of constraints is found to be inconsistent, then some constraint must be either relaxed or removed. A simple process-dependent removal strategy could be developed using the following logic [51]:

Algorithm 7.8 At each sampling instant, perform the following algorithm:

- Test for feasibility, if found infeasible then

1. relax (or remove) the least important soft constraint, test for feasibility
2. if the remaining set of constraints is feasible, pass on to the MPC algorithm and start optimization
3. else repeat the cycle again from step 1

- else pass on full set of constraints to the MPC algorithm and start the optimization.

Naturally, the algorithm relates to those constraints that are predicted to be violated, relaxing nonactive constraints will change nothing. The hope is that once enough soft constraints have been relaxed, the whole constraint set will become feasible and one can continue. The decision-making process is taken by a supervisory controller level, before the constraints are downloaded to the MPC algorithm. A more subtle variant is a *hierarchical strategy* [32], where the user is asked to assign a priority index to each group of constraints at the design stage. This index expresses the relative importance of a particular group. At every sample, the full set of constraints is checked for compatibility. At the time of infeasibilities, the supervisory controller level uses the priority indices to determine a set of low priority constraints, which must be removed to reestablish feasibility. These constraints are then reintroduced into the control law as soon as possible. An important modification is manipulation with the *lower constraint horizon*, where the removal of constraints is performed by increases made in the value of the horizon. At the time of infeasibilities, the conflict between constraints is resolved by defining a new value of the lower constraint horizon such that the set of constraints is feasible.

The technique of constraint softening involves removing certain constraints at times of infeasibilities and adding a term to the cost function that penalizes violations of temporarily discarded constraints. Similar to the removal strategy, a hierarchical constraint softening can be introduced.

The topic of feasibility and handling infeasible constraint configurations is an essential one, since the MPC algorithm is defined well only when constraints are feasible. More techniques and procedures concerning feasibility issues and the maintenance of feasibility can be found in works by Scokaert et al. and others [51, 53, 54, 57, 61].

References

1. Åstrom K, Furuta K (2000) Swinging up a pendulum by energy control. Automatica 36(2):287–295. doi:10.1016/S0005-1098(99)00140-5, http://www.sciencedirect.com/science/article/pii/S0005109899001405
2. Belavý C (2009) Teória Automatického Riadenia II: Návody na cvičenia, Slovenská vysoká škola technická v Bratislave: Strojnícka Fakulta, 1st edn. Bratislava, (Theory of automatic control II: seminar guide) in Slovak language
3. Bemporad A (1998) A predictive controller with artificial Lyapunov function for linear systems with input/state constraints. Automatica 34(10):1255–1260
4. Bemporad A, Chisci L, Mosca E (1994) On the stabilizing property of the zero terminal state receding horizon regulation. Automatica 30(12):2013–2015
5. Bertram JE, Kalman RE (1960) Control systems analysis and design via second method of Ljapunov. Trans ASME, J Basic Eng 82:371–400
6. Bitmead RR, Gevers M, Wertz V (1990) Adaptive optimal control: the thinking man's GPC. Prentice Hall, Englewood Cliffs
7. Bitsoris G (1988) On the positive invariance of polyhedral sets for discrete-time systems. Syst Control Lett 11(3):243–248
8. Blanchini F (1994) Ultimate boundedness control for uncertain discrete-time systems via set-induced Lyapunov functions. IEEE Trans Autom Control 39(2):428–433
9. Blanchini F (1999) Set invariance in control. Automatica 35(11):1747–1767
10. Blondel V, Tsitsiklis JN (1996) NP-hardness of some linear control design problems. SIAM J Control Optim 35:2118–2127
11. Blondel VD, Tsitsiklis JN (2000) A survey of computational complexity results in systems and control. Automatica 36(9):1249–1274. doi:10.1016/S0005-1098(00)00050-9, http://www.sciencedirect.com/science/article/pii/S0005109800000509
12. Boyd S, Ghaoui L, Feron E, Balakrishnan V (1994) Linear matrix inequalities in systems and control theory, 1st edn. Society for Industrial and Applied Mathematics, Philadelphia
13. Boyd S, Ghaoui LE, Feron E, Balakrishnan V (1994) Linear matrix inequalities in system and control theory. Studies in Applied Mathematics, SIAM, Philadelphia
14. Cannon M (2005) Model predictive control, lecture notes. Michaelmas Term 2005 (4 Lectures), Course code 4ME44. University of Oxford, Oxford
15. Chen H, Allgöver F (1998) A quasi-infinite horizon nonlinear model predictive control scheme with guaranteed stability. Automatica 34(10):1205–1217
16. Chen CC, Shaw L (1982) On receding horizon feedback control. Automatica 18:349–352
17. Clarke DW, Scattolini R (1991) Constrained receding-horizon predictive control. IEE Proc Part D 138(4):347–354
18. de Oliveira SL (1996) Model predictive control for constrained nonlinear systems. PhD thesis, California Institute of Technology, Pasadena
19. Furuta K (1992) Swing-up control of inverted pendulum using pseudo-state feedback. J Syst Control Eng 206(14):263–269. doi:10.1243/PIME_PROC_1992_206_341_02
20. Garcia CE, Prett DM, Morari M (1989) Model predictive control: theory and practice—a survey. Automatica 25(3):335–348
21. Gilbert EG, Kolmanovsky I (1995) Discrete-time reference governors and the non-linear control of systems with state and control constraints. Int J Robust Nonlinear Control 5:487–504
22. Gilbert EG, Kolmanovsky I (1999) Fast reference governors for systems with state and control constraints and disturbance inputs. Int J Robust Nonlinear Control 9:1117–1141
23. Gilbert EG, Tan KT (1991) Linear systems with state and control constraints: the theory and application of maximal output admissible sets. IEEE Trans Autom Control 36(9):1008–1020
24. Iwase M, Astom KJ, Furuta K, Akesson J (2006) Analysis of safe manual control by using Furuta pendulum. In: Computer aided control system design, 2006 IEEE international conference on control applications, 2006 IEEE international symposium on intelligent control, 2006 IEEE, pp 568–572. doi:10.1109/CACSD-CCA-ISIC.2006.4776708

25. Jirstrand M (1998) Constructive methods for inequality constraints in control. PhD thesis, Department of Electrical Engineering, Linköping University, Linköping
26. Karas A (2002) Stabilizujúce prediktívne riadenie systémov s obmedzeniami. PhD thesis, Slovak University of Technology in Bratislava, Bratislava (Stabilizing predictive control of systems with constraints.) in Slovak language
27. Karas A, Rohal'-Ilkiv B, Belavý C (2007) Praktické aspekty prediktívneho riadenia, 1st edn. Slovak University of Technology in Bratislava / Slovenská E-Akadémia, n.o., Bratislava (Practical aspects of predictive control) in Slovak language
28. Keerthi SS, Gilbert EG (1988) Optimal, infinite horizon feedback law for a general class of constrained discrete time systems: stability and moving-horizon approximations. J Optim Theory Appl 57:265–293
29. Kerrigan EC (2000) Robust constraint satisfaction: invariant sets and predictive control. PhD thesis, Control Group, Department of Engineering, University of Cambridge, Cambridge
30. Kouvaritakis B, Rossiter JA, Chang AOT (1992) Stable generalised predictive control: an algorithm with guaranteed stability. IEE Proc Part D 139(4):349–362
31. Kouvaritakis B, Cannon M, Karas A, Rohal'-Ilkiv B, Belavý C (2002) Asymmetric constraints with polyhedral sets in MPC with application to coupled tanks system. In: IEEE 2002 conference on decision and control, Las Vegas, pp 4107–4112. doi:10.1109/CDC.2002.1185011
32. Kuznetsov AG (1996) Constrained predictive control: a brief survey. Journal A (Benelux publication of the Belgian Federation of Automatic Control) 37(2):3–8
33. Kwon WH, Pearson AE (1978) On feedback stabilization of time-varying discrete linear systems. IEEE Trans Autom Control 23:479–481
34. Lofberg J (2004) YALMIP: a toolbox for modeling and optimization in MATLAB. In: Proceedings of the CACSD conference, Taipei
35. Lyapunov AM (1893) Problème général da la stabilité du mouvement. Academic Press, New York (reprinted in 1966)
36. Maciejowski JM (2000) Predictive control with constraints, 1st edn. Prentice Hall, Upper Saddle River
37. Mayne DQ, Michalska H (1990) Receding horizon control of non-linear systems. IEEE Trans Autom Control 35(5):814–824
38. Mayne DQ, Rawlings JB, Rao CV, Scokaert POM (2000) Constrained model predictive control: stability and optimality. Automatica 36(6):789–814
39. Michalska H, Mayne DQ (1993) Robust receding horizon control of constrained nonlinear systems. IEEE Trans Autom Control 38(11):1623–1633
40. Mosca E, Zhang J (1992) Stable redesign of predictive control. Automatica 28(6):1229–1233
41. Mosca E, Lemos JM, Zhang J (1990) Stabilising I/O receding-horizon control. In: Proceedings of 29th IEEE conference on decision and control, Honolulu
42. Nesterov Y, Nemirovskii A (1994) Interior-point polynomial methods in convex programming, Studies in applied mathematics. vol 13, SIAM, Philadelphia
43. Nevistić V, Primbs JA (1997) Finite receding horizon linear quadratic control: a unifying theory for stability and performance analysis. Technical report CIT-CDS 97-001, California Institute of Technology, Pasadena
44. Polak E, Yang TH (1993) Moving horizon control of linear systems with input saturation and plant uncertainty: part 1: robustness. Int J Control 58(3):613–638
45. Polak E, Yang TH (1993) Moving horizon control of linear systems with input saturation and plant uncertainty—part 2: disturbance rejection and tracking. Int J Control 58(3):639–663
46. Pólik I (2005) Addendum to the SeDuMi user guide version 1.1. Technical report, McMaster University, Advanced Optimization Lab, Hamilton, Ontario. http://sedumi.ie.lehigh.edu/
47. Primbs JA, Nevistić V (1997) Constrained finite receding horizon linear quadratic control. Technical report CIT-CDS 97-002, California Institute of Technology, Pasadena
48. Rawlings JB, Muske KR (1993) The stability of constrained receding horizon control. IEEE Trans Autom Control 38(10):1512–1516

49. Rohal'-Ilkiv B (2004) A note on calculation of polytopic invariant and feasible sets for linear continuous-time systems. Annu Rev Control 28:59–64
50. Rohal'-Ilkiv B, Belavý C, Karas A (2001) Stable infinite-horizon predictive control with amplitude and rate input constraints. In: 13th international conference on process control-process control '01, Štrbské Pleso, Vysoké Tatry, Slovak Republic, CD-ROM
51. Rossiter JA (2003) Model-based predictive control: a practical approach, 1st edn. CRC Press, Boca Raton
52. Rusko M (2004) A note to LQ / $\mathscr{H}_2$ optimal sector pole placement. In: 6th international scientific-technical conference on process control 2004, Kouty nad Desnou, conference CD, paper R251
53. Scokaert POM (1994) Constrained predictive control. Technical report OUEL 2023/94, Department of Engineering Science, Oxford University, Parks Road, Oxford
54. Scokaert POM, Clarke DW (1994) Stabilising properties of constrained predictive control. IEE Proc Part D 141(5):295–304
55. Scokaert POM, Rawlings JB (1996) Infinite horizon linear quadratic control with constraints. In: Proceedings of IFAC'96 world congress, San Francisco, vol 7a-04-1, pp 109–114
56. Scokaert POM, Rawlings JB (1998) Constrained linear quadratic regulation. IEEE Trans Autom Control 43(8):1163–1169
57. Scokaert POM, Rawlings JB (1999) Feasibility issues in model predictive control. AIChE J 45(8):1649–1659
58. Soeterboek R (1992) Predictive control-a unified approach. Prentice Hall, New York
59. Sturm JF (2001) SeDuMi 1.05 R5 user's guide. Technical report, Department of Economics, Tilburg University, Tilburg. http://sedumi.ie.lehigh.edu/
60. Sznaier M, Damborg MJ (1987) Suboptimal control of linear systems with state and control inequality constraints. In: Proceedings of the 26th IEEE conference on decision and control, pp 761–762
61. Vada J, Slupphaug O, Johansen TA, Foss BA (2001) Linear mpc with optimal prioritized infeasibility handling: application, computational issues and stability. Automatica 37:1835–1843
62. Wu SP, Vandenberghe L, Boyd S (1996) MAXDET-software for determinant maximization problems-user's guide. Information Systems Laboratory, Electrical Engineering Department, Stanford University
63. Xu Y, Iwase M, Furuta K (2001) Time optimal swing-up control of single pendulum. J Dyn Syst Meas Control 123(3):518–527. doi:10.1115/1.1383027, http://link.aip.org/link/?JDS/123/518/1
64. Zheng ZQ, Morari M (1995) Stability of model predictive control with mixed constraints. IEEE Trans Autom Control 40(10):1818–1823
65. Zhou K, Doyle J, Glover K (1994) Robust optimal control. Prentice-Hall, Englewood Cliffs
66. Ziegler MG (1995) Lectures on polytopes. Springer, New York

Chapter 8
Efficient MPC Algorithms

The time required to compute the control move in model predictive control (MPC) cannot exceed the sampling time of the controller. This is a clear and logical requirement, as the controller could not function otherwise: an input has to be applied to the feedback system exactly at the discrete sampling instant, thus it needs to be available before that time. In other words, the time necessary to evaluate the control law must be smaller than the sampling period. Simultaneously the sampling time strongly limits the choice of possible plants to control. One needs to choose a sampling period exceeding the dominant dynamic characteristics of the controlled system. A sampling period twice the bandwidth of the controlled system is often sufficient, however vibrating systems presenting oscillating motion often use 10–20 times the bandwidth of the largest controlled frequency—leaving room for higher order dynamics.

Initially, only predictive control algorithms without constraints have been utilized in industrial and laboratory applications. These did not have online computational time issues because no system constraints have been considered, thus making it possible to express the control law in a closed form [66]. Later constrained MPC formulations have appeared, significantly reducing the pool of possible applications because of the computationally demanding online optimization process. Early constrained MPC strategies have been implemented only on plants with slow dynamics such as chemical processes [41, 42]. From the viewpoint of the computational efficiency, we can consider any application *slow*, where the sampling time is measured in seconds or minutes. Stability guarantees such as the ones introduced in the previous chapter made the online constrained optimization process even more computationally demanding, thus calling for more efficient hardware platforms.

Systems with *fast* dynamics such as mechanical structures, automotive engines or vibration suppression require short sampling periods. The application of model predictive control on these systems is still a challenge, even with the advent of cheap and better hardware platforms. The online optimization of a multi-variable system model, with a long horizon and stability guarantees considering up-to-date standard hardware tools is still a formidable task. Considering a linear model, polyhedral

G. Takács and B. Rohal'-Ilkiv, *Model Predictive Vibration Control*,
DOI: 10.1007/978-1-4471-2333-0_8,

terminal sets (implying additional stability constraints) and a quadratic cost results in an optimization problem referred to as quadratic programming (QP). Quadratic programming-based MPC strategy (QPMPC) has been already successfully applied to systems with fast dynamics using a combination of quadratic programming solvers specialized for predictive control [22, 23, 26] and/or high performance hardware with optimized coding [89]. The proposed applications range from active vibration attenuation [90] to engine control [25]. Despite of the success of QPMPC in these applications, there is still a huge room for improvement. Some of the QPMPC demonstration examples which can be regarded as fast sampling [3, 25, 65, 86] can be in fact only considered as moderately fast sampling; especially when we take in account the possible applications with dominant dynamics measured in tens or even hundreds of kHz. Even if the sampling speeds in the aforementioned applications are fast enough for a wider array of mechatronic applications [25, 90], the featured MPC strategy usually does not guarantee stability a priori. The implementation of an MPC strategy with stability and feasibility guarantees would increase the complexity of the computations significantly, thus would lower the achieved maximal sampling speeds and the pool of possible applications further.

Since the advent of predictive controllers in practical applications, numerous formulations have been introduced which may improve the online computational efficiency of the optimization process. To use MPC in vibration control (or any other fast system), one may utilize the following approaches or their combination to speed up the online optimization process:

1. Increase of the computational power (e.g. use of DSP boards, high performance computers, field-programmable gate array (FPGA) architectures, etc.).
2. Low level optimization and clean-up of the existing algorithm implementation (e.g. transcript of the generic QP algorithm directly into machine language).
3. High level optimization of the algorithm—exploiting the specific structure of the MPC problem to make the QP algorithm more efficient (replacing generic QP with MPC optimized QP).
4. Sacrificing optimality for an alternate efficient optimizations (change of the original formulation, usually involving suboptimal albeit very efficient strategies such as Newton–Raphson-based MPC).
5. Pre-computing the QPMPC problem beforehand—transferring the online load into offline (multi-parametric explicit MPC).

We may safely state that the first three items in this short list can provide only a moderate speed gain. However, as numerous laboratory trials and timing analyses have proven, the latter two may cut down online computational demands enough to allow the use of MPC in vibration control [80, 84]. Nevertheless, the speed gain comes with a sacrifice, as alternate formulations of the MPC problem usually imply a loss of optimality levels while pre-computing the QPMPC problem can be overly complex with increasing problem dimensions.

This chapter evaluates computationally efficient MPC approaches, but it will not discuss the first two items on the list above. Instead of reviewing the same algorithm

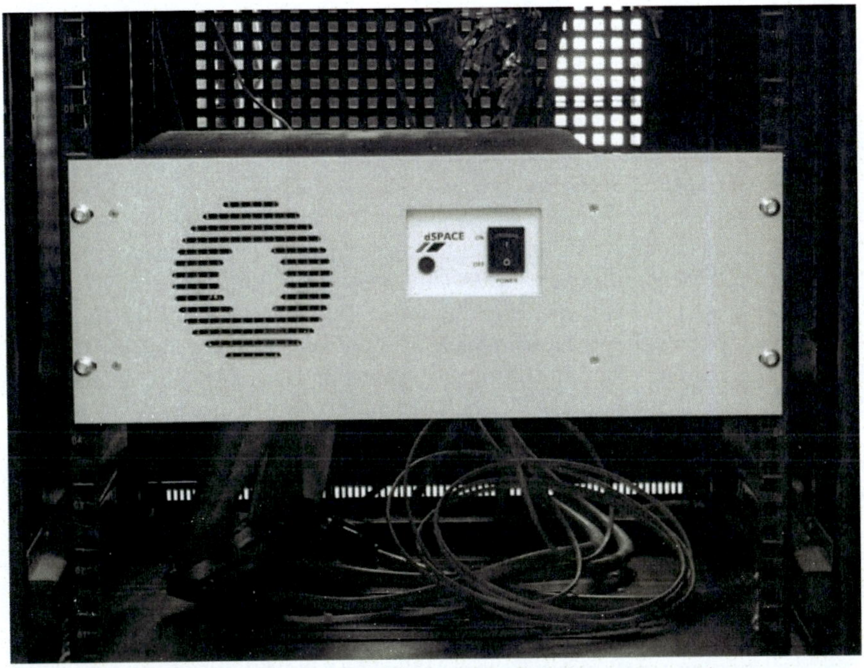

Fig. 8.1 Powerful rapid control prototyping systems with high memory capacity and computation speed are widespread in the automotive and aerospace industries. The proliferation of MPC controllers in practical applications is currently limited because of high computational costs. The use of model predictive control may be advanced not only by more powerful hardware, but better designed algorithms and strategies as well

with different hardware platforms, we will later use just one platform and analyze the efficiency of the algorithm formulation itself. While new products constantly come and go, an efficient algorithm formulation is ageless. One may easily select the best possible hardware for the job according to the available budget and manufacturer specifications. Rapid control system prototyping solutions are on the rise: powerful controller implementation platforms with high speeds and large memories are commonly used in high profile industrial projects, such as the automotive or the aerospace industries. Figure 8.1 illustrates a modern dSPACE control prototyping platform. Unfortunately, in many possible application fields expensive and powerful computation hardware is not an option, therefore the control algorithm must be able to work on implementation platforms with only modest possibilities.

Transcribing the QP algorithm directly into machine code or cleaning up the existing algorithm can be regarded as a last resort option to decrease computational times. This approach is time-consuming and its specifics will be not reviewed here, since it is more of a task for programming professionals. To those interested in the particulars of these two methods, Wills et al. have reached an impressive 5–25 kHz sampling period active vibration control using constrained MPC without stability guarantees in [89, 90].

Unlike the generic QP optimization problem as known in the field of mathematics, the MPC problem has a well defined, specific structure. Efficient MPC approaches capitalizing on the characteristic MPC structure belong to the third group listed previously. With certain techniques exploiting the structure of model predictive control, a computationally more efficient solver algorithm may be created which is specialized to MPC. Such an approach has been used by Ferreau to create a QP algorithm, which has been tailored for the needs of predictive control [24]. One of the major points to note here is that, despite the more efficient formulation, an MPC controller implemented through this modified QP algorithm still provides a fully optimal performance. The theoretical basics of this algorithm called qpOASES[1] will not be reviewed here, however, its implementation details are covered in Sect. 10.1 as qpOASES is used to benchmark other formulations later. To those interested in the theoretical particulars of qpOASES, we may recommend the works of Ferreau [22] and Ferreau et al. [26]. The technique listed as #3 in the list is also employed in combination with a suboptimal formulation (#4) in the work of Wang and Boyd [87, 88], briefly reviewed in Sect. 8.3 of this chapter. According to the findings of Wang and Boyd, if the variables of the MPC problem are reordered, the search direction of the interior point QP solver algorithm can be found by solving a block tridiagonal system of linear equations [88]. Using a generic QP algorithm the complexity of the optimization problem is cubic in dependence of the horizon n_c, while the above-mentioned approach[2] reduces to a linear complexity in the horizon. An alternative method making use of the structure of the QP problem has been proposed by Cannon et al. [16] and is also briefly discussed in Sect. 8.4 of this chapter.

The fourth item in the list points toward a group of approaches which sacrifice optimality in order to change the original optimal stabilized MPC formulation radically. Of the abundance of tricks used, complexity reduction usually involves the employment of simplified target sets such as the ones introduced in Sects. 7.5 and 7.6 of the previous chapter. Ellipsoidal target sets are often utilized to simplify the MPC formulation in the interest of reducing complexity [2, 12, 13]; and will also be at the center of attention of this work. The core of this chapter is presented in Sect. 8.1 where the theoretical basics of the computationally efficient and suboptimal algorithm of Kouvaritakis et al. is reviewed [14, 45, 46]. The essential elements of their approach are an augmented state-space system based on a pre-stabilized control loop and ellipsoidal stability and feasibility constraints. As it will be later outlined in Sect. 8.1, the elegant solution of Kouvaritakis et al. leads to an online formulation, which is simple and can be evaluated in very little time. The extension to increase optimality is reviewed in Sect. 8.1.2, while a formulation enlarging the region of attraction (thus the pool of possible controllable states) is given in Sect. 8.1.3.

One of the popular ways to decrease the computational time in MPC is expressing the solution of the QP beforehand as a piecewise-affine function of the state [5].

[1] The "qpOASES" solver software—Online Active Set Strategy is available as a free download at http://www.qpoases.org.

[2] The "Fast MPC" solver software based on this method is available as a free download at http://www.stanford.edu/~boyd/papers/fast_mpc.html.

This is a so-called explicit or multi-parametric MPC approach. The state-space is divided into controller regions offline. The online controller law is implemented as a static look-up table: the region in which the current state belongs is identified, so is the associated simple control law.[3] The drawback of this method is the long offline computational time and the complexity of the controller with state dimensions larger than $n_x = 2$ and long horizons. This approach belonging to the last, fifth group of our list is discussed in Sect. 8.2.1 of this chapter, while its suboptimal but efficient version is briefly reviewed as well in Sect. 8.2.2.

Although this book considers only linear MPC, we have to note that efficiency is even more crucial in nonlinear MPC (NMPC) formulations. To compute the inputs in NMPC, at each time step a nonlinear programming problem needs to be solved online [1]. Such optimization problems are in general nonconvex because the model itself is nonlinear. To make NMPC feasible for plants with moderately or very fast dynamics, both alternative formulations and better solvers are required [1].

8.1 Newton–Raphson MPC

This method allows for efficient online formulation thanks to the unique problem formulation first introduced by Kouvaritakis et al. in [45] based on earlier works such as [44, 76]. The strategy can be referred to as Newton–Raphson MPC (NRMPC), because its core online optimization algorithm is the Newton–Raphson (NR) root search method. NRMPC not only ensures a computationally efficient alternative to the traditional MPC methods, but also prevents uncertainty propagating through the predicted states. In order to guarantee the stability of the control course, NRMPC makes use of the tricks introduced previously: such as penalties on the terminal state [73], the dual-mode paradigm [64, 77–79] and invariant target sets [4, 17, 43, 62]. However, instead of using the maximal invariant target set as proposed by Gilbert and Tan [30], the method assumes simplified ellipsoidal targets [63, 91]. Another major difference is that the control loop is prestabilized by introducing *perturbations* from a nominal control law [75, 76].

NRMPC optimizes closed-loop predictions: a pre-stabilized loop is formed, where the predicted performance is not optimized over the system input u_k, rather a new free variable c_k is formed.[4] In the absence of constraints this pre-stabilized loop can be optimal in some sense, most commonly LQ optimal. (Although it is possible to choose any other criteria of optimality.) The free variable c_k is zero, whenever the constraints are inactive. During transients, this pre-stabilized loop is, no longer optimal and may not be even stabilizing. The role of the introduced free variable c_k called *perturbation* is to ensure the stability and feasibility of constraints. The

---

[3] The complex solver software suite "Multi-Parametric Toolbox (MPT)" based on this method is available as a free download at http://control.ee.ethz.ch/mpt/.

[4] Note that following our variable naming conventions, both u_k and c_k can be in general vectors, however for a single input–single output (SISO) system they are scalar values.

norm of c_k is minimized under the condition, that constraints must be satisfied on the whole control horizon for future predicted input variables and possibly states.

8.1.1 Basic NRMPC Formulation

Instead of using $\mathbf{u}_k$ and its elements u_{k+i} for $i = 0, 1, 2, \ldots, n_c - 1$ as an optimization variable, the new free variable c_k is introduced into the closed-loop formulation. The perturbations express input deviation from the fixed feedback optimal value. It is possible to express inputs at each sampling interval by [11, 45]:

$$u_k = \mathbf{K}x_k + c_k \tag{8.1}$$

where c_{k+i} for $i = 0, 1, 2, \ldots, n_c - 1$ expresses the degrees of freedom in the optimization problem. Thus instead of the well-known linear time-invariant (LTI) state-space equations the closed-loop system can be alternatively described by [45]: by

$$x_{k+1} = \boldsymbol{\Phi} x_k + \mathbf{B}c_k \tag{8.2}$$

where $\boldsymbol{\Phi}$ is the closed-loop dynamics of the LTI system. The dynamics of this *augmented* system can be then described by the autonomous state-space system:

$$\mathbf{z}_{k+1} = \boldsymbol{\Psi}\mathbf{z}_k \tag{8.3}$$

$$\boldsymbol{\Psi} = \begin{bmatrix} \boldsymbol{\Phi} & \mathbf{BE} \\ \mathbf{0} & \mathbf{T} \end{bmatrix} \tag{8.4}$$

Vector $\mathbf{z}_k$ consists of the state vector and the perturbation vector $\mathbf{f}_k$, and matrix $\mathbf{E} = [\mathbf{I}\,\mathbf{0}\ldots\mathbf{0}]$. The perturbation vector contains the perturbations for the actual sampling time up to the horizon length:

$$\mathbf{z}_k = \begin{bmatrix} x_k \\ \mathbf{f}_k \end{bmatrix} \tag{8.5}$$

where the vector of future perturbations is given by

$$\mathbf{f}_k = \begin{bmatrix} c_k \\ c_{k+1} \\ c_{k+2} \\ \vdots \\ c_{k+n_c-1} \end{bmatrix} \tag{8.6}$$

Matrix $\mathbf{T}$ acts as a shift matrix, as by multiplying with the perturbation vector it shifts its elements one sampling time forward. Its elements $\mathbf{0}_{n_u}$ are $n_u \times n_u$ matrices of zeros, where n_u is the system input size. $\mathbf{I}_{n_u}$ is an identity matrix of conforming dimension:

$$\mathbf{T} = \begin{bmatrix} \mathbf{0}_{n_u} & \mathbf{I}_{n_u} & \mathbf{0}_{n_u} & \dots & \mathbf{0}_{n_u} \\ \mathbf{0}_{n_u} & \mathbf{0}_{n_u} & \mathbf{I}_{n_u} & \dots & \mathbf{0}_{n_u} \\ \vdots & \vdots & \vdots & \ddots & \vdots \\ \mathbf{0}_{n_u} & \dots & \dots & \mathbf{0}_{n_u} & \mathbf{I}_{n_u} \\ \mathbf{0}_{n_u} & \mathbf{0}_{n_u} & \dots & \dots & \mathbf{0}_{n_u} \end{bmatrix} \tag{8.7}$$

The a priori stability guarantee of the NRMPC formulation is based on ellipsoidal invariant target sets, which were briefly introduced in Sect. 7.6 of the previous chapter. Let us recapitulate the basics of ellipsoidal target sets, and consider such a set E_x which is given by:

$$E_x = \left\{x | x^T \Gamma_{\mathbf{X}}^{-1} x \leq 1\right\} \tag{8.8}$$

where $\Gamma_{\mathbf{X}}^{-1} \geq 0$. Consider now a fixed state feedback control law $u_k = \mathbf{K}x_k$ for a linear time-invariant state-space system $x_{k+1} = (\mathbf{A} + \mathbf{BK})x_k$ for which we can create an invariant target set using ellipsoid E_x by simply stating that:

$$x_k \in E_x \Rightarrow x_{k+1} \in E_x \tag{8.9}$$

meaning that if the state x_k belongs to the ellipsoidal set at time (k) so must its next iteration x_{k+1}. From this, it is possible to show [9] that the condition of invariance will be:

$$\Gamma_{\mathbf{X}}^{-1} - \boldsymbol{\Phi}^T \Gamma_{\mathbf{X}}^{-1} \boldsymbol{\Phi} > 0 \tag{8.10}$$

where $\boldsymbol{\Phi}$ is the closed-loop matrix of this system $\boldsymbol{\Phi} = (\mathbf{A} + \mathbf{BK})$. Given only symmetric input constraints $\overline{u} = -\underline{u}$ the feasibility condition under the state feedback law $\mathbf{K}$ is given simply and logically as:

$$|\mathbf{K}x| \leq \overline{u} \tag{8.11}$$

For a state $x_k \in E_x$ the above feasibility condition can be rewritten equivalently in the following fashion [45]:

$$\|\mathbf{K}^T \Gamma_{\mathbf{X}}^{\frac{1}{2}}\| \leq \overline{u} \quad \Rightarrow \quad \overline{u}^2 - \mathbf{K}^T \Gamma_{\mathbf{X}} \mathbf{K} \geq 0 \tag{8.12}$$

The stability of the autonomous system (8.3) is guaranteed by the stability of Ψ. Similarly to the case of the LTI model itself without the augmentation as in (8.10), an elliptic invariant set must also exist for the augmented system. Let us denote this ellipse with E_z and define it in the following fashion:

$$E_z = \left\{\mathbf{z} | \mathbf{z}^T \Gamma_{\mathbf{Z}}^{-1} \mathbf{z} \leq 1\right\} \tag{8.13}$$

Because of matrix symmetry, it is possible to partition $\Gamma_{\mathbf{Z}}^{-1}$ similarly to the partitioning of $\mathbf{z}_k$ and get the blocks $\Gamma_x, \Gamma_f, \Gamma_{xf}$ and Γ_{fx}. It is worth noting that

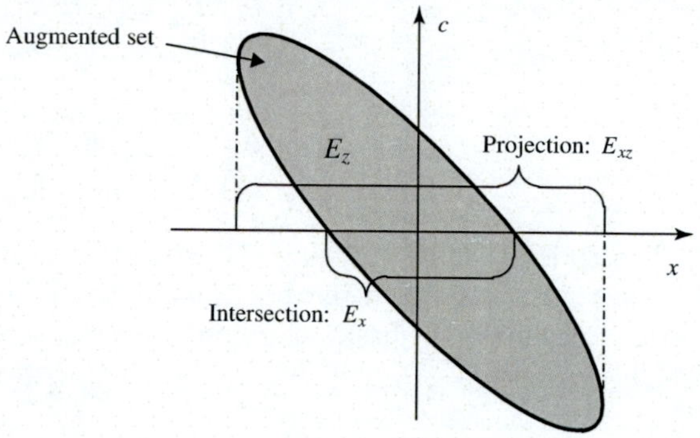

Fig. 8.2 Graphical representation of the augmented invariant set, its projection and intersection with the original state-space

$\Gamma_{xf} = \Gamma_{fx}{}^T$. This way it is possible to express the definition (8.13) by the following inequality:

$$x_k^T \Gamma_x x_k \leq 1 - 2\mathbf{f}_k^T \Gamma_{fx} x_k - \mathbf{f}_k^T \Gamma_f \mathbf{f}_k \tag{8.14}$$

In case where the perturbation vector $\mathbf{f}_k$ is not present, this inequality is satisfied by all possible $\mathbf{z}_k = \begin{bmatrix} \mathbf{x}_k & \mathbf{0} \end{bmatrix}^T$ for which $x_k \in E_x$ providing that $\Gamma_x = \Gamma_{\mathbf{X}}^{-1}$. A non-zero perturbation vector may be used to create an invariant ellipsoid, which is larger than the original $x_k \in E_x$. The maximizer of the right hand side of Eq. (8.14) after differentiation is $\mathbf{f}_k = -\Gamma_f{}^{-1}\Gamma_{fx}x_k$. If we substitute this back to the inequality, we obtain the following expression:

$$E_{xz} = \left\{ x | x^T \Gamma_{\mathbf{XZ}}{}^{-1} x \leq 1 \right\} \tag{8.15}$$

$$\Gamma_{\mathbf{XZ}} = \left[\Gamma_x - \Gamma_{xf} \Gamma_x{}^{-1} \Gamma_{fx} \right]^{-1} \tag{8.16}$$

Since $\Gamma_{\mathbf{XZ}}^{-1} \leq \Gamma_x$ and it is clear that $\Gamma_x = \Gamma_{\mathbf{X}}^{-1}$ it is obvious that the original invariant set is the subset of the new projection $E_x \subseteq E_{xz}$. It is possible to express matrix $\Gamma_{\mathbf{XZ}}$ in a more convenient way, using a matrix transformation:

$$\Gamma_{\mathbf{XZ}} = \mathbf{T}\Gamma_{\mathbf{Z}}\mathbf{T}^T \tag{8.17}$$

$$x_k = \mathbf{T}\mathbf{z}_k \tag{8.18}$$

It is difficult to illustrate the augmented invariant set E_z, its projection and intersection with the augmented state-space. In case the system in question is second order, the augmented state-space is already three-dimensional with only a one step ahead prediction. Therefore, Fig. 8.2 attempts to graphically illustrate a mere one-dimensional system with one step ahead prediction—that is one perturbation. E_z is the augmented invariant set, which has a dimension of $n_x + n_c$, where n_x is the system order and n_c is the prediction horizon or the size of the perturbation vector.

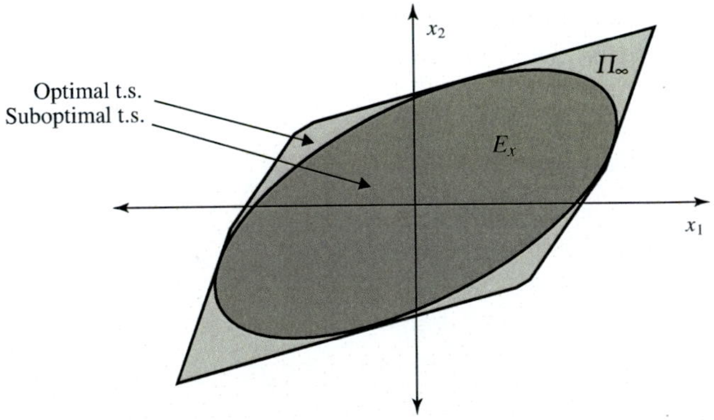

Fig. 8.3 Illustration of the optimal and suboptimal target set in x-space. The (hyper) ellipsoidal region defined by the NRMPC intersection is a subset of the ideal optimal target set. Abbreviation t.s. denotes target set

E_{xz} is the projection of E_z into the original state-space, E_x is the intersection with the state-space.

Figure 8.3 illustrates optimal and suboptimal target regions in the state-space. The intersection E_x of the augmented hyperellipsoid E_z with the state-space defined by NRMPC control law is just a subset of the ideal maximal invariant target set $\Pi_{n_a} = \Pi_\infty$. As the hyperellipsoid fills the polyhedral maximal invariant set, there is always some extra portion of the state-space that would approach the constraints more closely and thus improve on optimality. It is even harder to copy and fill the volume of a polyhedral maximal invariant set in higher dimensions, therefore the optimality of an NRMPC control law or in fact any MPC control law using ellipsoidal target sets will necessarily decrease with increasing state dimensions n_x.

Just as in any other MPC algorithm, invariance of the set and feasibility of the predictions has to be ensured. This may be carried out by an extension of the invariance conditions given previously by (8.10) to the augmented system:

$$\Psi^T \Gamma_{\mathbf{Z}}^{-1} \Psi - \Gamma_{\mathbf{Z}}^{-1} \leq 0 \tag{8.19}$$

This may be expressed in a form more convenient for semidefinite programming (SDP) [67], using Schur complements [9]:

$$\begin{bmatrix} \Gamma_{\mathbf{Z}} & \Gamma_{\mathbf{Z}} \Psi^T \\ \Psi^T \Gamma_{\mathbf{Z}} & \Gamma_{\mathbf{Z}} \end{bmatrix} \geq 0 \tag{8.20}$$

Unlike in the feasibility condition for x-space given by (8.12), now we have to account for the changes introduced in the calculation of the control input in (8.1); namely the presence of the perturbation value. The feasibility conditions for the augmented system will be therefore:

$$\left\| \begin{bmatrix} \mathbf{K}_i^T \, \mathbf{e}_i^T \end{bmatrix} \Gamma_{\mathbf{Z}}^{1/2} \right\| \leq \overline{u}_i \tag{8.21}$$

where $\mathbf{e}_i$ is the i-th column of the identity matrix, $\mathbf{K}_i$ is the i-th row of $\mathbf{K}$ and $i = 1, \ldots, n_u$ for a multiple input system. We may rewrite the feasibility condition of (8.21) and get a modified form given by:

$$\overline{u}_i^2 - \left[\mathbf{K}_i^T \; \mathbf{e}_i^T\right] \Gamma_{\mathbf{Z}} \left[\mathbf{K}_i^T \; \mathbf{e}_i^T\right]^T \geq 0 \tag{8.22}$$

The volume of the invariant hyperellipsoid E_{xz} is defined by $\det(\mathbf{T}\Gamma_{\mathbf{Z}}\mathbf{T}^T)$. To maximize the volume of the ellipsoid, while respecting feasibility and constraints one has to solve the following algorithm [45, 46]:

Algorithm 8.1 Maximize $\log\det(\mathbf{T}\Gamma_{\mathbf{Z}}\mathbf{T}^T)$ by respecting linear matrix inequalities defining invariance conditions (8.20) and feasibility conditions (8.22).

Since the LMI's (8.22) and (8.20) do not depend on the current state, this algorithm can be solved offline. It is assumed that there exists a feedback controller $\mathbf{K}$, a control horizon n_c , a matrix $\Gamma_{\mathbf{Z}}$ defining the (projection of) the hyperellipsoid in a way that $x_0 \in E_{xz}$.

For a sufficiently large control horizon the feasibility and the invariance of the problem can be handled by the degrees of freedom in $\mathbf{f}_k$. Therefore, $\mathbf{f}_k$ is responsible for:

- Ensuring that the states will remain in the projection of the augmented invariant set at all times: $x_k \in E_{xz}$
- Ensuring that constraints will not be exceeded but reached if necessary: $\underline{u} \leq u_k \leq \overline{u}$
- Using the remaining degrees of freedom to optimize the predicted performance

During the prediction the system described by (8.1) and (8.2) will remain linear. This means that constraints will leave it unaffected and $\mathbf{K}$ may be chosen without regard to them. $\mathbf{K}$ may be chosen by any conventional method making it optimal in an arbitrary sense. One may choose $\mathbf{K}$ to be the nominal LQ controller or for example to minimize worst-case cost. The overall algorithm will assume the following form [45, 46]:

Algorithm 8.2 *Offline procedure*: Calculate a feedback matrix $\mathbf{K}$, which is optimal in some sense, while ignoring constraints. Calculate $\Gamma_{\mathbf{Z}}$ to maximize the projection of the invariant hyperellipsoid E_{xz}, while respecting conditions for invariance and feasibility. *Online procedure*: At each sampling instant (k) perform the following minimization procedure:

$$\min_{f} \mathbf{f}_k^T \mathbf{f}_k \quad s.t. \quad \mathbf{z}_k^T \Gamma_{\mathbf{Z}}^{-1} \mathbf{z}_k \leq 1 \tag{8.23}$$

Only the first element of the calculated $\mathbf{f}_k$ is utilized and the procedure is repeated at the next sampling instant.

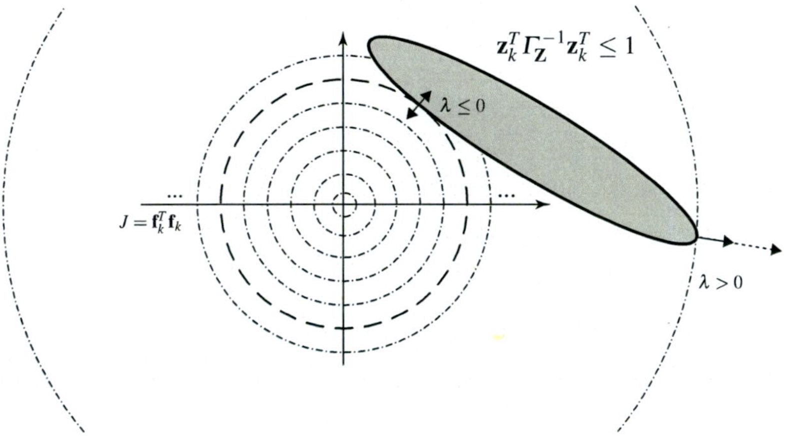

Fig. 8.4 Graphical illustration of the NRMPC optimization problem

Theorem 8.1 *Algorithm 8.2 is stabilizing.*

The detailed proof to this theorem can be found in the work of Kouvaritakis et al. [45]. Because $\mathbf{K}$ is stabilizing and the cost is decreasing, the state is advanced toward the equilibrium of the state-space. In addition, there will be a time when there is no need to perturb the sequence $u_k = \mathbf{K}x_k$ anymore and the solution for $\mathbf{f}_k$ will be zero.

The solution of the minimization problem (8.23) is actually the search for the shortest distance of an ellipsoid from the origin. This is a univariate optimization problem, graphically illustrated in Fig. 8.4. The circular iso lines with the center in the origin represent the function to be optimized min $\mathbf{f}_k^T \mathbf{f}_k$. The ellipse represents the constraints defined by $\mathbf{z}_k^T \Gamma_{\mathbf{Z}}^{-1} \mathbf{z}_k \leq 1$. One may deploy the method of Lagrange's constrained optimization problem to find the optimum, using the Lagrange multiplier λ. The gradient of the function is only negative at one point, because there is only one viable solution. In case the system origin is included in the ellipse defined by the constraints, there is no need to perform the optimization because the problem is already optimal.

If one utilizes the partitioning of $\Gamma_{\mathbf{Z}}$; may rewrite the optimization constraint

$$\mathbf{z}_k^T \Gamma_{\mathbf{Z}}^{-1} \mathbf{z}_k = x_k^T \Gamma_{fx} x_k + 2\mathbf{f}_k^T \Gamma_{xf} x_k + \mathbf{f}_k^T \Gamma_f \mathbf{f}_k \leq 1 \tag{8.24}$$

The solution of this optimization problem for the state at a given time (k) will be:

$$\mathbf{f}_k = \lambda \Theta \Gamma_{fx} x_k \tag{8.25}$$

$$\chi(\lambda) = \Gamma_{xf}[\Theta \Gamma_f^{-1} \mathbf{f}_k \Theta - \Gamma_f^{-1} \mathbf{f}_k] \Gamma_{fx} x_k + x_k^T \Gamma_x x_k - 1 = 0 \tag{8.26}$$

where Θ is defined as $\Theta = (\mathbf{I} - \lambda \Gamma_f)^{-1}$ and λ is the unique real root of $\chi(\lambda)$. The solution of this problem is trivial and the eigenvector–eigenvalue properties of Γ_f may be exploited to simplify the online procedure. It is necessary to ensure that the solution of (8.26) will converge when using the Newton–Raphson (NR) method to find λ.

Theorem 8.2 *For all $\lambda \leq 0$ all the derivatives of $\chi(\lambda)$ are positive, thus the Newton–Raphson method will converge quadratically when initialized with $\lambda = 0$ [46].*

This theorem utilizes the fact that for the i-th derivative of Θ we may express:

$$\frac{d\Theta^i}{d\lambda} = i\Theta^{i+1}\Gamma_f \quad \forall i \geq 1 \tag{8.27}$$

the proof of Theorem 8.2 is as follows:

Proof Using the relation expressed in (8.27) an the process of induction, it is possible to prove that [46]

$$\frac{d^i\chi(\lambda)}{d\lambda^i} = (i+1)!x_k^T\Gamma_{xf}\Delta\Lambda^{i-1}(\mathbf{I}-\lambda\Lambda)^{-(i+2)}\Delta^T\Gamma_{fx}x_k \tag{8.28}$$

where Λ and Δ is defined by the eigenvalue–eigenvector decomposition of Γ_f which is $\Gamma_f = \Delta\Lambda\Delta^T$. Since $\Gamma_{\mathbf{Z}}$ is positive definite and symmetric spectral matrix Λ will be also positive. The right hand side of (8.28) is positive for all $\lambda \leq 0$.

Since there is no need to evaluate complex constrained optimization tasks, the Newton–Raphson MPC (NRMPC) is very efficient. NRMPC presents an enormous computational advantage over MPC formulations utilizing quadratic programming or semidefinite programming. It is possible to show that utilizing modal decomposition of matrix Γ_f the computational time grows linearly with the prediction horizon.

8.1.2 Extension of the Newton–Raphson MPC

Since most of the computationally intensive operations can be performed in offline mode and one can utilize eigenvalue–eigenvector decomposition, the speed gain is a very important property of NRMPC. For exceptionally fast sampling applications like vibration damping, the advantages of NRMPC are evident. Also, the gain in computational speed is important for applications which require large prediction horizons to ensure feasibility properly. So what is the difference between NRMPC and the well-known MPC utilizing quadratic programming? While the formulation introduced in the previous section provides exceptional speed, it is still somewhat suboptimal. Although this is not very significant, QPMPC still has the advantage of using the largest admissible set Π_∞ and providing optimal solutions.

It is possible to extend NRMPC to match the optimality of QPMPC, as it has been demonstrated by Kouvaritakis et al. in [46]. There are possible perturbation vectors $\mathbf{f}_k$, which satisfy constraints over the prediction horizon and are from the maximal admissible set Π_∞ which ensure feasibility past the prediction horizon as well. This assumption and the affine dependence of control inputs u_{k+i} and the

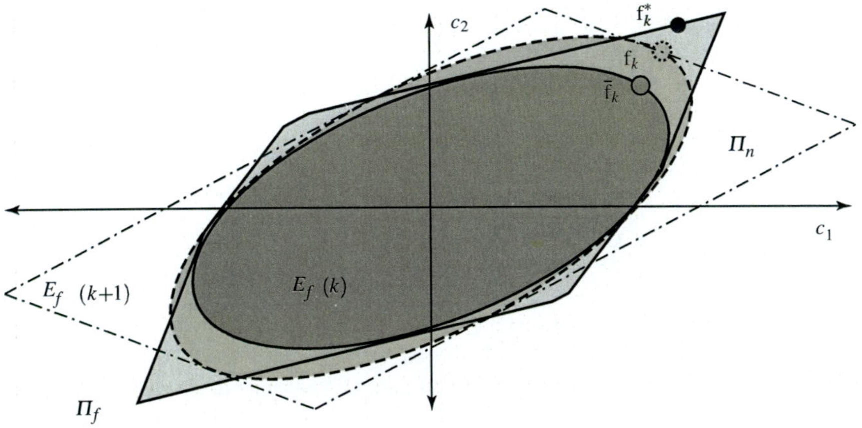

Fig. 8.5 Illustration of the perturbations in **f**-space. At time (k) there are perturbation vectors $\mathbf{f}_k < \bar{\mathbf{f}}_k$ which fall outside E_f. However, these vectors $\mathbf{f}_k$ still ensure feasibility past the prediction horizon and produce states which are a part of the maximal invariant set Π_∞. The low complexity set defined by the constraints is Π_n, $\bar{\mathbf{f}}_k$ is the original perturbation $\mathbf{f}_k$ is the new perturbation scaled by μ and $\mathbf{f}_k^*$ is the theoretically optimal perturbation vector

terminal state x_{n_c} implies a set of linear inequalities for the perturbation vector at the current sample time (k). The inequalities define a low complexity polyhedral set $\Pi_n(x_k)$. The choice of NRMPC for the perturbation vector is not necessarily the feasible solution of the minimum norm, because $E_f(x_k) \subseteq \Pi_n(x_k)$.

This concept is graphically illustrated in Fig. 8.5 where the set of perturbations E_f at time (k) is always only a subset of optimal perturbations Π_f allowed by the maximal allowable set $\Pi_\infty = \Pi_{n_a}$. In essence, it is possible to iterate the problem one step further, and search for a perturbation vector $\mathbf{f}_k$ which would be feasible at the next step and is smaller than the original $\bar{\mathbf{f}}_k$ by a factor of μ. This is illustrated by the iterated set of perturbations E_f at (k+1) and the low complexity set Π_n, which ensures that the new perturbation is still feasible. The set Π_n is a direct result of the constraints, where essentially one has to ensure that the new (augmented) state conforms to the constraints.

Let us denote the perturbation vector calculated by NRMPC with $\bar{\mathbf{f}}_k$. The idea is the scaling procedure of $\mathbf{f}_k$ subject to the constraint, defining that the feasibility at the current time sample will imply feasibility at the next one. The scaling scalar valuc is μ:

$$\mathbf{f}_k = \mu \bar{\mathbf{f}}_k, \quad \mu \leq 1 \tag{8.29}$$

Lemma 8.1 *Assuming the optimization* (8.23) *is feasible at the current time* (k) *and the minimizer is denoted as* $\mathbf{f}_k$, *the following inequality admits solutions* $0 \leq \mu \leq 1$ *according to* [46]:

$$a\mu^2 + b\mu + c \leq 0 \quad (8.30)$$

$$a = \bar{\mathbf{f}}_k^T \bar{\Gamma}_f \bar{\mathbf{f}}_k \quad (8.31)$$

$$b = 2x_k^T \bar{\Gamma}_{xf} \bar{\mathbf{f}}_k \quad (8.32)$$

$$c = x_k^T \bar{\Gamma}_x x_k \quad (8.33)$$

where $\bar{\Gamma}_x$, $\bar{\Gamma}_{xf}$, $\bar{\Gamma}_{fx}$, $\bar{\Gamma}_f$ are blocks generated by $\Gamma_{\mathbf{Z}}^{-1}$ at the time instant $(k+1)$ using $\Gamma_{\mathbf{Z}}^{-1}(k+1) = \bar{\Gamma}_{\mathbf{Z}}^{-1} = \Psi^T \Gamma_{\mathbf{Z}}^{-1} \Psi$.

Proof For the augmented state vector $z_k(\mu) = [x_k \; \mu\bar{\mathbf{f}}_k]^T$ the inequality (8.30) can be equivalently stated as $\mathbf{z}_{k+1}\Gamma_{\mathbf{Z}}^{-1}\mathbf{z}_{k+1} \leq 0$ for $\mathbf{z}_{k+1} = \Psi\mathbf{z}_k(\mu)$. This is true for $\mu = 1$ and we assume that E_z is invariant and $\mathbf{z}_k(\mu)$ belongs to it. The existence of scalers which are $0 \leq \mu \leq 1$ follows from the continuity [46]:

$$\mathbf{z}_k^T(\mu)\Gamma_{\mathbf{Z}}^{-1}\mathbf{z}_k(\mu) \Longrightarrow \mathbf{z}_k^T(\mu)\Psi^T\Gamma_{\mathbf{Z}}^{-1}\Psi\mathbf{z}_k(\mu) \quad (8.34)$$

The scaling value μ also has to satisfy the input and possibly the state constraints. Let us denote the smallest possible value which results this optimization μ_k^*.

Theorem 8.3 *According to Lemma 8.1, the new scaled perturbation vector $\mathbf{f}_k(\mu) = \mu\bar{\mathbf{f}}_k$ satisfies the condition* [46]:

$$\mathbf{f}_k^T \mathbf{f}_k \leq \bar{\mathbf{f}}_k^T \bar{\mathbf{f}}_k \quad (8.35)$$

and the control input utilizing the scaled perturbation $u_k = \mathbf{K}x_k + \mathbf{E}\mathbf{f}_k(\mu_k^*)$ ensures the feasibility of (8.24) *at the next sampling instant.*

The proof of this theorem is given in [46]. The algorithm for the extended Newton–Raphson MPC (ENRMPC) is as follows:

Algorithm 8.3

- Solve a semidefinite programming problem offline to maximize the projection E_{xz} defined by $\Gamma_{\mathbf{XZ}}$ of the extended invariant set E_z subject to constraints (8.20) and (8.22).
- Compute online the unique negative root of $\chi(\lambda)$ from (8.26) and successively compute $\bar{\mathbf{f}}_k$ from (8.25).
- Compute online the smallest possible scaling value μ_k according to relation (8.30) and the constraints then the final, scaled perturbation vector $\mathbf{f}_k$.
- Compute online and implement the first element of the control input vector. Perform the online optimization at the next time step all over again.

The extra calculation required to obtain the scaling factor μ^* can be performed explicitly and does not pose a significant online computational time increase. It is possible to compute μ^* explicitly, it involves the solution of one quadratic and a set of linear equations. If the suboptimality of NRMPC is acceptable, one can save the time for this, but in case optimality is an issue ENRMPC is the choice of control algorithm. According to the Monte Carlo simulations presented in [46] the performance of ENRMPC is practically indistinguishable from QPMPC, but with a significantly reduced computational cost. The use of this extension does not significantly improve the optimality of systems with lightly damped poles, such as models describing the vibration dynamics of very lightly damped structures [81, 83].

8.1.2.1 One Step Further...

The extension introduced previously uses scaling to increase optimality. This is done by presuming feasible solutions exist outside the invariant ellipsoid defined by the control law. It is nevertheless possible to conceive a further development on the original NRMPC extension, where invariance would be ensured $(k+2)$ steps and onward.[5]

Similarly to the original extension, let us assume the existence of a scaling variable according to (8.29). We may find solutions to μ by applying the following lemma:

Lemma 8.2 *Assuming the optimization (8.23) is feasible at the current time k and the minimizer is denoted* $\mathbf{f}_k$, *the following inequality admits solutions* $0 \leq \mu \leq 1$:

$$a\mu^2 + b\mu + c \leq 0 \tag{8.36}$$

$$a = \bar{\mathbf{f}}_k^T \tilde{\Gamma}_f \bar{\mathbf{f}}_k \tag{8.37}$$

$$b = 2x_k^T \tilde{\Gamma}_{xf} \bar{\mathbf{f}}_k \tag{8.38}$$

$$c = x_k^T \tilde{\Gamma}_x x_k \tag{8.39}$$

where $\tilde{\Gamma}_x, \tilde{\Gamma}_{xf}, \tilde{\Gamma}_{fx}, \tilde{\Gamma}_f$ *are blocks generated by* $\Gamma_{\mathbf{Z}}^{-1}$ *at the time instant* $(k+2)$ *using* $\Gamma_{\mathbf{Z}}^{-1}(k+2) = \tilde{\Gamma}_{\mathbf{Z}}^{-1} = \Psi^{2T}\Gamma_{\mathbf{Z}}^{-1}\Psi^2$.

Proof Similarly to the original extension, inequality (8.36) can be equivalently stated as $\mathbf{z}_{k+2}\Gamma_{\mathbf{Z}}^{-1}\mathbf{z}_{k+2} \leq 0$ for $\mathbf{z}_{k+2} = \Psi^2\mathbf{z}_k(\mu)$. This must be true for $\mu = 1$. The existence of scalers which are $0 \leq \mu \leq 1$ follows from the continuity:

$$\begin{aligned} \mathbf{z}_k^T(\mu)\Gamma_{\mathbf{Z}}^{-1}\mathbf{z}_k(\mu) &\Longrightarrow \mathbf{z}_k^T(\mu)\Psi^T\Gamma_{\mathbf{Z}}^{-1}\Psi\mathbf{z}_k(\mu) \\ &\Longrightarrow \mathbf{z}_k^T(\mu)\Psi^{2T}\Gamma_{\mathbf{Z}}^{-1}\Psi^2\mathbf{z}_k(\mu) \end{aligned} \tag{8.40}$$

8.1.3 *Optimizing Prediction Dynamics for the Newton–Raphson MPC*

It is possible to give a convex formulation for the optimization of prediction dynamics utilizing a nonlinear transformation of variables, as proposed by Cannon and

[5] As suggested by Dr. Mark Cannon in Oxford, November 2008.

Kouvaritakis in [14]. This approach leads to a generalization of prediction dynamics used in the previous subsections, which allows changing the dynamics of the controller depending on the predicted plant state. This method significantly enlarges the volume of the stabilizable set. Since changes take place in the offline formulation, computational time is not affected.

Given the invariance and feasibility conditions described by (8.20) and (8.22) one can compute the invariant set $\Gamma_{\mathbf{Z}}^{-1}$ offline via a convex optimization, given that matrices $\mathbf{T}$ and $\mathbf{E}$ are fixed. For this, the projection of E_z into x subspace is maximized as the optimization objective. This is acceptable if one can operate with a sufficiently large horizon n_c and a finite length perturbation vector. In practice, however, computational load grows linearly with increasing horizon. Also, for a given horizon n_c a larger feasible set is obtained if the "shift" matrices $\mathbf{T}$ and $\mathbf{E}$ are variables in the optimization of E_z. Unfortunately, optimizing for $\Gamma_{\mathbf{Z}}^{-1}$, $\mathbf{T}$ and $\mathbf{E}$ leads to a non-convex optimization problem.

Imsland et al. utilized optimized prediction dynamics to enlarge the region of attraction of the NRMPC control method, although their strategy involved a non-convex optimization approach with all its well-known drawbacks [35, 36]. At the same time, Cannon and Kouvaritakis have demonstrated that this problem can be avoided by reformulating the implied offline optimization to a convex LMI problem [14]. Let us define matrices $\mathbf{U}, \mathbf{V} \in \mathbb{R}^{n_x \times n_c}$, $\tilde{\mathbf{N}} \in \mathbb{R}^{n_x \times n_x}$, $\tilde{\mathbf{M}} \in \mathbb{R}^{n_u \times n_x}$ and also symmetric $\mathbf{X}, \mathbf{Y} \in \mathbb{R}^{n_x \times n_x}$ by the following identities:

$$\Gamma_{\mathbf{Z}}^{-1} = \begin{bmatrix} \mathbf{X}^{-1} & \mathbf{X}^{-1}\mathbf{U} \\ \mathbf{X}^{-1}\mathbf{U}^T & \bullet \end{bmatrix} \tag{8.41}$$

$$\Gamma_{\mathbf{Z}} = \begin{bmatrix} \mathbf{Y} & \mathbf{V} \\ \mathbf{V}^T & * \end{bmatrix} \tag{8.42}$$

$$\tilde{\mathbf{N}} = \mathbf{U}\mathbf{T}\mathbf{V}^T \tag{8.43}$$

$$\tilde{\mathbf{M}} = \mathbf{E}\mathbf{V}^T \tag{8.44}$$

where $\bullet$ and $*$ denotes the blocks of $\Gamma_{\mathbf{Z}}^{-1}$ and $\Gamma_{\mathbf{Z}}$ which are determined uniquely by $\mathbf{X}, \mathbf{Y}, \mathbf{U}, \mathbf{V}$ and $\Gamma_{\mathbf{Z}}^{-1}\Gamma_{\mathbf{Z}} = \mathbf{I}$. The identity $\Gamma_{\mathbf{Z}}^{-1}\Gamma_{\mathbf{Z}} = \mathbf{I}$ also implies:

$$\mathbf{U}\mathbf{V}^T = \mathbf{X} - \mathbf{Y} \tag{8.45}$$

From this, blocks $\bullet$ and $*$ of (8.41) may be expressed as:

$$\begin{bmatrix} \mathbf{X}^{-1} & \mathbf{X}^{-1}\mathbf{U} \\ \mathbf{X}^{-1}\mathbf{U}^T & \bullet \end{bmatrix} \begin{bmatrix} \mathbf{Y} & \mathbf{V} \\ \mathbf{V}^T & * \end{bmatrix} = \begin{bmatrix} \mathbf{I} & \mathbf{0} \\ \mathbf{0} & \mathbf{I} \end{bmatrix} \tag{8.46}$$

This can be evaluated by simple matrix algebra, to acquire the unknown matrix blocks:

$$\bullet = -\mathbf{V}^{-1}\mathbf{Y}\mathbf{X}^{-1}\mathbf{U} \tag{8.47}$$

$$* = -\mathbf{U}^{-1}\mathbf{V} \tag{8.48}$$

The following theorem will introduce equivalent invariance and feasibility conditions, which are convex in $\mathbf{X}, \mathbf{Y}, \tilde{\mathbf{N}}, \tilde{\mathbf{M}}$:

Theorem 8.4 *There exist* $\mathbf{T}, \mathbf{E}, \Gamma_{\mathbf{Z}}^{-1}$ *satisfying the invariance and feasibility conditions, only if the linear matrix inequalities in* $\mathbf{X}, \mathbf{Y}, \tilde{\mathbf{N}}, \tilde{\mathbf{M}}$ *are feasible* [14]:

$$\begin{bmatrix} \begin{bmatrix} \mathbf{Y} & \mathbf{X} \\ \mathbf{X} & \mathbf{Y} \end{bmatrix} & \begin{bmatrix} \boldsymbol{\Phi}\mathbf{Y} + \mathbf{B}\tilde{\mathbf{M}} & \boldsymbol{\Phi}\mathbf{X} \\ \tilde{\mathbf{N}} + \boldsymbol{\Phi}\mathbf{Y} + \mathbf{B}\tilde{\mathbf{M}} & \boldsymbol{\Phi}\mathbf{X} \end{bmatrix} \\ \begin{bmatrix} \boldsymbol{\Phi}\mathbf{Y} + \mathbf{B}\tilde{\mathbf{M}} & \boldsymbol{\Phi}\mathbf{X} \\ \tilde{\mathbf{N}} + \boldsymbol{\Phi}\mathbf{Y} + \mathbf{B}\tilde{\mathbf{M}} & \boldsymbol{\Phi}\mathbf{X} \end{bmatrix}^T & \begin{bmatrix} \mathbf{Y} & \mathbf{X} \\ \mathbf{X} & \mathbf{Y} \end{bmatrix} \end{bmatrix} \geq 0 \tag{8.49}$$

$$\begin{bmatrix} \overline{u}^2 & \begin{bmatrix} \mathbf{K}\mathbf{Y} + \tilde{\mathbf{M}} & \mathbf{K}\mathbf{X} \end{bmatrix} \\ \begin{bmatrix} \mathbf{K}\mathbf{Y} + \tilde{\mathbf{M}} & \mathbf{K}\mathbf{X} \end{bmatrix}^T & \begin{bmatrix} \mathbf{Y} & \mathbf{X} \\ \mathbf{X} & \mathbf{Y} \end{bmatrix} \end{bmatrix} \geq 0 \tag{8.50}$$

The feasibility of these conditions is necessary and sufficient for the feasibility of the original conditions (8.22), (8.20).

Proof Using Schur complements, the invariance condition (8.19) was equivalently expressed by (8.20). If both sides of this inequality are pre and post multiplied, we obtain:

$$\begin{bmatrix} \Sigma^T & \mathbf{0} \\ \mathbf{0} & \Sigma^T \end{bmatrix} \begin{bmatrix} \Gamma_{\mathbf{Z}} & \Gamma_{\mathbf{Z}}\Psi^T \\ \Psi^T\Gamma_{\mathbf{Z}} & \Gamma_{\mathbf{Z}} \end{bmatrix} \begin{bmatrix} \Sigma & \mathbf{0} \\ \mathbf{0} & \Sigma \end{bmatrix} \geq 0 \tag{8.51}$$

where Σ consists of

$$\Sigma = \begin{bmatrix} \mathbf{Y} & \mathbf{X} \\ \mathbf{V}^T & \mathbf{0} \end{bmatrix} \tag{8.52}$$

Using this with (8.45) actually yields (8.49). Similarly, the pre- and post-multiplication of condition (8.22) with

$$\begin{bmatrix} \mathbf{I} & \mathbf{0} \\ \mathbf{0} & \Sigma^T \end{bmatrix} \quad \text{and} \quad \begin{bmatrix} \mathbf{I} & \mathbf{0} \\ \mathbf{0} & \Sigma \end{bmatrix} \tag{8.53}$$

yields (8.50). The original feasibility and invariance conditions are fulfilled, if (8.49) and (8.50) is feasible. Matrices $\mathbf{U}$ and $\mathbf{V}$ can be assumed full rank, without the loss of generality. From the definitions of $\tilde{\mathbf{N}}$ and $\tilde{\mathbf{M}}$, solutions for the "shift" matrices $\mathbf{T}$ and $\mathbf{E}$ exist for given $\mathbf{K}, \tilde{\mathbf{M}}$, whenever $n_x \geq n_c$. For the case that $n_c = n_x$, we get unique solutions for the shift matrices:

$$\mathbf{T} = \mathbf{U}^{-1}\tilde{\mathbf{N}}\mathbf{V}^{-T} \tag{8.54}$$

$$\mathbf{E} = \tilde{\mathbf{M}}\mathbf{V}^{-1} \tag{8.55}$$

The offline optimization algorithm will change according to the previously introduced concepts:

Algorithm 8.4 The maximization of the projection of the invariant ellipsoid E_z into the x-subspace over $\mathbf{T}$, $\mathbf{E}$ and $\Gamma_{\mathbf{Z}}^{-1}$ is performed [14]:

- Solve the semidefinite programming problem (SDP):

$$\max[-(\log \det \mathbf{Y})] \tag{8.56}$$

  subject to conditions (8.49) and (8.50), in the optimization variables $\mathbf{X}$, $\mathbf{Y}$, $\tilde{\mathbf{N}}$, $\tilde{\mathbf{M}}$.
- Factorize $\mathbf{X}$ and $\mathbf{Y}$ to determine $\mathbf{U}$ and $\mathbf{V}$ according to relation (8.45).
- Using the relation defined by (8.43) for $\tilde{\mathbf{N}}$ and $\tilde{\mathbf{M}}$ solve for $\mathbf{T}$ and $\mathbf{E}$.

It is very important to note that the value of $\mathbf{Y}$ in the optimization problem (8.56) is independent of the control horizon n_c if $n_c \geq n_x$. If $n_c \leq n_x$ the optimization problem will be non-convex. A larger horizon than n_x will show no advantage at all. The maximal stabilizable set will not be bigger if we increase the horizon, it is sufficient to choose $n_c = n_x$.

Although this formulation ensures the maximal possible stabilizable set, it might cause issues of numerical nature.[6] The optimization may yield a hyperellipsoid with exaggerated proportions. There are several possibilities to avoid this situation, including the option of imposing bounds on the predicted cost along the trajectories of the autonomous system defined by (8.3). For a given bound γ the cost value $\tilde{J} \leq \gamma$ is ensured for all initial conditions of the autonomous system in E_z, if the invariance condition is replaced by:

$$\Gamma_{\mathbf{Z}}^{-1} - \Psi^T \Gamma_{\mathbf{Z}}^{-1} \Psi > \frac{1}{\gamma} \begin{bmatrix} \mathbf{C}^T & \mathbf{K}^T \\ \mathbf{0} & \mathbf{E}^T \end{bmatrix} \mathscr{D} \begin{bmatrix} \mathbf{C} & \mathbf{0} \\ \mathbf{K} & \mathbf{E} \end{bmatrix} \tag{8.57}$$

where $\mathscr{D} = \text{diag}\,(\mathbf{I}, \mathbf{R})$. By a suitable transformation, this condition can be equivalently expressed in $\mathbf{X}$, $\mathbf{Y}$, $\tilde{\mathbf{N}}$, $\tilde{\mathbf{M}}$ using the following LMI:

$$\begin{bmatrix} \gamma\mathbf{I} & \mathbf{0} & \mathscr{D}^{1/2} \begin{bmatrix} \mathbf{CY} & \mathbf{CX} \\ \mathbf{KY} + \tilde{\mathbf{M}} & \mathbf{KX} \end{bmatrix} \\ * & \begin{bmatrix} \mathbf{Y} & \mathbf{X} \\ \mathbf{X} & \mathbf{Y} \end{bmatrix} & \begin{bmatrix} \boldsymbol{\Phi}\mathbf{Y} + \mathbf{B}M & \boldsymbol{\Phi}\mathbf{X} \\ \tilde{\mathbf{N}} + \boldsymbol{\Phi}\mathbf{Y} + \mathbf{B}M & \boldsymbol{\Phi}\mathbf{X} \end{bmatrix} \\ * & * & \begin{bmatrix} \mathbf{Y} & \mathbf{X} \\ \mathbf{X} & \mathbf{Y} \end{bmatrix} \end{bmatrix} \geq 0 \tag{8.58}$$

This LMI can be included in the optimization (8.56) in the place of (8.49).

[6] This problem turned out to be quite significant, see Sect. 11.3 for details.

8.1.4 Warm Starting and Early Termination

Unfortunately, the online NRMPC approach is not suitable for warm starting or early termination. This can be attributed to the fact that the (transformed[7]) cost $J = \mathbf{f}_k^T \mathbf{f}_k$ is not necessarily non-increasing during the iterations of the Newton–Raphson algorithm, neither is the feasibility of $\mathbf{f}_k$ guaranteed before the algorithm converges [47]. Stability guarantees can only be given if the perturbation $\mathbf{f}_k$ is feasible, and the cost at time (k) does not exceed the cost of the tail predicted at time $(k-1)$ associated with $\mathbf{E}\mathbf{f}_{k-1}$. Terminating the NRMPC algorithm early would mean that these conditions are not met, losing the advantages gained by the application of stability guarantees.

Based on the already computationally efficient NRMPC algorithm; Kouvaritakis, Li and Cannon recently proposed a modified method in [47], where both early termination and warm starting is possible. The approach called autonomous augmented MPC (AAMPC) uses a formulation, which instead of the dual optimization variables—i.e. the Lagrange multiplier λ—works with primal variables. Similarly to NRMPC, the AAMPC approach uses an univariate optimization procedure to compute the control inputs. While the NRMPC approach uses a dual optimization variable which can only be initialized from zero at sample time (k), the primal formulation of AAMPC allows warm starting utilizing the optimization results from the previous step $(k-1)$. Moreover, even though the cost is guaranteed to be decreasing in NRMPC by successive time samples $k, k+1, \ldots$, this is not true for the iterations of the NR optimization process at a given sample. The AAMPC method justifies this shortcoming by a formulation, where early termination is made possible by using a cost, which is decreasing even within the iterations for a given time sample. Instead of the Newton–Raphson procedure, the AAMPC method uses a line search algorithm for optimization.

By evaluating 100 random fifth order state-space systems in simulation, Kouvaritakis et al. reported that the AAMPC algorithm performed slightly worse in terms of optimality compared to NRMPC algorithm, being about 6% suboptimal with warm starting and 7% suboptimal when cold starting. The higher suboptimality of this novel algorithm is compensated by its speed gain, where AAMPC needed 56% of the computation time of NRMPC with warm starting and 87% without warm starting. The suboptimality issue is turned around when instead of a cost-based termination criterion we consider early termination of the algorithms based on the number of iterations. If AAMPC is given only one iteration and a (slightly modified) NRMPC algorithm is given 3 iterations, the NRMPC algorithm is about 30% less optimal than AAMPC due to the non-monotonically decreasing cost, while requiring almost twice as much computation time.

The novel AAMPC algorithm is certainly an interesting addition to the family of computationally efficient MPC strategies. Nevertheless, the approximately twofold efficiency increase is only possible by the expense of an even higher degree of suboptimality than NRMPC. As it will be later demonstrated, suboptimality proved

[7] See Sect. 10.3.2 for more details on cost transformation.

to be the weakest point of the NRMPC approach on lightly damped vibrating systems, especially with higher order prediction models. The online task execution times achieved with second-order models were already excellent, suggesting that there is room for much higher sampling times and more complicated models. Because the efficiency of NRMPC was already acceptable and the suboptimality of NRMPC was a fundamental issue, the AAMPC algorithm has not been tested in the simulations and experiments featured in the upcoming chapters.

8.2 Multi-Parametric MPC

8.2.1 Optimal Multi-Parametric MPC

The multi-parametric programming approach in MPC may be regarded as a relatively recent development of control theory. This approach proposed by Bemporad et al. [5, 6] takes advantage of the fact that MPC is a constrained linear piecewise-affine (PWA) problem, for which explicit solutions can be readily calculated. The PWA control laws for the MPC problem are pre-computed offline, which means that at the implementation stage only a piecewise-linear function is evaluated. This approach ensures fast online calculation, however, its main drawback is that offline computational time and memory requirements grow fast with increasing problem dimensions. Generally, if the number of constraints is exceeding 10, it is not practical to employ multiparametric programming-based MPC method [61]. In addition to various procedures aimed at reducing computational complexity, current research is also focused on formulating robust MPMPC controllers able to handle systems with built-in uncertainties [58], adaptive MPMPC formulations able to serve systems with varying parameters [51] and MPMPC laws for systems modeled by fuzzy laws [52, 53, 57].

The scope of applications, where multi-parametric programming-based MPC is currently employed or proposed is broad, ranging from marine applications [39], chemical processing [31], paper production [7], engine control [28], fuel-cell management [69], spacecraft attitude control [34] and even insulin delivery to diabetic patients [21, 70].

Using multi-parametric programming, the constrained quadratic programming problem is solved offline. The solution assumes the form of a piecewise-affine state feedback control law, which may be represented in a form [5, 48, 71]:

$$u(x) = \begin{cases} \mathbf{O}_1 x + \mathbf{o}_1 & \text{if } x_k \in \mathscr{X}_1 \\ \mathbf{O}_2 x + \mathbf{o}_2 & \text{if } x_k \in \mathscr{X}_2 \\ \mathbf{O}_3 x + \mathbf{o}_3 & \text{if } x_k \in \mathscr{X}_3 \\ \vdots & \vdots \\ \mathbf{O}_N x + \mathbf{o}_N & \text{if } x_k \in \mathscr{X}_N \end{cases} \tag{8.59}$$

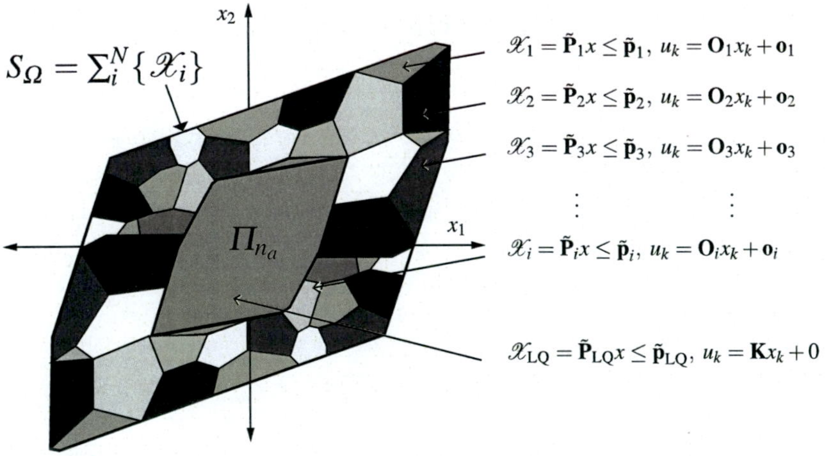

Fig. 8.6 Illustration of the regions and associated control laws of a conceptual MPMPC controller in two-dimensional state-space. The region of attraction is sub-divided into polyhedral regions $\mathscr{X}_i$, each of which has an associated control law defining u_k in case the actual state x_k belongs to that region

where x_k is a state, acting as input to the controller function. Matrices $\mathbf{O}_i$ and vectors $\mathbf{o}_i$ define a fixed feedback and a constant for the given control law. The current measured state is a part of a polyhedral set $\mathscr{X}_i$, the sum of these sets forms a polyhedral partition $S_\Omega = \mathscr{X}_1, \mathscr{X}_2, \mathscr{X}_3, \ldots, \mathscr{X}_N$ in state-space, while N is the total number of regions defining the region of attraction. The concept of polyhedral regions with associated control laws is graphically illustrated by an example in two-dimensional state-space in Fig. 8.6. The polyhedral regions $\mathscr{X}_i$ are characterized by intersections of half spaces in the hyperspace:

$$\mathscr{X}_i = \left\{ x \in \mathbb{R}_x^n | \tilde{\mathbf{P}}_i x \leq \tilde{\mathbf{p}}_i \right\} \tag{8.60}$$

These sets intersect only at the boundaries, therefore in case the problem is feasible, the current state x_k unambiguously belongs to one of the partitions, and can be associated with a PWL control law. Such a controller is defined by the following data: $\left\{ \mathbf{O}_i, \mathbf{o}_i, \tilde{\mathbf{P}}_i, \tilde{\mathbf{p}}_i \right\}_{i=1}^N$.

It is also possible to use a vertex representation instead of a half-space description of the polyhedral regions. This is an equivalent interpretation, though requiring some changes both in the offline and online algorithmic implementation. Vertex representation can be mathematically characterized as:

$$\mathscr{X}_i = \mathrm{conv}(\mathbf{h}_i^1, \mathbf{h}_i^2, \ldots, \mathbf{h}_i^N) \tag{8.61}$$

where *conv* denotes a convex hull of vertices $\mathbf{h}_i$.

The offline multi-parametric optimization process for acquiring the explicit solution of an MPC process may be summarized by the following concept algorithm [72, 74]:

Algorithm 8.5 To find the solution of the MPMPC problem offline, given a linear time-invariant system and process settings, perform the following task once:

- For all feasible active sets, define a polyhedral region $\mathscr{X}_i$ such, that in case a given state $x_k \in \mathscr{X}_i$, than the control course $u_k = \mathbf{O}_i x_k + \mathbf{o}_i$ is optimal and feasible. The sum of regions $\mathscr{X}_i$ is the region of attraction $\Pi_{n_a} = \Pi_\infty$.
- Reduce regions $\mathscr{X}_i$ to prevent overlaps and duplications.
- Store the region look-up table or binary tree $\mathscr{X}_i$ and the corresponding PWL functions for the online controller.

There are several possible methods finding the polyhedral set corresponding to a state at a given sampling instant. The most straightforward way is to sequentially search through all the regions. This is easily implementable, albeit computationally inefficient. In the worst case, N matrix operations must be performed in a form $\tilde{\mathbf{P}}_i x - \tilde{\mathbf{p}}_i \leq 0$. Given that the usual explicit MPC controller contains several tens of thousands of regions, direct search can seriously impair computational efficiency.

The other major method is the use of a binary search tree, which utilizes a much more efficient strategy than simple direct search. The drawback of the binary search tree method is the necessary time to calculate it, adding significant amount of computational time to the offline process.

In the online control process, at each sampling interval the set $\mathscr{X}_i$ defining a region corresponding to the actual state is found. The next task is to utilize the function (8.59) corresponding to the polyhedral set $\mathscr{X}_i$ in calculating the control output. Control output is simply the function of the current state $u_k = f(x_k)$. This is a very efficient process, assuming low problem dimensionality current hardware is capable of high sampling speeds. The online algorithm may be summarized as [38]:

Algorithm 8.6 To find the solution of the MPMPC problem online, at each sampling instant perform the following tasks:

- Measure or observe current state of the system.
- Identify the index i of the polyhedral region, such that $x_k \in \mathscr{X}_i$.
- Evaluate the PWL function corresponding to the index found at the previous step: $u_k = \mathbf{O}_i x_k + \mathbf{o}_i$.

For high bandwidth applications with small dimensionality, the MPMPC approach may yield manageable small number of controller regions within tractable offline computational times. The pre-computed MPC controller law has the great advantage of being readily transferable to mass-produced cheap microprocessor applications. However, a large region of attraction in constrained MPMPC with guaranteed feasibility may require lengthy horizons [82]. Generally speaking a problem of dimensionality above 10—including the prediction horizon necessary to ensure a given region of attraction—becomes difficult to manage [61].

8.2.1.1 Complexity Reduction

The complexity and the computation cost of the multi-parametric MPC approach can be reduced by one of two methods [55, 59], which either involves the reduction of the size of the look-up table [19, 20, 29, 32, 37] or improves the speed of the online algorithm [18, 40, 50, 54, 85]. Although these methods do reduce the memory requirements on the online implementation platform or improve the search speed of the online run, they require additional operations in the offline regime. The computational cost of performing simplification operations such as optimal region merging (ORM) or generating a binary search tree (BST) may be prohibitively high for systems requiring lengthy prediction horizons.

A very interesting development has been proposed to reduce the complexity of MPMPC controllers by Kvasnica et al. in [49] and later in [60]. The complexity reduction procedure suggested by Kvasnica et al. may be especially relevant to the control of underactuated lightly damped vibrating systems, where the control outputs are likely to be saturated for most part of the state-space. The number of regions is reduced by a method called *clipping* which is also maintaining closed loop properties of the original system, thus if the original system was providing guaranteed stability and feasibility, so will the clipped version. The idea behind clipping is based on the fact that for some states x_k the outputs u_k are saturated and certain constraints are active. In clipping the unsaturated regions are extended in a way that they also cover the saturated ones [56]. In the online regime, the original search procedure is performed, but the function values are also passed to a so-called *clipping function*, which ensures the equivalence with the original function for all feasible initial states [56].

Given a prerequisite of having a saturated continuous MPMPC law (or a PWA function in general) computed according to the framework of Algorithm 8.5, consisting of a set of individual regions described by $\mathscr{X}_i = \left\{x \in \mathbb{R}^n_x | \tilde{\mathbf{P}}_i x \leq \tilde{\mathbf{p}}_i\right\}$, the offline post-processing algorithm which extends the unsaturated regions to the saturated ones can be described as follows [49, 56]:

Algorithm 8.7 Extract the indices of unsaturated regions $\mathscr{X}_u$ and the indices of their adjacent regions, then for each unsaturated region $\mathscr{X}_i \in \mathscr{X}_u$ perform the following procedure in arbitrary order:

- Find the indices of the half spaces over which the adjacent regions of $\mathscr{X}_i$ are saturated regions $\mathscr{X}_s$.
- Form a new region $\tilde{\mathscr{X}}_i$ by removing the indexed half space from the unsaturated region $\mathscr{X}_i$ (extend the $\mathscr{X}_i$ to cover the saturated region).
- The new region from the previous step $\tilde{\mathscr{X}}_i$ is intersected with the domain of definition, in other words the region of attraction to obtain $\tilde{\mathscr{X}}_i = \tilde{\mathscr{X}}_i \cap S_\Omega$.
- If the new set $\tilde{\mathscr{X}}_i$ also intersects with other unsaturated regions $\mathscr{X}_j$ from $\mathscr{X}_j \in \mathscr{X}_u$ then the new region will change to the set difference of the new region $\tilde{\mathscr{X}}_i$ and the other unsaturated regions: $\tilde{\mathscr{X}}_i = \tilde{\mathscr{X}}_i \backslash \mathscr{X}_j$.

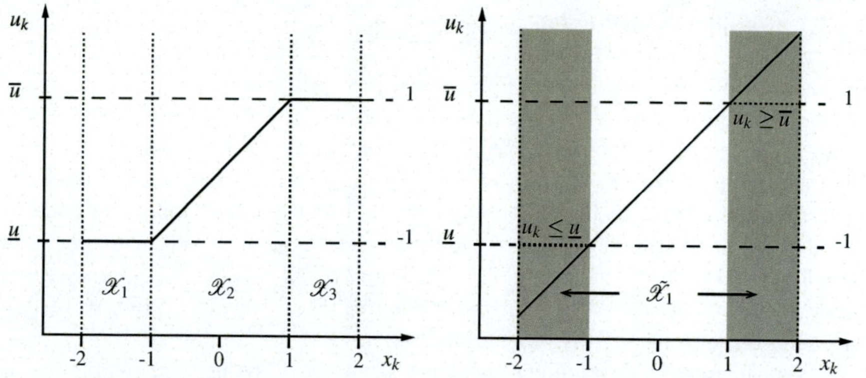

(a) Original MPMPC regions **(b)** MPMPC regions after clipping

Fig. 8.7 Conceptual illustration of the clipping operation on a one-dimensional example. The three original MPMPC regions in (**a**) are reduced to one in (**b**) by extending the unsaturated region to the saturated ones and the application of a clipping function

- Store the new region $\tilde{\mathscr{X}}_i$ and update the original definition matrices $\tilde{\mathbf{P}}_i$ and $\tilde{\mathbf{p}}_i$ as well.
- Move to a new unsaturated region and start from the beginning, until all regions are evaluated.

In the online regime a clipping function must be implemented, which is based on the online procedure of Algorithm 8.6. In addition to the original MPMPC online computation task, after u_k is computed it is passed through the following clipping function and updated according to [49, 56]:

$$u_k = \begin{cases} \overline{u} & \text{if} \quad u_k \geq \overline{u} \\ \underline{u} & \text{if} \quad u_k \leq \underline{u} \\ u_k & \quad\text{otherwise} \end{cases} \tag{8.62}$$

Let us illustrate this complexity reduction procedure by a simple one-dimensional example, featured in Fig. 8.7. The output is constrained by an upper and a lower constraint defined by $\overline{u} = -\underline{u} = 1$, while the PWL law can be expressed by three regions. One of those regions $\mathscr{X}_2 = \{-1 \leq x \leq -1\}$, $u_k(x) = 1x_k + 0$ is unsaturated, while the other two $\mathscr{X}_1 = \{-2 \leq x \leq -1\}$, $u_k(x) = 0x_k - 1$ and $\mathscr{X}_3 = \{1 \leq x \leq 2\}$, $u_k(x) = 0x_k + 1$ are saturated. The definition domain of the PWL law is $S_\Omega = \{-2 \leq x \leq 2\}$. Starting by the one and only unsaturated region $\mathscr{X}_2$ we can determine that both adjacent regions $\mathscr{X}_1$ and $\mathscr{X}_3$ are saturated and the half spaces $x \leq -1$ and $x \geq 1$ can be removed from the definition of $\mathscr{X}_2$, to form a new region $\tilde{\mathscr{X}}_1$. We take the intersection of this new region with the domain of definition, thus the new region will be defined over $\tilde{\mathscr{X}}_1 = \{-2 \leq x \leq 2\}$. This new region is not intersecting with any other unsaturated region, thus it is stored as the new clipped PWL region.

The number of regions after performing this simplification procedure will be in general significantly smaller than the number of regions before. In fact, most of the time it will be close to the number of the unsaturated regions [56]. After performing tests with randomized systems, Kvasnica et al. reported an average of eight times smaller number of regions, which is a significant improvement. For SISO systems, the clipping method reduced the number of regions to the number of unsaturated regions for 90% of the cases. According to these randomized tests, clipping not only reduced the complexity of the regions more than ORM, but also needed less computational effort in the offline regime while being suitable for systems with a high number of regions.

As it will be demonstrated later, the number of regions in the MPMPC control of lightly damped vibrating systems may be very high due to the long prediction horizons needed to ensure a large region of attraction. This region of attraction is mostly made up of saturated regions, which can be eliminated using the above introduced method. Nevertheless, the clipping procedure is a post-processing operation, which needs the original PWL law to operate. Clipping is aimed at reducing online storage memory and search times, but it does not reduce the offline computation cost. In case the original MPMPC offline computation procedure is deemed intractable for a given system, clipping will not solve the issue.

8.2.2 Multi-Parametric Programming-Based Minimum Time Suboptimal MPC

The minimum-time MPMPC approach replaces the cost function usually considered in MPC with a much simpler alternative. Instead of using the weighted sum of state and input contributions over an n_c steps long horizon, minimum-time MPMPC minimizes the number of steps necessary to enter a given terminal set for an initial state x_k. As a result, only a single terminal set is considered at each iteration. The algorithms introduced by Grieder et al. in [33] and also featured in the work of M. Kvasnica [48] not only reduce the complexity of the online solution, but considerably reduce offline computation time as well. The offline minimum time MPMPC problem may be summarized according to the following algorithm [48]:

Algorithm 8.8 A sequence of one step long horizon multi-parametric programs are solved backwards in time:

- Design an invariant target Π_0 set around the origin.
- Solve an $n_c = 1$ problem with Π_0 as the terminal set.
- Store Π_1 along with its regions $\mathscr{X}_{1j}$ and the associated PWA feedback laws.
- Repeat using Π_i as a terminal set and solving one-step problems until convergence is achieved.

The resulting controller consists of all such partitions for the individual iterations. The simplified online minimum time multi-parametric MPC problem may be summarized by the following algorithm:

Algorithm 8.9 Perform the following procedure at each online sampling instant:

- Observe or measure actual system state at sample x_k.
- Identify the partition Π_i which both contains x_k and it has a minimal cost, e.g. it steers the state to the origin in minimum time.
- Identify the region number $\mathscr{X}_{ij}$ containing x_k in the previously identified partition Π_i, and retrieve the associated j-th feedback law.
- Calculate u_k by evaluating the PWA feedback law.
- Apply u_k to the system and repeat at the next instant.

8.3 Approximate Primal-Barrier Interior Point Method-Based MPC

An efficient MPC approach has been proposed by Wang and Boyd in [87] and subsequently in [88] by using an approximate primal-barrier interior point method (APB-IPM). This seemingly complex term refers to a collection of ideas and methods, which using its structural properties increases the efficiency of the well-known quadratic programming-based MPC problem by a factor of several tens or hundreds.

To briefly summarize what is APB-IPM, we may list the following ideas [87, 88]:

- Modified QP optimization: exploit the structure of MPC and use variable re-ordering to significantly simplify the online interior point optimization.
- Warm-starting: use the predictions computed in the previous time step, to initialize the optimization process in the current time step.
- Early termination: in order to save on computational time, the optimization process is terminated early. This yields an optimality decrease of 2–3% when compared to truly optimal solvers.

In the terms of efficiency, the APB-IPM has a lot to offer. According to the tests performed using randomized examples in [88], a system with $n_x = 4$ states, $n_u = 2$ inputs and a $n_p = 30$ steps long horizon, the computation time has been listed as 2.79 ms. This is an almost 150 fold increase in speed when compared to the modest 400 ms computation time of the SDPT3-based optimization process. The timing data is based on a C implementation of the APM-IPM algorithm, running on a Linux-based 3 GHz AMD Athlon computer. More complex examples listed in [88] have an increased state dimension, however the maximal prediction horizon used in the demonstration examples is $n_c = 30$ steps. A randomized MPC problem with $n_x = 30$ states, $n_u = 8$ inputs and a $n_p = 30$ steps long horizon has been tackled by the APB-IPM algorithm a little under 26 ms, this is a 130 fold speed increase compared to the poor 3400 ms computation time of SDPT3.

While the simulations demonstrate an impressive speed increase over fully optimal (and generic) QP, we have to note that this method is suboptimal as well. The optimality loss depends on the algorithm settings, such as the number of maximal iterations. Roughly speaking, the two magnitudes speed increase compared to QP may be maintained with a suboptimality of 2–3%.

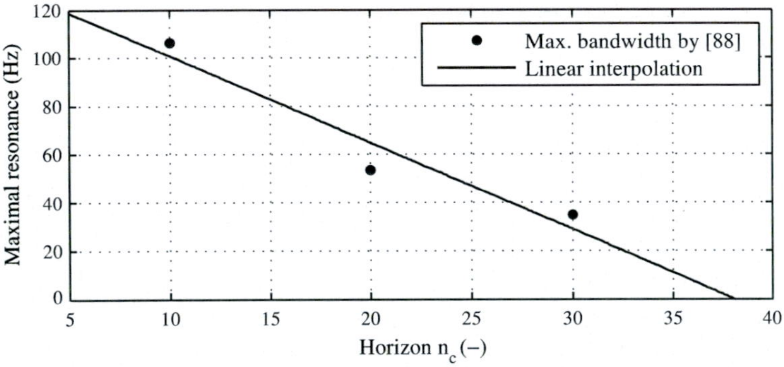

Fig. 8.8 The approximation of the maximal possible resonance frequency of the mechanical system depending on the prediction horizon for a fourth order model with the IPB–IPM method

Let us now take these results and consider them in a vibration attenuation example. A fourth order example containing two resonant frequencies and two points of actuation with a $n_c = 30$ steps horizon may be sampled by at most 350 Hz (2.79 s). Taking ten samples over period gives a 35 Hz maximum for the second resonant frequency—which in fact assumes a very flexible mechanical system, with low resonant frequencies.

Using the interior point method with appropriate variable re-ordering one may assume that to solve the problem it is enough to perform $O(n_c(n_x + n_u)^3)$ instead of $O(n_c^3(n_x + n_u)^3)$ as in the case of generic QP. In other words, the problem dimension depends on the horizon linearly, instead of cubically. Using this assumption and the test data featured in [88], we may interpolate for larger problem horizons. For the above described system explicitly controlling and modeling the first two resonances, we may approximate the maximal useful sampling frequency as demonstrated in Fig. 8.8. Taking into account that this is the maximal achievable sampling period for a fourth order system, the maximal resonant frequency of the controlled mechanical system is approximately an order of magnitude lower than the sampling times in Fig. 8.8.

Although the original simulation data presented in [88] does not contain a second order test example, it is an interesting question how it would compare to other methods or optimal QP with a long prediction horizon. A possible issue may be, that the computational overhead associated with calling the APB-IPM algorithm dominates over the optimization procedure itself, therefore its advantages would not be that obvious.

8.4 Efficient MPC Based on Pontryagin's Minimum Principle

In conventional MPC approaches the predicted input trajectories are the optimization variables, while the state variables are eliminated from the problem. This means that the number of variables of the optimization problem will grow linearly with the horizon length. However, the QP problem will require solving matrix factorizations, which will grow cubically with the horizon length.

A novel approach is described by Cannon et al. in [16]. The matrix factorizations used in general-purpose quadratic programming algorithms are replaced by recursions of state and co-state variables, based on Pontryagin's minimum principle. This way it is possible to develop solvers which have computational complexity only linearly depending on the horizon length.

An optimal control problem for input constrained linear systems using the Euler-Lagrange approach is developed in [68]. The optimal control problem in [16] is developed for input constrained linear system using the Euler-Lagrange approach. For a given active set of constraints, Pontryagin's minimum principle is used to calculate input and state trajectories as functions of initial and terminal states. The number of optimization variables is reduced to plant order, since the multipliers of input constraints are eliminated from the problem. An active set method successively solves equality constrained problems in a reduced space. In [16] it was shown that this method converges to the solution in a finite number of operations.

Let us consider a state-space model of a system, with input constraints, terminal state constraints $\mathbf{V}x_{n_c} \leq 1$. The actual plant state is assumed to be measured at each sampling instant, then the problem can be formulated as a QP problem:

$$\min_{x,\mathbf{u}} J(x_k, \mathbf{u}_k) \tag{8.63}$$

subject to the following constraints:

$$x_{k+1} = \mathbf{A}x_k + \mathbf{B}u_k, \qquad k = 0, \ldots, n_c - 1 \tag{8.64}$$

$$\underline{u} \leq u \leq \bar{u}, \qquad k = 0, \ldots, n_c - 1 \tag{8.65}$$

$$\mathbf{V}x_{n_c} \leq 1 \tag{8.66}$$

$$x_0 = x(0) \tag{8.67}$$

where $x(0)$ denotes the measured or observed state. A receding horizon control law is assumed, where only the first element of the input vector is implemented and the optimization procedure is repeated at each sampling instant. Stability is ensured by a suitable choice of terminal weight:

$$\frac{1}{2}x_{n_c}^T \mathbf{P}_f x_{n_c} = \frac{1}{2}\sum_{i=n_c}^{\infty}(x_{k+i}^T \mathbf{Q}x_{k+i} + r u_{k+i}^T u_{k+i}) \tag{8.68}$$

$$u_{k+i} = \mathbf{K}x_{k+i} \quad \forall \ i \geq n_c \tag{8.69}$$

It is assumed that the problem (8.63) is feasible for the given plant state x_0. The terminal state constraints are assumed to be constructed in a way that they ensure the feasibility of input constraints in case that $k \leq n_c$. After introducing Lagrange multipliers $\lambda = [\lambda_0, \ldots, \lambda_{n_c-1}]$ and also $\iota = [\underline{\iota}_0, \bar{\iota}_0, \ldots, \underline{\iota}_{n_c-1}, \bar{\iota}_{n_c-1}]$, the Karush-Kuhn-Tucker (KKT) conditions for the optimality of (8.63) can be expressed:

$$x_{k+1} = \mathbf{A}x_k + \mathbf{B}u_k, \qquad k = 0, \ldots, n_c - 1 \tag{8.70}$$

$$\lambda_k = \mathbf{A}^T \lambda_{k+1} + \mathbf{Q}x_{k+1}, \qquad k = 0, \ldots, n_c - 2 \tag{8.71}$$

$$u_k = -r^{-1}\mathbf{B}^T \lambda_k - r^{-1}\bar{\iota}_k + r^{-1}\underline{\iota}_k \qquad k = 0, \ldots, n_c - 1 \tag{8.72}$$

$$\bar{\iota}_k \leq 0, \; \bar{\iota}_k^T(\bar{u} - u_k) = 0, \quad \bar{u} - u_k \leq 0, \qquad k = 0, \ldots, n_c - 1 \tag{8.73}$$

$$\underline{\iota}_k \leq 0, \; \underline{\iota}_k^T(u_k - \underline{u}) = 0, \quad u_k - \underline{u} \leq 0, \qquad k = 0, \ldots, n_c - 1 \tag{8.74}$$

with the additional terminal and initial conditions:

$$\lambda_{n_c-1} = \mathbf{P}_f x_{n_c} + \mathbf{V}^T \nu \tag{8.75}$$

$$\nu \leq 0, \; \nu^T(\mathbf{V}x_{n_c} - 1) = 0, \; \mathbf{V}x_{n_c} - 1 \leq 0 \tag{8.76}$$

where $\nu = [\nu_1, \ldots, \nu_{n_\nu}]$ are Lagrange multipliers for the terminal condition. Traditionally an active set approach is deployed to solve (8.63) with the constraints, which transforms the problem into having only equality constraints. Let us denote the current active set as:

$$\mathbf{a} = [a_0, \ldots, a_{n_c-1}, a_{n_c}, \ldots, a_{n_c+n_\nu-1}] \tag{8.77}$$

We can express the corresponding equality problem (EP) as follows:

$$\min_{x,\mathbf{u}} J(x_k, \mathbf{u}_k) \tag{8.78}$$

$$x_{k+1} = \mathbf{A}x_k + \mathbf{B}u_k, \qquad k = 0, \ldots, n_c - 1 \tag{8.79}$$

$$u_k = \bar{u} \text{ if } a_k = 1, \qquad k = 0, \ldots, n_c - 1 \tag{8.80}$$

$$u_k = \underline{u} \text{ if } a_k = -1, \qquad k = 0, \ldots, n_c - 1 \tag{8.81}$$

$$\mathbf{e}_i^T \mathbf{V}x_{n_c} = 1 \text{ if } a_{n_c-1+i} = 1, \qquad i = 1, \ldots, n_\nu \tag{8.82}$$

where $\mathbf{e}_i$ is the i-th column of the identity matrix. A general-purpose active set solver, which is used in QPMPC uses the solution x_k, u_k, λ, ι to choose a new active set $\mathbf{a}$ from all of the active sets, for which the above defined problem admits a solution [8, 10]. This is done iteratively until a solution is obtained for the original problem (8.63). The conventional approach eliminates the state variables from the optimization structure. This way the number of variables is reduced to the horizon length times the input DOF. The disadvantage of this approach is that if the structure grouping cost and constraints is removed, the computational burden grows cubically.

It was shown in [16], that using Pontryagin's minimum principle the equality problem (8.78) can be expressed as a two-point boundary value problem. For a case of one input DOF:

$$x_{k+1} = \mathbf{A}x_k + \mathbf{B}u_k \qquad k = 0, \ldots, n_c - 1 \tag{8.83}$$

$$u_k = \bar{u},\ \bar{\iota}_k = -\mathbf{B}^T\lambda_k - r\bar{u},\ \underline{\iota}_k = 0 \qquad \text{if } a_k = 1 \tag{8.84}$$

$$u_k = \underline{u},\ \underline{\iota}_k = \mathbf{B}^T\lambda_k + r\underline{u},\ \bar{\iota}_k = 0 \qquad \text{if } a_k = -1 \tag{8.85}$$

$$u_k = -r^{-1}\mathbf{B}^T\lambda_k,\ \bar{\iota}_k = \underline{\iota}_k = 0 \text{if } a_k = 0 \tag{8.86}$$

$$\lambda_{n_c-1} = \mathbf{P}_f x_{n_c} + \mathbf{V}^T \nu \tag{8.87}$$

$$\mathbf{e}_i^T \mathbf{V} x_{n_c} = 1,\ \text{if } a_{n_c-1+i} = 1 \qquad i = 1, \ldots, n_v \tag{8.88}$$

$$\nu_i = 0,\ \text{if } a_{n_c-1+i} = 0 \qquad i = 1, \ldots, n_v \tag{8.89}$$

It is also assumed that matrix $\mathbf{A}$ is non-singular and invertible, so the state at the current sampling instant may be expressed by:

$$x_k = \mathbf{A}^{-1}x_{k+1} - \mathbf{A}^{-1}\mathbf{B}u_k \tag{8.90}$$

For a given terminal state x_{n_c} and ν one can determine x_k, u_l, λ, ι satisfying (8.83)–(8.87) by simulating it through the n_c step long control horizon.

Let $\eta \in \mathbb{R}^{n_x}$ denote the available degrees of freedom in x_{n_c} and ν which are satisfying both (8.88) and also (8.89). The active rows of $\mathbf{V}$ will be expressed by $\mathbf{V}_A \in \mathbb{R}^{n_a \times n_x}$—this corresponds to the active terminal constraints. The non-zero entries of ν are expressed by $\nu_A \in \mathbb{R}^{n_a}$. It is possible to express x_{n_c} and ν_A in terms of η:

$$x_{n_c} = [\mathbf{V}_A^{\perp}\ \mathbf{0}]\eta + \mathbf{V}_A^{+}\mathbf{1} \tag{8.91}$$

$$\nu_A = [\mathbf{0}\ \mathbf{I}]\eta \tag{8.92}$$

where $\mathbf{V}_A^{+}$ is a right inverse of $\mathbf{V}_A$. The matrix $\mathbf{V}_A^{\perp}$ spans the null space of $\mathbf{V}_A$. The null space is the kernel of $\mathbf{V}_A$ and it is valid that $\mathbf{V}_A\mathbf{V}_A^{\perp} = \mathbf{0}$ [27]. Equation (8.87) defines λ_{n_c-1} as an affine function of η:

$$\lambda_{n_c-1} = [\mathbf{P}_f\mathbf{V}_A^{\perp}\mathbf{V}_A^{T}]\eta + \mathbf{P}_f\mathbf{V}_A^{+}\mathbf{1} \tag{8.93}$$

From this it follows that the trajectories of x_k, u_k, λ, ι defined by Eqs. (8.83) through (8.87) are each affine functions of η. For example:

$$x_k = \Theta_{pk}\eta + \theta_{pk},\ k = 0, \ldots, n_c \tag{8.94}$$

where $\Theta_p \in \mathbb{R}^{n_x \times n_x}$ and $\theta_p \in \mathbb{R}^{n_x \times n_x}$ depend on the active set. The solution of the EP can be characterized by η. The minimizing argument (x, u) of the equality problem (8.78) is generated by $\eta = \Theta_{p0}^{-1}(x_0 - \theta_{p0})$ in (8.91), (8.92) and (8.83)–(8.87).

The solution of the equality problem (8.78) is feasible for (8.70) if and only if the inequality constraints on u, ι, x_{n_c}, ν are satisfied. The feasible region for η such that (8.73), (8.74) and (8.76) hold true is a convex polytope. This follows from the linearity of the constraints and the affine dependence of variable η. The convex polytope may be expressed as:

$$\mathscr{Y}(\mathbf{a}) = \{\eta : x(\eta), \lambda(\eta), u(\eta), \iota(\eta)\} \tag{8.95}$$

while satisfying (8.73–8.74) and (8.76). Since (8.70) defines the conditions of optimality for (8.63), an η optimal for the equality problem with a given active set $\mathbf{a}$ and $\eta \in \mathscr{Y}(\mathbf{a})$ is the minimizing argument of (8.63). Cannon et al. [15] proposes to successively solve the EP and update the active set via a line search. The following algorithm describes the single shooting approach [15, 16]:

Algorithm 8.10 Initialize with $\eta^{(0)}$ and an active set $\mathbf{a}^{(0)}$, such that η is in the polytopic set defined by $\mathscr{Y}(\mathbf{a}^{(0)})$, and with iteration number $k = 0$

- Compute a new x_{n_c} and ν_A corresponding to the current active set $\mathbf{a}^{(k)}$.
- Calculate Θ_{p0} and θ_{p0} and the inequalities defining $\mathscr{Y}(\mathbf{a}^{(k)})$ by simulating (8.83–8.87) backwards through the horizon $k = n_c - 1, \ldots, 0$.
- Solve the current EP by computing $\eta = \Theta_{p0}^{-1}(x_0 - \theta_{p0})$ and find a new search direction $\mathbf{j}^{(k)} = \eta - \eta^{(k)}$.
- Perform a line search: $\alpha_l^{(k)} = \max_{\alpha_l \le 1} \alpha_l$ subject to $\eta^{(k)} + \alpha_l \mathbf{j}^{(k)} \in \mathscr{Y}(\mathbf{a}^{(k)})$.
- If $\alpha_l^{(k)} < 1$ set $\eta^{(k+1)} = \eta^{(k)} + \alpha_l^{(k)}\mathbf{j}^{(k)}$ and increase $k = k + 1$, update the active set $\mathbf{a}^{(k)}$ and the new $\eta^{(k)}$ based on the new set of active constraints. Return to the initial set.
- Else set $\eta^* = \eta^{(k)} + \mathbf{j}^{(k)}$.

It is possible to show that the above defined algorithm converges to the solution of (8.63) in finite time. The initial point of the algorithm may also be derived from an estimate of the current optimal solution based on the previous step. This can help to "hot start" the algorithm.

This was the so-called single shooting approach. Although it is computationally efficient, it has a major drawback: the equality problem is solved via a backward recursion. Depending on the horizon length n_c and the eigenvalues of the matrix $\mathbf{A}$, the predicted trajectories can be inaccurate. State and input variables are characterized by only one variable η, state and co-state recursions are generally in unstable directions.

An additional approach is derived in [15, 16] based on Ricatti recursions, which improves the numerical robustness of the algorithm. By using the so-called sweep method, the optimal control sequence for the current EP is derived from a feedback law based on the current state x_k. The feedback law is computed at each point of the horizon via a backward recursion (8.83). It is possible to express x_k, u_k, λ, ι as explicit functions of x_0 by simulating over the prediction horizon. Mathematical ill conditioning is alleviated, since this forward recursion generates optimal trajectories for x_k, u_k with respect to the equality problem and the optimal trajectories are necessarily stable.

References

1. Agachi PS, Nagy Z, Cristea MV, Imre-Lucaci A (2006) Model based control: case studies in process engineering, 1st edn. Wiley-VCH, Weinheim
2. Antwerp JGV, Braatz RD (2000) Model predictive control of large scale processes. J Process Control 10(1):1–8. doi:10.1016/S0959-1524(99)00050-5, http://www.sciencedirect.com/science/article/B6V4N-3XFKHC8-1/2/18dea51e69b5a1869ec7b653f1615d9c
3. Arnold E, Neupert J, Sawodny O, Schneider K (2007) Trajectory tracking for boom cranes based on nonlinear control and optimal trajectory generation. In: IEEE international conference on control applications: CCA 2007, pp 1444–1449. doi: 10.1109/CCA.2007.4389439
4. Bemporad A, Chisci L, Mosca E (1994) On the stabilizing property of the zero terminal state receding horizon regulation. Automatica 30(12):2013–2015
5. Bemporad A, Bozinis NA, Dua V, Morari M, Pistikopoulos EN (2000) Model predictive control: a multi-parametric programming approach. In: Pierucci S (ed) European symposium on computer aided process engineering-10, Computer aided chemical engineering, vol 8. Elsevier, pp 301–306. doi:10.1016/S1570-7946(00)80052-8, http://www.sciencedirect.com/science/article/B8G5G-4NK5JX8-1V/2/76240158054cdb0fb1454f6e0eaa5dfe
6. Bemporad A, Morrari M, Dua V, Pistikopoulos EN (2000) The explicit solution of model predictive control via multiparametric quadratic programming. In: Proceedings of the American control conference, Chicago, pp 872–876
7. Bemporad A, Bernardini D, Cuzzola FA, Spinelli A (2010) Optimization-based automatic flatness control in cold tandem rolling. J Process Control 20(4):396–407. doi:10.1016/j.jprocont.2010.02.003, http://www.sciencedirect.com/science/article/B6V4N-4YH9YDM-1/2/d0e76c0bfed87a92f529fcf8061a96de
8. Boyd S, Vandenberghe L (2004) Convex Optimization. Cambridge University Press, Cambridge
9. Boyd S, Ghaoui LE, Feron E, Balakrishnan V (1994) Linear matrix inequalities in system and control theory: studies in applied mathematics. SIAM, Philadelphia
10. Camacho EF, Bordons C (1995) Model predictive control in the process industry, Advances in industrial control, vol 104. Springer, London
11. Cannon M (2005) Model predictive control, lecture notes, Michaelmas Term 2005 (4 lectures). Course code 4ME44. University of Oxford, Oxford
12. Cannon M, Kouvaritakis B (1997) Ellipsoidal bounding in stable predictive control. In: Proceedings of the 5th IEEE Mediterranean conference on control and systems
13. Cannon M, Kouvaritakis B (1998) Fast suboptimal predictive control with guaranteed stability. Syst Control Lett 35(1):19 – 29. doi: 10.1016/S0167-6911(98)00018-8, http://www.sciencedirect.com/science/article/B6V4X-3VCV2K5-T/2/e26cc7439ebfb7bebdbba41a20d48a16
14. Cannon M, Kouvaritakis B (2005) Optimizing prediction dynamics for robust MPC. IEEE Trans Autom Cont 50(11):1892–1897. doi:10.1109/TAC.2005.858679
15. Cannon M, Liao W, Kouvaritakis B (2006) Efficient MPC optimization using Pontryagin's minimum principle. In: IEEE conference on decision and control, San Diego, pp 5459–5464
16. Cannon M, Liao W, Kouvaritakis B (2007) Efficient MPC optimization using Pontryagin's minimum principle. Int J Robust Nonlinear Control 18(8):831–844
17. Chen CC, Shaw L (1982) On receding horizon feedback control. Automatica 18:349–352
18. Christophersen FJ, Kvasnica M, Jones CN, Morari M (2007) Efficient evaluation of piecewise control laws defined over a large number of polyhedra. In: Tzafestas SG, Tzafestas PA (eds) Proceedings of the European control conference ECC '07, pp 2360–2367
19. Cychowski M, O'Mahony T (2005) Efficient approximate robust mpc based on quad-tree partitioning. In: Control applications: CCA 2005: proceedings of 2005 IEEE conference on, p 239. doi: 10.1109/CCA.2005.1507131
20. Cychowski M, Ding B, O'Mahony T (2005) An orthogonal partitioning approach to simplify robust model predictive control. In: Intelligent control, 2005: proceedings of the 2005 IEEE

international symposium on, Mediterranean conference on control and automation, p 877. doi: 10.1109/.2005.1467130
21. Dua P, Kouramas K, Dua V, Pistikopoulos E (2008) MPC on a chip—recent advances on the application of multi-parametric model-based control. Comput Chem Eng 32(4–5):754–765. doi: 10.1016/j.compchemeng.2007.03.008, http://www.sciencedirect.com/science/article/B6TFT-4N98825-1/2/d6fda698cfbc90a5034664731405dced (festschrift devoted to Rex Reklaitis on his 65th birthday)
22. Ferreau HJ (2006) An online active set strategy for fast solution of parametric quadratic programs with applications to predictive engine control. Master's thesis, University of Heidelberg
23. Ferreau HJ (2011) qpOASES—online active set strategy. Leuven. http://www.qpoases.org
24. Ferreau HJ (2011) qpOASES user's manual. Optimization in Engineering Center (OPTEC) and Department of Electrical Engineering, Leuven
25. Ferreau HJ, Ortner P, Langthaler P, del Re L, Diehl M (2007) Predictive control of a real-world diesel engine using an extended online active set strategy. Annu Rev Cont 31(2):293–301. doi:10.1016/j.arcontrol.2007.09.001, http://www.sciencedirect.com/science/article/B6V0H-4R05C2B-2/2/23db757b09f804365ba12dc1844992d9
26. Ferreau HJ, Bock HG, Diehl M (2008) An online active set strategy to overcome the limitations of explicit MPC. IEEE Trans Autom Cont 18(8):816–830
27. Fletcher R (2000) Practical methods of optimization. Wiley, New York
28. García-Nieto S, Martínez M, Blasco X, Sanchis J (2008) Nonlinear predictive control based on local model networks for air management in diesel engines. Control Eng Pract 16(12):1399–1413. doi: 10.1016/j.conengprac.2008.03.010, http://www.sciencedirect.com/science/article/B6V2H-4SJR2B9-1/2/074f553c85e17d19dbc9885761d63047
29. Geyer T, Torrisi FD, Morari M (2008) Optimal complexity reduction of polyhedral piecewise affine systems. Automatica 44(7):1728 – 1740. doi:10.1016/j.automatica.2007.11.027, http://www.sciencedirect.com/science/article/pii/S0005109807004906
30. Gilbert EG, Tan KT (1991) Linear systems with state and control constraints: the theory and application of maximal output admissible sets. IEEE Trans Autom Cont 36(9):1008–1020
31. Grancharova A, Johansen TA, Kocijan J (2004) Explicit model predictive control of gas-liquid separation plant via orthogonal search tree partitioning. Comput Chem Eng 28(12):2481–2491. doi:10.1016/j.compchemeng.2004.06.010, http://www.sciencedirect.com/science/article/B6TFT-4CYWJH6-1/2/017c01ad65b6d18bf88e42ce347091d0
32. Grieder P, Wan Z, Kothare M, Morari M (2004) Two level model predictive control for the maximum control invariant set. In: American control conference, Boston
33. Grieder P, Kvasnica M, Baotic M, Morari M (2005) Stabilizing low complexity feedback control of constrained piecewise affine systems. Automatica 41(10):1683–1694
34. Hegrenæs Ø, Gravdahl JT, Tøndel P (2005) Spacecraft attitude control using explicit model predictive control. Automatica 41(12):2107–2114. doi: 10.1016/j.automatica.2005.06.015, http://www.sciencedirect.com/science/article/B6V21-4H99JGM-2/2/d96885a9c443d0f2cebee658396ae01b
35. Imsland L, Rossiter JA (2005) Time varying terminal control. In: 16th IFAC world congress in Prague, Prague, p TP/19
36. Imsland L, Bar N, Foss BA (2005) More efficient predictive control. Automatica 41(8):1395–1403. doi:10.1016/j.automatica.2005.03.010, http://www.sciencedirect.com/science/article/B6V21-4G7NT35-1/2/52a9590bfe1ccc2a9561165c3fbdf872
37. Johansen T, Grancharova A (2003) Approximate explicit constrained linear model predictive control via orthogonal search tree. IEEE Trans Autom Cont 48(5):810–815. doi:10.1109/TAC.2003.811259
38. Johansen TA, Jackson W, Schreiber R, Tøndel P (2007) Hardware synthesis of explicit model predictive controllers. IEEE Trans Control Syst Technol 15(1):191–197
39. Johansen TA, Fuglseth TP, Tøndel P, Fossen TI (2008) Optimal constrained control allocation in marine surface vessels with rudders. Control Eng Pract 16(4):457–464.

doi:10.1016/j:conengprac.2007.01.012, http://www.sciencedirect.com/science/article/B6V2H-4NBR3R5-1/2/2b78779171ca8833add555fffbb7b731 (Special section on manoeuvering and control of marine craft)

40. Jones C, Grieder P, Rakovic S (2005) A logarithmic-time solution to the point location problem for closed-form linear MPC. In: IFAC world congress, Prague
41. Karas A (2002) Stabilizujúce prediktívne riadenie systémov s obmedzeniami. PhD thesis, Slovak University of Technology in Bratislava, Bratislava, (Stabilizing predictive control of systems with constraints—in Slovak language)
42. Karas A, Rohal'-Ilkiv B, Belavý C (2007) Praktické aspekty prediktívneho riadenia, 1st edn. Slovak University of Technology in Bratislava/Slovenská E-Akadémia, n.o., Bratislava (Practical aspects of predictive control) in Slovak language
43. Keerthi SS, Gilbert EG (1988) Optimal, infinite horizon feedback law for a general class of constrained discrete time systems: stability and moving-horizon approximations. J Optim Theory Appl 57:265–293
44. Kouvaritakis B, Rossiter JA, Chang AOT (1992) Stable generalised predictive control: an algorithm with guaranteed stability. IEE Proc Part D 139(4):349–362
45. Kouvaritakis B, Rossiter J, Schuurmans J (2000) Efficient robust predictive control. IEEE Trans Autom Cont 45(8):1545–1549. doi:10.1109/9.871769
46. Kouvaritakis B, Cannon M, Rossiter J (2002) Who needs QP for linear MPC anyway? Automatica 38:879–88. doi:10.1016/S0005-1098(01)00263-1, http://www.sciencedirect.com/science/article/pii/S0005109801002631
47. Kouvaritakis B, Li S, Cannon M (2010) A line search improvement of efficient MPC. Automatica 46(11):1920–1924. doi:10.1016/j.automatica.2010.07.003, http://www.sciencedirect.com/science/article/B6V21-50NH0BX-3/2/0b4491d922a7d04d1b0315edae0e8944
48. Kvasnica M (2009) Real-time model predictive control via multi-parametric programming: theory and tools, 1st edn. VDM, Saarbrücken
49. Kvasnica M, Fikar M (2010) Performance-lossless complexity reduction in explicit mpc. In: Proceedings of the 49th IEEE conference on decision and control 2010, pp 5270–5275
50. Kvasnica M, Christophersen FJ, Herceg M, Fikar M (2008) Polynomial approximation of closed-form MPC for piecewise affine systems. In: Proceedings of the 17th world congress of the international federation of automatic control, Seoul, pp 3877–3882
51. Kvasnica M, Herceg M, Čirka L', Fikar M (2009) Robust adaptive minimum-time control of piecewise affine systems. In: Proceedings of the 48th IEEE conference on decision and control, pp 2454–2459
52. Kvasnica M, Herceg M, Čirka L', Fikar M (2009) Time optimal control of fuzzy systems: a parametric programming approach. In: Proceedings of the 28th IASTED conference on modelling, identification and control, pp 640–805
53. Kvasnica M, Herceg M, Čirka L', Fikar M (2009) Time-optimal control of Takagi–Sugeno fuzzy systems. In: Proceedings of the 10th European control conference, Budapest, pp 916–921
54. Kvasnica M, Christophersen FJ, Herceg M, Fikar M (2010) Polynomial approximation of closed-form MPC for piecewise affine systems. In: Selected topics in modelling and control, vol 6, Slovak University of Technology Press, pp 105–112
55. Kvasnica M, Rauová I, Fikar M (2010) Automatic code generation for real-time implementation of model predictive control. In: Proceedings of the 2010 IEEE international symposium on computer-aided control system design, Yokohama, pp 993–998
56. Kvasnica M, Fikar M, Čirka L', Herceg M (2011) Complexity reduction in explicit model predictive control. In: Selected topics on constrained and nonlinear control: STU Bratislava–NTNU Trondheim, pp 241–288
57. Kvasnica M, Herceg M, Čirka L', Fikar M (2011) Explicit minimum-time controllers for fuzzy systems. In: Selected topics on constrained and nonlinear control. Preprints, STU Bratislava–NTNU Trondheim, pp 287–292

58. Kvasnica M, Herceg M, Čirka L', Fikar M (2011) Robust explicit time-optimal control of PWA systems with parametric uncertainties. In: Selected topics on constrained and nonlinear control. Preprints, STU Bratislava–NTNU Trondheim, pp 295–300
59. Kvasnica M, Rauová I, Fikar M (2011) Real-time implementation of model predictive control using automatic code generation. In: Selected topics on constrained and nonlinear control. Preprints, STU Bratislava–NTNU Trondheim, pp 311–316
60. Kvasnica M, Rauová I, Fikar M (2011) Separation functions used in simplification of explicit MPC feedback laws. In: Selected topics on constrained and nonlinear control. Preprints, STU Bratislava–NTNU Trondheim, pp 303–308. http://www.kirp.chtf.stuba.sk/publication_info.php?id_pub=1067
61. Maciejowski JM (2000) Predictive control with constraints, 1st edn. Prentice Hall, Upper Saddle River
62. Mayne DQ, Michalska H (1990) Receding horizon control of non-linear systems. IEEE Trans Autom Cont 35(5):814–824
63. Mayne DQ, Rawlings JB, Rao CV, Scokaert POM (2000) Constrained model predictive control: stability and optimality. Automatica 36(6):789–814
64. Michalska H, Mayne DQ (1993) Robust receding horizon control of constrained nonlinear systems. IEEE Trans Autom Cont 38(11):1623–1633
65. Morger A (2010) Synthesis, design and control of a tendon-driven robot platform for vestibular stimulation during sleep. Master's thesis, ETH Zurich
66. Mosca E (1994) Optimal, Predictive and adaptive control, 1st edn. Prentice Hall, Upper Saddle River
67. Nesterov Y, Nemirovskii A (1994) Interior-point polynomial methods in convex programming, Studies in applied mathematics, vol 13. SIAM, Philadelphia
68. Ohtsuka T (2004) A continuation/GMRES method for fast computation of nonlinear receding horizon control. Automatica 40(4):563–574. doi:10.1016/j.automatica.2003.11.005, http://www.sciencedirect.com/science/article/pii/S0005109803003637
69. Panos C, Kouramas KI, Georgiadis MC, Brandon N, Pistikopoulos EN (2010) Modelling and explicit MPC of PEM fuel cell systems. In: Pierucci S, Ferraris GB (eds) 20th European symposium on computer aided process engineering, Elsevier, vol 28, pp 517–522. doi:10.1016/S1570-7946(10)28087-2, http://www.sciencedirect.com/science/article/B8G5G-505XT0T-36/2/4e263f053c6a882ca5aaece78b07e431
70. Percival M, Wang Y, Grosman B, Dassau E, Zisser H, Jovanovic L, Doyle FJ III (2011) Development of a multi-parametric model predictive control algorithm for insulin delivery in type 1 diabetes mellitus using clinical parameters. J Process Control 21(3):391–404. doi:10.1016/j.jprocont.2010.10.003, http://www.sciencedirect.com/science/article/B6V4N-51BP6G6-2/2/33f4ab368180bad868c94b831fb10c17
71. Pistikopoulos EN, Georgiadis MC, Dua V (eds) (2007) Multi-parametric model-based control, vol 2, 1st edn. Wiley-VCH, Weinheim
72. Pistikopoulos EN, Georgiadis MC, Dua V (eds) (2007) Multi-parametric programming, vol 1, 1st edn. Wiley-VCH, Weinheim
73. Rawlings JB, Muske KR (1993) The stability of constrained receding horizon control. IEEE Trans Autom Cont 38(10):1512–1516
74. Rossiter JA (2003) Model-based predictive control: a practical approach, 1st edn. CRC Press, Boca Raton
75. Rossiter JA, Kouvaritakis B (1993) Constrained stable generalized predictive control. IEE Proc Part D 140(4):243–254
76. Rossiter JA, Kouvaritakis B, Rice MJ (1998) A numerically robust state-space approach to stable-predictive control strategies. Automatica 34(1):65–73
77. Scokaert POM, Rawlings JB (1996) Infinite horizon linear quadratic control with constraints. In: Proceedings of IFAC'96 world congress, San Francisco, vol 7a-041, pp 109–114
78. Scokaert POM, Rawlings JB (1998) Constrained linear quadratic regulation. IEEE Trans Autom Cont 43(8):1163–1169

79. Sznaier M, Damborg MJ (1987) Suboptimal control of linear systems with state and control inequality constraints. In: Proceedings of the 26th IEEE conference on decision and control, pp 761–762
80. Takács G (2009) Efficient model predictive control applied on active vibration attenuation. PhD thesis, Institute of Autmation, Measurement and Applied Informatics, Faculty of Mechanical Engineering, Slovak University of Technology in Bratislava, submitted in fulfillment of the requirement for the degree of Doctor of Philosophy, pp 257
81. Takács G, Rohal'-Ilkiv B (2009) Implementation of the Newton–Raphson MPC algorithm in active vibration control applications. In: Mace BR, Ferguson NS, Rustighi E (eds) Proceedings of the 3rd international conference on noise and vibration: emerging methods, Oxford
82. Takács G, Rohal'-Ilkiv B (2009) MPC with guaranteed stability and constraint feasibility on flexible vibrating active structures: a comparative study. In: Hu H (ed) Proceedings of The eleventh IASTED international conference on control and applications, Cambridge
83. Takács G, Rohal'-Ilkiv B (2009) Newton–Raphson based efficient model predictive control applied on active vibrating structures. In: Proceedings of the European control conference, Budapest
84. Takács G, Rohal'-Ilkiv B (2009) Newton–Raphson MPC controlled active vibration attenuation. In: Hangos KM (ed) Proceedings of the 28th IASTED international conference on modeling, identification and control, Innsbruck
85. Tøndel P, Johansen TA, Bemporad A (2003) Evaluation of piecewise affine control via binary search tree. Automatica 39(5):945–950. doi:10.1016/S0005-1098(02)00308-4, http://www.sciencedirect.com/science/article/pii/S0005109802003084
86. Van den Broeck L, Diehl M, Swevers J (2009) Time optimal MPC for mechatronic applications. In: Proceedings of the 48th IEEE conference on decision and control, Shanghai, pp 8040–8045
87. Wang Y, Boyd S (2008) Fast model predictive control using online optimization. In: Proceedings of the 17th world congress the international federation of automatic control, Seoul, pp 6974–6979
88. Wang Y, Boyd S (2010) Fast model predictive control using online optimization. IEEE Trans Control Syst Technol 18(2):267–278
89. Wills A, Bates D, Fleming A, Ninness B, Moheimani R (2005) Application of MPC to an active structure using sampling rates up to 25 kHz. In: 44th IEEE conference on decision and control, 2005 and 2005 European control conference. CDC-ECC '05, pp 3176–3181. doi:10.1109/CDC.2005.1582650
90. Wills AG, Bates D, Fleming AJ, Ninness B, Moheimani SOR (2008) Model predictive control applied to constraint handling in active noise and vibration control. IEEE Trans Control Syst Technol 16(1):3–12
91. Zhou K, Doyle J, Glover K (1994) Robust optimal control. Prentice-Hall, Englewood Cliffs

Part III
Model Predictive Vibration Control

Chapter 9
Applications of Model Predictive Vibration Control

This chapter will briefly review some of the existing applications of model predictive control for vibration attenuation or its closely related fields. The application of model predictive control as a vibration reduction strategy is not common, and there are only a handful of available publications related to this field. The existing literature is mostly based on well-established, however slightly outdated theoretical findings such as GPC and DMC; see for example [27, 33, 78, 102, 154] and others. At this time the only available published work featuring model predictive vibration control *with system constraints* and online optimization is presented by Wills et al. in [142, 143], which has inspired the authors of this book to further investigate the topic and include stability constraints in [126–128, 125]. Since predictive control for active vibration attenuation especially with constraints and stability guarantees is a nearly non-existent field [10, 33], a further research of its properties and application possibilities is warranted.

Of course, as it was already noted in the preceding chapters, one may expect that systems and structures with multiple-inputs and multiple-outputs (MIMO) are especially suited for model predictive vibration control. Unlike in the case of simple single-input and single-output (SISO) systems, the advantages of MPC methods and the increased performance of vibration attenuation over other control approaches will be evident in more complex application scenarios, involving multi-point sensing and actuation. MPC controlled systems will always perform better than saturated strategies, as these controllers are either overly aggressive because they have been tuned without constraints in mind (potentially leading to unstable control course), or are too conservative because their performance is meant to reflect the saturation limits (not using the full potential of the closed-loop system).

In addition to reviewing suitable applications of MPC as a vibration reduction strategy, the properties and issues of the current implementations utilizing other control approaches are also discussed in this chapter. The few existing works on MPC or related vibration attenuation applications will be presented, along with a discussion of the application of model predictive vibration control in fields where so far only traditional industrial feedback control has been applied.

G. Takács and B. Rohal'-Ilkiv, *Model Predictive Vibration Control*,
DOI: 10.1007/978-1-4471-2333-0_9, © Springer-Verlag London Limited 2012

Fig. 9.1 The tuned mass damper on display in "Taipei 101" [22] is a prime example of the potential of both passive and active vibration control systems designed to save lives in civil engineering structures

We will begin the first chapter of Part III by presenting simple demonstration examples in Sect. 9.1: cantilever beams, plates, disks and other structures. These laboratory applications may seem elementary, but in many cases adequately emulate the dynamic behavior of more complicated real-life systems. Academic literature already lists a few successful applications of various MPC algorithms for the vibration damping of these structures, some of which we will discuss here. The following section examines current examples of vibration control in the field of robotic manipulators and other similar systems. Structural attenuation in optical systems is introduced in Sect. 9.3, while the difficult application scenario of active noise control is briefly reviewed afterward. The automotive industry driven by the desire to fulfill customer needs has always been keen on introducing innovations—including different forms of vibration attenuation and damping—in order to increase passenger comfort. The possible application fields of model predictive control in the transport industry are investigated in Sect. 9.5. Vibrations induced by earthquakes are especially dangerous to human life; therefore, any measure against its effects is extremely important. In addition to some of the already existing passive counter-measures, as the tuned mass damper of the Taipei 101 illustrated in Fig. 9.1, Sect. 9.6 prevalently deals with active or semi-active control systems in the field of civil engineering, which may benefit from the application of model predictive control. After a brief excursion into the world of vibration insulated platforms and machine tools, the chapter is finished by a section devoted to the aerospace industry and spaceflight.

9.1 Concept Demonstration Examples

The seemingly simple cantilever beam structure is at the focus of many research publications, as it can represent the basic dynamic behavior of a broad range of physical systems, such as wing surfaces in fixed and rotary wing aircraft, civil engineering structures, space structures and others. From the viewpoint of the control engineer, the dynamics of such a simple experimental laboratory structure fully emulate the problems encountered with real lightly damped structures such as fast dynamics, the need for constraints and others.

9.1.1 Cantilever Beams

The active vibration control of a cantilever beam using model predictive control has been considered by Kim and Washington in [48]. In this work, predictive control is only utilized to enhance the properties of sliding mode control. Generalized predictive control has been featured in [102] to suppress the vibrations of a piezoelectric patch-driven flexible clamped beam. The resultant GPC law is in a closed form without constraints. The authors compared GPC to positive position feedback, citing inconclusive performance advantages of GPC over PPF.

Adaptive GPC has been implemented for a piezoelectric device-driven cantilevered beam in [77] and later in [78]. The data acquisition of this system has been set at 200 Hz with the adaptive controller updates performed at third of the sampling rate. The adaptive feature certainly requires increased computational efforts; however, for an unconstrained predictive controller, such sampling speeds are not very impressive and severely limiting the effective bandwidth of the vibration attenuation system. In addition, the GPC approach may be considered slightly outdated by today's prevailing research trends. The lack of system constraints allows for stability guarantees, which are given in a more straightforward fashion than in constrained MPC, thus avoiding any problems with limited regions of attraction.

A model predictive controller more consistent with current research trends applied to the structural vibration control of an experimental active cantilever is presented by Wills et al. in [142] and later in [143]. Here a linear time-invariant state-space system is used to model the vibration dynamics of the cantilever. The need for system constraints is emphasized on the grounds of the always-present physical actuator limitations. The model utilized by Wills et al. is of a relatively high order; that is $n_x = 12$ states to cover 5 transversal vibration modes in the 0–500 Hz bandwidth. This implementation assumes system constraints, thus uses a QP-based optimization procedure in the online regime. Given the fairly large model and the QP algorithm, the work has demonstrated impressive sampling speeds up to 5 kHz with a $n_c = 12$ steps long horizon. The high sampling speeds have been reached by using a digital signal processor board with a customized active-set QP algorithm. The question of stability has been treated by using a dual-mode MPC algorithm; however, without the application of a constraint checking horizon to ensure the feasibility of the predicted input tail. Although this dual-mode concept is a big step up from the generic finite horizon MPC approach, it can still not guarantee stability a priori. The findings by Wills et al. confirm that constrained MPC outperforms a saturated (clipped) LQ controller in terms of vibration damping performance.

In previous works by the authors of this book a computationally efficient, suboptimal constrained MPC approach has been applied to damp cantilever beam vibrations in [128, 127], where stability guarantees have been given a priori. Later this suboptimal approach has been compared to multi-parametric MPC in [126], suggesting issues with the implementability of stabilized constrained MPC on lightly damped systems. The vibration damping performance and implementation properties of different MPC algorithms with constraints and guaranteed stability have been analyzed as well. Based on this, the application of constrained stabilized MPC to lightly damped vibrating structures will be introduced in detail in the upcoming chapters.

The active vibration control of simple cantilevered beams through various control strategies other than MPC is discussed in numerous publications. A neural-networks approach is used in [46], fuzzy control is utilized in [66, 140], genetic algorithms are used for actuator placement and feedback gain optimization in [105, 148]. Electrorheological fluid-based mounts are used to damp a cantilever semi-actively in [43]. Several other works examine the different aspects of clamped cantilever beam vibrations [76, 133].

9.1.2 Plates and Shells

Another article considering an adaptive-predictive control approach based on the GPC controller concept is aimed at suppressing plate vibrations [27]. Similarly to the previously introduced works, Eure applied a finite horizon predictive controller without stability guarantees. The adaptive GPC controller has been able to cover a bandwidth of several kHz with corresponding high order models, which would been very difficult to perform with a quadratic optimization-based constrained MPC algorithm.

The vibration control of plates and shells, especially cantilever plates is also a highly researched topic. Some of the works focusing on the topic are for example [20, 98, 101]. Robust vibration control of circular shaped plates is investigated in [38].

9.1.3 Others

A less traditional active vibration control demonstration device is used by VandenBroeck at al. in [135, 136]. The authors utilize a two degree of freedom mass-spring-damper system and apply a novel time optimal MPC approach to this system.

The two DOF system is actuated by a hydraulic piston. The time optimal MPC approach of VandenBroeck steers the system into equilibrium in the shortest possible time [135], without concentrating on the effort of minimizing the cost function. This novel and alternative interpretation of the MPC law can be beneficial for other vibrating systems as well.

9.2 Manipulators in Robotics

The increasing efficiency of robots and manipulators demands the constant increase of operational speeds while at the same time due to cost effectiveness and weight optimization the manipulating arms become increasingly flexible. Due to the increased flexibility, the vibrational response of such systems cannot be ignored anymore. Lightweight manipulators are of special interest for spacecraft.

Model predictive control of a flexible link manipulator mechanism is often investigated in the literature. The problem is recurring and is not strictly limited to manipulators, as these mechanisms have the same generic dynamical behavior as other lightly damped structures, such as solar panels, antenna systems, truss structures in space or for that matter simple cantilever beams. Even though the majority of works dealing with manipulators do not explicitly focus on vibration issues, some do allow for the attenuation of vibration dynamics or at least consider the possibility to use an MPC controller for this purpose.

Unconstrained predictive vibration control of elastic manipulators appears as early as 1996 in [151] by Yim. The author uses an unconstrained formulation of the predictive control law with a slightly modified quadratic cost function to arrive at a closed control law. The stability of the system is investigated by inspecting pole locations of the resulting closed-loop system.

An unconstrained MPC method has been applied to the control of vibrations of a flexible link manipulator in [157] by Zmeu and Shipitko. The closed form of the controller utilized a model based on an offline artificial neural network learning process. The work did not treat the stability of the system explicitly.

A multivariable model predictive control based vibration reduction system for a flexible link mechanism has been introduced in [33]. Much like other available publications considering predictive vibration control, Hassan et al. assumes a predictive controller based on early formulations such as FIR- or FSR-based methods. Constraints or stability issues have not been explicitly treated in this work either.

Boscariol et al. considered a constrained MPC control of a flexible link manipulator in [10] and compared control performance to more traditional approaches. The work is aimed both at position and vibration control and utilizes a linearized state-space model of the dynamics. Even though the robustness of the controller is discussed and tested in simulation by introducing uncertainties, a priori stability guarantees are not ensured in this work, neither is the question of stability thoroughly discussed. An unstable MPC controlled manipulator arm may pose serious issues in many critical applications: both the vibration control and the position control of the arm may go out of hand, and risk mission objectives and possibly even human life.

The vibration control of flexible link manipulators is discussed, for example, in [44, 67, 106, 121, 144] with control strategies other than MPC.

9.3 Optical Systems

The vibration control of optical systems enhances image quality that is deteriorated due to the vibration of the optical system or its components. A well-known example of optical image stabilizers can be found in high-end photography equipment such as camera bodies and lenses.

The other good example of vibration control systems in optics is telescopes in astronomy. It is very difficult to cast mirrors larger than 7 m in diameter from a single piece of glass, therefore future reflectors shall be constructed from an array of optical systems [62, 99]. The problem with such a multi-mirror setup is that the positioning of the individual elements has to be precise enough to mimic the properties of a monolithic mirror even in the presence of outside disturbances. Moreover, in ground-based astronomical observatories the source of disturbance of the images come not only from mechanical sources, but the mirrors are significantly affected by atmospheric conditions as well.

The application of MPC in the vibration control of optical systems is manyfold and the need for actuator constraints is important in this situation as well. Unlike with the lightly damped systems presented before, the actuator and disturbance asymmetry is small in optics, therefore the need of long horizon MPC to ensure a proper sized region of attraction is not likely. The implementation potential of constrained MPC with stability guarantees as a vibration reduction technique in optics is high and only limited by the bandwidth of the disturbance and computational efficiency of the algorithm, and not the region of attraction.

Optical jitter has been attenuated experimentally by a real-time implementation of an adaptive GPC algorithm in [78]. The sampling rate has been set at 600 Hz citing that the jitter occurs at half this frequency. While it is questionable whether a sampling speed which is the double of the upper bandwidth is satisfactory enough, an unconstrained predictive controller could surely do better than this speed.

Model predictive control has been suggested to control the vibrations of rear-view mirrors in luxury or heavy vehicles by Larchez in [58]. The image quality of rear-view mirrors affects driver comfort due to increased eyestrain and blurred images may cause a road safety hazard as well. Disturbances contributing in increased blurriness due to vibration are limited to the 5–200 Hz frequency range. The need for predictive control has been justified because of the delays in the hardware loop. Larchez utilized a type of a simple adaptive-predictive controller, based on the filtered-x least mean square approach. The work demonstrated a significant vibration attenuation capability in experimental tests, nevertheless lacked system constraints. Even though the sampling frequencies were relatively high, there was no need to use long prediction horizons since neither stability, nor feasibility guarantees have been given. From the control engineering viewpoint, the lack of system constraints removes most of the implementation difficulties. Because real actuators are always constrained by saturation, this kind of approach also introduces questions on the stability or the real optimality of the algorithm. To be fair, a simpler algorithm requires simpler and cheaper hardware, which is essential in the cost sensitive automotive industry. In this setting, the theoretical questions of stability or optimality are irrelevant as an actively attenuated rear-view mirror can pose a significant image improvement, and thus a competitive advantage over a conventional one.

The majority of works in active vibration control of optical systems utilize feed-forward approaches such as digital filters. A digital filter-based system has been introduced in [90] to correct the tracking error in automotive DVD drives due to road vibrations, while sliding mode control is used in [155] for a similar task. Another field of application for AVC in optics is the stabilization and tracking control of scanning probes in atomic force microscopes [18, 23]. An active vibration control system for an airplane mounted optical bench effectively attenuating sub 0.1 Hz frequencies is presented in [92]. Other works concerned with AVC in optical systems are, for example, [15, 62, 71, 86].

9.4 Active Noise Control

Active noise control (ANC) is a closely related field to active vibration control and it is concerned with attenuating sound waves. Sound is a pressure wave which may be actively attenuated by structure-integrated actuators or noise-cancellation speakers.

From the viewpoint of model predictive control, ANC is a field where the practical implementation of predictive controllers is very difficult. This is caused by the high frequency and wide bandwidth excitation that is usually encountered in acoustics. Audible sound is limited to frequencies between 20 Hz and 20 kHz. The sampling of the MPC algorithm covering acoustic frequencies has to exceed the highest expected frequency by approximately ten times. The implementation of a constrained MPC controller with online QP optimization in the range of the upper limits of the human hearing is burdensome with currently available hardware. However, an unconstrained controller is more likely especially if the disturbance is only limited to narrow bandwidths, thus requiring small model orders.

The situation is somewhat relieved by the fact that the sound energy of acoustic disturbance is unlikely to exceed actuator capabilities. Unlike in the case of lightly damped structures, a constrained MPC controller with stability guarantees does not require excessive horizons to ensure a region of attraction (a feasible set of states) covering all conceivable disturbances.

Adaptive dynamic matrix control, an older form of model predictive control based on the finite step response of the controlled system is utilized in [154]. As with other publications using simple predictive control formulations, Zhang and Gal did not consider the inclusion of system constraints into their formulation. The approach utilized in [154] seems to be both practical and as the results show functional, however not up-to-date with the modern findings of model predictive control theory such as constraint inclusion, state-space models or stability and feasibility guarantees.

The adaptive GPC predictive control strategy has been utilized in [77, 78] to minimize sound pressure in a closed experimental noise control test bed. Compared with the unattenuated system, the adaptive GPC algorithm was effective up to approximately 200 Hz. This has been reached by the implementation of a predictive controller without system constraints, essentially avoiding the usual implementation issues related with sampling speeds in MPC.

One of the applications where noise is not merely a comfort factor is submarines, where increased noise levels may cause the detection and elimination of the vessel by the enemy. Piezoelectric stack actuators configured counter phase in a T-shaped active stiffener have been considered in [91] and subsequently in [92]. The control strategy used in these works is not MPC based; rather it is founded on the equation of motion of the hull at the stiffeners, where the control moment is essentially calculated from minimizing the undesired deflections or pressures. While both displacement and radiated pressure minimization is considered in numerical examples with 40–90% noise level reduction [92], the inclusion of a state-space model-based MPC control approach could be certainly beneficial. The calculated resonant modes are located at 12, 24 and 35 Hz [91], therefore the real-time application of a MPC controller is feasible. Although the displacement effect of the piezoelectric actuators is still somewhat smaller than the expected deformations of a large submarine hull, the region of attraction in stabilized and feasible MPC is not a similarly serious issue as it is with cantilever beam-like structures. This in practice means the use of shorter horizons and the possibility to apply a broader range of MPC algorithms.

There are several works investigating the active control of acoustic noise using various traditional feedback control strategies. Active noise control in acoustic cavities such as trains or aircraft fuselages is presented using positive position feedback control techniques in [17]. A distributed control approach has been chosen in [35] to drive piezoelectric actuators anti-phase to minimize sound radiation. The positioning of control sources in three-dimensional noise control settings is discussed in detail in [56]. Li et al. performed a simulation study investigating active noise control of a medical MRI scanner in [63]. There is a wide selection of available publications discussing active noise control using different traditional control strategies, for example [12, 49, 54, 60].

9.5 Automotive Industry

Because the suspension transfers the force between the vehicle body and the road, a well-designed active suspension may enhance driving comfort, handling, vehicle service time and road safety. For commercial heavy vehicles, the aim of active or semi-active suspensions is to lessen dynamic tire forces to protect road surfaces and to protect cargo integrity. These objectives can be effectively met by the use

Fig. 9.2 Body vibration of a passenger car due to road excitation, measured by FKFS, Stuttgart, Germany [34]. The *darker* shades denote less vibration, while the *lighter* shades found mostly at the center of large panels indicate increased vibration levels

of MPC [75]. The body vibration of a passenger car resulting from typical road excitation is illustrated in Fig. 9.2, where the darker shades indicate lower vibration levels, corresponding to the stiffer areas of the body. The higher vibration levels are indicated by the lighter shades, those tend to be the areas inside large panels and surfaces.

Vehicle suspension systems essentially consist of a fixing mechanism (usually a wishbone or similar), a spring and a damper. This damper may be exchanged to a semi-active device, which provides a variable damping force based on the decisions of some control algorithm and feedback measurements. The (semi-)active dampers provide the variable damping force based on viscosity change as in MR dampers, or through a variable orifice valve, which can set the fluid flow conditions inside the damper. Model predictive control-based "preview" enhanced semi-active suspensions have been already considered for the HMMWV military vehicle [75].

A modified constrained model predictive control algorithm is proposed to control switching shock absorbers on a trailer semi-truck in [51]. The controller applied by Kok et al. is not predictive in a classical sense; however, it preserves most of its characteristic properties. The predictions in this work are generated by the observed disturbances measured at the front axle, while the control inputs are applied on the rear axle of the semi-truck—referred to as a control preview. In spite of a relatively powerful hardware configuration for the time, the real-time experimental implementation of this system was not feasible. A similar approach is featured in simulation in [75], where the optimization horizon was 16 steps with a 100 Hz sampling. Neither of the aforementioned works treated the question of stability arising from the nonlinear control law.

The MPC-based controller design for an electromagnetic motor-driven suspension actuator is presented in [39]. The controller is responsible for supplying current to the coil of the electromagnet and thus ensuring a given position. The constraints arise from both the available current limits (input) and the useful working space as well (output). A sampling speed of 100 Hz has been used with a relatively long $n_c = 40$ steps prediction horizon, given a system with three states. These parameters are within the limits of feasible implementation on current hardware, using generic QP algorithms. The work arrives at the conclusion that vibration damping capability of the constrained MPC system exceeding saturated LQ. Stability of the constrained MPC control law is not treated by Huang and Zhang.

The difficulty in designing a semi-active suspension is the hysteretic nonlinear behavior of MR dampers. Some authors employ soft-computing techniques such as neural networks to overcome this difficulty [152]. Other authors utilized a hybrid fuzzy—sliding mode controller for an ER-based automotive suspension system [124]. Classical feedback and various feedforward control-based semi-active and active vehicle suspension systems are discussed by many works, for example [26, 29, 74, 150, 152].

Other vibration control applications closely related to the automotive industry are for example the active vibration damping of seat suspension systems [123], to increase ride comfort and eliminate certain health concerns. The controller applied by Sun et al. is a $\mathscr{H}_\infty$ with a finite frequency response. The bandwidth of the controller is tuned to human comfort, around 4–8 Hz or the resonant range of internal organs. Sun et al. discusses the need to adjust for finite actuator stroke, therefore both the implied discrete sampling frequency and the requirement for constraints suggests the use of model predictive control for the vibration control of active seating systems.

Semi-active engine suspensions based on squeeze-mode ER fluid actuators are suggested in [141], while an electromagnetic inertia-mass actuator is utilized in [9] for the same task. A review of active vibration and noise canceling techniques is given in an earlier work by Shoureshi and Knurek in [109] and later in [110]. Since the advent of cheap microcontrollers, a wide spectrum of publications have appeared on the active vibration control and noise suppression in the automotive industry including works such as [41, 52, 122] and others.

In addition to the automotive industry, the vibration control of railway cars is considered by Kozek et al., where a heavy metro railway car is actuated by piezo stack actuators [53]. The combined artificial neural network and PI controller-based vibration control of magnetic levitation (MAGLEV) trains is investigated in [149].

9.6 Civil Engineering

Active control technologies are valuable as they may save human life and financial property during earthquakes. The integrity of civil engineering structures such as high-rise buildings, bridges, towers and infrastructure is not only jeopardized by seismic activity but by wind as well. Aerodynamic forces may induce vibrations similar to the case of wing flutter in aviation and severely disturb or destroy structures. The case of the infamous Tacoma Narrows Bridge collapse[1] still remains as a cautionary tale on the power of mechanical vibrations [7]. In addition to optimizing the structural design of buildings in areas of high seismic activity, passive means of damping are often employed, such as tuned mass dampers (See Figs. 9.1 and 9.5 for an illustration.), lead rubber bearings, friction pendulum bearings and others.

[1] See the Fig. 1.2 on p. 3 for an illustration.

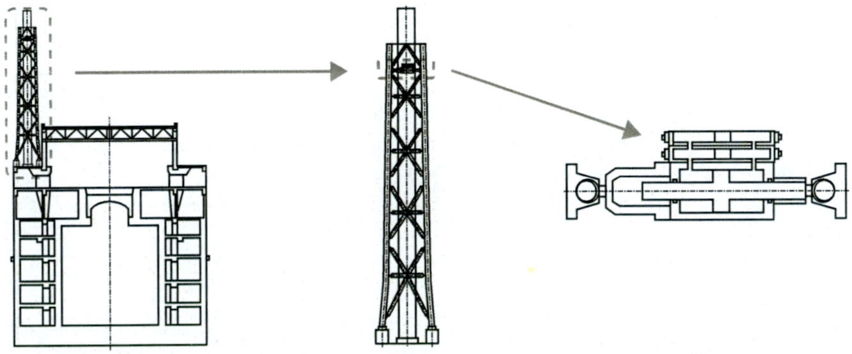

Fig. 9.3 A passive vibration control device is implemented at a stack of the Kashiwazaki Kariwa Nuclear Power Station located at a seismically sensitive area of Japan [146]. The image shows the cross-section of the reactor buildings, a detail of the stack and the passive vibration isolation device

In addition to passive technologies, semi-active and active seismic isolation systems have started to appear both in academic studies and in real buildings as well.

Although the major nuclear disaster at the Fukushima Daiichi power plant following the $M_w = 9.0$ magnitude [134] Tōhoku earthquake has not been the direct result of vibration phenomena but rather of equipment failure due to the tsunami wave [129, 145], engineers and safety experts are pushed even more to implement additional safety measures. The reactor buildings withstood the earthquake with an acceleration magnitude somewhat above their design limit, but the critical external power supply infrastructure that could potentially power the cooling equipment of the reactors was destroyed [107]. Figure 9.3[2] illustrates the passive seismic control system implemented for a stack of the reactor buildings six and seven at the Kashiwazaki Kariwa Nuclear Power Station in Japan [146], prior to the more recent Tōhoku earthquake. Because this power station has been previously affected by an earthquake in 2007, vibration insulation systems are steadily implemented as a way to enhance the seismic safety of this highly sensitive building. Active systems can improve on seismic safety even further; however, precautions must be made to ensure closed-loop stability.

In addition to input constraints arising from the physical limits of actuators, another set of constraints may be essential for an earthquake prevention technology. Output constraints in a form of maximal horizontal deflections in the building directly relate to the preservation of structural integrity. Model predictive control could be effectively used to incorporate these needs into the control law, while stability guarantees are also essential in this application field [50, 103]. As a potentially unstable controller may render the controlled plant unstable in the presence of a disturbance, so could an unstable control algorithm make the effects of a minor earthquake even worse by exciting the building into its resonance. This is mostly true for

2 Courtesy of the Japan Society of Maintenology.

Fig. 9.4 Building models on a shake table at the testing facility of the University of California, San Diego. The *left* building is fixed directly to the table surface, while the one on the *right* is equipped with passive vibration insulation [111]

earthquake systems with active actuators. Since the physical effect of an actuator used in earthquake vibration control is small compared to the expected disturbance, an MPC controller with guaranteed stability utilized in such an application could possibly require a very large region of attraction. This actuator-disturbance asymmetry suggests implementation problems, which are similar to the case presented in the following chapters.

A semi-active magnetorheological damper-enabled earthquake control system is suggested in [147]. The response of the simulated high-rise building is nonlinear, however for design purposes Yan et al. considered linear MPC. The MPC control law featured in this work is unconstrained and has been expressed in its closed form. The authors debate the inherent stability of this system; however, this is not due to the controller design. On the contrary, even if the stability of an unconstrained MPC law could be expressed, here it is implied by the use of MR dampers that cannot add energy to the system, thus cannot render it unstable. This is up to debate, as the control system and even a badly designed passive system can still indirectly alter the resonant frequencies of a structure and shifting them closer to disturbance. The performance advantage of the closed form MPC law in comparison with LQ and saturated LQ was inconclusive, as different controllers performed better in different aspects and situations. Figure 9.4 illustrates the models of multi-story buildings tested on a shake table, while Fig. 9.5 shows the passive tuned mass damper of the Taipei 101. Both of these methods are designed as earthquake protection measures—their passive nature guarantees that no energy can be introduced to the building via improperly designed active control systems.

Active vibration damping systems with traditional industrial control methods are suggested by many publications, for example [13, 31, 32, 61, 68, 146]. Marzbanrad et al. essentially combines a classical LQ feedback loop with a feedforward loop

Fig. 9.5 Illustration of the scale of the tuned mass damper in Taipei101 [118]. To achieve better damping performance with smaller devices, passive tuned mass dampers may be replaced by active vibration control systems. Controllers must guarantee system stability and constraint feasibility at all times

acting as a sort of preview for the excitation to come [73]. In addition, residential and commercial buildings, vibration damping technologies have been suggested for bridges as well [89, 100]. State-feedback $\mathscr{H}_\infty$ optimal control is described in [45, 100] to implement earthquake protection in buildings. Asymptotic stability of the controllers is investigated and guaranteed in the work of Karimi et al. Piezoelectric actuators are uncommon in seismic vibration control because of the questionable range, however they are featured with a positive position (acceleration) feedback strategy in [103]. A SMA-based earthquake protection system is suggested in [130] by Torra et al. Fuzzy control with stability guarantees is considered in [50] suggesting that researchers in earthquake engineering do acknowledge the importance of a priori stability guarantees as well.

Other civil engineering applications include the vibration control of disturbances created by impulsive loads such as blasts and explosions. El Wahed et al. proposed both an ER and an MR fluid-based blast resistant structure for applications in structures such as offshore drilling platforms in [137]. According to the abstract, the vibration control of an elevator is solved by multi-parametric progamming in the Chinese language paper by Ping and Ju [94].

9.7 Manufacturing, Machinery and Tools

Vibration in manufacturing and machinery contributes directly to financial losses and products with decreased quality. For example, the lateral vibration of the spindle in high-speed lathe machines causes manufacturing errors and significantly contributes to machine failure.

This section reviews active vibration control applications in the field of rotor systems, active mounts and power tools. Of these works by Bai and Ou and Shi et al. utilizes traditional predictive control formulation in a closed form [6, 108], while [19] utilizes a constrained explicit MPC control scheme. Another interesting application of MPC in machinery is the low frequency load-sway attenuation in cranes presented by Neupert et al. in [87]. Neither active vibration isolated platforms nor machines have a known use of the model predictive control strategy yet.

9.7.1 Rotor Systems

Vibrations in rotating systems such as shafts may appear due to imbalance, misalignment and outside disturbances. Linear voice coil motors are utilized in [6] to attenuate the transverse vibrations of a shaft. The work of Bai and Ou utilizes older FIR-based predictive control and GPC concepts without constraint handling. Because of the lack of constraints, the predictive control law is derived directly in a close form. The question of stability is treated by the experimental variation of tuning parameters; however, an a priori stability guarantee is not given. The authors pointed out that the GPC algorithm outperformed the FIR-based predictive controller.

Dynamic matrix control—a traditional step response-based algorithm—is used in [108] in cascade with a PI controller for the control of a two-mass rotational drive system. Although this work is not specifically for vibration control, its results apply to this field as well. The unconstrained MPC control law in a closed form does not utilize the full potential of the contemporary results of predictive control theory.

A considerably more up-to-date approach is utilized for the same physical problem by Cychowski and Szabat in [19]. The work implements an explicit, MP-based MPC controller in real-time on a two-mass drive system with good results. The implemented MPMPC control law is constrained, however stability guarantees are not given and the subject is not treated in this work. The sampling time selected for this work is 500 μs and with the number of regions not exceeding 90 the worst-case computational time is less than the half of the sampling period. We have to note that a potentially larger region of attraction and a resulting longer prediction horizon would be needed with stability guarantees—resulting in a MPMPC controller with more regions. Given the nature and dynamics of the two-mass drive system, an explicit MP-based MPC approach is a good choice.

In addition to practically eliminating friction, active magnetic bearings (AMB) also control the vibration levels of rotating machinery. Such a system is discussed for example in [42], where the prevailing control strategy to use is $\mathscr{H}_\infty$. AMB systems can be regarded as hard to control because the underlying dynamics is unstable, multivariable, coupled and nonlinear with uncertainties [42]. Moreover, due to the general physical configuration of such rotor systems, often resonant frequencies over 500 Hz are to be expected, which make MPC implementation difficult.

The asymmetry between actuation capabilities and disturbances is not significant, therefore much shorter prediction horizons are expected than in the case of flexible, lightly damped systems. Magnetic systems are not the only way to damp rotor vibrations: a magnetorheological fluid-based semi-active system is featured in [156] while an ER fluid-based rotor vibration damper is presented in [11].

9.7.2 Active Mounts and Production Systems

The increased interest of the scientific and manufacturing community on micro-electromechanical systems (MEMS) demands manufacturing platforms, which are virtually free of vibrations. Such outside disturbances include micro-vibrations from the ground, instrument movement in the laboratory, people etc.

An effective way to overcome these vibrations is again active or semi-active vibration control. Various traditional control strategies are common for platforms, however to the knowledge of the authors no MPC-based vibration insulation platform has been presented to this date.

Magnetorheological damper-based semi-active vibration control mount platforms are presented in [16, 43]. Even though such a physical system is inherently stable because of the inability of the MR damper to inject energy into the structure, the presence of constraints may warrant a control system with stability guarantees. In other works, magnetostrictive actuators are used for the vibration control of a micromachining platform [153]. The use of piezoceramics is also very common for vibration damping in micromachining platforms.

Automatic manufacturing and production systems have an advantage over other applications mentioned here because the source of disturbance causing oscillations and vibrations is constant, periodic and well defined. Moreover, the source of disturbance is usually limited to the drive system and intermediate mechanisms [40, 97], therefore an active vibration control system is not necessary nor is it recommended. As the feedback sensing of these systems is often very expensive or not practical, traditional approaches may be recommended instead of MPC. Control strategies without feedback for systems such as filling of open containers with liquids in a production cycle may be very effective in reducing spillage [96]—as it has been demonstrated by experiments with liquid filled containers, displaced by a strategy with acceleration input shaping [40, 95].

9.7.3 Anti-Sway Control for Cranes

To make the operation of boom cranes more efficient, the loading and unloading time of cargo has to be shortened. This can be achieved by performing faster maneuvers by the crane operator, which in turn may result in significant unwanted oscillation of the load. This load oscillation is referred to as sway and it is easy to find the analogy with vibrations. Traditionally the load sway is compensated by the operator, however due to the need for the further decrease of loading times, automatic compensating techniques have found their way into this industry as well. Such commercial systems are already implemented: offering reduced sway, increased velocity, turnover and safety [64].

Hubinský proposed a strategy to control load sway and residual oscillations in cranes by feeding the drive system with acceleration signals subject to an input shaping strategy [40]. The slow oscillation times of loads in this application suggest that the sensing of the vibration levels cannot be realized by piezoceramic transducers [40], instead in the interest of simple practical realization a direct approach without feedback has been suggested. Because the load sway in cranes is caused mainly by the movement of the drive system itself, input shaping of drive acceleration profile may reduce load oscillations dramatically [40]. We have to note, that this strategy does not take into account outside disturbances such as wind, imbalance of the load due to non-uniform mass distribution and dynamic changes in inertia.

To tackle the load swaying issue, an MPC-based concept controller has been proposed by Neupert et al. in [87]. After creating a simplified linearized model of the load and crane dynamics, the tracking controller is formulated as a finite horizon, constrained, linear[3] MPC control problem. Natural constraints also arise by the configuration of the physical space, in which the load can be manipulated and transferred safely. These boundaries are then transferred and defined as actuator constraints in the predictive control problem. Additionally, the controller features velocity and acceleration (in other words rate constraints), which have been designed to prevent resonance created by over-aggressive actuator action.

The physical problem on which this concept controller has been evaluated is a large harbor crane, the LIEBHERR Harbour Mobile Crane LHM 400, illustrated[4] in Fig. 9.6. Since the crane features a rope length of up to 90 m, the swaying periods can extend to 19 s or roughly 0.05 Hz. In terms of vibration attenuation, this is a system with very slow dynamics. Referring to practical experience, a control horizon of $n_c = 10$ steps has been utilized in [87], with a sampling of $T_s = 1$ s.

The problem of controller stability in the work by Neupert et al. has been treated by a zero state terminal constraint. This fixed equality constraint eventually has been discarded and replaced by an extension of the quadratic cost function with a terminal penalty term. Neupert et al. has abandoned the idea of using terminal constraints because under non-nominal conditions such as plant-model mismatch and with short horizons these constraints would render the MPC problem infeasible.

Instead of the above-mentioned solution, the feasibility and stability of the MPC controller could be ensured by using the well-known dual-mode infinite horizon MPC formulation with a polytopic terminal set. Feasibility of the constraints and stability beyond the prediction horizon is ensured by the addition of additional constraints determined by the length of a constraint checking horizon. However, this formulation would also create a problem similar to the one already discussed: states would be

---

3 Due to the nonlinearity of the load behavior, the authors utilize linearization of the actual dynamics.

4 Courtesy of Liebherr.

(a) LHM 400 crane (b) LHM 400 in operation

Fig. 9.6 LIEBHERR Harbour Mobile Crane LHM 400 [65]. A drawing of the crane is shown in (**a**), while (**b**) illustrates the LHM 400 in operation

contained within the region of attraction, while the size of this region has to cover all the expected conditions, otherwise the optimization problem may become infeasible.

Let us look at the given harbor crane problem in detail and assess whether there is an analogy between lightly damped and under-actuated vibrating systems: The controller should cover the highest resonant frequency of the system, which has considerable effect on dynamics. For a vibrating mechanical system it is often enough to consider the first dominant resonant frequency to approximate the dynamics. However, in the case of a boom crane, we have parametric vibrations—with the cable length l_c being our governing parameter. Let us represent the swaying load with a mathematical pendulum. For small amplitudes the oscillation frequency approximated by

$$f = \frac{1}{2\pi}\sqrt{\frac{g}{l_c}} \tag{9.1}$$

where f is the approximate oscillation frequency, g the local gravitational constant and l_c the rope length. A maximal rope length of 90 m produces a very slow 0.05 Hz oscillation, but as the rope gets shorter, this oscillation frequency increases. For a 10 m rope the oscillation increases to 0.16 Hz, while a 1 m rope causes swaying motion with 0.5 Hz motion (Fig. 9.7).

A boom crane actuated by powerful hydraulic pistons and motors is nowhere near as underactuated as a mechanical structure damped by piezoelectric patches. In the case of a crane we may safely assume that for the slowest sway motion of 19 s it takes about the whole period to return the system near to its equilibrium, that is to get the system state back to the target set. To account for the shorter end of the rope length spectrum, one would require 10 Hz sampling, that is 0.1 seconds. To create

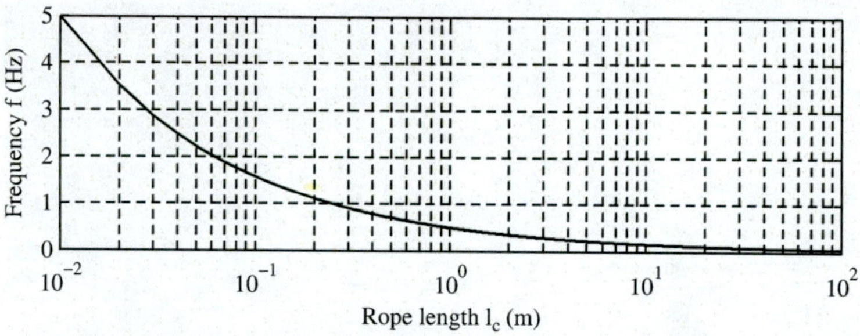

Fig. 9.7 Resonance frequency depending on actual rope length l_c on a crane

a controller capable of keeping up with the fastest resonance, while having a safely exaggerated volume of the region of attraction we would need a prediction horizon in the order of $n_c = 100$–200 steps. To compute such a controller online with a 0.1 s sampling may, or may not be an issue—that depends mainly on the model order.

From the above consideration, it is clear that implementing a model predictive controller with polytopic stability and feasibility guarantees may not be trivial for this system, which resembles many of the characteristic issues of under-damped flexible systems.

9.7.4 Machine Tools

Chatter suppression is solved by using an ER fluid based semi-active system for a boring machine in [138]. Albizuri et al. solves the chatter suppression of center-less grinding machines in [3]. Certain common household items such as washing machines can also benefit from semi-active or active vibration control [119]. The authors of this book have not been able to find an example of an MPC strategy-based vibration reduction application for industrial or household machines.

The power tool manufacturing company DeWALT markets its line of power tools—mainly rotary hammers and demolition hammers—with the (marketing) statement that they are enabled with "Active Vibration Control". Although these tools do have a form of vibration control, it is certainly not active: currently a spring loaded tuned mass is utilized for damping in the aforementioned product line [21]. According to the company, this passive approach decreases handle vibrations in the work axis by 70% while the overall vibration levels are decreased by 30% thus increasing work comfort and component lifetime. Another power tool manufacturer Hilti also markets some of its products with "Active Vibration Reduction" since 1998. Again, while this technology can effectively improve working conditions and prevent hand-arm vibration related injuries, it is not active in the scientific sense. The approach in Hilti products rests on mechanism and tool shape optimization, passive vibration isolation and tuned mass dampers [36].

9.8 Vibration Control in Aircraft and Spacecraft

Increased government spending in the military and space sectors has always fueled active research of new technologies. This desire for military and scientific superiority ensures that new ideas are readily implemented and accepted in highly specific fields such as space flight and military aviation.

To the knowledge of the authors, there are currently no implementations of model predictive control for spacecraft or aircraft vibration control. This holds true for both (known) practical implementations and for academic literature. This section will briefly characterize some of the currently available works on the vibration control of spacecraft and aircraft and introduce some basic concepts and terms used in the application field.

9.8.1 Aircraft Vibration

Active vibration control may enhance the flight properties of fixed and rotary wing aircraft, reduce weight and increase passenger comfort. Since structures in aviation are not completely rigid, some interactions will always occur between the inertial and elastic properties of aircraft structures and aerodynamic forces. This phenomenon in short is referred to as *aeroelasticity*. In the presence of outside aerodynamic forces, the aircraft structure will deform. This deformation is then a ground for increased force interactions, thus creating a self-feeding deformation and disturbance cycle. In other words, the aeroelastic phenomena may be understood as a type of unintentional positive feedback process. There are two types of aeroelasticity: static and dynamic, of which the latter much resembles the characteristics of vibrating structures.

The aerodynamic forces may excite the structure to vibrate in one or more of its resonant modes, creating a potentially dangerous and destructive situation. This effect is referred to as *flutter* in aviation and although it is not identical to the forced resonant phenomena presented in Sect. 2.3—its attenuation requires similar tools and methods. Another common dynamic effect is called *buffeting* and it refers to the excitation of the structure by random impulse-like forces due to the separated flow surfaces. The periodic vibration of aircraft structures is undesired and can be effectively damped using active technologies. Active aeroelastic control of fixed-winged aircraft through a transfer function based approach is considered for example in [14]. The control systems of aircraft responsible for setting control surfaces may also contribute to flutter; this is referred to as *aeroservoelasticity*. Unlike the control strategies utilized currently, MPC may enhance flutter avoidance by explicitly handling constraints arising from actuator limitations and the mechanical properties of the structure. Active aeroelastic surfaces in fixed wing aircraft are have been actively researched

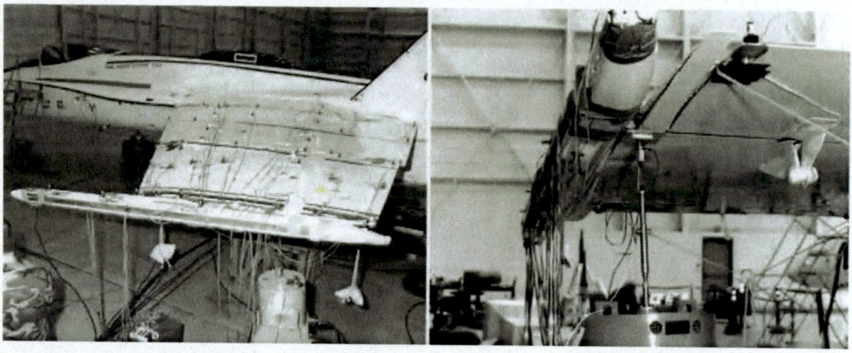

(a) Modified F/A-18A with visible accelerometers **(b)** Shaker test of the active aeroelastic wing

Fig. 9.8 The upper wing surfaces of the Active Aeroelastic Wing F/A-18 test aircraft are covered with accelerometers and other sensors during ground vibration tests at NASA Dryden Flight Research Center in (**a**) [132], while (**b**) shows a large shaker connected to the instrumented wing surface in a dynamic test [131]

in the not so distant past: Fig. 9.8[5] shows the modified F/A-18A research test aircraft with an active aeroelastic wing developed by NASA [57, 131].

As the dominant resonances occur near 20 Hz in aerospace applications, from the viewpoint of computing efficiency it is likely that model predictive control is a feasible implementation choice. Wing tip surfaces are flexible and embedded actuators such as piezoelectric patches cannot effectively match the energy introduced by the disturbances. This in practice means, that a controller with stability guarantees must have a large region of attraction—as it will be demonstrated in the next chapters using the simple cantilever beam demonstration system.

The vibration control of the tail section of military fighter aircraft are of special importance. Fighter aircraft spend a substantial amount of the flying hours in high angles of attack, in which the dynamic loads on the tail are especially high [25]. Vibrations of the twin-tail section of an F-15 military aircraft have been damped by a velocity based feedback control law in [25] using piezoelectric actuators. Another work employing the velocity feedback strategy for aircraft tail sections is featured in [24] also providing stability analysis of the suggested system. Parametric stability of a nonlinear dynamic model of a twin aircraft tail has been investigated with a closed-loop feedback controller later in [4]. In addition to the control of wing surfaces to prevent flutter and other aeroelastic effects, similarly to the automotive industry there is an interest in the application of MR dampers in landing gear [5].

The noise from civilian and military aircraft is also a problem, especially in densely populated areas. Jet engines produce two types of noise: a low-frequency component dominates at takeoff and climb, while a higher pitched sound is audible at landing maneuvers [104]. Passive counter-measures like high-bypass-ratio jet engines offer a significant noise reduction, which could be further improved with active strategies.

[5] Courtesy of NASA.

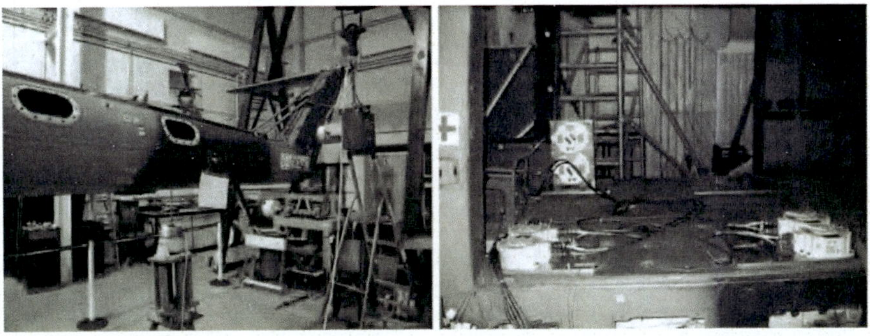

(a) Shaker test on a helicopter tail **(b)** Force actuators shown in the fuselage

Fig. 9.9 Electrodynamic shakers are connected to the tail section of a military helicopter in (**a**), while (**b**) shows the magnetic force actuators mounted in the fuselage [88]

In addition to the jet engine noise, passenger comfort may be increased by using active panels and linings inside the fuselage.

Fixed-winged aircraft is not the only area of application of active vibration control in the aviation industry. Rotary wing aircraft—or helicopters are subject to strong dynamic disturbances as well. Vibration damping in helicopters can be divided into three basic approaches [8]:

- Vibration control of the fuselage, actuators on the fuselage
- Vibration control of the fuselage, actuators on the rotor system or individual blades
- Vibration control of the rotor system or individual blades

An illustration of a dynamic test performed on a helicopter tail section is featured in Fig. 9.9.[6] Here the outside disturbance caused by the tail rotor is simulated by large shakers and the vibration levels inside the fuselage are attenuated using force actuators.

The flight speed record in helicopters has just recently been broken by the Sikorsky X2 helicopter prototype [117]. The enhanced horizontal flight and hovering capabilities and the flight comfort of the aircraft is partly due to the implemented active vibration control system [116, 117]. Other commercial aircraft, for example the Sikorsky 76D or the Sikorsky S-92 already features a type of active vibration control system, implemented through a nose or rotor hub mounted pair of force actuators [113, 114]. Military aircraft like the Sikorsky UH-60M (Blackhawk) technical sheets also list active vibration suppression systems, however implementation details of these technologies are proprietary [115]. The upcoming active rotor hub mounted vibration system may feature semi-active magnetorheological actuator based damping. Although there is not much information on the technology, recent press releases suggest the involvement of an outside contractor specialized in magnetorheological dampers [112].

[6] Courtesy of the Noise & Vibration Control Ltd.

Other major helicopter manufacturers and airspace or defense contractors have also successfully implemented experimental vibration reduction systems. Vibrations are reduced below 0.05 G at 4/rev and 8/rev speeds in a Kawasaki BK117 helicopter [47]. Fuselage vibrations are semi-actively damped in the HAL Dhruv Advanced Light Helicopter, where four isolation elements are mounted between the main gearbox and the fuselage [30, 70]. The same magnetorheological damper based technology is used in the Eurocopter EC225/EC725 [70].

Shape memory alloy materials have been used in experimental rotor blade systems to control stiffness, blade angle twist, natural frequency and damping properties in [59]. Vibration control of individual rotor blades is discussed in [8]. Coupled fuselage-rotor modes are damped in a rotary wing unmanned aerial vehicle (UAV) using positive position feedback in [2]. A very interesting idea is presented by Lu and Meng in [71], where the authors suggest the use of an ER fluid filled composite plate. Such novel materials can possibly shift frontiers on the active vibration insulation of air and spacecraft.

The flutter control of helicopter rotors has a related problem area in the field of power generation: TingRui and YongSheng have investigated the aerolastic behavior of wind turbine blades in [69].

9.8.2 Spacecraft Vibration

The vibration resistance of space structures ranging from rockets, orbiters, satellites, space telescopes is an important design factor. These engineering structures are subjected to rigorous vibration testing at their design stage to investigate their dynamic response. For example Fig. 9.10[7] shows the now retired space shuttle and its components undergoing vibration testing [79, 83, 84], while Fig. 9.11[8] shows a space station component and a probe in vibration testing as an illustration [82, 85]. Spacecraft undergo weeks of intensive thermal and vibration testing to imitate the temperature and dynamic forces encountered at launch and spaceflight [85]. Any measure increasing the vibration resistance of spacecraft can improve the safety of both manned and unmanned spaceflight and decrease the cost of cargo transportation into space. Active vibration control may have a role in the design of future economic and reusable spacecraft, which became an actual issue after the recent termination of the space shuttle program and the last flight of the orbiter Atlantis STS-135 into outer space.

The worst-case loading scenario for a payload such as a satellite is its launch. Payloads, for example the previously mentioned satellites, have to be designed for this worsened condition, adding considerable cost to both the manufacturing process and the launch itself. This cost increase is mainly due to the passive vibration insulation and damping precautions, which add weight to the space vehicle and the payload.

[7] Courtesy of NASA.

[8] Courtesy of NASA.

(a) Orbiter Enterprise in test stand installation

(b) Removal of orbiter Enterprise from test stand

(c) Mobile launcher platform and solid rocket boosters

Fig. 9.10 The photograph (**a**) shows the now retired orbiter (space shuttle) Enterprise in its liftoff configuration undergoing a dynamic vibration test, its removal from the stand in (**b**) and the ground vibration test of a mobile launcher platform with the solid rocket boosters in (**c**) [79, 83, 84]

(a) Apollo telescope mount

(b) Cassini Saturn probe

Fig. 9.11 The Apollo telescope mount, one of four major components is undergoing horizontal vibration test in (**a**), while (**b**) shows the Cassini Saturn probe in a similar vibration and thermal testing scenario [82, 85]

(a) Space truss under vibration test **(b)** Passive damping element in space truss

Fig. 9.12 A space truss undergoing vibration tests on a trust-boom test hardware is presented in (**a**) [80] while (**b**) [81] shows a passive damping element within the space truss

The payload within the capsule is secured using the so-called payload adapter fitting (PAF). Traditionally PAF have been predominantly rigid structures absorbing little of the launch vibrations. Recently passive approaches have emerged and successfully flown to space, considering vibration damping when designing PAF. The other approach is to replace the PAF with an actively controlled structure.

A whole spacecraft-based active vibration isolation scheme is proposed in [28], where MPC is used to track pressure for a pneumatic actuator. In fact, this approach uses MPC not to directly control the vibration levels, but as an inner loop of a two-controller cascade—therefore cannot be considered as an example of MPC-based vibration suppression scheme. The MPC law is responsible to track desired pressure levels in the actuator, while the vibration control itself is taken care of by a simple rate-feedback controller. The MPC law used in [28] does not assume constraints; therefore, it has been expressed as a closed feedback law. No computational issues arise from such an implementation, which essentially ignores real actuator limits and the question of controller stability.

The vibration of a smart grid structure resembling the configuration and dynamic behavior of solar panel structures is actively controlled using simple position feedback in [55]. Similarly, classical position feedback control is utilized in [37] for the vibration control of flexible spacecraft. The vibration suppression for inter-satellite communication links is presented in [72].

An application field closely related to space flight are robotic manipulators discussed in Sect. 9.2. In addition to the manipulators discussed earlier without an explicit intention to use in microgravity environment, for example [139], deals with a fuzzy controlled manipulator with piezoelectric patches for space.

The vibration attenuation of space borne optical interferometers is a great interest of the scientific community [99] as well. Such systems are based on large flexible systems in the range of 10 m while the optical stabilization has to be carried out at a 10 nm level. Such a vibration attenuation system is presented, for example, in [86]. The vibration attenuation of any space borne truss structure is of great interest in aeronautics. Figure 9.12[9] shows a trust-boom structure undergoing vibration testing with a passive damping element [80, 81].

Large flexible antenna systems in space are also an exciting field for novel MPC applications. A vibration control system for such an application has been suggested by Agrawal and Bang in [1]. The wind-induced vibrations of a ground-based parabolic radio telescope are damped by ER fluid-based actuators in a simulation study by Su et al. in [120].

References

1. Agrawal BN, Bang H (1996) Adaptive structures for large precision antennas. Acta Astronaut 38(3):175–183. doi:10.1016/0094-5765(96)00062-8, http://www.sciencedirect.com/science/article/B6V1N-3VTW8Y7-3/2/a53f7c4acb3ee1541568e0db4062d985
2. Ahmed B, Pota H (2011) Dynamic compensation for control of a rotary wing UAV using positive position feedback. J Intell Rob Syst 61:43–56. doi:10.1007/s10846-010-9487-7, http://dx.doi.org/10.1007/s10846-010-9487-7
3. Albizuri J, Fernandes M, Garitaonandia I, Sabalza X, Uribe-Etxeberria R, Hernández J (2007) An active system of reduction of vibrations in a centerless grinding machine using piezoelectric actuators. Int J Mach Tools Manuf 47(10):1607–1614. doi:10.1016/j.ijmachtools.2006.11.004, http://www.sciencedirect.com/science/article/B6V4B-4MR1K43-1/2/aa1014dd203b27a44c75cef37a2adf09
4. Amer Y, Bauomy H (2009) Vibration reduction in a 2DOF twin-tail system to parametric excitations. Commun Nonlinear Sci Numer Simul 14(2):560–573. doi:10.1016/j.cnsns.2007.10.005, http://www.sciencedirect.com/science/article/B6X3D-4PYP723-2/2/b9d5375168fadb0b4e67857e92948bfc
5. Aslam M, Xiong-Liang Y, Zhong-Chao D (2006) Review of magnetorheological (MR) fluids and its applications in vibration control. J Mar Sci Appl 5(3):17–29
6. Bai M, Ou KY (2002) Experimental evaluation of adaptive predictive control for rotor vibration suppression. IEEE Trans Control Syst Technol 10(6):895–901. doi:10.1109/TCST.2002.804124
7. Billah KY, Scanlan RH (1991) Resonance, Tacoma Narrows Bridge failure, and undergraduate physics textbooks. Am J Phys 59(2):118
8. Bittanti S, Cuzzola FA (2002) Periodic active control of vibrations in helicopters: a gain-scheduled multi-objective approach. Control Eng Pract 10(10):1043–1057. doi:10.1016/S0967-0661(02)00052-7, http://www.sciencedirect.com/science/article/B6V2H-45KSPJJ-3/2/9647861ce849d131c7d4b90cdb964751
9. Bohn C, Cortabarria A, Härtel V, Kowalczyk K (2004) Active control of engine-induced vibrations in automotive vehicles using disturbance observer gain scheduling. Control Eng Pract 12(8):1029–1039 (in Special Section on Emerging Technologies for Active Noise and Vibration Control Systems). doi:10.1016/j.conengprac.2003.09.008, http://www.sciencedirect.com/science/article/B6V2H-49Y3VWS-1/2/dd7bcefd1618f3820896ddbd6dce7430
10. Boscariol P, Gasparetto A, Zanotto V (2010) Modell predictive control of a flexible links mechanism. J Intell Rob Syst 58:125–147. doi:10.1007/s10846-009-9347-5, http://dx.doi.org/10.1007/s10846-009-9347-5
11. Bouzidane A, Thomas M (2008) An electrorheological hydrostatic journal bearing for controlling rotor vibration. Comput Struct 86(3–5):463–472. (Smart Struc-

[9] Courtesy of NASA.

tures) doi:10.1016/j.compstruc.2007.02.006, http://www.sciencedirect.com/science/article/B6V28-4NDVGTG-1/2/32b829850e8db109469179cdb7d7d4f6

12. Camino J, Arruda J (2009) $\mathscr{H}_2$ and $\mathscr{H}_\infty$ feedforward and feedback compensators for acoustic isolation. Mech Syst Sig Process 23(8):2538–2556. doi:10.1016/j.ymssp.2009.04.006, http://www.sciencedirect.com/science/article/B6WN1-4W7J0YN-2/2/918091cd3d7b23193d5b3637eb2342ce
13. Carotti A, Lio G (1991) Experimental active control: bench tests on controller units. Eng Struct 13(3):242–252. doi:10.1016/0141-0296(91)90036-C, http://www.sciencedirect.com/science/article/B6V2Y-4829VWB-CG/2/4414a8cb4321f4e346ca04468e610264
14. Cavagna L, Ricci S, Scotti A (2009) Active aeroelastic control over a four control surface wing model. Aerosp Sci Technol 13(7):374–382. doi:10.1016/j.ast.2009.06.009, http://www.sciencedirect.com/science/article/B6VK2-4X315M3-1/2/e145579962804cd5026e72e011405013
15. Chang CS, Liu TS (2007) LQG controller for active vibration absorber in optical disk drive. IEEE Trans Magn 43(2):799–801. doi:10.1109/TMAG.2006.888417
16. Choi SB, Hong SR, Sung KG, Sohn JW (2008) Optimal control of structural vibrations using a mixed-mode magnetorheological fluid mount. Int J Mech Sci 50(3):559–568. doi:10.1016/j.ijmecsci.2007.08.001, http://www.sciencedirect.com/science/article/B6V49-4PD4XHC-1/2/c491dc4a4a881e38b0e20ceef7206dec
17. Creasy M, Leo D, Farinholt K (2008) Adaptive positive position feedback for actively absorbing energy in acoustic cavities. J Sound Vib 311(1–2):461–472. doi:10.1016/j.jsv.2007.09.013, http://www.sciencedirect.com/science/article/B6WM3-4R2HKR0-3/2/e9d3c9817e3b4c302a861a4a3bb6fcb1
18. Cunningham M, Jenkins D, Clegg W, Bakush M (1995) Active vibration control and actuation of a small cantilever for applications in scanning probe instruments. Sens Actuators A 50(1–2):147–150. doi:10.1016/0924-4247(96)80099-9, http://www.sciencedirect.com/science/article/B6THG-3YVM62R-15/2/ea100dfeea242e7471472799494a5b93
19. Cychowski M, Szabat K (2010) Efficient real-time model predictive control of the drive system with elastic transmission. IET Control Theory Appl 4(1):37–49. doi:10.1049/iet-cta.2008.0358
20. Darus IM, Tokhi M (2005) Soft computing-based active vibration control of a flexible structure. Eng Appl Artif Intell 18(1):93–114. doi: 10.1016/j.engappai.2004.08.017, http://www.sciencedirect.com/science/article/B6V2M-4DFT21W-2/2/0e01e702eebed40a2e2dbd2925feed
21. DeWALT Industrial Tool Co (2007) DeWALT active vibration control. http://www.dewalt.co.uk/vibration/powertool-selection/avc/
22. du Plessis A (2010) Tuned mass damper on display in Taipei 101. Photograph licensed under the Creative Commons Attribution 3.0 Unported license. http://commons.wikimedia.org/wiki/File:Taipei_101_Tuned_Mass_Damper_2010.jpg
23. Eielsen A, Fleming A (2010) Passive shunt damping of a piezoelectric stack nanopositioner. In: American control conference (ACC) 2010, pp 4963–4968
24. Eissa M, Bauomy H, Amer Y (2007) Active control of an aircraft tail subject to harmonic excitation. Acta Mech Sin 23:451–462. doi:10.1007/s10409-007-0077-2
25. El-Badawy AA, Nayfeh AH (2001) Control of a directly excited structural dynamic model of an F-15 tail section. J Franklin Inst 338(2–3):133–147. doi:10.1016/S0016-0032(00)00075-2, http://www.sciencedirect.com/science/article/B6V04-42HNMDV-3/2/e3bf6f797834c8e8638324be88fb78f7
26. Eski I, Yildirim S (2009) Vibration control of vehicle active suspension system using a new robust neural network control system. Simul Modell Pract Theory 17(5):778–793. doi:10.1016/j.simpat.2009.01.004, http://www.sciencedirect.com/science/article/B6X3C-4VHSDJ4-1/2/d2fe946695b369279d2e1229f15a61bd
27. Eure KW (1998) Adaptive predictive feedback techniques for vibration control. Doctoral dissertation, Virginia Polytechnic Institute and State University, Blacksburg

28. Fei H, Zheng G, Liu Z (2006) An investigation into active vibration isolation based on predictive control: Part I: Energy source control. J Sound Vib 296(1–2):195–208. doi:10.1016/j.jsv.2006.02.021, http://www.sciencedirect.com/science/article/B6WM3-4K18VTN-5/2/80551dd9ac1e1a29dac9c9e25601ea6e
29. Fischer D, Isermann R (2004) Mechatronic semi-active and active vehicle suspensions. Control Eng Pract 12(11):1353–1367 (Mechatronic Systems). doi:10.1016/j.conengprac.2003.08.003, http://www.sciencedirect.com/science/article/B6V2H-49V1CR4-2/2/0dd89d1b7760e7303a32b5bdd2cbbf9b
30. GlobalSecurity (2010) Dhruv advanced light helicopter. http://www.globalsecurity.org/military/world/india/alh.htm
31. Guclu R (2006) Sliding mode and PID control of a structural system against earthquake. Math Comput Modell 44(1–2):210–217. doi:10.1016/j.mcm.2006.01.014, http://www.sciencedirect.com/science/article/B6V0V-4JP9FV5-1/2/0900f85ba6e764d746c054ac040aff77 (Advances in business modeling and Decision technologies, pp 1–95)
32. Guclu R, Yazici H (2008) Vibration control of a structure with ATMD against earthquake using fuzzy logic controllers. J Sound Vib 318(1–2):36–49. doi:10.1016/j.jsv.2008.03.058, http://www.sciencedirect.com/science/article/B6WM3-4SM0XJT-1/2/fe8f6a66297ad6e12f0791a83e4eed36
33. Hassan M, Dubay R, Li C, Wang R (2007) Active vibration control of a flexible one-link manipulator using a multivariable predictive controller. Mechatronics 17(1):311–323
34. Helfer M (2006) Body vibration due to road excitation. Released to the public domain by the copyright holder. FKFS, Stuttgart. http://commons.wikimedia.org/wiki/File:Body_vibration_due_to_road_excitation.jpg
35. Henrioulle K, Dehandschutter W, Sas P (1998) Design of an active noise control system using a distributed actuator. Flow, Turbulence and Combustion 61:189–209. doi:10.1023/A:1009901115843
36. HILTI Corporation (2010) Active vibration reduction. Lower vibration for higher productivity. Whitepaper. http://www.hilti.com/fstore/holcom/LinkFiles/Hilti_HS_Active%20Vibration%20Reduction_1.pdf
37. Hu Q (2009) A composite control scheme for attitude maneuvering and elastic mode stabilization of flexible spacecraft with measurable output feedback. Aerosp Sci Technol 13(2–3):81–91. doi:10.1016/j.ast.2007.06.007, http://www.sciencedirect.com/science/article/B6VK2-4P96269-2/2/5fbc47249fdd3f1963c5ba856f071c55
38. Hu YR, Ng A (2005) Active robust vibration control of flexible structures. J Sound Vib 288(1–2):43–56. doi:10.1016/j.jsv.2004.12.015, http://www.sciencedirect.com/science/article/B6WM3-4FJTP4X-4/2/7d5773e39fd6ea0a80496e691131c32f
39. Huang K, Yu F, Zhang Y (2010) Model predictive controller design for a developed electromagnetic suspension actuator based on experimental data. In: 2010 WASE international conference on information engineering (ICIE), vol 4, pp 152–156. doi:10.1109/ICIE.2010.327
40. Hubinský P (2010) Riadenie mechatronických systémov s nízkym tlmením, 1st edn. Slovenská technická univerzita v Bratislave, Nakladateľstvo STU (Control of mechatronic systems with low damping), Bratislava (in Slovak language)
41. Jahromi A, Zabihollah A (2010) Linear quadratic regulator and fuzzy controller application in full-car model of suspension system with magnetorheological shock absorber. In: 2010 IEEE/ASME international conference on mechatronics and embedded systems and applications (MESA), pp 522–528. doi:10.1109/MESA.2010.5552010
42. Jastrzebski RP, Hynynen KM, Smirnov A (2010) $\mathscr{H}_\infty$ control of active magnetic suspension. Mech Syst Sig Process 24(4):995–1006. doi: 10.1016/j.ymssp.2009.10.008, http://www.sciencedirect.com/science/article/B6WN1-4XJP3XR-2/2/51b0222180b2610516135c196f226
43. Jung WJ, Jeong WB, Hong SR, Choi SB (2004) Vibration control of a flexible beam structure using squeeze-mode ER mount. J Sound Vib 273(1–2):185–199. doi:

10.1016/S0022-460X(03)00478-4, http://www.sciencedirect.com/science/article/B6WM3-49DFFMM-1/2/1255ad59eca53b0c021632de61aef0b8

44. Kang B, Mills JK (2005) Vibration control of a planar parallel manipulator using piezoelectric actuators. J Intell Rob Syst 42:51–70. doi:10.1007/s10846-004-3028-1
45. Karimi H, Zapateiro M, Luo N, Rossell J (2010) Feedback vibration control of a base-isolated building with delayed measurements using $\mathscr{H}_\infty$; techniques. In: American control conference (ACC) 2010, pp 750–755
46. Kawabe H, Tsukiyama N, Yoshida K (2006) Active vibration damping based on neural network theory. Mater Sci Eng A 442(1–2):547–550, Proceedings of the 14th international conference on internal friction and mechanical spectroscopy. doi:10.1016/j.msea.2006.02.234, http://www.sciencedirect.com/science/article/B6TXD-4KPFKNH-2/2/51634002bdd85fe7ee55df4b6b28e7e4
47. Kawasaki Heavy Industries, Ltd (2011) AVR (Active Vibration Reduction) system for BK117 Helicopter. http://www.khi.co.jp/english/rd/tech/136/ne136s01.html
48. Kim B, Washington GN (2008) Active vibration control of a cantilevered beam using model predictive sliding mode control. In: 49th AIAA/ASME/ASCE/AHS/ASC structural dynamics and materials conference, Schaumburg, pp 2008–2038
49. Kim I, Kim YS (2009) Active vibration control of trim panel using a hybrid controller to regulate sound transmission. Int J Precis Eng Manuf 10:41–47. doi:10.1007/s12541-009-0007-2
50. Kim Y, Langari R, Hurlebaus S (2009) Seismic response control of a large civil structure equipped with magnetorheological dampers. In: IEEE international conference on fuzzy systems 2009. FUZZ-IEEE 2009, pp 215–220. doi:10.1109/FUZZY.2009.5277045
51. Kok J, van Heck J, Huisman R, Muijderman J, Veldpaus F (1997) Active and semi-active control of suspension systems for commercial vehicles based on preview. In: Proceedings of the 1997 American control conference, vol 5, pp 2992–2996. doi:10.1109/ACC.1997.612006
52. Kowalczyk K, Karkosch HJ, Marienfeld PM, Svaricek F (2006) Rapid control prototyping of active vibration control systems in automotive applications. In: Computer aided control system design, 2006 IEEE international conference on control applications, 2006 IEEE international symposium on intelligent control, 2006 IEEE, pp 2677–2682. doi:10.1109/CACSD-CCA-ISIC.2006.4777062
53. Kozek M, Benatzky C, Schirrer A, Stribersky A (2009) Vibration damping of a flexible car body structure using piezo-stack actuators. Control Eng Pract 19:298–310. doi:10.1016/j.conengprac.2009.08.001, http://www.sciencedirect.com/science/article/B6V2H-4X3MR4Y-2/2/3ef1d868e70c2b6f10fd9412f9c8c1de
54. Krishnaswamy K, Rajamani R, Woo J, Cho Y (2005) Structural vibration control for broadband noise attenuation in enclosures. J Mech Sci Technol 19:1414–1423. doi:10.1007/BF03023900
55. Kwak MK, Heo S (2007) Active vibration control of smart grid structure by multi-input and multioutput positive position feedback controller. J Sound Vib 304(1–2):230–245. doi:10.1016/j.jsv.2007.02.021, http://www.sciencedirect.com/science/article/B6WM3-4NH6N96-2/2/ca7b43602b9d052e388f4b2a28f1ebae
56. Kwon OS, Kim BK, Ih JG (1994) On the positioning of control sources in active noise control of three-dimensional interior space. J Mech Sci Technol 8:283–292. doi:10.1007/BF02953357
57. Landis T (2001) NASA Dryden Flight Research Center (NASA-DFRC). Full scale dynamic model of the EOS-AM1 satellite. Image ID: EC01-0288-2
58. Larchez A (2007) Finite element modelling of piezoelectric structures. PhD thesis, School of Electrical, Computer and Telecommunication Engineering, University of Wollongong, Wollongong
59. Lau K, Zhou L, Tao X (2002) Control of natural frequencies of a clamped-clamped composite beam with embedded shape memory alloy wires. Compos Struct 58(1):39–47. doi:10.1016/S0263-8223(02)00042-9, http://www.sciencedirect.com/science/article/B6TWP-45XTP9W-N/2/07b9a065ac866d8869a4240deb918851
60. Lee J, Kim J, Cheong C (1999) Piezoelectric smart structures for noise reduction in a cabin. J Mech Sci Technol 13:451–458. doi:10.1007/BF02947714

61. Lee JH, Su RK, Lee PK, Lam LC (2002) Semi-active damping device for vibration control of buildings using magnetorheological fluid. In: Anson M, Ko J, Lam E (eds) Advances in building technology, Elsevier, Oxford, pp 969–976. doi:10.1016/B978-008044100-9/50122-4, http://www.sciencedirect.com/science/article/B858K-4PCJRKH-47/2/1c6a74db22e114e2cbddec5d173950f8
62. Li K, Kosmatopoulos E, Ioannou P, Boussalis H, Mirmirani M, Chassiakos A (1998) Control techniques or a large segmented reflector. In: Proceedings of the 37th IEEE conference on decision and control 1998, vol 1, pp 813–818. doi:10.1109/CDC.1998.760789
63. Li M, Lim TC, Lee JH (2008) Simulation study on active noise control for a 4-T MRI scanner. Magn Reson Imaging 26(3):393–400. doi: 10.1016/j.mri.2007.08.003, http://www.sciencedirect.com/science/article/B6T9D-4R8KT3W-2/2/2797c565f329cf6cd1e567eefb69607e
64. Liebherr-Elektronik GmbH (2011) Liebherr electronics. Catalog. 10559768-04/2011
65. Liebherr-International Deutschland GmbH (2010) Data sheet LHM 400—LIEBHERR mobile harbour crane LHM 400. http://www.liebherr.com/MCP/en-GB/products_mcp.wfw/id-11612-0/tab-1295_1527
66. Lin J, Liu WZ (2006) Experimental evaluation of a piezoelectric vibration absorber using a simplified fuzzy controller in a cantilever beam. J Sound Vib 296(3):567–582. doi:10.1016/j.jsv.2006.01.066, http://www.sciencedirect.com/science/article/B6WM3-4K0FG0H-2/2/e4fad7e52e98cf46123aa869cf780b65
67. Lin LC, Lee TE (1997) Integrated PID-type learning and fuzzy control for flexible-joint manipulators. J Intell Rob Syst 18:47–66. doi:10.1023/A:1007942528058
68. Liu SJ, Huang ZH, Chen YZ (2004) Automobile active suspension system with fuzzy control. J Cent South Univ Technol 11:206–209. doi:10.1007/s11771-004-0042-1
69. Liu T, Ren Y (2011) Vibration and flutter of wind turbine blade modeled as anisotropic thin-walled closed-section beam. Science China Technol Sci pp 1–8. doi:10.1007/s11431-010-4230-y
70. LORD Corporation (2010) Active vibration control (AVC) systems. Cary. http://www.lord.com/Products-and-Solutions/Vibration-and-Motion-Control/Active-Vibration-Control-%28AVC%29-Systems.xml
71. Lu H, Meng G (2006) An experimental and analytical investigation of the dynamic characteristics of a flexible sandwich plate filled with electrorheological fluid. Int J Adv Manuf Technol 28:1049–1055. doi:10.1007/s00170-004-2433-8
72. Luo T, Hu Y (2002) Vibration suppression techniques for optical inter-satellite communications. In: IEEE 2002 international conference on communications, circuits and systems and west sino expositions, vol 1, pp 585–589. doi:10.1109/ICCCAS.2002.1180687
73. Marzbanrad J, Ahmadi G, Jha R (2004) Optimal preview active control of structures during earthquakes. Eng Struct 26(10):1463–1471. doi: 10.1016/j.engstruct.2004.05.010, http://www.sciencedirect.com/science/article/B6V2Y-4CYNR00-1/2/271b4c49fa053fb1a95d5df632c701c
74. Mcmanus SJ, Clair KAS, Boileau P, Boutin J, Rakheja S (2002) Evaluation of vibration and shock attenuation performance of a suspension seat with a semi-active magnetorheological fluid damper. J Sound Vib 253(1):313–327. doi: 10.1006/jsvi.2001.4262, http://www.sciencedirect.com/science/article/B6WM3-45Y1C16-N/2/33a165ac8f2fe7d8fad7bd83d9484957
75. Mehra R, Amin J, Hedrick K, Osorio C, Gopalasamy S (1997) Active suspension using preview information and model predictive control. In: Proceedings of the 1997 IEEE international conference on control applications, pp 860–865. doi:10.1109/CCA.1997.627769
76. Moon SJ, Lim CW, Kim BH, Park Y (2007) Structural vibration control usinglinear magnetostrictive actuators. J Sound Vib 302(4–5):875–891. doi: 10.1016/j.jsv.2006.12.023, http://www.sciencedirect.com/science/article/B6WM3-4N2M6HH-5/2/417522adfca8640acfa76e890ae0533c
77. Moon SM, Clark RL, Cole DG (2005) The recursive generalized predictive feedback control: theory and experiments. J Sound Vib 279(1-2):171–199.

doi:10.1016/j.jsv.2003.12.034, http://www.sciencedirect.com/science/article/B6WM3-4C005WR-2/2/4580a0865591eaa5ca1bf02e09dedcb7
78. Moon SM, Cole DG, Clark RL (2006) Real-time implementation of adaptive feedback and feedforward generalized predictive control algorithm. J Sound Vib 294(1–2):82–96. doi:10.1016/j.jsv.2005.10.017, http://www.sciencedirect.com/science/article/B6WM3-4HYMY76-1/2/50d98047187533ebe9d3ea8310446e77
79. NASA Kennedy Space Center (NASA-KSC) (2003) Ground vibration test of a mobile launcher platform with the solid rocket boosters. Image ID: KSC-03PD-3153
80. NASA Langley Research Center (NASA-LaRC) (2002) Passive viscous damping struts have been fabricated. Image ID: EL-2002-00103
81. NASA Langley Research Center (NASA-LaRC) (2002) Truss-boom test hardware. Image ID: EL-2002-00100
82. NASA Marshall Space Flight Center (NASA-MSFC) (1971) Vibration testing of the Apollo telescope mount. Image ID: MSFC-7019987
83. NASA Marshall Space Flight Center (NASA-MSFC) (1978) Orbiter Enterprise test stand installation. Image ID: MSFC-7992411
84. NASA Marshall Space Flight Center (NASA-MSFC) (1978) Orbiter Enterprise test stand installation. Image ID: MSFC-7992452
85. NASA Marshall Space Flight Center (NASA-MSFC) (1996) NASA Headquarters–Greatest Images of NASA (NASA-HQ-GRIN). Image ID: GPN-2000-001982
86. Neat G, Melody J, Lurie B (1998) Vibration attenuation approach for spaceborne optical interferometers. IEEE Trans Control Syst Technol 6(6):689–700. doi:10.1109/87.726529
87. Neupert J, Arnold E, Schneider K, Sawodny O (2010) Tracking and anti-sway control for boom cranes. Control Eng Pract 18(1):31–44. doi: 10.1016/j.conengprac.2009.08.003, http://www.sciencedirect.com/science/article/B6V2H-4XHT48W-1/2/06713de1b8ba53b8f60bc0598692008d
88. Noise & Vibration Control Ltd Co (2009) Electro-dynamic shakers connected to structure for active vibration control experiment. http://www.nvcontrol.com/sitebuildercontent/sitebuilderpictures/CRW_2225_2.JPG
89. Ok SY, Kim DS, Park KS, Koh HM (2007) Semi-active fuzzy control of cable-stayed bridges using magneto-rheological dampers. Eng Struct 29(5):776–788. doi:10.1016/j.engstruct.2006.06.020, http://www.sciencedirect.com/science/article/B6V2Y-4KM46VD-4/2/1c85c3a0d12e30e2d5afddaa590f7059
90. Pan MC, Chien C (2010) Adaptive hybrid tracking-error control for DVD drives in vehicular systems. Microsyst Technol 16:279–286. doi:10.1007/s00542-009-0856-8
91. Pan X, Tso Y, Juniper R (2008) Active control of low-frequency hull-radiated noise. J Sound Vib 313(1–2):29–45. doi:10.1016/j.jsv.2007.11.022, http://www.sciencedirect.com/science/article/B6WM3-4RDB90B-6/2/396dfb897d90f7dc0555b249fef3848d
92. Pan X, Tso Y, Juniper R (2008) Active control of radiated pressure of a submarine hull. J Sound Vib 311(1–2):224–242. doi:10.1016/j.jsv.2007.09.001, http://www.sciencedirect.com/science/article/B6WM3-4PYJF3F-2/2/6ab6ad0f79e7f751265db011ef2e7a15
93. Patton R (1994) Design of an active vibration control system for isolation of an optical bench. In: Proceedings of the 26th southeastern symposium on system theory 1994, pp 43–46. doi:10.1109/SSST.1994.287913
94. Ping H, Ju Z (2008) Explicit model predictive control system and its application in active vibration control of mechanical system of elevator. In: Control and decision conference 2008. CCDC 2008. Chinese, pp 3738–3742. doi:10.1109/CCDC.2008.4598029
95. Pospiech T, Hubinský P (2009) Beschleunigungsprofil für Abfüllanlagen. Resonanzfreies Positionieren von schwingungsfähigen Systemen am Beispiel von offenen Behältern mit Flüssigkeiten. Automatisierungstechnische Praxis (Acceleration profile for bottling. Resonance-free positioning of oscillatory systems using the example of open containers with liquids) 51(6):19–21 (in German language)

96. Pospiech T, Hubinský P (2009) Input shaping for slosh-free moving containers with liquid. Int J Mech Control 9(02):13–20
97. Pospiech T, Hubinský P (2009) Schwingungsfähige Systeme resonanzfrei Positionieren. Elektrotechnik + Automation (Resonance-free positionoing of vibrating systems) 9:46–49 (in German language)
98. Pradhan S (2005) Vibration suppression of FGM shells using embedded magnetostrictive layers. Int J Solids Struct 42(9–10):2465–2488. doi:10.1016/j.ijsolstr.2004.09.049, http://www.sciencedirect.com/science/article/B6VJS-4F6SSGN-1/2/b6f9e2e6ffc65bfc0c4af5083e37df0b
99. Preumont A (2002) Vibration control of active structures, 2nd edn. Kluwer Academic, Dordrecht
100. Preumont A, Seto K (2008) Active control of structures, 3rd edn. Wiley, Chichester
101. Qiu ZC, Wu HX, Ye CD (2009) Acceleration sensors based modal identification and active vibration control of flexible smart cantilever plate. Aerosp Sci Technol 13(6):277–290. doi:10.1016/j.ast.2009.05.003, http://www.sciencedirect.com/science/article/B6VK2-4WB3NH7-2/2/e7bef32fa0e1ef301516f9b393ea8a97
102. Richelot J, Bordeneuve-Guibe J, Pommier-Budinger V (2004) Active control of a clamped beam equipped with piezoelectric actuator and sensor using generalized predictive control. In: 2004 IEEE international symposium on industrial electronics, vol 1, pp 583–588. doi:10.1109/ISIE.2004.1571872
103. Ríos-Gutiérrez M, Silva-Navarro G (2010) Suppression of mechanical vibrations in a building like structure by means of a piezoelectric patch actuator and positive acceleration feedback. In: 2010 7th international conference on electrical engineering computing science and automatic control (CCE), pp 452–457. doi:10.1109/ICEEE.2010.5608581
104. Rossing TD, Moore RF, Wheeler PA (2001) The science of sound. 3rd edn. Addison Wesley, San Francisco
105. Roy T, Chakraborty D (2009) Optimal vibration control of smart fiber reinforced composite shell structures using improved genetic algorithm. J Sound Vib 319(1–2):15–40. doi:10.1016/j.jsv.2008.05.037, http://www.sciencedirect.com/science/article/B6WM3-4T0X2NT-1/2/6e02883f5e6352192210eb9b36700538
106. Shan J, Liu HT, Sun D (2005) Slewing and vibration control of a single-link flexible manipulator by positive position feedback (PPF). Mechatronics 15(4):487–503. doi:10.1016/j.mechatronics.2004.10.003, http://www.sciencedirect.com/science/article/B6V43-4DR87K7-4/2/2dd311fdd61308e1415cd45c1edc3076
107. Shenoi S (2011) Editorial. Int J Crit Infrastruct Prot 4(1):1–2. doi:10.1016/j.ijcip.2011.03.004, http://www.sciencedirect.com/science/article/pii/S18745482110 00084
108. Shi P, Liu B, Hou D (2008) Torsional vibration suppression of drive system based on DMC method. In: 7th world congress on intelligent control and automation 2008. WCICA 2008, pp 4789–4792. doi:10.1109/WCICA.2008.4593699
109. Shoureshi R, Knurek T (1996) Automotive applications of a hybrid active noise and vibration control. IEEE Control Syst Mag 16(6):72–78. doi:10.1109/37.546272
110. Shoureshi R, Gasser R, Vance J (1997) Automotive applications of a hybrid active noise and vibration control. In: Proceedings of the IEEE international symposium on industrial electronics 1997. ISIE '97, vol 3, pp 1071–1076. doi:10.1109/ISIE.1997.648888
111. Shustov / WikiMediaorg (2006) Concurrent experiments with two kinematically equivalent to a real prototype building models on a shake-table. Photograph. Online, the file is licensed under the creative commons attribution-share alike 3.0 unported license. http://commons.wikimedia.org/wiki/File:Kinematically_equivalent_building_models_on_a_shake-table.jpg

112. Sikorsky Aircraft Corporation (2010) Sikorsky innovations completes testing of hub mounted vibration system. http://www.sikorsky.com/About+Sikorsky/News/Press+Details?pressvcmid=297bee22ced7b210VgnVCM1000004f62529fRCRD\&keyword=active%20vibration%20control\&dateFrom=null\&dateTo=null\&model=null\&business=null\&matchCriteria=null\&matchKeyword=all\&fromSearchPage=true\&modelIndex=0\&businessIndex=0\&page=1
113. Sikorsky Aircraft Corporation (2010) Sikorsky S-76D helicopter technical information. Technical data sheet. http://www.sikorsky.com/StaticFiles/Sikorsky/Assets/Attachments/Mission%20Downloads/S76-105a_S76D_VIP_TI.pdf
114. Sikorsky Aircraft Corporation (2010) Sikorsky UH-60M BLACK HAWK helicopter—United States Army multi-mission combat assault helicopter. Technical data sheet. http://www.sikorsky.com/StaticFiles/Sikorsky/Assets/Attachments/Mission%20Downloads/S92-056a_TI_SAR.pdf
115. Sikorsky Aircraft Corporation (2010) Sikorsky UH-60M BLACK HAWK helicopter—United States Army multi-mission combat assault helicopter. Technical data sheet. http://www.sikorsky.com/StaticFiles/Sikorsky/Assets/Attachments/Mission%20Downloads/A-144_UH60M_Brochure.pdf
116. Sikorsky Aircraft Corporation (2010) X2 TECHNOLOGY demonstrator. http://www.sikorsky.com/vgn-ext-templating-SIK/v/index.jsp?vgnextoid=40c96eb78fa78110VgnVCM10
117. Sikorsky Aircraft Corporation (2010) X2 technology demonstrator achieves 225 knots, sets new top speed for helicopter—target milestone of 250 knots looms in Q3 2010. http://sikorsky.com/About+Sikorsky/News/Press+Details?pressvcmid=a4a2962fa4f0a210VgnVCM1000004f62529fRCRD
118. Someformofhuman / WikiMedia Commons (2008) The tuned mass damper in Taipei 101. Illustration licensed under the creative commons attribution—share alike 3.0 unported, 2.5 generic, 2.0 generic and 1.0 generic license. http://commons.wikimedia.org/wiki/File:Taipei_101_Tuned_Mass_Damper.png
119. Spelta C, Previdi F, Savaresi SM, Fraternale G, Gaudiano N (2009) Control of magnetorheological dampers for vibration reduction in a washing machine. Mechatronics 19(3):410–421. doi:10.1016/j.mechatronics.2008.09.006, http://www.sciencedirect.com/science/article/B6V43-4TT1G22-1/2/3d8e5bd1cc63e7181272ef848f15508c
120. Su YX, Duan BY, Wei Q, Nan RD, Peng B (2002) The wind-induced vibration control of feed supporting system for large spherical radio telescope using electrorheological damper. Mechatronics 13(2):95–110. doi:10.1016/S0957-4158(01)00042-3, http://www.sciencedirect.com/science/article/B6V43-46WPHMS-2/2/eca7cd44909e99a1f8c6ad76a4fd4f19
121. Sun D, Mills JK, Shan J, Tso SK (2004) A PZT actuator control of a single-link flexible manipulator based on linear velocity feedback and actuator placement. Mechatronics 14(4):381–401. doi:10.1016/S0957-4158(03)00066-7, http://www.sciencedirect.com/science/article/B6V43-49DN5K4-1/2/fa21df547f182ad568ccfb2ddf3a6352
122. Sun J, Yang Q (2007) Automotive suspension system with an analytic fuzzy control strategy. In: IEEE international conference on vehicular electronics and safety 2007. ICVES, pp 1–4. doi:10.1109/ICVES.2007.4456375
123. Sun W, Li J, Zhao Y, Gao H (2010) Vibration control for active seat suspension systems via dynamic output feedback with limited frequency characteristic. Mechatronics (in Press). doi:10.1016/j.mechatronics.2010.11.001, http://www.sciencedirect.com/science/article/B6V43-51KH6DW-1/2/9f06f9d31ca4a47bf3b8e034ba8c6150
124. Sung KG, Han YM, Cho JW, Choi SB (2008) Vibration control of vehicle ER suspension system using fuzzy moving sliding mode controller. J Sound Vib 311(3–5):1004–1019. doi:10.1016/j.jsv.2007.09.049, http://www.sciencedirect.com/science/article/B6WM3-4R2H1TN-4/2/b3a297765c3ac7767b2d64fda7a6a3d7

125. Takács G, Rohal'-Ilkiv B (2009) Implementation of the Newton–Raphson MPC algorithm in active vibration control applications. In: Mace BR, Ferguson NS, Rustighi E (eds) Proceedings of the 3rd international conference on noise and vibration: emerging methods, Oxford
126. Takács G, Rohal'-Ilkiv B (2009) MPC with guaranteed stability and constraint feasibility on flexible vibrating active structures: a comparative study. In: Hu H (ed) Proceedings of the eleventh IASTED international conference on control and applications, Cambridge
127. Takács G, Rohal'-Ilkiv B (2009) Newton–Raphson based efficient model predictive control applied on active vibrating structures. In: Proceedings of the European control control conference, Budapest
128. Takács G, Rohal'-Ilkiv B (2009) Newton-Raphson MPC controlled active vibration attenuation. In: Hangos KM (ed) Proceedings of the 28th IASTED international conference on modeling, identification and control, Innsbruck
129. The Japan Times (2011) Disaster analysis you may not hear elsewhere. http://search.japantimes.co.jp/cgi-bin/fl20110320x2.html, Retrieved 21 June 2011
130. Torra V, Isalgue A, Martorell F, Terriault P, Lovey F (2007) Built in dampers for family homes via SMA: an ANSYS computation scheme based on mesoscopic and microscopic experimental analyses. Eng Struct 29(8):1889–1902. doi:10.1016/j.engstruct.2006.08.028, http://www.sciencedirect.com/science/article/B6V2Y-4MFKD84-1/2/8742fa675c346a7b34f395d9422cbc22
131. Tschida T (2002) NASA Dryden Flight Research Center (NASA-DFRC) Active aeroelastic wing F/A-18 research aircraft during a ground vibration testing. Image ID: EC02-0203-55
132. Tschida T, NASA Dryden Flight Research Center (NASA-DFRC) (2002) The upper wing surface of the active aeroelastic wing F/A-18 test aircraft. Image ID: EC02-0203-46
133. Tzou H, Chai W (2007) Design and testing of a hybrid polymeric electrostrictive/piezoelectric beam with bang-bang control. Mech Syst Sig Process 21(1):417–429. doi:10.1016/j.ymssp.2005.10.008, http://www.sciencedirect.com/science/article/B6WN1-4HR75KY-1/2/73701e5908a2ea598fa7bec1ce093563
134. United States Geological Survey (2011) Magnitude 9.0—near the east coast of Honshu, Japan. http://earthquake.usgs.gov/earthquakes/eqinthenews/2011/usc0001xgp/, Retrieved 21 June 2011
135. Van den Broeck L, Diehl M, Swevers J (2009) Time Optimal MPC for mechatronic applications. In: Proceedings of the 48th IEEE conference on decision and control, Shanghai, pp 8040–8045
136. Van den Broeck L, Swevers J, Diehl M (2009) Performant design of an input shaping prefilter via embedded optimization. In: Proceedings of the 2009 American control conference, St. Louis, pp 166–171
137. Wahed AKE, Sproston JL, Schleyer GK (2002) Electrorheological and magnetorheological fluids in blast resistant design applications. Mater Des 23(4):391–404. doi:10.1016/S0261-3069(02)00003-1, http://www.sciencedirect.com/science/article/B6TX5-450HD50-1/2/0da443f054d99983150525d47bf17aeb
138. Wang M, Fei R (1999) Chatter suppression based on nonlinear vibration characteristic of electrorheological fluids. Int J Mach Tools Manuf 39(12):1925–1934. doi:10.1016/S0890-6955(99)00039-5, http://www.sciencedirect.com/science/article/B6V4B-3X7N8GJ-7/2/6cc38d51af69b4fbb0aa1135681b5356
139. Wei JJ, Qiu ZC, Han JD, Wang YC (2010) Experimental comparison research on active vibration control for flexible piezoelectric manipulator using fuzzy controller. J Intell Rob Syst 59:31–56. doi:10.1007/s10846-009-9390-2
140. Wenzhong Q, Jincai S, Yang Q (2004) Active control of vibration using a fuzzy control method. J Sound Vib 275(3–5):917–930. doi: 10.1016/S0022-460X(03)00795-8, http://www.sciencedirect.com/science/article/B6WM3-49P82Y8-3/2/4041c663559fb530f34deadda058c82d

141. Williams E, Rigby S, Sproston J, Stanway R (1993) Electrorheological fluids applied to an automotive engine mount. J Non-Newtononian Fluid Mech 47:221–238. doi: 10.1016/0377-0257(93)80052-D, http://www.sciencedirect.com/science/article/B6TGV-44V49DV-75/2/a6f4db8ffcb810f6167c845a984dd93f
142. Wills A, Bates D, Fleming A, Ninness B, Moheimani R (2005) Application of MPC to an active structure using sampling rates up to 25 kHz. In: 44th IEEE conference on decision and control 2005 and 2005 European control conference. CDC-ECC '05, pp 3176–3181. doi:10.1109/CDC.2005.1582650
143. Wills AG, Bates D, Fleming AJ, Ninness B, Moheimani SOR (2008) Model predictive control applied to constraint handling in active noise and vibration control. IEEE Trans Control Syst Technol 16(1):3–12
144. Wilson DG, Robinett RD, Parker GG, Starr GP (2002) Augmented sliding mode control for flexible link manipulators. J Intell Rob Syst 34:415–430. doi:10.1023/A:1019635709331
145. World Nuclear News (2011) Massive earthquake hits Japan. http://www.world-nuclear-news.org/RS_Massive_earthquake_hits_Japan_1103111.html, Retrieved 21 June 2011
146. Yamashita K (2009) Efforts toward enhancing seismic safety at Kashiwazaki Kariwa nuclear power station. E-J Adv Maint 1(3):GA7. http://www.jsm.or.jp/ejam/Vol.1.No.3/GA/7/article.html
147. Yan G, Sun B, Lü Y (2007) Semi-active model predictive control for 3rd generation benchmark problem using smart dampers. Earthq Eng Eng Vib 6:307–315. doi:10.1007/s11803-007-0645-2
148. Yang Y, Jin Z, Soh CK (2005) Integrated optimal design of vibration control system for smart beams using genetic algorithms. J Sound Vib 282(3–5):1293–1307. doi:10.1016/j.jsv.2004.03.048, http://www.sciencedirect.com/science/article/B6WM3-4DJBPM1-6/2/944b2e30a1b99c969b56adbf527d9b1c
149. Yau J (2009) Vibration control of maglev vehicles traveling over a flexible guideway. J Sound Vib 321(1–2):184–200. doi:10.1016/j.jsv.2008.09.030, http://www.sciencedirect.com/science/article/B6WM3-4TWSWP3-1/2/c2ef06bef3677e1ed29b82857a322d58
150. Yildirim S (2004) Vibration control of suspension systems using a proposed neural network. J Sound Vib 277(4–5):1059–1069. doi:10.1016/j.jsv.2003.09.057, http://www.sciencedirect.com/science/article/B6WM3-4BM6CCP-4/2/0db857f0580d634772e8d782485e76bf
151. Yim W (1996) Modified nonlinear predictive control of elastic manipulators. In: Proceedings of the 1996 IEEE international conference on robotics and automation, vol 3, pp 2097–2102 (Special Issue: Inverse Problems). doi:10.1109/ROBOT.1996.506180
152. Zapateiro M, Luo N, Karimi H, Vehí J (2009) Vibration control of a class of semiactive suspension system using neural network and backstepping techniques. Mech Syst Sig Process 23(6):1946–1953. doi:10.1016/j.ymssp.2008.10.003, http://www.sciencedirect.com/science/article/B6WN1-4TTMJRM-1/2/b6b45074716201902e0b01b664ebbeb9
153. Zhang CL, Mei DQ, Chen ZC (2002) Active vibration isolation of a micro-manufacturing platform based on a neural network. J Mater Process Technol 129(1–3):634–639. doi:10.1016/S0924-0136(02)00671-4, http://www.sciencedirect.com/science/article/B6TGJ-46V46C0-4P/2/8e8228760a4ac6759cef159e6fcb7606
154. Zhang QZ, Gan WS (2004) A model predictive algorithm for active noise control with online secondary path modelling. J Sound Vib 270(4–5):1056–1066. doi: 10.1016/S0022-460X(03)00516-9, http://www.sciencedirect.com/science/article/B6WM3-49D6XFX-4/2/805d0549ca0e60339bb5e2c798de7264
155. Zhou Y, Steinbuch M, Kostic D (2002) Estimator-based sliding mode control of an optical disc drive under shock and vibration. In: Proceedings of the 2002 international conference on control applications, vol 2, pp 631–636. doi:10.1109/CCA.2002.1038673

156. Zhu C (2005) A disk-type magneto-rheological fluid damper for rotor system vibration control. J Sound Vib 283(3–5):1051–1069. doi:10.1016/j.jsv.2004.06.031, http://www.sciencedirect.com/science/article/B6WM3-4F4H9R2-1/2/48abebbf8d1230fcd80eee7d19fe52fa
157. Zmeu K, Shipitko E (2005) Predictive controller design with offline model learning for flexible beam control. In: Proceedings of the 2005 international conference on physics and control, pp 345–350. doi:10.1109/PHYCON.2005.1514005

Chapter 10
MPC Implementation for Vibration Control

This chapter introduces the implementation details of three different model predictive control (MPC) strategies applied to active vibration control (AVC). All methods have their own advantages and disadvantages; therefore, it is difficult to recommend just one strategy for AVC. The implementation of predictive control strategies without constraints is not different from the implementation of the very common linear quadratic controller (LQ), therefore will not be considered here. An MPC strategy without constraints is a fixed feedback law in a closed form and its software implementation only takes a formulation of a state observer and a matrix operation to evaluate inputs. MPC strategies without constraints have known implementation examples in AVC; however, the lack of system constraints raises the question of their advantage and justification over a simple LQ strategy [3, 13, 21, 45, 46, 56, 58, 75–78]. The inclusion of constraints requires a completely different approach in implementation, as online optimization or its equivalents are needed to compute control inputs. The use of constrained MPC in active vibration control is much less common due to this difficulty and existing implementations are few [6, 11, 22, 27, 44, 73, 74]. None of the MPC-based AVC strategies available in the literature consider the question of algorithm stability. As the authors believe that an MPC strategy shall incorporate a priori stability guarantees, all the model predictive vibration control strategies introduced in this chapter and the subsequent ones will consider this feature. To the knowledge of the authors, currently the only known implementation of constrained, stable MPC algorithms in AVC are the articles that inspired the writing of this book [63–65].

Traditional stabilized infinite horizon dual-mode quadratic programming based model predictive control (QPMPC) is the basis and benchmark of all MPC algorithm variations; therefore, we will use it as a base of comparison to evaluate the performance and computational efficiency of other algorithms. Its merits are clear: it provides optimal response, but with a price of intense online computational

DOI: 10.1007/978-1-4471-2333-0_10, © Springer-Verlag London Limited 2012

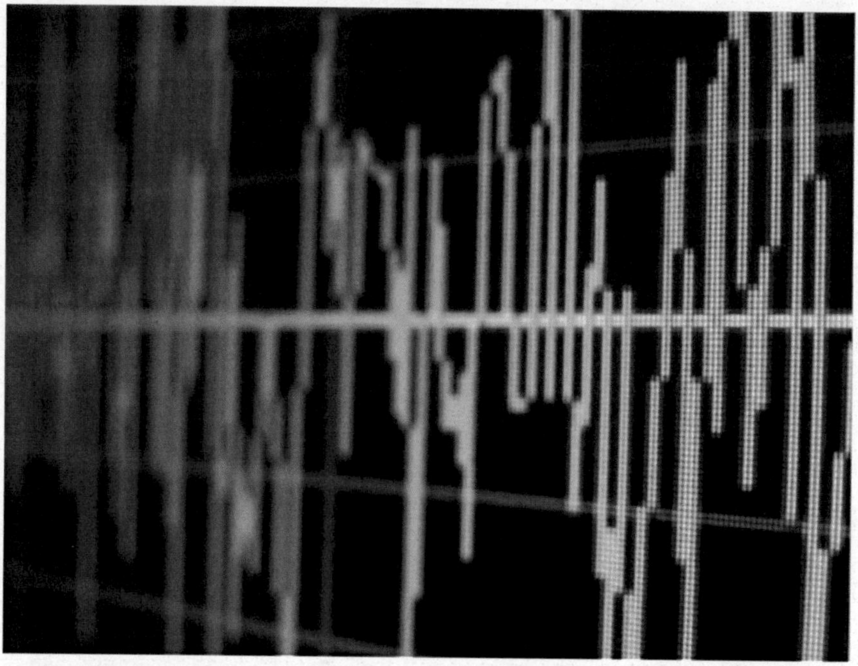

Fig. 10.1 The image illustrates a signal rendered on a virtual scope of the xPC Target rapid software prototyping system. In order to assess the timing properties of MPC algorithms correctly, all strategies used identical experimental implementations in the xPC Target environment

requirements. It can be recommended to those AVC applications which do not require fast sampling times.

More typical is the AVC application with fast sampling times. If the discrepancy between actuator capabilities and expected disturbance levels is not large—that is the structure is either well damped or with aggressive actuation—multi-parametric programming based MPC (MPMPC) can be recommended. Examples of such systems are among others narrow-band optical systems, manipulators, automotive dampers and others. This recommendation holds only in the case if the system dynamics are to be modeled in a very narrow band, containing one or at most two resonant modes thus restraining the size of the prediction model. The MPMPC controller can provide optimal performance just as QPMPC, but it also may gain some off and online computational advantage by sacrificing performance. One of the common examples of such a suboptimal MPMPC method is minimum-time MPMPC. According to the experience of the authors, the performance loss is not justified by the modest computational advantage gained using this suboptimal MPMPC implementation.

Flexible systems with weak actuation such as piezoelectric patches, electrochemical actuators, electrostrictive actuators and others have a large difference between the static effect of the actuating elements and the expected range of disturbances. As it has been previously noted, examples of such systems are helicopter rotor beams, wing

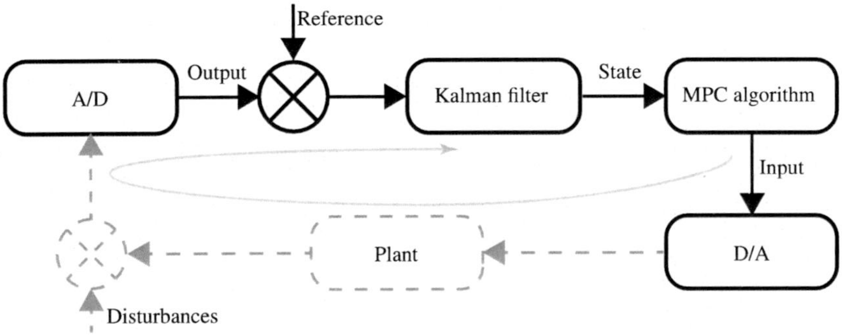

Fig. 10.2 Schematic representation of the model predictive control algorithm. The illustration shows the conceptual connection between distinct parts of the control scheme, such as analog-digital and digital-analog conversion, state observation and the MPC algorithm itself

surfaces, large manipulators, antennae, etc. To ensure the stable constrained model predictive vibration control of these systems a method such as Newton–Raphson MPC (NRMPC) can be recommended. Its merits are computational efficiency and the large region of attraction; however, this comes with a price of efficiency reduction, especially with larger model orders.

All MPC algorithms—in fact all state-based control schemes—require some common basic building blocks to create a working real-time implementation (Fig. 10.1). Such control schemes need to read the analog data from the sensors using an analog to digital (A/D) converter. This signal or signals are then processed by a state observer, usually a Kalman filter, and the states of the system are estimated. The states are fed into the MPC algorithm or any other state-based algorithm, which then computes the inputs. The inputs are passed to the controlled system and digital signals are converted back to analog with a digital to analog converter (D/A). The conceptual connection of these basic elements of a control system is illustrated in Fig. 10.2.

Section 10.1 introduces the implementation basics of traditional quadratic programming-based MPC with stability guarantees. Both the overall control loop and the quadratic programming solver are featured in this section with a recommendation on solver software. The next section, that is Sect. 10.2 presents a simple practical implementation of both optimal and minimum-time MPMPC through an off-the-shelf and readily available software toolbox. Properties of both the optimal MPMPC controller and its suboptimal minimum-time implementation are discussed. The controller implementation for the AVC demonstrator is used to illustrate the MPMPC controller law for lightly damped systems including the controller regions, inputs in three dimensions and the cost function. The final section of this chapter introduces the implementation details of the Newton–Raphson MPC approach. As this model predictive control strategy is not as widely known as the previous two, the most attention will be devoted to it. The offline SDP optimization problem is discussed in Sect. 10.3.1. The important cost transformation procedure is introduced next which is followed by some considerations on the Newton–Raphson root search routine in Sect. 10.3.3. A brief review of the implementation of the NRMPC extension is given in Sect. 10.3.4 while the section is finished by a few passages on code implementation details.

Appendix B is an important addition to this chapter for those who are seeking details on the code implementation properties of various MPC strategies. Appendix B contains well-commented code segments for the QPMPC, MPMPC and the NRMPC implementation. Offline QPMPC problem setup in Matlab and simulation pointers are discussed in Sect. B.1 while Sect. B.2 is devoted to the implementation of MPMPC through the MPT Toolbox. Since the implementation of NRMPC is not commonly available in the literature, the final Sect. B.3 discusses both the offline and the online code in detail.

10.1 Implementation of the QPMPC Algorithm

The quadratic programming based MPC controller considered in this book rests on the traditional infinite horizon dual-mode MPC formulation with stability and feasibility guarantees via terminal constraints. The theoretical basis for this particular type of MPC strategy has been introduced earlier in Chaps. 6 and 7 with the a priori stability guarantees presented in Sects. 7.3 and 7.4. Further reference for those interested in the theoretical particulars of traditional QPMPC formulation can be found in the books by Maciejowsky and Rossiter [42, 57] or other books written on the general topic of model predictive control, for example [4, 7, 23, 72] and others.

The problem setup, creating prediction and cost prediction matrices is carried out in the Matlab scripting environment. Stability in the QPMPC formulation is guaranteed through suitably formulated state feedback and terminal cost function matrices. The deployment of dual-mode predictions is the part of the formulation as well: the first mode considers n_c free control moves, while the second mode assumes the LQ control law [10, 43]. The feasibility of process constraints is ensured beyond the prediction horizon by the inclusion of a constraint checking horizon.

All controllers featured in the simulation and experimental studies of this book utilized identical settings. The QPMPC controller, which is implemented on the AVC demonstrator, uses a $T_s = 0.01$ s sampling, roughly 10 times faster than the highest measured and modeled frequency, thus avoiding aliasing effects and taking into account Shannon's sampling theorem [26]. Input penalty is set at $\mathbf{R} = r = 1\text{E-}4$, state penalty matrix is $\mathbf{Q} = \mathbf{C}^T\mathbf{C}$. Input constraints $\overline{u} = \underline{u} = \pm\, 120$ V are identical to the other cases as well, while state constraints have been inactive.

Note that we have characterized the MPC problem in the previous chapters in general: that is, we have included both input and state constraints in the formulation. In fact, one of the enormous advantages of MPC is the possibility to create and enforce state constraints, which in turn can be easily transformed into output constraints. Nevertheless, it is very important to keep in mind that state or output constraints are not the only reason to choose the constrained MPC formulation over other alternatives. As it has been briefly explained in Sect. 6.7.1, input saturation limits are not the same as the advanced constraint handling ability of MPC. The simulation studies and experimental tests featured in the upcoming chapters of this book will not include state or output constraints, but their inclusion is not a fundamental problem and is not essential in discussing the issues related to the implementation of MPC in active vibration control. Even if only input constraints are included, a saturated LQ controller is very different from a constrained MPC controller, which unlike the clipped LQ, ensures optimal and stable closed-loop control.

The only remaining yet important setting is the prediction horizon. In simulation studies the horizon has been set at very high values, in the order of $n_c = 150$ steps to match the region of attraction of the NRMPC controller. However, in experimental tests the QPMPC acted as benchmark and was the weakest link from the viewpoint of computational power. The prediction horizon therefore was set at lower values, which produced a computational period almost on the verge of crashing the controller and close to the sampling period. With a second order example and the given prototyping PC this horizon was in the order of $n_c = 70$ steps which allows only a beam tip deflection of ± 10 mm.

Online quadratic programming optimization is performed via the 2.0 version of the *qpOASES*[1] open-source C++ active set strategy by Ferreau et al. [14, 15, 18]. This quadratic programming solver implements numerous theoretical features, which make it especially suitable for MPC applications. qpOASES has been chosen as a comparison platform used in the upcoming chapters for its convenient implementation and the fact that unlike generic QP solvers, it has been fine tuned for predictive control applications. Note that although the QPMPC algorithm has been considered as the slowest of all the approaches considered here, the qpOASES active set implementation is still an efficient MPC strategy. Unlike in the other cases, computational efficiency lies in the online algorithm and not in the MPC formulation.

qpOASES has been considered for the time optimal MPC control of mechatronic systems by Van den Broeck et al. in [69, 70]. Van den Broeck et al. [69] utilized a novel time optimal MPC approach to control a two DOF mass-spring-damper system, using a fifth order state-space system sampled at $T_s = 0.01$ Hz containing two pairs of complex conjugate poles. The average computational time for this system was 6–7 ms while the mean time was approximately 1 ms.[2] The qpOASES active set solver has also been used successfully for the control of the exhaust gas recirculation (EGR) valve and variable turbocharger geometry in diesel engines with an 50 ms sampling and five step horizon by Ferreau et al. in [17]. An MPC-based trajectory optimization for boom cranes[3] using qpOASES has been suggested by Arnold et al. [2], while the control of robotic arms has been solved via a qpOASES supported solution as well [47, 55].

The qpOASES quadratic programming solver is loaded and compiled via its Simulink interface, then prediction and constraint matrices are passed on to it from the Matlab workspace. The QPMPC controller algorithm is included in the same block scheme as used for the other investigated controllers.

---

[1] Download at http://www.qpoases.org.

[2] Van den Broeck et al. used a computer with a Mobile Pentium 2 GHz processor with 2 GB of RAM.

[3] See more on the topic of Sect. 9.7.3.

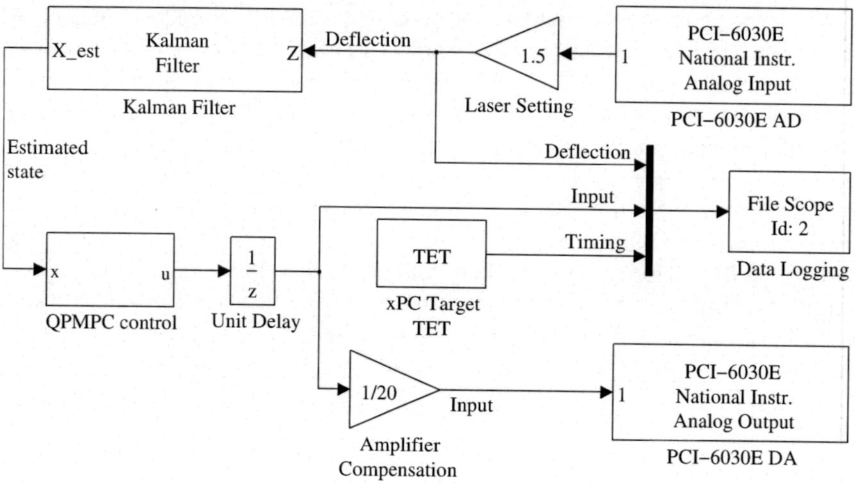

Fig. 10.3 Block scheme of the QPMPC controller in Simulink, for the application of model predictive vibration control on the demonstrator hardware

The QPMPC controller loop, which is identical to the MPMPC and NRMPC case except for the controller strategy itself, is featured in Fig. 10.3. The controller loop contains means for A/D and D/A conversion, state observation, data logging and others. The block containing the QPMPC controller takes the actual observer state x_k as an input, and outputs controller voltage u_k. For diagnostics and design purposes, it may be required to output variables such as iteration data or the value of the cost function.

The Simulink scheme implementing the controller gains its feedback signal directly from a block reading the analog input of the measuring card. The output from this block is scaled according to the current settings of the laser head, so it is expressed directly in millimeters. Practical measurements mostly involve a 1.5 mm/V gain. This signal can be compensated for drift and reference imprecision manually. The measurement results are saved to a file, thus the feedback signal is connected to the File Scope block. The File Scope block ensures real-time data logging onto the xPC Target PC hard drive, which can be later re-used for measurement processing.

The reference feedback is then compared to the desired setting, which of course in the case of vibration damping is zero. The feedback measurement passes through a product default Kalman filter, a part of the Signal Processing Toolbox [67]. The Kalman filter is always enabled, and the number of filters is set to one. Initial condition is assumed to be zero. The state transition matrix and measurement matrix is read from the script file, making use of the state transition and measurement matrices **A** and **C** also used for controller computation. Initial condition for estimated covariance is set to the identity matrix **I**, measurement covariance is set to one.

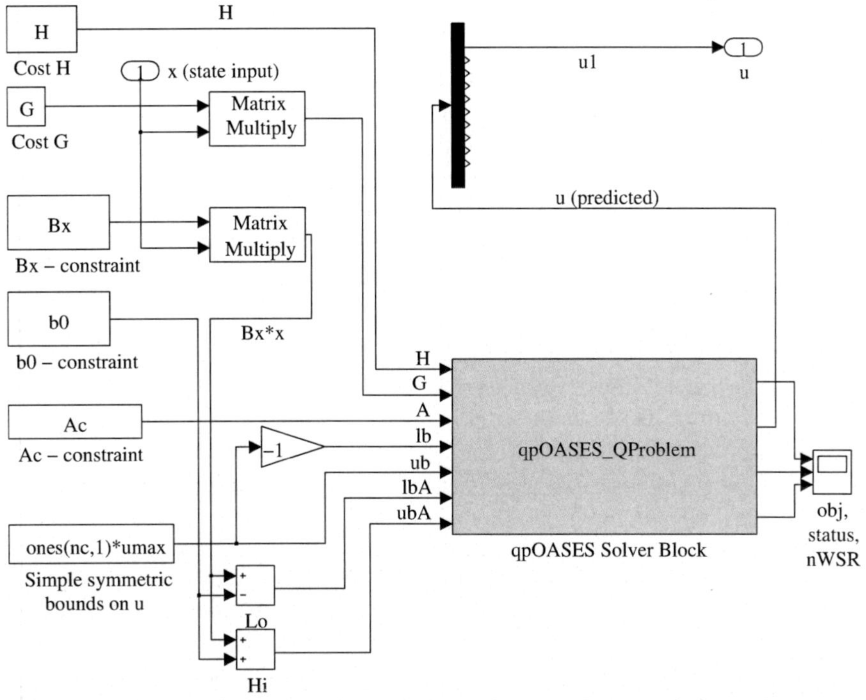

Fig. 10.4 Parsing the MPC problem to the qpOASES interface. The solver takes the measured state x_k combined with the prediction matrices $\mathbf{G}, \mathbf{H}$ and constraints $\mathbf{A}_c, \mathbf{b}_0, \mathbf{B}_x$ as parameters and computes the input sequence $\mathbf{u}_k$ and input u_k

The subsystem marked as QPMPC in Fig. 10.3 is featured in its full detail in Fig. 10.4. Here it is visible how the predicted cost matrices and constraints[4] calculated in the problem setup stage are parsed onto the qpOASES interface. The qpOASES solver block[5] takes the cost prediction matrices **H** and **G** as its input and in addition to that, the reformulated constraints combined with the actual state measurement x_k. The main output of this block is the vector of predicted inputs of which only the first element is required and selected. The qpOASES solver block outputs diagnostics values such as the cost function value J_k, a status indicator containing information whether the optimization was feasible and the number of iterations. For further details on the use of qpOASES one shall refer to the documentation [16].

---

[4] The illustration in Fig. 10.4 shows the variable naming conventions familiar from earlier chapters. The code implementation in Appendix B uses the following notation for variables: $\mathbf{H} =$ Hqp, $\mathbf{G} =$ F, $\mathbf{B}_x =$ BxQP, $\mathbf{b}_0 =$ b0QP and $\mathbf{A}_c =$ AcQP.

[5] The particular solver module featured in Fig. 10.4 is *qpOASES_QProblem*, which is recommended for problem formulations assuming nominal systems and fixed constraints. In case a sequential quadratic programming problem is required by the MPC formulation (e.g. in adaptive systems), the resulting optimization task can be solved more efficiently by using the sequential *qpOASES_SQProblem* module of qpOASES [16]. The timing analysis featured in Sect. 12.5 utilizes the sequential qpOASES solver, but with constant parameter and constraint matrices.

A detailed description of the offline QPMPC controller computation is given in Sect. B.1 of Appendix B. This section lists working code with comments and concentrates on the offline formulation details, such as the computation of prediction matrices, constraint formulation and offline simulation. By the aid of the qpOASES solver the online part can be easily realized in the Simulink simulation and rapid prototyping environment.

10.2 Implementation of the MPMPC Control Algorithm

The multi-parametric programming based model predictive control strategy introduced in Sect. 8.2.1 of Chap. 8 is an excellent candidate for the predictive control of vibrating systems. This is mainly valid in the case of strong actuation and narrow-bandwidth disturbance, not so much for lightly damped systems or broadband actuation. The use of MPMPC as an AVC strategy seldom appears in the academic literature, there are only a few examples [11, 50]. An MPMPC controller has been considered in simulation for the AVC demonstrator by Polóni et al. in [54]. Later a more detailed simulation example followed by a practical implementation has been featured in Takács et al. [64], along with an investigation into the properties of the offline controller computation process. For those interested in a more detailed treatment of the theoretical and practical properties of multi-parametric programming based MPC we may recommend the books by Pistikopoulos et al. [51, 52] or by Kvasnica [31].

The simulation and experimental studies featured in the upcoming chapters make use of both optimal MPMPC and suboptimal minimum-time MPMPC to test the performance, off and online computational requirements and other properties of this strategy for lightly damped structures. Keeping in mind the need of engineering practitioners for readily available solutions, the authors have decided to utilize an off-the-shelf solution for the formulation of the MPMPC controllers. This section will briefly introduce some of the implementation properties of MPMPC controllers in the AVC of lightly damped structures through the Multi-Parametric Toolbox (MPT) by Kvasnica et al. which is a freely available Matlab toolbox for the design, analysis and rapid deployment of piecewise-affine (PWA) controllers [31, 33–35]. In addition to this, the MPT toolbox provides software tools for visualization, computational geometry and parametric optimization—all closely related to the problem area.

MPMPC controller computation can be set up using linear, quadratic or minimum-time performance objectives or custom objective functions. In addition to the typical constraints such as input, output or state, it is possible to add custom ones: contraction, collision avoidance and others. The resulting multi-parametric control laws are not only usable in Matlab: it is possible to deploy them onto rapid software prototyping platforms using the Real-Time Workshop. A stand-alone C code can be generated and implemented on custom control hardware. It is possible to use the MPT Toolbox through command line, although a user-friendly graphical interface is also available. The overall experience with the toolbox has been extremely pleasant. With a general knowledge of MPMPC basics, one may become familiar with the toolbox in hours. Building up a working practical knowledge with multi-parametric controllers and conducting simulations is possible in a short time. Likewise, exporting controllers into stand-alone C and deployment onto hardware has been a hassle free process. A detailed account of multi-parametric MPC controllers with a particular focus on the utilization of the MPT toolbox in engineering practice is given in the book by Kvasnica for those interested [31].

This work assumes the latest available release (Version 2.6.2) of the toolbox, available online.[6] Release 7.5.0.342[7] of Matlab used throughout this work proved to be incompatible with the latest release of MPT toolbox, therefore an earlier version 7.2.0.232[8] of Matlab was utilized for multi-parametric controller calculation and simulations. However, after the experimental controller has been calculated, it has been implemented under the newer Matlab release and its corresponding tools such as Simulink and xPC Target kernel. Similarly to the controllers used in simulations assessing offline calculation times and other properties of the MPMPC method in Sect. 11.2, the experimental multi-parametric MPC controller deployed in real-time has been computed using the MPT Toolbox as well.

A detailed account of the MPMPC controller implementation is given in Sect. B.2 of Appendix B. Here systematic instructions are given for the implementation of multi-parametric MPC through well-commented code segments. In addition to offline and online controller deployment, a short account is given on the computation of minimum-time MPMPC controllers through the MPT Toolbox.

10.2.1 Optimal Multi-Parametric Programming-Based MPC

A second order linear time-invariant state-space model of the AVC demonstrator with a sampling rate of 100 Hz (0.01 s) has been used to calculate the experimental MPMPC controller [64]. In fact, this model is identical to that used for the QPMPC or NRMPC controller. The control objective was set to regulation toward origin, since the beam equilibrium is located at the origin of the state-space. The cost function was set to quadratic (2-norm).

To approximate the region of attraction offered by the experimental NRMPC controller presented as a simulation example in Sect. 11.6 and in order to cover a sufficient allowable deflection range, the prediction horizon was chosen as $n_c = 152$ steps. This horizon allows an approximate maximal deflection of $\sim\pm 18$ mm at the beam tip. Guaranteed feasibility of constraints and controller invariance was a requirement just as with the other MPC strategies. Due to the long prediction horizon, this controller

[6] Software package and extended documentation is available at: http://control.ee.ethz.ch/~mpt/.

[7] Also known as *R2007b*.

[8] Also known as *R2006a*.

contains a very large number of regions, that is exactly 46,849 polyhedral partitions with the corresponding control laws. Offline computational time was matching the complexity of the controller, exceeding 18 h. Region merging or binary table calculation [20, 37, 68] was not performed, due to the expected heavy—possibly intractable—computational load. Experimental tests utilized MPMPC controllers with shorter $n_c = 70$ steps long horizons and the corresponding smaller number of controller regions.

To create a basis for a fair comparison of different MPC controller strategies, penalization and constraints shall be identical to the QPMPC or NRMPC case: input penalty has been set to $\mathbf{R} = r = 1\text{E-}4$, state penalty matrix was set to $\mathbf{Q} = \mathbf{C}^T\mathbf{C}$. Input constraints were set to $\overline{u} = \underline{u} = \pm 120$ V and output or state constraints were not engaged.

The MPT Toolbox features rapid code deployment functionality [36, 38]. The controller stored as a Matlab multi-field variable can be exported as a standalone C code by using the command *mpt_exportc(ctrl)*, where *ctrl* is the controller name. This C code can be then integrated into a given application. The size of the resulting C source code for this particular example is 10 MB. Aiding the simple use of MPC in practical applications, future versions of the MPT Toolbox will feature a command to automatically generate the MPMPC code for industrial programmable logic controllers (PLC) [36, 38].

The experimental MPMPC algorithm has been implemented in Matlab Simulink, and loaded onto a target computer running the xPC Target kernel. Software versions and control hardware was identical to that used during the NRMPC tests, in order to allow for an unbiased comparison of results. Except the MPMPC block, the control loop is in fact identical to the one presented for the QPMPC controller and it is illustrated in Fig. 10.5. For the interest of rapid code deployment, the MPMPC controller has been integrated into a custom S-Function block in this work.

The straightforward online controller logic presented in Sect. 8.2.1 simply involves finding the region index corresponding to the actual state measurement, then calculating the feedback according to the fixed feedback law associated to it. This is repeated at each sampling interval and implemented in the source file *mpt_getInput.c*, a part of the MPT Toolbox. The controller look-up tables are included in a separate file declared within the function source itself.

Despite the large number of regions and file size, the MPMPC controller performs remarkably well on the given hardware. Considering this hardware performance, the average execution times achieved with this complex MPMPC controller are notable. The C code export feature of the Multi-Parametric Toolbox proved to be a valuable and excellent tool, making possible the implementation of the controller an easy and problem free process.

10.2.1.1 Properties of the Optimal MPMPC Controller

A very important controller analysis tool offered by the MPT Toolbox is the extended *plot(ctrl)* function, where *ctrl* is the Matlab variable containing the controller

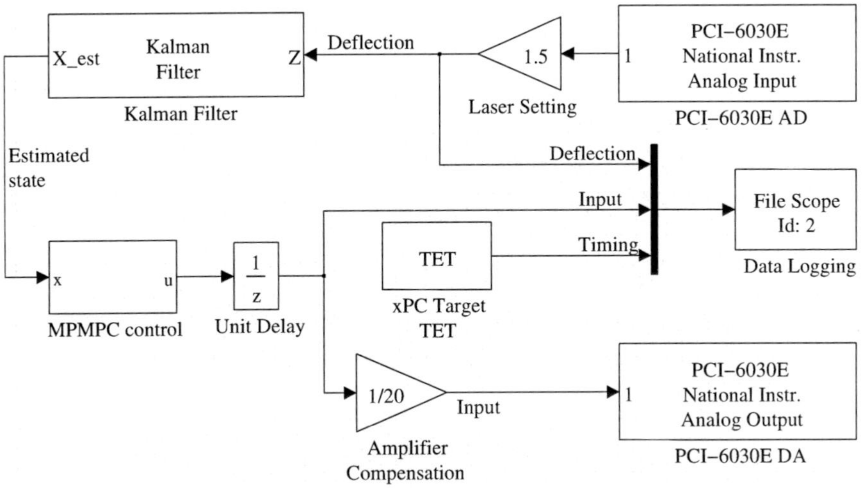

Fig. 10.5 Block scheme of the MPMPC controller in Simulink, for the active vibration cancellation application. Other than the controller itself the scheme is identical to the QPMPC or NRMPC implementation

structure. This command plots the regions of polyhedral partition defining the PWA control law [32]. The toolbox also assesses stability and invariance properties for the given controller. Values of control actions in relation to polyhedral regions can be plotted by the *plotu(ctrl)* command. Cost function in relation to the PWA controller regions may be plotted using the command *plotj(ctrl)*. Among many other excellent analysis tools, visualization of reachable sets and set of initial conditions is implemented as well.

Figure 10.6 shows multi-parametric MPC controller regions plotted in two-dimensional state-space. Controller partitions shown on the image belong to a controller based on the second order model of the AVC experimental device. Prediction horizon here is 140 steps, the region of attraction does not allow initial deflection levels over $\sim \pm 16$ mm. The region of attraction is divided into 39,777 polyhedral regions with associated control laws, showing partitions before merging.

MPMPC controller action expressed in Volts is plotted against the state-space in two dimensions in Fig. 10.7a. This plot shows the control output for the same second order controller producing 39,777 partitions for a 140 steps long horizon. From a practical engineering standpoint, there is one very important aspect to look for in this graph. One would logically expect an almost pure switching behavior from the controller, if the deflections exceed the maximal static effect of the piezoelectric transducers. As it will be later demonstrated, this is precisely how the exact MPC controllers generate their output signal. Figure 10.7a shows two major regions for either extreme case: the controller produces the upper or lower constrained value, depending on the combination of states. In between the two outermost possibilities, there is only a minimal transition—sharply dividing the state-space to two major regions.

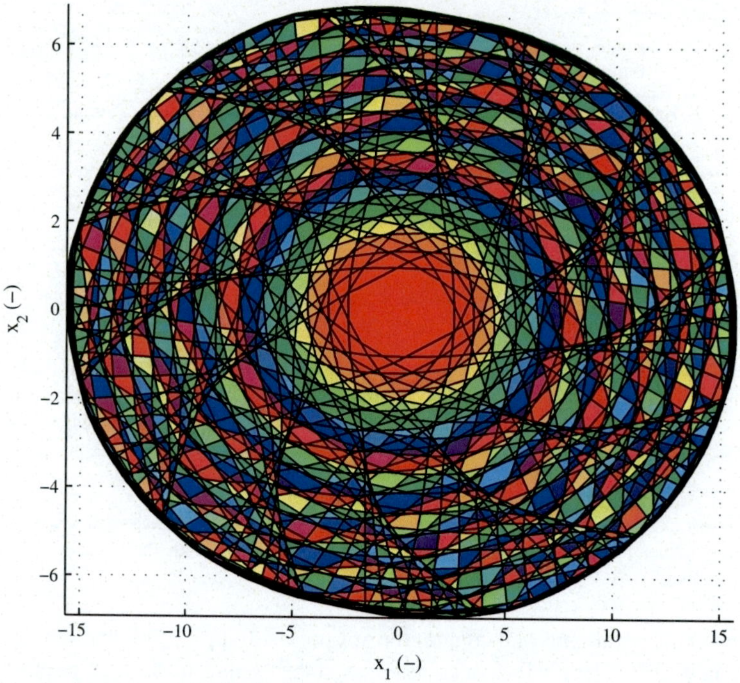

Fig. 10.6 MPMPC controller regions plotted in two-dimensional state-space. Figure shows 39,777 polyhedral regions for second order model with a $n_c = 140$ steps long control horizon

In fact, this essential phenomenon is applied and made use to create controllers in many active vibration damping applications. Examining the explicit optimal solution of this problem presents possibilities to approximate exact solutions with much simpler interpretations. Mechanical vibration damping based on heuristic shunt switching usually exploits properties of the state-space, where for example the sign of the state variables serves as a decision basis for the control actions. In a more sophisticated approach, D. Niederberger calculated and analyzed the explicit solution for the MPC control problem of a vibration damping system similar to the one used in this work [49]. Then the author designed an electrical, logical circuit emulating the MPMPC control law. Here the transition line between the two extreme switching positions as seen in Fig. 10.7a served as a basis to create hardware implemented control logic. A real-life practical application could benefit from this method by eliminating the need of relatively expensive computational equipment.

Figure 10.7b shows the plot of the quadratic cost function value $J(x)$, related to the polytopic controller regions in state-space. This plot has also been created using the controller example used later in experiments, but with a slightly shorter prediction horizon. The optimal controller utilized in the experimental comparison of

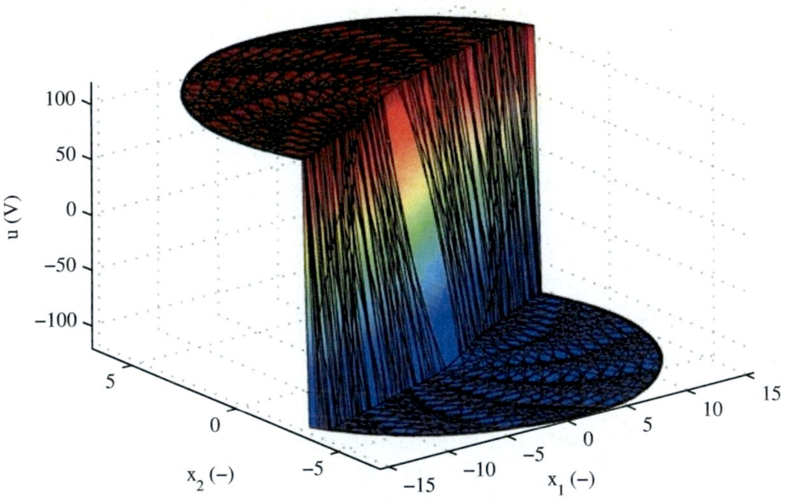

(a) MPMPC controller action

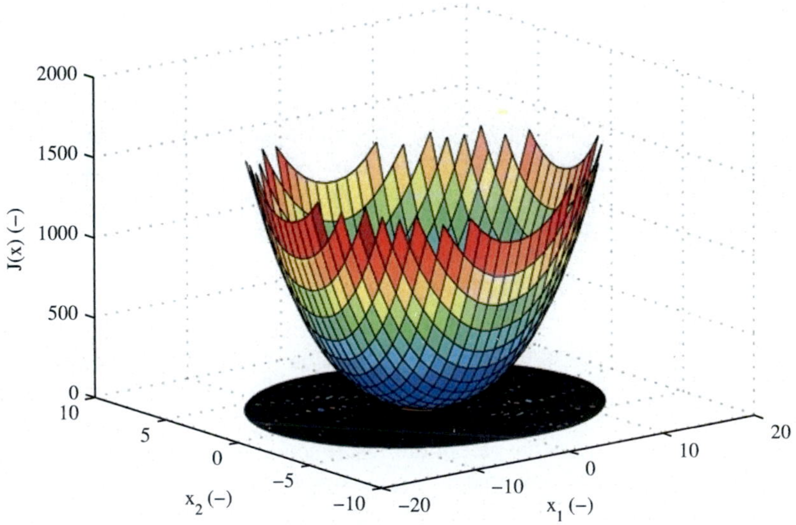

(b) MPMPC cost function

Fig. 10.7 Optimal MPMPC controller action in Volts is plotted in relation to states in two dimensions in (**a**), while (**b**) illustrates the MPMPC cost function $J(x)$ over the two-dimensional state-space

MPC methods uses a much shorter $n_c = 70$ step long horizon. The controller has been computed in 2818 s, which is over 46 min. This stable and invariant controller is defined over 10,099 polytopic regions.

10.2.2 Multi-Parametric Programming-Based Minimum-Time Suboptimal MPC

The suboptimal, minimum-time multi-parametric programming based MPC controller has been computed utilizing the Multi-Parametric Toolbox as well. The only difference between the optimal explicit controller is a setting enabling minimum-time controller computation and an inclusion of an output constraint. This additional symmetric output constraint $-\underline{y} = \overline{y}$ is necessary for the offline algorithm to converge on the same horizon and region of attraction as the optimal controller.

In case the minimum-time controller is compared to other stabilized constrained MPC strategies, the regions of attraction can be matched through a simple loop. Given a maximal implied deflection $|y_{\max}|$, the prediction horizon in optimal multi-parametric MPC or QPMPC is increased until the desired deflection level is reached. The online implementation of the minimum-time MPMPC controller is identical to the optimal MPMPC law introduced previously.

The explicit minimum-time controller featured in the experiments of Chap. 12 is defined over 1000 regions in two dimension, and required an offline computation time just over 65 s. This is a controller which ensures a region of attraction equivalent to those used with the optimal MPMPC and QPMPC controllers with the given horizon n_c. As it is evident, the minimum-time controller contains over an order of magnitude smaller amount of controller regions than its optimal counterpart and has been computed in a 40 times shorter time. This controller is also stable and invariant.

The regions defining the minimum-time controller are plotted in Fig. 10.8. The regions in the controller look different from the ones plotted for the optimal controller. It is interesting to see the two diagonal lines in the middle, which denote the sharp jumps in between the two extreme output levels. One portion is for the upper input level limit $\overline{u}$, while the other is for the lower $\underline{u}$. The band of regions stretching in the middle is the actual transient region with non-extreme input values. A better view of this effect is presented in Fig. 10.9a where the regions are plotted against the voltage input of the system as provided by the controller. Here the upper surface is the upper input constraint of $\overline{u} = 120$ V while the lower is $\underline{u} = -120$ V.

Finally, Fig. 10.9b shows the plot of the cost function. Comparing this plot to the optimal MPMPC cost function, one may see a significant difference. As it has been explained earlier in Sect. 8.2.2, the cost function is not the weighted sum of state and input contributions, rather the number of steps it takes to reach a terminal set. Therefore, the upper and largest area is the set of states from which it takes $J = n_c$ steps (or less) to reach the terminal set. Of course, if there is a lower cost layer for the given state, the corresponding law will be evaluated accordingly. One may refer to the book by Kvasnica for details on minimum-time MPMPC and its more advanced extensions [31].

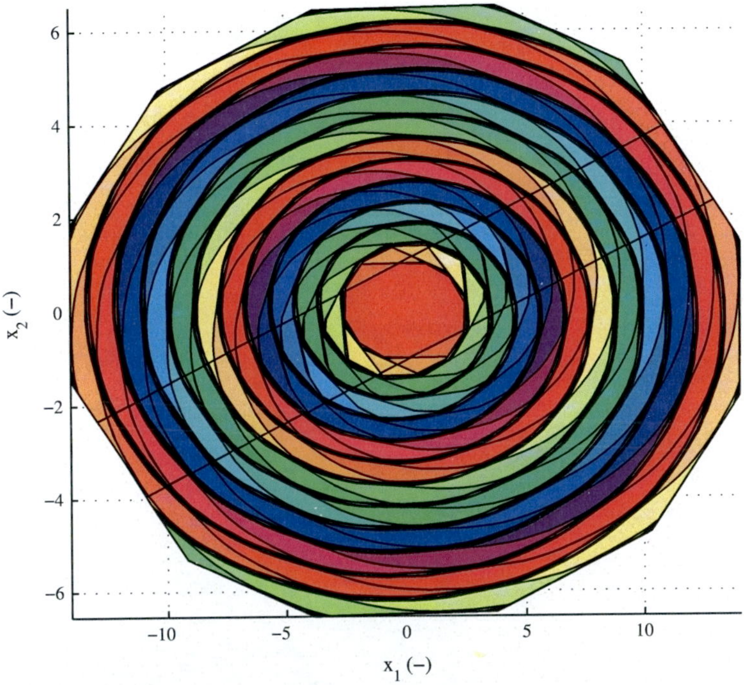

Fig. 10.8 Regions defining the minimum-time MPMPC controller for the vibrating system

10.3 Implementation of the NRMPC Control Algorithm

The Newton–Raphson MPC approach (NRMPC) has been selected as the third major predictive control strategy to be implemented on the AVC demonstrator. This particular MPC method has been first introduced by Kouvaritakis et al. in [28] and in addition to the brief review featured in Sect. 8.1 of this book one may find additional theoretical details in the publications by Kouvaritakis and Cannon et al. [28–30], Cannon and Kouvaritakis [8], Li et al. [39] and Imsland et al. [24, 25].

Up to now, NRMPC has not been implemented on any physical system[9] and its performance and timing properties have only been tested on randomly generated examples. The NRMPC algorithm has been first suggested for the active vibration control of lightly damped mechanical systems in [61, 62], later experimental studies confirmed its functionality in [63–66].

The inherent computational efficiency of NRMPC and the increased volume of the region of attraction by the additional optimization introduced by Cannon and Kouvaritakis in [8] render it as a very good controller candidate for the model predictive vibration control of lightly damped systems and this particular demonstration

[9] At the time of writing the book manuscript and according to the literature research performed by the authors.

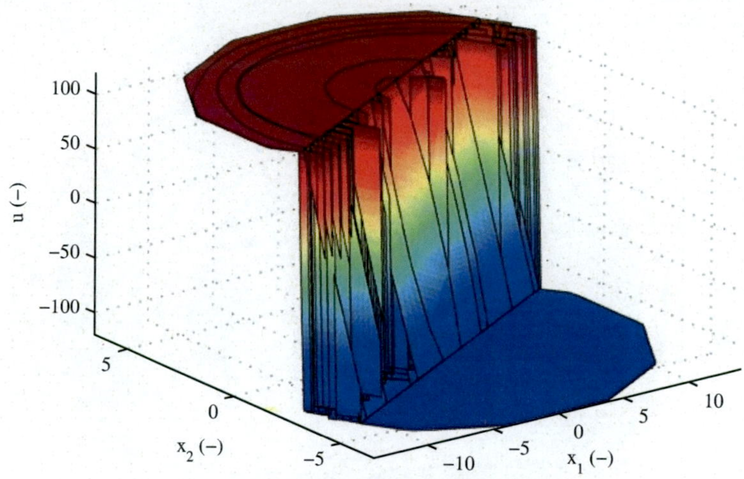

(a) MT MPMPC controller action

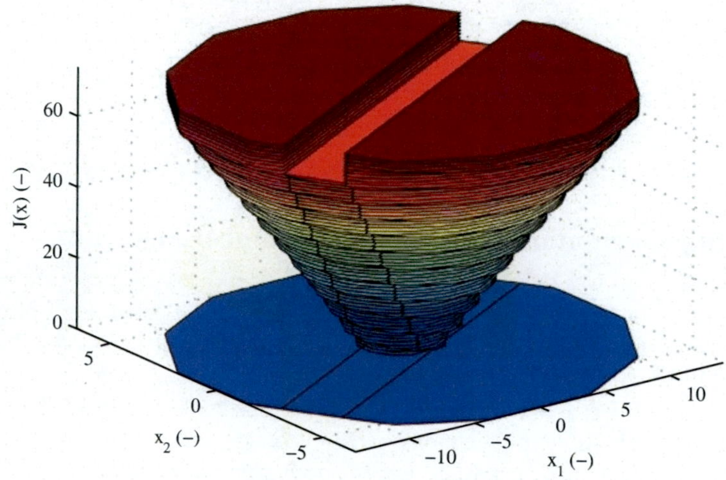

(b) MT MPMPC cost function

Fig. 10.9 Minimum-time MPMPC controller action in Volts is plotted in relation to states in two dimensions in (**a**), while (**b**) illustrates the minimum-time MPMPC cost function $J(x)$ over the two-dimensional state-space

hardware. As described in Sect. 8.1, NRMPC provides unmatched online computational savings by transferring time intensive operations into the offline stage. Although Monte Carlo simulations performed with the NRMPC algorithm in [29] showed that the theoretically suboptimal algorithm could be on par with truly optimal infinite-horizon dual-mode quadratic programming based MPC, flexible vibrating systems proved to be a more difficult case. The reader is kindly referred to the simulation results in Sects. 11.3, 11.4 and 11.5 for more details on the possible issues with the use of NRMPC for vibration control.

In addition to the information presented in this section, Chap. B contains a detailed account on the implementation details of NRMPC with well-commented code segments for both the problem setup and the real-time deployment. A summary of the offline portion of the NRMPC code implemented as a Matlab m-file is given in Sect. B.3.1 while the C language online code and a method of running a pre-compiled matrix algebra library on the real-time rapid prototyping software is discussed in Sect. B.3.2. The section devoted to NRMPC in Appendix B is finished by a treatment of different SDP optimization software choices suitable for the offline NRMPC optimization task in Sect. B.3.3.

10.3.1 SDP Problem Formulation and Solution

Generally speaking, a semidefinite optimization problem minimizes or maximizes a linear objective function, subject to linear equality or inequality constraints on the symmetric variable matrix $\mathbf{Z}$ and positive semidefinite constraint on $\mathbf{Z}$. These inequality constraints are known as linear matrix inequalities. The objective function in NRMPC is defined as the maximization of the volume of an invariant hyper-ellipsoid projection and intersection, constrained by the feasibility and invariance conditions. This is in fact a semidefinite programming problem and has to be solved utilizing a suitable SDP solver. In addition to the application of SDP in control theory, it is also used in combinatorial optimization, operations research, quantum chemistry and financial applications. Many problems may be modeled or approximated as a semidefinite programming problem. It is a relatively recent field and subject of active research. The general mathematical definition of the SDP can be summarized as follows [48]:

$$\text{SDP}: \begin{cases} \text{minimize} \sum_{i=1}^{m} \mathbf{G}_i \mathbf{w}_i & \\ \text{s.t.} \quad \mathbf{Z} = \sum_{i=1}^{m} \mathbf{A}_{\mathbf{e}i} \mathbf{w}_i - \mathbf{b}_{\mathbf{e}} \\ \qquad\quad \mathbf{Z} \geq 0 \end{cases} \tag{10.1}$$

where $\sum_{i=1}^{m} \mathbf{G}_i \mathbf{w}_i$ is the linear cost function of a variable $\mathbf{w}_i$ in general form and $\mathbf{Z} = \sum_{i=1}^{m} \mathbf{A}_{\mathbf{e}i} \mathbf{w}_i - \mathbf{b}_{\mathbf{c}}$ expresses optimization constraints in a linear matrix inequality form with the additional requirement that $\mathbf{Z}$ is to be positive semidefinite [71].

The semidefinite programming problem formulated by offline NRMPC optimization requirements is passed to the solver through YALMIP toolbox [40]. YALMIP is a freely available and distributed modeling language for defining and solving optimization problems in Matlab. Although most of the SDP solvers are also directly callable in Matlab, the main advantage of YALMIP is its ease of use. The objective function and constraints defined by LMI can be set up in a syntax and formulation conventions similar to that normally encountered in Matlab. YALMIP implements many modeling tricks that allow the user to concentrate on high-level modeling instead of taking the time to set up intricate problem definitions for a given solver [41]. For the interest of rapid code development and easy problem reconfiguration, the YALMIP interface has been considered throughout this work.

Although invoking solvers through a parser decreases computational efficiency, the scale of problems considering the usual complexity of NRMPC allows for the use of YALMIP without noticeable increase in evaluation time. However, in case the fast reconfiguration of the offline algorithm is not necessary anymore, parsing the problem directly to the SDP solver is a more desirable option. Furthermore, a possible adaptive implementation of the NRMPC algorithm, or a version where the constraints are to be re-set and changed on the fly, would benefit from an offline process and solver callable directly through as a C language routine.[10]

Several SDP solvers were evaluated as candidates for obtaining a solution for the optimization problem in the offline NRMPC algorithm. This work utilized YALMIP for problem formulation: thus, only solvers compatible with this parser were considered. Program efficiency was not an issue because of the problem scale; therefore, the solvers were not selected according to this criterion. Optimality of the solution was of much greater importance. Certain solvers were incompatible with the given Matlab release, while others encountered unknown errors or crashed during optimization. No exact measures of optimality have been used, since with certain optimization software the provided solution was so far from optimal that even constraints were violated at the online simulation stage. According to the analysis given in Sect. B.3.3, finally SeDuMi by Sturm et al. [59] has been chosen as the ideal candidate for the NRMPC offline optimization task. SeDuMi is a second order cone and semidefinite programming solver commonly used with the YALMIP parser. The development of SeDuMi is ongoing with many platform versions and a freely distributed source. The tests of the offline NRMPC code performed with SeDuMi showed no serious numerical issues and the resulting controller parameters were suitable for the real-time code implementation. For those interested, the offline problem formulation is explored in detail in Sect. B.3. One may find information on the use of Yalmip and SeDuMi in [40, 41, 53, 59, 60].

10.3.2 Cost Transformation

The cost to be minimized in the online NRMPC run has been expressed in (8.23) as $J_{\mathrm{NRMPC}} = \mathbf{f}^T\mathbf{f}$ which can be only true in the case the augmented states are transformed to make the cost equivalent with the original MPC formulation. To express this mathematically in a simple way, one may state [8]:

[10] See 12.7.2.6 for details.

$$J_\infty = J_{\mathrm{NRMPC}} = \sum_{i=0}^{\infty}\left(x_{k+i}^T \mathbf{Q} x_{k+i} + u_{k+i}^T \mathbf{R} u_{k+i}\right) \tag{10.2}$$
$$= \mathbf{z}_k^T \mathbf{L} \mathbf{z}_k$$
$$= \begin{bmatrix} x_k \\ \mathbf{f}_k \end{bmatrix}^T \begin{bmatrix} \mathbf{L}_x & \mathbf{L}_{xc} \\ \mathbf{L}_{cx} & \mathbf{L}_c \end{bmatrix} \begin{bmatrix} x_k \\ \mathbf{f}_k \end{bmatrix}$$

where x_k is a vector of states u_k[11] is generally a vector of inputs, $\mathbf{Q}$ is a state penalization matrix, $\mathbf{R}$ is an input penalization matrix, $\mathbf{z}_k$ is augmented state $\mathbf{f}_k$ is a perturbation vector and $\mathbf{L}$ is the cost transformation matrix.

In case shift matrix $\mathbf{C}_0$, penalization $\mathbf{Q}$, $\mathbf{R}$ and the closed-loop fixed feedback $\mathbf{K}$ are known, the transformation matrix $\mathbf{L}$ can be calculated as follows:

$$\mathbf{L} - \Psi^T \mathbf{L} \Psi^T = \begin{bmatrix} \mathbf{Q} & 0 \\ 0 & 0 \end{bmatrix} + \begin{bmatrix} \mathbf{K}^T \\ \mathbf{C}_0^T \end{bmatrix} \begin{bmatrix} \mathbf{L}_x & \mathbf{L}_{xc} \\ \mathbf{L}_{cx} & \mathbf{L}_c \end{bmatrix} \begin{bmatrix} \mathbf{K} & \mathbf{C}_0 \end{bmatrix} \tag{10.3}$$

where Ψ is the closed-loop matrix of the augmented system. Individual partitions $\mathbf{L}_x$ and $\mathbf{L}_c$ of the cost matrix may be computed using the following relations:

$$\mathbf{L}_x - \Phi^T \mathbf{L}_x \Phi = \mathbf{Q} + \mathbf{K}^T \mathbf{R} \mathbf{K} \tag{10.4}$$

where Φ is the closed-loop matrix $\Phi = (\mathbf{A} + \mathbf{B}\mathbf{K})$ and

$$\mathbf{L}_c - \mathbf{A}_0^T \mathbf{L}_c \mathbf{A}_0 = \mathbf{C}_0^T \mathbf{B}^T \mathbf{L}_x \mathbf{B} \mathbf{C}_0 + \mathbf{C}_0^T \mathbf{R} \mathbf{C}_0 \tag{10.5}$$

The previous two Lyapunov equations are solved successively in the offline code, to obtain the partitions of the cost transformation matrix.

NRMPC formulation expects to minimize the function $\mathbf{f}^T\mathbf{f}$ at each sampling instant with respect to constraints. This simple minimization criterion is essential in formulating the efficient structure of this algorithm. In order to minimize an equivalent transformed cost and still having the same simple function, the augmented states $\mathbf{z}_k$ have to be transformed in the online optimization task. This can be expressed by the following equation:

$$\bar{\mathbf{z}}_k = \begin{bmatrix} \mathbf{I} & 0 \\ 0 & \mathbf{L}_c^{-1/2} \end{bmatrix} \mathbf{z}_k \tag{10.6}$$

where $\bar{\mathbf{z}}_k$ is the transformed augmented state vector and $\mathbf{I}$ is an identity matrix of conforming dimensions.

To prevent numerical problems occurring at the stage of creating a negative matrix square root of $\mathbf{L}_c$, as defined in (10.6) an alternative approach is utilized in practice [8]. An eigenvalue–eigenvector decomposition of the cost transformation partition $\mathbf{L}_c$ is

---

[11] Remember that these are in general vectors, the boldface type is reserved for the vector of predicted inputs $\mathbf{u}_k$ to simplify notation.

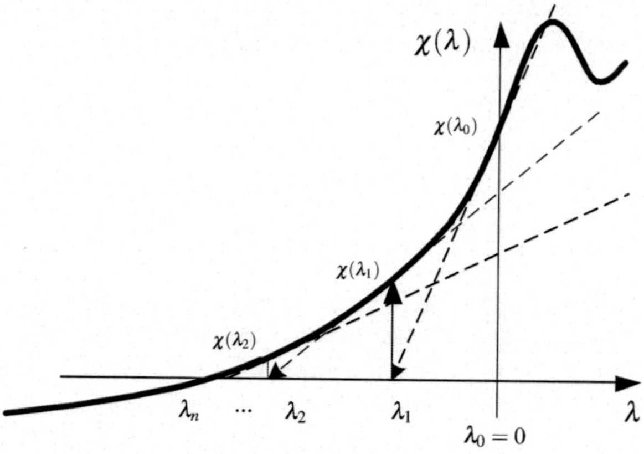

Fig. 10.10 Illustration of the Newton–Raphson procedure, deployed to find the root of the function $\chi(\lambda)$

computed, and the maximum diagonal element is selected.[12] The final transformation matrix is calculated according to the procedure described in Sect. B.3.1.5.

10.3.3 The Newton–Raphson Procedure

Compared to the classical quadratic programming approach, the online optimization task performed in NRMPC is significantly simpler. In practice one can simplify the performed operations further, thus saving as much time as possible. The first problem is to find λ according to Eq. (8.26). This is actually the part when one needs to use the Newton–Raphson procedure. The underlying concept is very simple and in case of the NRMPC is [1, 9, 19]:

$$\frac{d\chi(\lambda)}{d\lambda} = \frac{0 - \chi(\lambda)}{\lambda_{n-1} - \lambda_n} \tag{10.7}$$

The procedure itself is also trivial and is represented by the following algorithm:

Algorithm 10.1 At each sampling instant initialize with $\lambda = 0$ and perform an iteration which calculates:

$$\lambda_{n+1} = \lambda_n - \frac{\chi(\lambda)}{d\chi(\lambda)d\lambda} \tag{10.8}$$

until the change in λ is smaller than the pre-determined tolerance.

---

12 There is an additional tolerance value. **L** is supposed to be positive definite, although sometimes may contain negative elements too.

Subscript n denotes the iterations of the Newton–Raphson procedure and the process itself is illustrated in Fig. 10.10. According to this geometric representation at initialization time the algorithm computes a tangent line to the function $\chi(\lambda)$ and the intersection of this tangent line with the λ axis. This is the next iteration point where the process is repeated. The algorithm will eventually close in on the root of the function and terminate.

One may take advantage of expressing the matrix Γ_f as an eigenvalue/eigenvector decomposition. Fortunately, in addition the decomposition is also valid for the powers and inverse of Γ_f [29]:

$$\Gamma_f^i = \Delta \Lambda^i \Delta^T \tag{10.9}$$

where Δ, Λ is defined by the eigenvalue–eigenvector decomposition of Γ_f and i is the i-th power or inverse. Equation (8.26) contains the expression $\Theta = (\mathbf{I} - \lambda \Gamma_f^{-1})$. Using the decomposition (10.9) we may denote this as [29]:

$$\Theta = \Delta \text{diag}(1./(1 - \lambda \mathbf{S}_v))\Delta^T \tag{10.10}$$

where "*diag*" denotes a diagonalization operation, $\mathbf{S}_v$ is a vector of eigenvalues gained from Λ and ./ denotes the piecewise division operation. This is a very useful feature for the online computation since it substitutes the inversion of the full matrix expression $(\mathbf{I} - \lambda \Gamma_f^{-1})$. Furthermore, this expression occurs not only in evaluating the perturbation vector $\mathbf{f}_k$, but also in the function $\chi(\lambda)$ and its first derivative. Let $\mathbf{m}_v = (1 - \lambda \mathbf{S}_v)$ and $\mathbf{m}^i = \text{diag}(1./\mathbf{m}_v^{(-i)})$, then we may express the function (8.26) and its derivative as:

$$\chi(\lambda) = x_k^T \mathbf{W}_1 \mathbf{m}^2 \mathbf{W}_2 \mathbf{x}_k + \mathbf{W}_3 \tag{10.11}$$

$$\frac{d\chi(\lambda)}{d\lambda} = 2x_k^T \mathbf{W}_4 \mathbf{m}^3 \mathbf{W}_2 \mathbf{x}_k \tag{10.12}$$

Equation (8.25) yielding the perturbation vector $\mathbf{f}_k$ will successively transform to

$$\mathbf{f} = \lambda \Delta \mathbf{m}^1 \mathbf{W}_4 x_k \tag{10.13}$$

Matrices $\mathbf{W}_1$,$\mathbf{W}_2$,$\mathbf{W}_3$ can be calculated offline, therefore saving some time avoiding unnecessary multiplications at every NR iteration and sample time. Matrix $\mathbf{W}_4$ can be calculated before the NR process for the actual sample time initiates:

$$\begin{aligned}
\mathbf{W}_1 &= \Gamma_{xf} \Delta \\
\mathbf{W}_2 &= \mathbf{W}_1^T \\
\mathbf{W}_3 &= \mathbf{W}_1 \Lambda^{-1} \\
\mathbf{W}_4 &= -x_k^T \mathbf{W}_3 \mathbf{W}_2 + x_k^T \Gamma_x x_k - 1
\end{aligned}$$

When initialized with $\lambda = 0$ the algorithm shall converge after a finite number of iterations. Just to be on the safe side, one shall implement a safety feature, which will detect too many iterations, or iterations that would result in exceeding the sample time. In this case, the calculated control input should somehow make up for the error. One may take for example the control input calculated at the previous instant, or the prediction for the actual time step. Another choice would be the use of a suboptimal λ. Further details on the Newton–Raphson operation can be found in Sect. B.3.2.3.

10.3.4 NRMPC Extension

To search outside the augmented invariant ellipsoid described in Sect. 8.12, we have to perform the optimization problem described by (8.30) while still fulfilling the input constraints. If after evaluation of the quadratic equation (8.30), the determinant is negative, the optimization failed and $\mu = 1$. The perturbation values shall remain unchanged. In case the equation yields two values, the lower one μ_L and the higher one μ_H there are two possibilities for a failed optimization:

- If $\mu_H < 0$ then the lower one is also negative. Optimization failed, set $\mu = 1$.
- If $\mu_L > 0$ then the higher value is also larger than one. Optimization failed, set $\mu = 1$.

In case there are two real roots of which $\mu_L \in <0, 1>$, then it will be true that $\mu_H > 1$. The possibility that μ_H is between μ_L and 1 is outruled, since the quadratic equation already defines that the problem shall be feasible at $k + 1$, therefore $\mu = 1$ in the worst case. The optimal $\mu^* = 1$ has to be from the interval $<\mu_L, 1>$. To fulfill the symmetric input constraints we evaluate

$$\bar{\mu}_L = -\left(\frac{u}{|\mathbf{Ef}|} + \frac{\mathbf{K}x}{\mathbf{Ef}}\right) \tag{10.14}$$

Out of all the values $\bar{\mu}_L$ we select the maximal one to get the desired optimal μ^*.

This extension has not been utilized in the simulations or the experiments featured in this book. The increased computational complexity is not justified by the negligible performance gain for this particular application.

10.3.5 Code Implementation

The offline part of the NRMPC controller has been implemented using the Matlab high-level scripting language, *m-scripts*. Controller structure is obtained by evaluating linear matrix inequalities defined by the invariance and feasibility conditions. LMI are parsed using YALMIP [40] to the SeDuMi [59] optimization software, capable of evaluating the semidefinite programming problem defined in the offline NRMPC algorithm.

Matlab scripting environment has been chosen for the offline code because of its simplicity, fast reconfiguration and the possibility of concentrating on the high-level problem instead of programming details. This is sufficient for a laboratory setup, however in a conceivable practical application tuning parameters and constraints need to be often reconfigured on site. A version of the offline NRMPC code applicable to mass produced hardware or even suitable for adaptive applications is not considered in this book.

As in the case of the QPMPC and MPMPC controllers, the real-time implementation of the online NRMPC algorithm is realized in Matlab Simulink, utilizing software tools provided by the Real Time Workshop toolbox and custom coding. In addition to the S-Function containing the C code of the NRMCP algorithm, the controller scheme contains essential blocks for input/output and other functions. The real-time control application is running on a dedicated machine under the xPC Target kernel.

Fortunately, the online part of the NRMPC controller is simple and does not even contain full matrix inversions, only essential piecewise inversions. This is not only desirable because of computational efficiency, but it also made the implementation straightforward. Not much more is required to transcribe the algorithm than basic matrix algebra, program decision points, cycles and other fundamental operations. The code is introduced in Sect. B.3.2 for those interested in a detailed explanation of the implementation steps.

The online part of the controller code requires matrix–matrix, matrix–vector and other mathematical operations not implemented as default in the C programming environment. Although it would be possible to create custom written subprograms for these procedures in C, readily available universal solutions exist. Basic linear algebra subprograms (BLAS) can be used as basic programming units in algorithms requiring mathematical processing [12]. BLAS operations may be divided into different levels:

- Level 1—scalar, vector and vector-vector operations
- Level 2—matrix vector operations
- Level 3—matrix-matrix operations

The implementation of the NRMPC controller requires all of the above. BLAS routines may be divided also according to required precision levels to single, double, complex and double complex. There are also different routines for the same mathematical task, making use of the properties of matrices—for example simmetricity. These properties help to maximize computational efficiency of these operations.

In addition to the fact that there is no need to make custom subprograms for each matrix algebra operation, these packages have been developed and perfected over the years by researchers in academia and industry.[13] Another very important aspect of this collection of algorithms is that they are available optimized for platform specific environments. Shared memory and multi-threading versions are also accessible.

[13] Matlab is in fact based on BLAS and LAPACK.

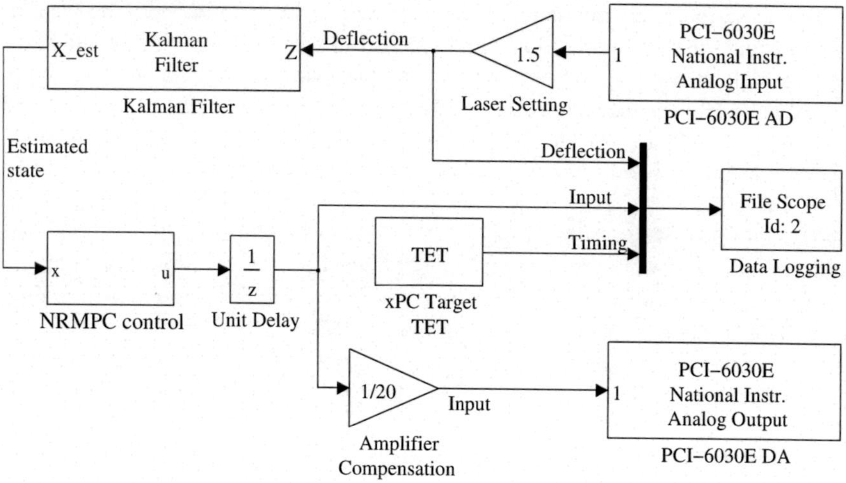

Fig. 10.11 The block scheme of the NRMPC controller in Simulink. Except the controller itself, this block is identical to the ones presented earlier

The BLAS package has been originally developed in the FORTRAN77 programming language. Although a C interface is available for BLAS [5], it is possible to call FORTRAN77 subroutines directly from C. This is utilized in the online implementation of the NRMPC algorithm. Pre-compiled versions of the BLAS package are available online for different platforms and BLAS is included in distributions of Matlab. In fact, all the essential algebraic operations are performed via BLAS in Matlab: the command multiplying two matrices is translated into the proper BLAS routine for full efficiency. All mathematical operations in the online code have been reviewed, unnecessary operations merged or concatenated to further increase efficiency.

The NRMPC controller code is implemented within a custom S-Function block. The feedback control system implemented to the AVC demonstrator is in other aspects identical to the one used with the QPMPC or MPMPC controllers. The controller loop is shown in Fig. 10.11. The estimated state, conforming to the dimensions of model considered when creating the offline NRMPC controller, is passed onto a custom S-Function block implementing the online NRMPC code in C language. This block takes Θ, $\mathbf{T}$, $\mathbf{S}_m$, Γ_{xf}, Γ_x, $\mathbf{K}$ and $\mathbf{C}_0$ as its working parameters in addition to the observed state. All of these parameters are the result of running the Matlab script responsible for initialization and offline NRMPC algorithm evaluation. The S-Function block has the core C program and two custom functions specified as modules. The C program is contained within this S-Function, which has been created using the S-Function Builder as shown in Fig. 10.12. In addition to the specified parameters including their data type, input and output ports are declared. Here the input is the estimated state, and output the scalar controller variable. Discrete sampling time depends on the application. The output of this block is also passed onto monitoring and data logging blocks. Obviously, the controller output is connected to the physical

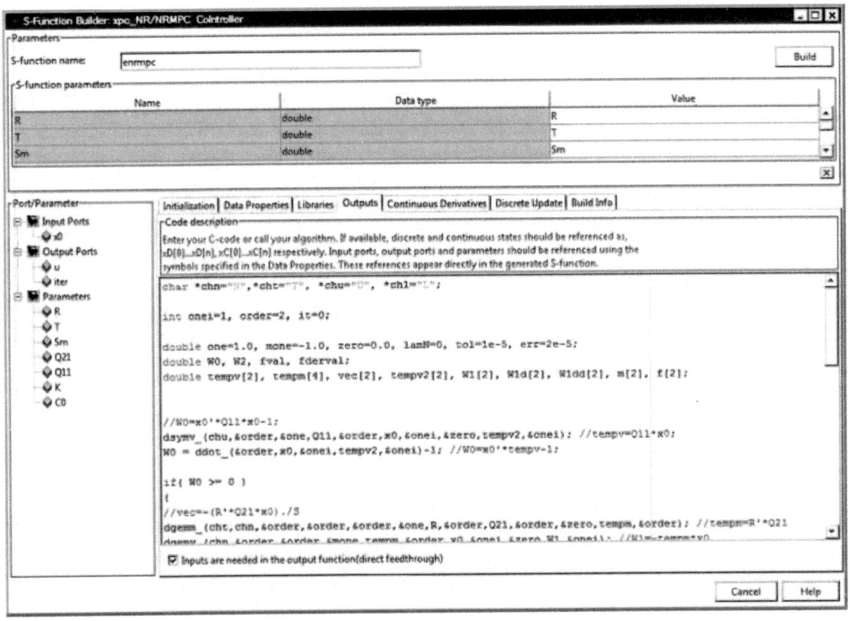

Fig. 10.12 Screenshot from the S-Function Builder implementing the online NRMPC code. The figure features a tab showing a portion of the output code in C

analog voltage output of the measuring card. The NRMPC block provides voltage output directly, therefore has to be compensated by the amplifier gain. In this case, this is a 20× decrease.

References

1. Anstee R (2006) The Newton–Raphson Method. Lecture notes. Whitepaper, The University of British Columbia, Vancouver. http://www.math.ubc.ca/~anstee/math104/104newtonmethod.pdf
2. Arnold E, Neupert J, Sawodny O, Schneider K (2007) Trajectory tracking for boom cranes based on nonlinear control and optimal trajectory generation. In: IEEE international conference on control applications CCA, pp 1444–1449. doi:10.1109/CCA.2007.4389439
3. Bai M, Ou KY (2002) Experimental evaluation of adaptive predictive control for rotor vibration suppression. IEEE Trans Control Syst Technol 10(6):895–901. doi: 10.1109/TCST.2002.804124
4. Baocang D (2009) Modern predictive control, 1st edn. Chapman and Hall / CRC, Boca Raton
5. Basic Linear Algebra Subprograms Technical (BLAST) Forum (2001) Basic Linear Algebra Subprograms Technical (BLAST) Forum Standard. Available: http://www.netlib.org/blas/blast-forum/blas-report.pdf
6. Boscariol P, Gasparetto A, Zanotto V (2010) Model predictive control of a flexible links mechanism. J Intell Rob Syst 58:125–147. doi: 10.1007/s10846-009-9347-5

7. Camacho EF, Bordons C (2007) Model predictive control. 2nd edn. Springer, London
8. Cannon M, Kouvaritakis B (2005) Optimizing prediction dynamics for robust MPC. IEEE Trans Autom Control 50(11):1892–1897. doi:10.1109/TAC.2005.858679
9. Chapra SC (2004) Applied numerical methods with MATLAB for engineers and scientists. 1st edn. McGraw-Hill Science, New York
10. Chen H, Allgöver F (1998) A quasi-infinite horizon nonlinear model predictive control scheme with guaranteed stability. Automatica 34(10):1205–1217
11. Cychowski M, Szabat K (2010) Efficient real-time model predictive control of the drive system with elastic transmission. IET Control Theory Appl 4(1):37–49. doi: 10.1049/iet-cta.2008.0358
12. Dongarra J (2002) Basic linear algebra subprograms technical forum standard. Int J High Perform Appl Supercomput 16:1–111
13. Eure KW (1998) Adaptive predictive feedback techniques for vibration control. Doctoral dissertation, Virginia Polytechnic Institute and State University, Blacksburg
14. Ferreau HJ (2006) An online active set strategy for fast solution of parametric quadratic programs with applications to predictive engine control. Master thesis, University of Heidelberg
15. Ferreau HJ (2011) qpOASES—online active set strategy. Leuven. Available http://www.qpoases.org
16. Ferreau HJ (2011) qpOASES User's Manual. Optimization in engineering center (OPTEC) and department of electrical engineering, K. U. Leuven
17. Ferreau HJ, Ortner P, Langthaler P, del Re L, Diehl M (2007) Predictive control of a real-world diesel engine using an extended online active set strategy. Annu Rev Control 31(2):293–301. doi:10.1016/j.arcontrol.2007.09.001, http://www.sciencedirect.com/science/article/B6V0H-4R05C2B-2/2/23db757b09f804365ba12dc1844992d9
18. Ferreau HJ, Bock HG, Diehl M (2008) An online active set strategy to overcome the limitations of explicit MPC. Int J Robust Nonlinear Control 18(8):816–830
19. Fletcher R (2000) Practical methods of optimization. Wiley, New York
20. Geyer T, Torrisi FD, Morari M (2008) Optimal complexity reduction of polyhedral piecewise affine systems. Automatica 44(7):1728–1740. doi:10.1016/j.automatica.2007.11.027, http://www.sciencedirect.com/science/article/pii/S00051098070 04906
21. Hassan M, Dubay R, Li C, Wang R (2007) Active vibration control of a flexible one-link manipulator using a multivariable predictive controller. Mechatronics 17(1):311–323
22. Huang K, Yu F, Zhang Y (2010) Model predictive controller design for a developed electromagnetic suspension actuator based on experimental data. In: WASE international conference on information engineering (ICIE) 4:152–156. doi:10.1109/ICIE.2010.327
23. Woo HK, Soo HH (2005) Receding horizon control: model predictive control for state models. 1st edn. Springer, London
24. Imsland L, Rossiter JA (2005) Time varying terminal control. In: 16th IFAC world congress in Prague, p TP/19
25. Imsland L, Bar N, Foss BA (2005) More efficient predictive control. Automatica 41(8):1395–1403. doi:10.1016/j.automatica.2005.03.010, http://www.sciencedirect.com/science/article/B6V21-4G7NT35-1/2/52a9590bfe1ccc2a9561165c3fbdf872
26. Inman DJ (2007) Engineering vibrations, 3rd edn. Prentice Hall, NJ
27. Kok J, van Heck J, Huisman R, Muijderman J, Veldpaus F (1997) Active and semi-active control of suspension systems for commercial vehicles based on preview. In: Proceedings of the 1997 American control conference 5:2992–2996. doi:10.1109/ACC.1997.612006
28. Kouvaritakis B, Rossiter J, Schuurmans J (2000) Efficient robust predictive control. IEEE Trans Autom Control 45(8):1545–1549. doi:10.1109/9.871769
29. Kouvaritakis B, Cannon M, Rossiter J (2002) Who needs QP for linear MPC anyway?. Automatica 38:879–884. doi:10.1016/S0005-1098(01)00263-1, http://www.sciencedirect.com/science/article/pii/S0005109801002631
30. Kouvaritakis B, Li S, Cannon M (2010) A line search improvement of efficient MPC. Automatica 46(11):1920–1924. doi:10.1016/j.automatica.2010.07.003, http://www.sciencedirect.com/science/article/B6V21-50NH0BX-3/2/0b4491d922a7d04d1b0315edae0e8944

31. Kvasnica M (2009) Real-time model predictive control via multi-parametric programming: Theory and tools, 1st edn. VDM Verlag, Saarbrücken
32. Kvasnica M (2011) Selected Topics on constrained and nonlinear control. workbook, STU Bratislava—NTNU Trondheim, Chap Multi-Parametric Toolbox, pp. 101–170
33. Kvasnica M, Grieder P, Baotić M (2004) Multi-Parametric Toolbox (MPT). Online, Available: http://control.ee.ethz.ch/
34. Kvasnica M, Grieder P, Baotic M, Morari M. (2004) Multi-parametric toolbox (MPT). In: Alur R, Pappas GJ (eds) Hybrid systems: computation and control, Lecture notes in computer science, Springer Berlin / Heidelberg, 2993:121–124
35. Kvasnica M, Grieder P, Baotic M, Christophersen FJ (2006) Multi-parametric toolbox (MPT). Extended documentation, Zürich, http://control.ee.ethz.ch/~mpt/docs/
36. Kvasnica M, Rauová I, Fikar M (2010) Automatic code generation for real-time implementation of model predictive control. In: Proceedings of the 2010 IEEE international symposium on computer-aided control system design, Yokohama pp 993–998
37. Kvasnica M, Fikar M, Čirka Ľ, Herceg M (2011) Selected Topics on constrained and nonlinear control. Textbook, STU Bratislava—NTNU Trondheim, Chap Complexity reduction in explicit Model predictive control, pp 241–288
38. Kvasnica M, Rauová I, Fikar M (2011) Real-time implementation of model predictive control using automatic code generation. In: Selected topics on constrained and nonlinear control. Preprints, STU Bratislava—NTNU Trondheim, pp 311–316
39. Li S, Kouvaritakis B, Cannon M (2010) Improvements in the efficiency of linear MPC. Automatica 46(1):226–229. doi:10.1016/j.automatica.2009.10.010, http://www.sciencedirect.com/science/article/B6V21-4XGCHXB-3/2/20a93fa6dd4fb88469638ac3bc2fe729
40. Lofberg J (2004) YALMIP: A toolbox for modeling and optimization in MATLAB. In: Proceedings of the CACSD conference, Taipei
41. Lofberg J (2009) YALMIP wiki. Available: http://control.ee.ethz.ch/~joloef/wiki/pmwiki.php?n=Main.HomePage
42. Maciejowski JM (2000) Predictive control with constraints, 1st edn. Prentice Hall, New Jersey
43. Mayne DQ, Rawlings JB, Rao CV, Scokaert POM (2000) Constrained model predictive control: Stability and optimality. Automatica 36(6):789–814
44. Mehra R, Amin J, Hedrick K, Osorio C, Gopalasamy S (1997) Active suspension using preview information and model predictive control. In: Proceedings of the 1997 IEEE international conference on control applications, pp 860–865, doi:10.1109/CCA.1997.627769
45. Moon SM, Clark RL, Cole DG (2005) The recursive generalized predictive feedback control: theory and experiments. J Sound Vib 279(1–2):171–199. doi:10.1016/j.jsv.2003.12.034, http://www.sciencedirect.com/science/article/B6WM3-4C005WR-2/2/4580a0865591eaa5ca1bf02e09dedcb7
46. Moon SM, Cole DG, Clark RL (2006) Real-time implementation of adaptive feedback and feedforward generalized predictive control algorithm. J Sound Vib 294(1–2):82–96. doi:10.1016/j.jsv.2005.10.017, http://www.sciencedirect.com/science/article/B6WM3-4HYMY76-1/2/50d98047187533ebe9d3ea8310446e77
47. Morger A (2010) Synthesis, design and control of a tendon-driven robot platform for vestibular stimulation during sleep. Master's thesis, ETH Zurich
48. Nesterov Y, Nemirovskii A (1994) Interior-point polynomial methods in convex programming, Studies in Applied Mathematics, vol 13. SIAM, Pennsylvania
49. Niederberger D (2005) Hybrid systems: computation and control, vol 3414, Springer / Heidelberg, Berlin, Design of optimal autonomous switching circuits to suppress mechanical Vibration, pp 511–525
50. Ping H, Ju Z (2008) Explicit model predictive control system and its application in active vibration control of mechanical system of elevator. In: Control and decision conference, CCDC . Chinese, pp 3738–3742. doi:10.1109/CCDC.2008.4598029
51. Pistikopoulos EN, Georgiadis MC, Dua V (eds) (2007) Multi-parametric model-based control, vol 2, 1st edn. Wiley

52. Pistikopoulos EN, Georgiadis MC, Dua V (eds) (2007) Multi-parametric programming, vol 1, 1st edn. Wiley
53. Pólik I (2005) Addendum to the SeDuMi user guide version 1.1. Technical report, McMaster University, Advanced optimization lab, Hamilton, Ontario. Available: http://sedumi.ie.lehigh.edu/
54. Polóni T, Takács G, Kvasnica M, Rohal'-Ilkiv B (2009) System identification and explicit control of cantilever lateral vibrations. In: Proceedings of the 17th international conference on process control, Štrbské Pleso
55. Rauter G, Zitzewitz JV, Duschau-Wicke A, Vallery H, Riener R (2010) A tendon-based parallel robot applied to motor learning in sports. In: Proceedings of the IEEE international conference on biomedical robotics and biomechatronics
56. Richelot J, Bordeneuve-Guibe J, Pommier-Budinger V (2004) Active control of a clamped beam equipped with piezoelectric actuator and sensor using generalized predictive control. In: 2004 IEEE international symposium on industrial electronics, vol. 1, pp 583–588 vol. 1. doi:10.1109/ISIE.2004.1571872
57. Rossiter JA (2003) Model-based predictive control: a practical approach, 1st edn. CRC Press, Boca Raton
58. Shi P, Liu B, Hou D (2008) Torsional vibration suppression of drive system based on DMC method. In: 7th World Congress on intelligent control and automation. WCICA, pp 4789–4792. doi:10.1109/WCICA.2008.4593699
59. Sturm JF (1999) Using SeDuMi 1.02, a MATLAB toolbox for optimization over symmetric cones. Optimization methods and software—special issue on interior point methods 11–12:625–653
60. Sturm JF (2001) SeDuMi 1.05 R5 user's guide. Technical report, department of economics, Tilburg university, Tilburg, Available: http://sedumi.ie.lehigh.edu/
61. Takács G (2008) Efficient model predictive control applied on active vibration control. Written report for PhD qualification exam, Slovak Technical University, Faculty of mechanical engineering, Bratislava
62. Takács G, Rohal'-Ilkiv B (2008) Model predictive control in vibration attenuation. In: Proceedings of the 2nd international conference education research and innovation, Bratislava
63. Takács G, Rohal'-Ilkiv B (2009) Implementation of the Newton-Raphson MPC algorithm in active vibration control applications. In: Mace BR, Ferguson NS, Rustighi E (eds) Proceedings of the 3rd international conference on noise and vibration: emerging methods, Oxford
64. Takács G, Rohal'-Ilkiv B (2009) MPC with guaranteed stability and constraint feasibility on flexible vibrating active structures: a comparative study. In: Hu H (ed) Proceedings of the 11th IASTED international conference on control and applications, Cambridge
65. Takács G, Rohal'-Ilkiv B (2009) Newton-Raphson based efficient model predictive control applied on active vibrating structures. In: Proceedings of the European control conference, Budapest
66. Takács G, Rohal'-Ilkiv B (2009) Newton–Raphson MPC controlled active vibration attenuation. In: Hangos KM (ed) Proceedings of the 28th IASTED international conference on modeling, identification and control, Innsbruck
67. The Mathworks (2007) Matlab signal processing blockset v6.6 (R2007b). Software. The MathWorks Inc., Natick
68. Tøndel P, Johansen TA, Bemporad A (2003) Evaluation of piecewise affine control via binary search tree. Automatica 39(5):945–950. doi:10.1016/S0005-1098(02)00308-4, http://www.sciencedirect.com/science/article/pii/S00051098020 03084
69. Van den Broeck L, Diehl M, Swevers J (2009) Time optimal MPC for mechatronic applications. In: Proceedings of the 48th IEEE conference on decision and control, Shangai, pp 8040–8045
70. Van den Broeck L, Swevers J, Diehl M (2009) Performant design of an input shaping prefilter via embedded optimization. In: Proceedings of the 2009 American control conference, St. Louis, pp 166–171

71. Vandenberghe L, Boyd S (1996) Semidefinite programming. SIAM Review 38(1):49–95
72. Wang L (2009) Model predictive control system design and implementation using MATLAB, 1st edn. Springer, London
73. Wills A, Bates D, Fleming A, Ninness B, Moheimani R (2005) Application of MPC to an active structure using sampling rates up to 25 kHz. In: 44th IEEE conference on decision and control, 2005 and 2005 European control conference. CDC-ECC '05, pp 3176–3181. doi:10.1109/CDC.2005.1582650
74. Wills AG, Bates D, Fleming AJ, Ninness B, Moheimani SOR (2008) Model predictive control applied to constraint handling in active noise and vibration control. IEEE Trans Control Syst Technol 16(1):3–12
75. Yan G, Sun B, Lü Y (2007) Semi-active model predictive control for 3rd generation benchmark problem using smart dampers. Earthquake engineering and engineering vibration 6:307–315. doi:10.1007/s11803-007-0645-2
76. Yim W (1996) Modified nonlinear predictive control of elastic manipulators. In: Proceedings of the 1996 IEEE international conference on robotics and automation, vol. 3, pp 2097–2102. doi:10.1109/ROBOT.1996.506180
77. Zhang QZ, Gan WS (2004) A model predictive algorithm for active noise control with online secondary path modelling. J Sound Vib 270(4–5):1056–1066. doi:10.1016/S0022-460X(03)00516-9, http://www.sciencedirect.com/science/article/B6WM3-49D6XFX-4/2/805d0549ca0e60339bb5e2c798de7264
78. Zmeu K, Shipitko E (2005) Predictive controller design with offline model learning for flexible beam control. In: Proceedings of the 2005 international conference on Physics and control, pp 345–350. doi:10.1109/PHYCON.2005.1514005

Chapter 11
Simulation Study of Model Predictive Vibration Control

This chapter presents the results of simulations performed with various model predictive control (MPC) strategies applied to the state-space model of the active vibration control (AVC) demonstration device, representing a class of lightly damped mechanical systems. The simulations have been aimed at evaluating the implementation properties of MPC algorithms and investigating the issues caused by the lightly damped nature of the controlled plant, numerical limitations, optimality problems and others.

The active vibration suppression of lightly damped flexible mechanical systems is a uniquely difficult control task, in case model predictive control is considered with stability and feasibility guarantees. As the numerical study in Sect. 11.1 implies, this is caused by the long horizons necessary to create a sufficiently large region of attraction for the control law. Using the state-space model of the AVC demonstrator, minimal necessary horizons are compared with the largest achievable deflections at the tip under stabilized infinite horizon dual-mode quadratic programming-based MPC (QPMPC). Because of the several hundred steps long prediction horizons—necessary to drive the system state from an initial state caused by a large deformation—the application of classical QPMPC on lightly damped mechanical systems with fast sampling is very unlikely. Although multi-parametric MPC is computationally more efficient than QPMPC, the long prediction horizons may prohibit practical implementation because of the intractable offline computation times. Amongst others, the time required to evaluate an MPMPC controller is related to the achieved horizon length in Sect. 11.2. The number of polyhedral regions in the controller and the size of the executable to be loaded onto the hardware running the online MPMPC routine is also evaluated here.

Simulation evaluation of the Newton–Raphson MPC (NRMPC) algorithm shed light on some serious numerical problems, which may occur in real-time implementations. Imprecision at the offline optimization stage caused violation of the invariance condition and by that indirectly the violation of system constraints as demonstrated in Sect. 11.3. In addition to exhibiting the nature of this difficulty, the effects of imposing a prediction performance bound are evaluated. As the simulation results

G. Takács and B. Rohal'-Ilkiv, *Model Predictive Vibration Control*,
DOI: 10.1007/978-1-4471-2333-0_11,

imply, invariance condition violation is clearly caused by numerical obstacles and can be partly remedied by increasing solver precision.

Suboptimality is a natural drawback of the NRMPC approach. However, it is well illustrated in Sect. 11.4 that simulations performed with higher order lightly damped state-space models are suboptimal beyond all expectations. Both the evolution of controller outputs in the time domain and state trajectories point out the deficiencies of NRMPC in comparison with optimal methods. As the trials indicated in Sect. 11.4 suggest, lowering controller output penalization may improve the situation slightly and allows the actuators to use their full potential. The use of a certain penalization value is rationalized for second order models of the vibration attenuation example in this work, and compared to the much lower settings suitable for higher order examples.

To see the optimality improvement promised by the discussion in Sect. 8.1.2.1, simulation trials were performed with the alternate NRMPC extensions. Closed-loop costs are compared for a simple example in Sect. 11.5, and despite their drawbacks, the several steps ahead extensions show significant improvement in process optimality for certain problem classes. The evolution of perturbation scaling variables acquired via evaluating for the future membership and feasibility conditions are also shown. Although promising for certain applications, the simulation carried out with a fourth order model of the vibrating system shows no significant improvement in closed-loop cost and performance. The use of alternative NRMPC extensions is not justified for the problem considered in this work.

Finally, Sect. 11.6 compares the vibration damping performance of quadratic programming, multi-parametric programming and based MPC strategies. The results of this trial show no surprising facts which could not have been deducted from the theoretical discussion. The NRMPC algorithm respects constraints and behaves as expected while the QPMPC- and MPMPC-based controllers provide a faster and better damped response to the same initial condition.

11.1 On the Horizon Length of Stable MPC

A simple second order prediction model for the experimental demonstration device described in Chap. 5 with a 100 Hz rate would be suitable to use with a quadratic programming based MPC controller. The optimization task is likely to be tractable, as higher sampling rates have already been successfully implemented for vibration control [71, 72]. Neither the model order, nor the sampling rate requires algorithmic or special hardware requirements. However, this is only true if constraints are considered—but no guarantees on stability or feasibility are given.

If system constraints are considered, the set of allowable states or initial conditions is generally limited to a subset of space given by those constraints—for example box constraints on the state. The use of an MPC formulation with an a priori stability guarantee further limits the size (volume) of this set of allowable states. This limited subset of state-space is called the region of attraction. If the disturbance takes the

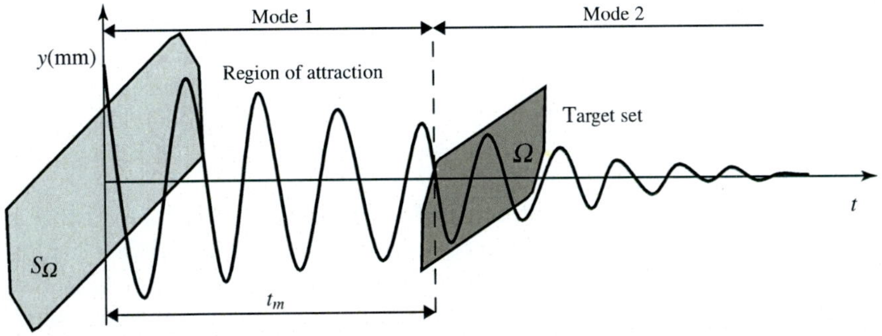

Fig. 11.1 Conceptual illustration of the time t_m required to enter the target set, when starting from a given initial condition in the region of attraction

system state outside this region, the optimization is simply infeasible. Therefore, the volume of the region of attraction shall be made as large as possible.

Amongst others, the size or volume of the region of attraction depends on the given MPC strategy. In case it is an optimal strategy such as dual-mode stabilized QPMPC or optimal MPMPC, the region is a polyhedral set in hyperspace. In case it is a suboptimal MPC strategy such as NRMPC, or target sets like the ones in Sect. 7.5 or 7.6 are used, the region of attraction is only a subset of the maximal possible region of attraction. In addition to the type of MPC strategy, the volume of the region of attraction is impacted by the choice of state and input penalization matrices **Q** and **R**, the character of the controlled system itself and the length of the prediction horizon.

The dynamic properties of the controlled system have a great influence on the relative size of the region of attraction and the target set. For certain systems with very light damping, such as the AVC demonstrator example, the effect of the actuators is very small compared to the range of disturbances one might reasonably expect. In other words, a large region of attraction has to be computed to allow the state to migrate inside the target set by the end of the prediction horizon. In this light we may alternatively state that the necessary horizon length for stabilized MPC control may be understood as the number of steps necessary to drive the system state from a given initial condition into the terminal set.

This concept is graphically illustrated in Fig. 11.1. The slowly decaying waveform suggests a lightly damped behavior of the beam tip or possibly other systems where the settling time is very long in comparison with sampling rate. Such systems have been discussed earlier in Sect. 5.1.1, where it has been suggested that a class of engineering problems such as manipulators [8, 19, 26, 42, 61, 63, 73, 74, 76], helicopter rotors [7, 40, 43], wing surfaces [1, 14, 15] and space structures [20, 39, 44, 51, 57] may have very similar dynamic behavior to the simple vibrating cantilever from the viewpoint of predictive control.

The initial condition of these systems has to be contained within the region of attraction, shown in a conceptional manner as the large polyhedron. There is a certain point in the course of behavior, where the state of the system enters the given terminal set—illustrated as the smaller polyhedron on the figure. The minimal necessary prediction horizon for stable MPC with constraint feasibility guarantees is then the time t_m in samples. Note that here the invariant sets are only for illustration purposes, and the concepts of state-space and control output course in the time domain are mixed purely for demonstration.

This indeed suggests that all systems where the effect of the control action is limited in comparison with the effort necessary to drive the trajectory into equilibrium will require lengthy control horizons. The displacement effect of piezoelectric actuators is small compared to the range they need to cover, when used for the vibration suppression of very flexible structures. This effect is not limited to piezoelectric actuators as active vibration attenuation through electrostrictive or magnetostrictive materials [4, 9, 49, 56, 69], electrochemical materials [2, 30, 46, 58] and possibly other actuators would create dynamic models with similar properties and issues for MPC implementation.

The number of samples required to settle the system from its initial condition gives an indication of extremely long control horizons. In the case of the laboratory demonstration model running saturated linear quadratic (LQ) control—the time necessary to reach near equilibrium from an initial deflection of 15 mm is approximately around 2 s. This divided by the sampling rate suggests a necessary minimal prediction horizon of 200 steps if dual-mode QPMPC with polytopic terminal sets is considered as a controller. The issue is the same with MPMPC, only instead of the online computational issues, the implementation difficulties would be transferred to the offline controller computation stage. The majority of the MPC implementations available in the literature considered no constraints, which neither limits the available state-space nor creates computational issues [16, 19, 50, 59, 75]. Other researchers considered constrained MPC that requires online optimization, thus creating certain implementation issues, however failed to address the question of stability [8, 13, 21, 27, 48, 71, 72]. Unlike these cases the authors of this book considered MPC implementations with stability guarantees in preliminary works such as in [64–67] and this will also be the assumption throughout the simulations and experiments presented in this and the upcoming chapter.

Let us get back to the case of vibrating systems and estimate the necessary prediction horizon based on the settling time of the system. If an exponential decay of vibration amplitudes is presumed due to the damping, we may easily approximate the prediction horizon necessary to ensure a feasible and stable MPC run. In case the system has been let to vibrate freely starting from a given initial condition, amplitudes d_t at a given time can be approximated by the following relation [6, 17, 23]:

$$d_t = d_0 e^{-\zeta \omega_n t} \tag{11.1}$$

where d_0 is the initial deflection, t is the time in seconds since the initial conditions had affected the system, ζ is the damping ratio and ω_n is the first or dominant natural frequency of the vibrating system. If MPC control of the system is presumed with guaranteed stability, there must be an amplitude level under which the system enters the target set. In case we denote this level with d_{ts}, then by utilizing relation (11.1), the minimal prediction horizon n_{min} for initial deflection d_0 and smaller can be approximated by:

$$n_{min} = \frac{-\lg(\frac{d_{ts}}{d_0})}{2\pi \zeta f_0 T_s} \tag{11.2}$$

where f_0 is the first or dominant mechanical eigenfrequency, T_s the sampling rate considered for control and the rest of the variables as defined for (11.1).

11.1.1 Simulating Necessary Horizon Lengths

Two simulations have been designed to determine the relationship between maximal allowable deflection at the beam tip and minimal associated prediction horizon. The first algorithm is based on the traditional dual-mode QPMPC formulation [3, 45, 60, 70], with stability and feasibility guarantees ensured by terminal constraints [12, 47]. The mathematical model used in the simulation and its settings corresponded to the physical laboratory device and the implementation has been carried out assuming the general workflow introduced earlier in Sect. 10.1.

The choice of method to resolve the minimal control horizon for a given deflection was to evaluate the online quadratic programming problem. In case the QP problem was infeasible at any of the steps, the horizon has been increased until a full successful run has been achieved. This has been repeated for increasing values of tip deflection and corresponding states. The summary of the algorithm goes as follows:

Algorithm 11.1

1. Increase initial deflection d and corresponding state x_k.
2. Perform the minimization $\min_{\mathbf{u}} J(\mathbf{u}_k, x_k)$ subject to input and terminal constraints.
3. If the minimization is infeasible, increase prediction horizon n_c and repeat algorithm from step 2.
4. Else note deflection value d and corresponding horizon n_c and repeat algorithm from step 1.

This method is inexact, though it is time efficient compared to the computationally more intensive search for the extreme points of the maximal admissible set given the expected horizon lengths.

An alternative algorithm to determine the minimal horizons to ensure a given beam tip deflection is based on utilizing successive computations of multi-parametric MPC controllers with varying deflection constraints. The multi-parametric controllers have been iteratively computed using the MPT Toolbox [31, 34, 35]. This method is slightly more computationally intensive than Algorithm 11.1, but also more exact:

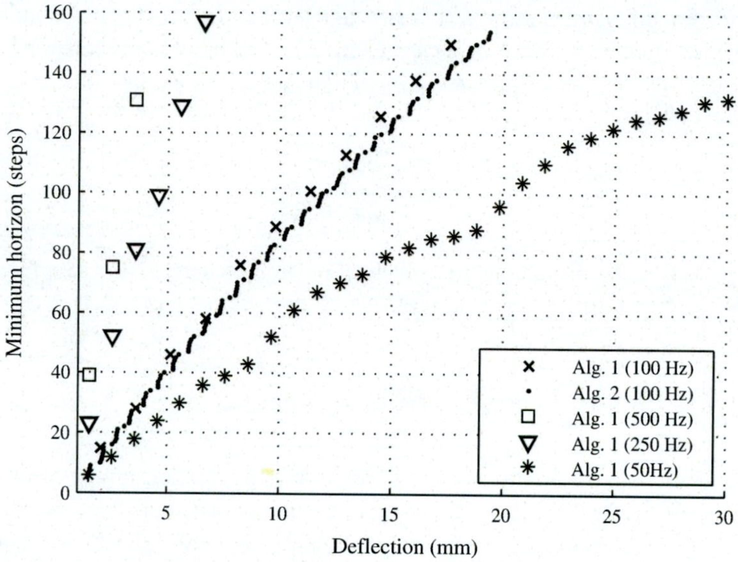

Fig. 11.2 Maximal allowable beam tip deflection with corresponding minimal stable dual-mode QPMPC horizon length for different model orders n_x

Algorithm 11.2

1. Increase initial deflection d and set deflection constraints to $\overline{y} = -\underline{y} = d$.
2. Compute minimum time multi-parametric MPC controller.
3. If the computed horizon n_c is equal to the one in the previous step, repeat from step 1.
4. Else denote prediction horizon n_c and the corresponding deflection d, then repeat from 1.

Results of simulating minimal prediction horizon lengths using the formerly introduced Algorithms 11.1 and 11.2 are featured in Fig. 11.2. As it is evident from this analysis, a fairly small deflection measured at the beam end requires a very long prediction horizon. For the second order prediction models with a 100 Hz sampling and a modest $y_{max} = 15$ mm maximal allowable deflection is approaching $n_c = 150$ steps. If one includes an excess range reserve in order to prevent infeasible states to enter the optimization procedure, this value is even higher. Considering a higher order prediction model allowing the control of faster dynamics, one requires a faster sampling rate. As an example we demonstrate that a 500 Hz sampled model of the beam would require approximately a $n_c = 500$ steps ahead prediction horizon to include states resulting from the same deflection.

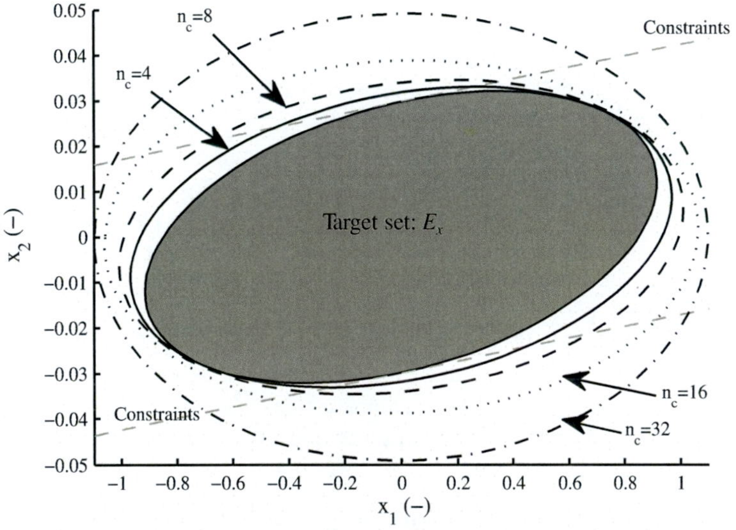

Fig. 11.3 Region of attraction in relation to horizon length, plotted for the NRMPC control law based on a second order model of the physical system. The innermost ellipse drawn with a *solid* line denotes $n_c = 4$ steps the *dash-dot* outer $n_c = 32$ steps ahead prediction

11.1.2 NRMPC and Horizon Length

The original development of the NRMPC algorithm presented by Kouvaritakis et al. in [28] and [29] assumes a fixed shift matrix **T** and **E**. The role of vector **E** is merely to select the first perturbation value from the vector of perturbations. This development of NRMPC assumes a fixed and pre set prediction horizon. Naturally, enlarging the horizon here implies a larger region of attraction, too.

While it is possible to use NRMPC in its original formulation on certain systems, this may not be the case with the given active vibration attenuation example and possibly other under-damped physical systems. The size of the region of attraction is very small for typical prediction horizons—in fact beyond the point of practical use. This is clearly illustrated in Fig. 11.3. A second order model of the physical system has been used to plot the region of attraction of the NRMPC law for different horizons. The intersection of the augmented ellipsoid with the state-space, the target set is shown as the shaded area.[1] The volume of the augmented set projection—which is actually the region of attraction is shown as the ellipse outlines with growing volume. The innermost ellipse shows projection assuming 4 steps ahead prediction. Horizon length was then increased by a factor of two, up to the value of 32 steps shown as the outermost ellipse. Even the relatively large 32 steps horizon would only allow a maximal initial beam tip deflection of ±0.5 mm, which is indeed unreasonably small.

[1] Simulations show that the intersection (target set) size shrinks with increasing prediction horizon, although in this case with a visually indistinguishable rate—thus not shown on the figure.

In the case of NRMPC, the region of attraction can be only the approximation of the polytopic set created by the exact QPMPC control law. The volume of the set of stabilizable states will be actually smaller than that of QPMPC. As it has been demonstrated in Sect. 11.1.1, very long horizons are needed to include the necessary range of deflections into the vibration control problem, which is also true for the NRMPC controller.

The volume of the region of attraction can be maximized by optimizing prediction dynamics as suggested by Cannon and Kouvaritakis in [11] and in a slightly different approach by Imsland et al. in [22]. In the method also introduced in Sect. 8.1.3 the matrices **T** and **E** will be full and unknown variables in the offline optimization problem. The horizon will be fixed and equal to the system order. However, this approach may cause numerical problems when calculating the ellipsoids acting as parameters for the online algorithm. This effect has been noted when using underdamped systems, such as the model for the experimental device.

11.2 Properties of MPMPC for Active Vibration Control

A promising stable and computationally efficient MPC control strategy for vibrating systems is the proven and actively researched MPMPC method [5], briefly introduced in Sect. 8.2.1. In MPMPC, the controller is formulated as a set of regions in the state-space with associated affine and fixed structures stored in a lookup table [53]. However, it has been suggested by the simulation results presented in the previous Sect. 11.1 that the size of the necessary region of attraction for this application may require very lengthy prediction horizons. As the computational burden of MPMPC is transferred into the offline computation of the controller regions, generally speaking systems with more than four states and horizons in the range of ten steps cannot be recommended for MPMPC. Although numerous efforts have been made to limit the size of the online lookup table and to shorten search times [24, 36], the offline computation effort required to realize the simplification procedures themselves may be prohibitive in some cases.

Narrow-band excitation and small discrepancy between the capabilities of actuators and the disturbance render MPMPC as a good controller choice. Nevertheless, one must not forget that lightly damped vibrating systems with weak static actuation are very different. MPMPC has been suggested to use for a vibrating cantilever by Polóni et al. without stability guarantees in [55] while this work was later expanded to include a priori stability guarantees in [65] shedding line on the inconvenient offline properties of MPMPC with lightly damped systems.

It has been long known that the main drawback of the MPMPC approach is its extensive offline computational need [31, 52]. In this section, several simulation trials are introduced using the MPMPC approach to evaluate the offline computational properties of the algorithm. Instead of using generic systems with "nice" dynamics we rather assume the characteristic dynamics of lightly damped underactuated vibrating systems. Since this work is focused on the application of efficient and stable MPC on systems of vibration suppression, practical questions like whether the offline calculation time is tractable using current computer hardware were analyzed. In addition to the very important issue of offline calculation times, this section investigates the number of regions and controller executable sizes necessary to deploy stable MPMPC on a real system subject to narrow-band excitation.

11.2.1 MPMPC Computation Time

As it has been implied in Sect. 11.1, in order to cover a broad range of deformations on a flexible structure, the computationally efficient MPMPC formulation will still require the same long horizons as QPMPC. While the online computational time is significantly reduced by the use of the MPMPC formulation, large horizons are prohibitive for other reasons. Several hundred steps long prediction horizons make the calculation of the controller intractable, especially with higher order models. Real life applications necessitate eventual system reconfiguration; changes in the constraints also call for repeated and extensive controller recomputation.

Simulations have been carried out to determine the properties of MPMPC control, using a second and fourth order state-space model of the laboratory device. Controller parameters and requirements have been identical to the QPMPC case presented in Sect. 11.1. Most importantly the region of attraction of MPMPC matched that of the QPMPC set. In other words, the controllers were required to cover the same range and produce the same online response, however with a different method of algorithmic implementation.

Explicit controllers with growing prediction horizon were evaluated using the MPT Toolbox [31–33], on a personal computer conforming to current hardware configuration standards.[2] The maximal prediction horizon that could be reliably computed for a second order system in the offline procedure was 162 steps. Over this horizon, the solver ran into problems, most likely due to memory issues. The offline computational times required to perform various tasks in MPMPC controller evaluation are shown in Fig. 11.4. A real-time implementation necessitates the calculation of the controller and its compilation from source code; therefore, the minimal realization time is the sum of these. Using extrapolation from the simulation data, the time required to evaluate an explicit controller allowing maximal deflections would take approximately 7 days.

Figure 11.4 also shows the time necessary to perform two non-essential tasks aimed at complexity reduction in MPMPC. One of them is merging, used to simplify the controller in order to reduce the number of regions, thereby lowering file size and search times. This merging procedure is implemented in the MPC Toolbox [38] based on the *optimal region merging* method of Geyer et al. [18]. It essentially involves a complexity reducing procedure, where the regions defining the same control law are

---

[2] AMD Athlon X2 DualCore 4400+ @ 2.00 GHz, 2.93 GB of RAM.

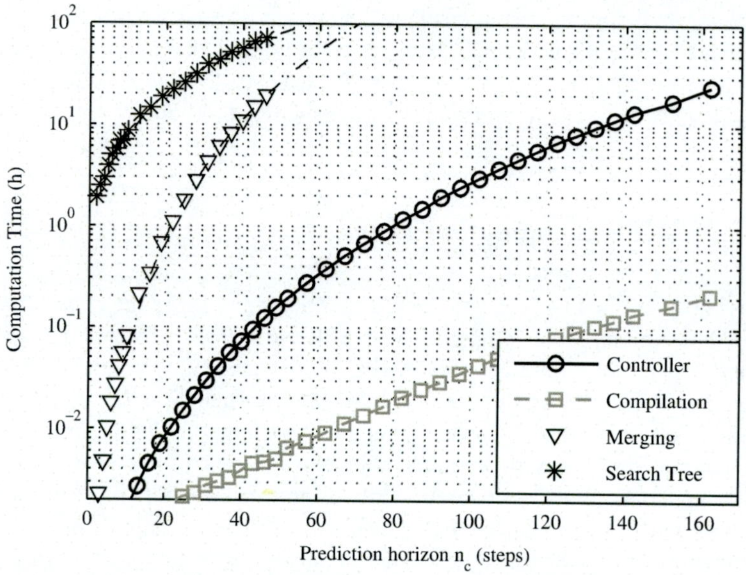

Fig. 11.4 Horizon length related to the offline controller computational time combined with executable compilation time in hours ($n_x = 2$)

unified. Using simulation results to extrapolate to higher initial deflections, the time to simplify regions for a maximal allowable deflection of ± 30 mm is estimated to 500 days. In fact, it is well known that optimal region merging performed as a post-processing operation is prohibitive for systems with state dimensions above $n_x = 2$ due to excessive offline computational demands [35, 37].

The other non-essential task is to use a binary search tree, instead of direct region searching algorithms [25]. This approach has an advantage of decreasing online search times, albeit it requires additional memory to store the precompiled binary search tree and data structure. As pointed out previously, the significant offline computational load is the main drawback in this case—not the online performance. The MPT Toolbox utilizes the method of Tøndel et al. [68] to generate the binary search tree. The calculation of this tree proved to be the most computationally intensive task, therefore it has been evaluated only up to horizons $n_c = 46$ steps ahead, taking over 70 h to complete. Figure 11.4 also illustrates the time necessary to generate this search tree for different horizons and corresponding maximal deflections. Extrapolation suggests extreme calculation times necessary to cover the whole operating region, similarly to the case of merging.

When a fourth order system has been considered, the multi-parametric calculation of the controller regions failed when the horizon exceeded a mere $n_c = 16$ steps forward. With this relatively short horizon and low model order, the controller computational time combined with the executable compilation time approaches 20 h. An MPMPC controller is clearly intractable for model predictive vibration attenuation applications with broadband excitation.

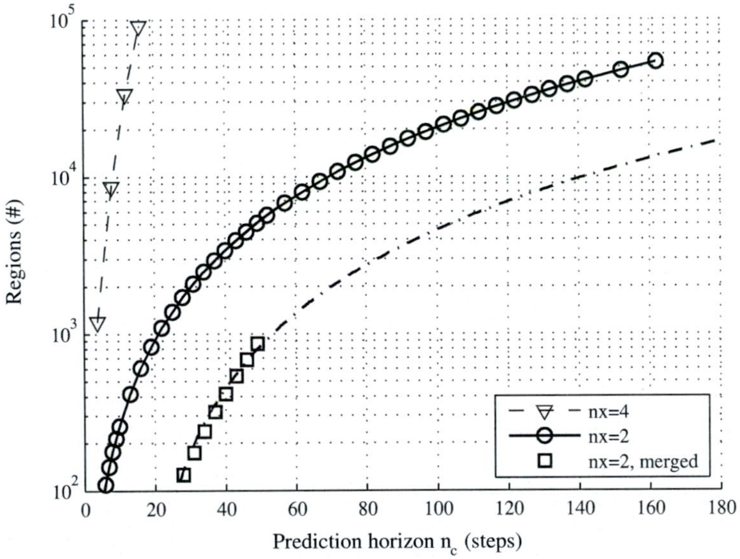

Fig. 11.5 Number of polyhedral MPMPC regions for a second order model, related to growing prediction horizon

11.2.2 MPMPC Regions

The explicitly computed MPC controller will contain a significant number of polytopic regions, thus implying issues with on the memory requirements of the controller hardware and increased search times. Region simplification (merging) reduces the number of regions and the controller executable size, but also adds to the computational time.

Figure 11.5 relates the control horizon and thus the maximal allowable tip deflection to the number of controller regions, which increases at an exponential rate. As illustrated in the figure, the number of regions is even more rapidly increasing in case a fourth order system is used. Merging reduces the amount of polytopic controller regions; however, with increasing horizons computation times become intractable as illustrated before.

If one considers covering the second eigenfrequency by the controller, the model order changes to four and at the same time sampling rates will have to increase to include higher frequency dynamics. Faster sampling requires even longer horizons, and this in fact will show the drawbacks of MPMPC in such and similar applications.

Simulations performed using the fourth order state-space model of the laboratory device confirmed this. While the computation of the controller already fails over $n_c = 16$ due to a solver crash attributed to memory issues, this horizon only yields a largest allowable deflection of a little over 1 mm. To allow ± 20 mm deflections a 400-step ahead horizon is required. This clearly is intractable considering the given conditions.

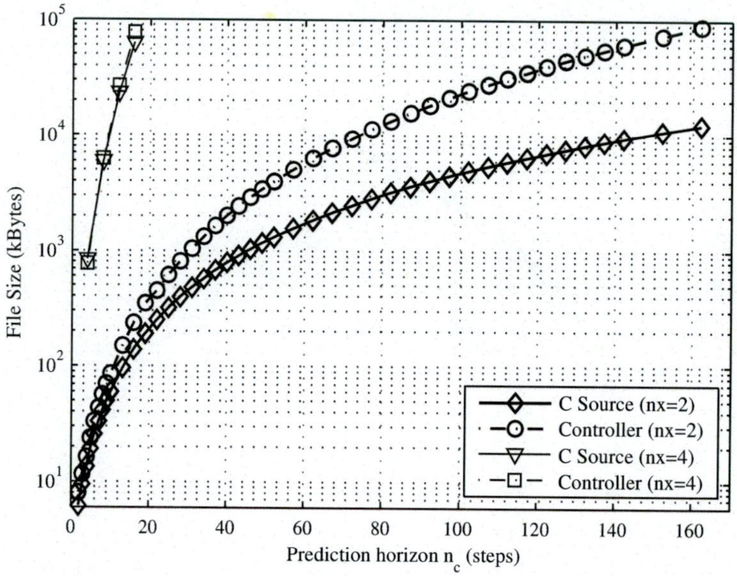

Fig. 11.6 MPMPC controller and compiled executable file size related to prediction horizon

11.2.3 MPMPC Controller Size

The quantity of regions has a direct effect on the controller size as well. The size of the file containing the raw controller and the C source code of the search tables is indicated in Fig. 11.6 for a second and fourth order system. When considering a fourth order system with a horizon length of $n_c = 16$ steps, the executable size grows up to 60 MBytes.

11.3 Issues with NRMPC Invariance

The dual-mode control law assumes that a fixed feedback law will steer the state into equilibrium beyond the prediction horizon. Both the feasibility of process constraints and from this indirectly the stability of the process is guaranteed by enforcing process constraints beyond the horizon. Enforcing process constraints beyond the horizon will create a target set which in nature will be invariant, or self-contracting.

In simple terms, invariance of the target set in an MPC strategy with guaranteed stability means that once the system state enters the target set it cannot leave it again. This is assuming the state is steered into equilibrium without further outside disturbance. Naturally, in a real MPC implementation the state may leave the target set since disturbances may occur at any point of the control course, and thus the controller

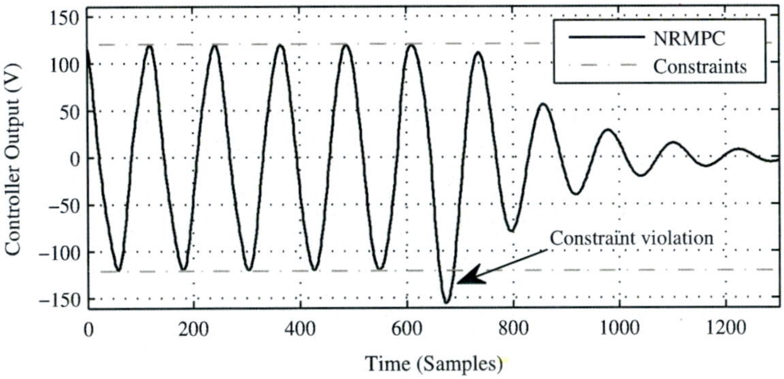

Fig. 11.7 NRMPC simulation run showing violation of the lower output constraints

strategy never "switches" permanently to the LQ mode. Given that the formulation of the MPC strategy and its implementation is correct, the violation of the invariance condition in simulation may indicate numerical errors in the implementation. These problems appear in the state-space representation as the state trajectory entering and subsequently leaving the target set, but from the viewpoint of input sequences and possibly outputs, this may materialize as a violation of system constraints.

Violation of both the invariance condition and process constraints has been observed in numerous simulations involving the NRMPC algorithm with optimized prediction dynamics. Figure 11.7 shows the evolution of control outputs for a second order linear time-invariant state-space model of the active vibrating beam. Here too the constraints were set to the physical limits of the transducers, ±120 V. Input penalization was set to $\mathbf{R} = r =$ 1E-4, which is based on simulations involving simple LQ controllers. State penalty has been fixed as $\mathbf{Q} = \mathbf{C}^T\mathbf{C}$. An initial condition emulating the deflection of the beam has been set, as a disturbance at the beginning of the simulation. No further disturbance has been assumed in this test.

To illustrate the issues with invariance better, this NRMPC algorithm implementation was slightly atypical. Normally the algorithm would decide between engaging NRMPC portion or pure LQ code according to whether current state at sample k is part of the target set[3] or not. Here however, following the essential principles of invariance, the controller switched permanently to LQ mode in case the state entered the target set. This approach assumes that simulations involve initial conditions and no other disturbances are present during the control course. Naturally, this assumption is unacceptable in an experimental implementation, although it serves a clear diagnostic purpose if invariance is questionable.

The evolution of outputs in Fig. 11.7 seems to be normal until a certain point, where the lower process constraints are clearly violated. After this, the output diagram seems to be ordinary again, albeit serious issues are suggested by the fact that the LQ controller produced outputs exceeding preset limits. Constraint violation occurs

[3] Coincident with the intersection of the augmented ellipsoid with original state-space.

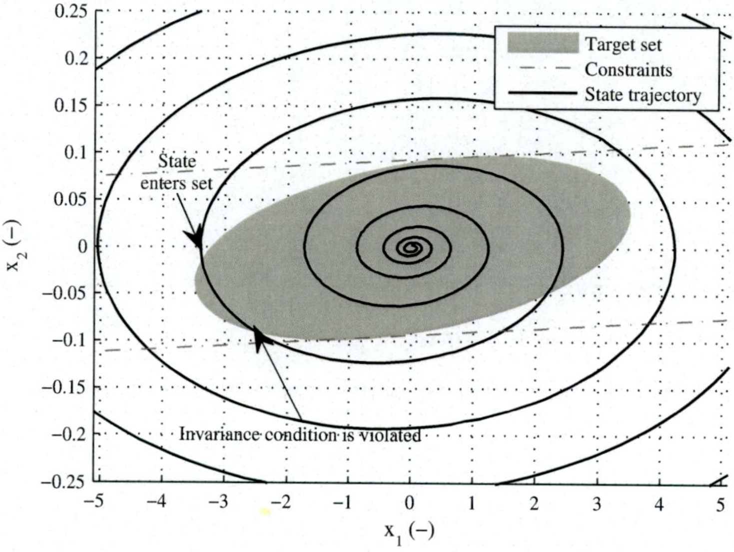

Fig. 11.8 NRMPC simulation run showing the evolution of states in x space. Violation of the invariance condition occurs as the states leave the target set

after the LQ controller is engaged, but for some reason the states do not remain in the target set where a pure linear-quadratic controller could take care of the rest of the control course.

Figure 11.8 shows a clearer picture on the nature of the problem. Here state-space is depicted in two dimensions, along with the evolution of the state trajectory. States are spiraling towards the origin from the initial condition. This is how oscillations at the output are represented using the state variables. At a certain point, the state trajectory enters the target set, depicted as the gray shaded ellipse in the middle of the figure. There is a problem though, after the state entered the target set it leaves it, which is a clear violation of the invariance condition.

This behavior has been only observed when the dynamics optimization principle as presented by Cannon and Kouvaritakis in [11] is implemented into the NRMPC offline optimization algorithm. In case the original NRMPC formulation is considered, none of these peculiar problems occur. Many signs point to the fact that this problem is of numerical nature, and can be solved by modifying the offline optimization algorithm or solver settings. Variables of the offline NRMPC optimization contain both extremely large and small numbers; simple matrix multiplications may be erroneous and are prone to numerical problems. The optimization process itself may also cause issues with invariance.

11.3.1 Performance Bounds

Optimized prediction dynamics in NRMPC makes it possible to recover the maximal region of attraction under any given fixed feedback law, while only requiring the horizon equal to that of the model order [11, 22]. While this is theoretically possible, the previous example demonstrates the practical limits of this approach.

A partial remedy to the issue with invariance violation is setting a boundary on the cost function value [11]. This unfortunately also directly affects the projection size, thus limiting the region of attraction. There is a performance boundary or cost limit, which redefines the invariance condition according to (8.57).[4] The bound of the predicted cost is then set by γ, which is enforced for all initial conditions of the autonomous system in E_z:

$$\tilde{J} \leq \gamma \tag{11.3}$$

Sacrificing the size of region of attraction may not be an issue with certain model classes. However, simulations performed using the LTI model of the vibration damping application showed that the range of allowable vibration deflections was severely compromised in the interest of numerical stability. A performance bound, deemed low enough to ensure numerical integrity of the optimization limited the deflection range under ~10 mm. This clearly defeats the purpose of optimizing prediction dynamics, and makes the practical use of NRMPC in this application questionable.

Using a second order model of the physical system, several tests were performed to approximate the ideal compromise between the size of augmented ellipsoidal set projection and numerical stability. For a given performance bound γ these tests varied the initial condition, changing the first element of the state vector from 0 to decreasing negative values according to $x_0 = [x_{11} \quad 0]^T$. For each initial condition, the evolution of control outputs was calculated, and its maximal value plotted against the given starting state. The tests utilized penalties $\mathbf{R} = r = 1\text{E-}4$, $\mathbf{Q} = \mathbf{C}^T\mathbf{C}$ and output has been constrained within $\overline{u} = -\underline{u} = 120$ V.

As visible in Fig. 11.9, the unrestrained cost bound and maximal region of attraction produces erroneous results for certain initial conditions, while others involve evolution of controller output where the maximal value stays within constraints. This approach is a rough estimation, and has its obvious limitations. Naturally it does not imply that a given γ necessarily ensures a numerically stable controller for all initial conditions, but only for the given search direction. The resolution of this simulation is an additional weak point, where it is possible to imagine a situation where invariance violation occurs exactly between steps changing initial condition.

Despite the shortcomings of this approach, Fig. 11.9 shows that the given controller is numerically adequate below a $\gamma = 1\text{E}5$ performance bound, and suggests

---

[4] For detailed mathematical description please refer to [11] or the relevant section in this work: See: Sect. 8.1.3

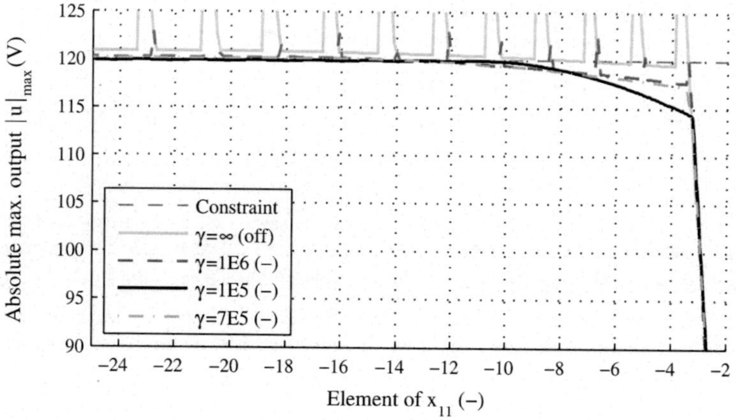

Fig. 11.9 Maximal absolute controller output for a simulation run with different performance limits γ, related to the initial condition. Maximal output values exceeding the constraints indicate invariance issues

that the optimal value might be somewhere around $\gamma = 7\text{E}5$ which provides a control process preserving invariance, but still maximizing the range of allowable initial conditions. Therefore, there must be a performance bound gamma γ over which the numerical integrity of the offline process is optimized.

An approximate method to estimate this level and to ensure a large region of attraction let us consider the following algorithm:

Algorithm 11.3
To estimate the maximal level of performance bound γ where the offline NRMPC optimization remains numerically viable, perform the following algorithm:

- Create a prediction model and evaluate offline NRMPC problem for the current performance bound γ. Determine bounds on the set defining the region of attraction.
- For a given direction in state-space, choose an initial condition and run a simulation. If no constraint violation is detected, repeat with a given resolution of initial conditions until the edge of region of attraction is reached.

If constraint violation is detected decrease γ and start over, otherwise increase γ. Stop if pre defined resolution is reached.

In this case, invariance problems were detected according to the previously introduced method: by checking for constraint violation during a given simulation run. A more sophisticated technique could be determining whether the state trajectory leaves the target set after entering. Despite the approximate nature of the algorithm, a good estimate on the performance bound ensuring the maximal safe region of attraction may be computed.

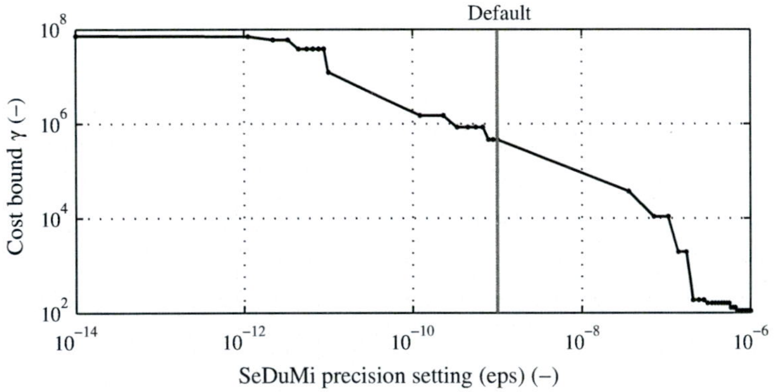

Fig. 11.10 Approximate maximal performance bound γ before invariance violation occurs, depending on solver precision settings

11.3.2 Solver Precision and Invariance

The violation of the invariance condition experienced throughout simulation trials of the NRMPC algorithm clearly has a numerical character. This suggests opportunity to fine-tune the SDP solver parameters in order to increase precision. Although several possible SDP solvers were considered for the implementation of the offline NRMPC problem in Sect. 10.3.1 none of them was deemed to be suitable[5] for this application except SeDuMi.

It is possible to redefine some of the default solver parameters in SeDuMi [62]. There are three variables controlling numerical tolerance, although the exact role of these is unlisted in the manual and customizing them is not recommended [54]. The desired solver accuracy is influenced by setting the *pars.eps* command structure to a smaller value. The default numerical accuracy is set to $eps = 1\text{E-}9$, when this value is reached the optimization terminates. Setting this value smaller means more precision, although optimization will take longer. Fortunately, given the typical problem dimensionality in NRMPC, this is not an issue. Setting parameter *eps* to 0 means that the optimization will continue as long as the solver can make progress.

Figures 11.10 and 11.11 show the results of simulations searching for the connection between solver precision settings and the size of region of attraction. The simulations have been performed using the NRMPC algorithm, utilizing a second order LTI model of the laboratory device. Input penalty was maintained at $\mathbf{R} = r = 1\text{E-}4$ and state penalty has been set to $\mathbf{Q} = \mathbf{C}^T\mathbf{C}$. Inputs were constrained to $\overline{u} = -\underline{u} = 120$ V, which agrees with the piezo transducer physical limits of ± 120 V. Algorithm 11.3 has been used to approximate the maximal possible performance bound and the corresponding volume of the region of attraction before violation of the invariance condition occurs.

[5] See B.3.3 for details.

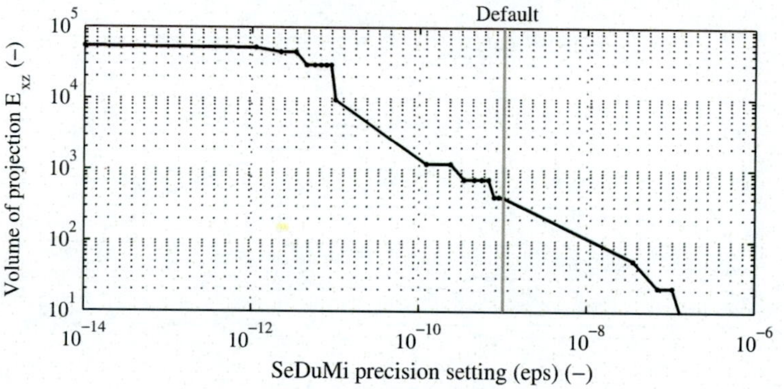

Fig. 11.11 Approximate maximal volume of the region of attraction depending on solver precision settings

Solver precision is plotted against the approximate maximal safe performance bound γ in Fig. 11.10. The corresponding volume of region of attraction for the given example and settings is shown in Fig. 11.11. The default precision is indicated on the figures as the vertical line at the 1E-9 mark. Superseding the default tolerance settings and algorithm precision to the obtainable maximum in SeDuMi increased the level of performance bound γ by more than two orders of magnitude. The resulting growth in the volume of the region of attraction has been similarly more than two orders of magnitude. In the light of the practical application, the allowable deflection range increased about an order of magnitude. A maximal tip deflection of ± 3 mm on the laboratory device is hardly exploitable, however increasing this to ± 30 mm allows the controller to perform its task under any mechanically viable situation.

These simulation and findings refer to the case with a second order model of the vibrating beam. We have to note that with other examples, especially higher order models, this improvement was not so significant. To preserve invariance and prevent numerical problems, the size of the initial stabilizable set was sacrificed significantly.

11.4 Issues with NRMPC Optimality

Simulations performed using a second order mathematical model of the vibrating structure showed the viability of using NRMPC for the model predictive vibration control of lightly damped structures. The optimality difference between the performance of QPMPC experienced both in simulations and experiments proved to be minor enough to produce an indistinguishable vibratory response. In fact, the Monte Carlo simulations described by Kouvaritakis et al. in [29] showed no more than 2% increase in closed-loop cost when compared to QPMPC. The paper suggests that for randomly generated examples with second and fifth order examples and the

NRMPC extension implemented, closed-loop cost remained only 1% worse in 97% of the cases.

Unfortunately, the lightly damped example used in this work proves to be a more difficult case for NRMPC. This is especially evident if the model order is increased anything above $n_x = 2$. Although from a practical engineering standpoint prediction dynamics and a controller based on a second order model is satisfactory, one might argue that a more complex prediction model could also explicitly include higher order dynamics. This is valid in particular for controllers covering more than one vibratory modes and a broadband excitation.

A fourth order model considered for the application on the vibrating beam could explicitly include first and second mode dynamics, thus in this case cover the bandwidth of approximately 0–80 Hz. Such a model has been prepared by using the experimental identification method described in Sect. 5.2. Simulations examining optimality were performed with this model using 250 Hz sampling, and characterized by the following linear time-invariant state-space system:

$$\mathbf{A} = \begin{bmatrix} 0.97778 & -0.52189 & -0.13501 & 0.60563 \\ 0.069689 & 0.94384 & -0.4819 & -0.45066 \\ -0.0027596 & 0.10094 & 0.29177 & 1.2812 \\ -0.0098709 & -0.047971 & -0.66151 & 0.33772 \end{bmatrix} \tag{11.4}$$

$$\mathbf{B} = \begin{bmatrix} 0.00061339 & -0.00046246 & -0.00043512 & -0.00015075 \end{bmatrix}^T$$

$$\mathbf{C} = \begin{bmatrix} -0.38171 & -0.5632 & -0.48193 & 0.40628 \end{bmatrix}$$

and with $\mathbf{D} = 0$.

The problems caused by suboptimality of the NRMPC method are clearly illustrated in Fig. 11.2, where the evolution of ENRMPC and NRMPC controller outputs u_k are plotted in time domain and compared to the truly optimal QPMPC controlled system. Here an initial condition $x_0 = [0.75\ 0\ 0\ 0]^T$ has been considered, which is equivalent to an initial deflection 1.5 mm at the beam tip. Every effort was made to create similar circumstances for both controllers. The prediction horizon of the QPMPC controller was set to $n_c = 36$ steps, the smallest possible for the considered initial condition. The horizon allowed engaging constraints without requiring too lengthy computations. Input penalty has been set to $\mathbf{R} = r = 1\text{E-}4$ and this time the state penalty $\mathbf{Q}$ was equivalent to the identity matrix of conforming dimensions for both QPMPC and NRMPC controllers. Process constraints have been engaged only on inputs, restraining them to ± 120 V. Simulations showed no significant improvement when using the extension with the NRMPC controller; neither did optimized prediction dynamics greatly affect the control outputs.

It is evident that a QPMPC-based controller evolution produces the expected switching behavior, while NRMPC outputs resemble a smoother sinusoidal curve. What is more important, the NRMPC controller outputs are far from the constraints thus not utilizing the full possible potential of the actuators. This is obviously noticeable in the damping performance of the controller too, although tip vibrations are irrelevant to the point and not shown here. Using the extension (ENRMPC) as suggested by Kouvaritakis et al. in [29] does not provide satisfactory improvement

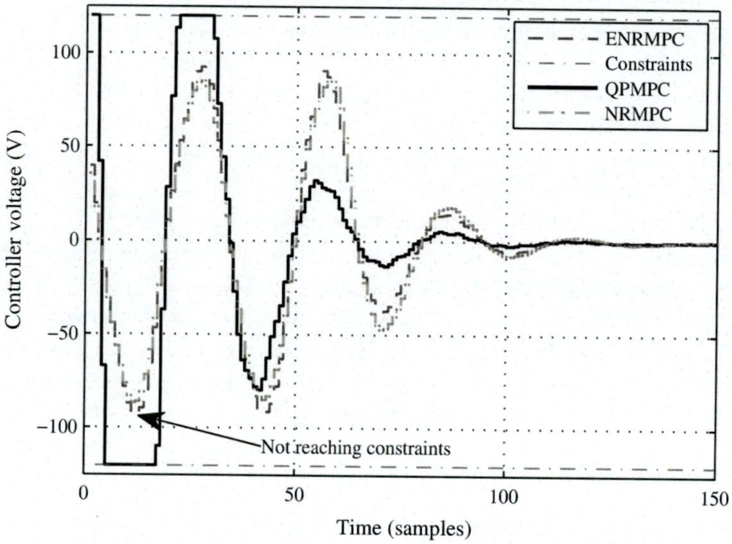

Fig. 11.12 Evolution of controller outputs for QPMPC and NRMPC controlled vibrations on a fourth order system model. Note the high degree of suboptimality for NRMPC

either. This simulation did not make use of optimized prediction dynamics, however it would not affect the outcome in any way. Due to the minimal difference between the ENRMPC and NRMPC control outputs, the following state trajectories will not differentiate between them. The trajectory marked as NRMPC utilizes the extension, thus presenting the slightly better case.

Figure 11.3 illustrates the projection of the control trajectory in state-space into the two-dimensional plane defined by x_1, x_2 and $x_3 = 0$, $x_4 = 0$. Here the cut of the NRPMPC target set is shown as a shaded ellipsoid, and the cut of QPMPC target set as the slightly larger polyhedral region. Intersection of the multidimensional plane defined by constraints with the above-mentioned coordinate system is also depicted. As suggested by the development of control outputs in Fig. 11.12, the QPMPC control trajectory is spiraling toward desired equilibrium at a much faster pace than NRMPC. Furthermore, it is important to note that the NRMPC target set is significantly smaller than that for QPMPC. In addition, the set is not approaching the region bounded by the half-spaces defined by constraints close enough.

Given the difficulties visualizing state behavior in a multidimensional system, projections of trajectories and cuts of target sets defined by the rest of the state coordinates are also depicted in Fig. 11.14. The volume difference of QPMPC and NRMPC target sets is significant in each view. It is worth noting that states x_1 and x_2 are most dominant in the trajectory, the rest of the components play a less vital role in the overall outcome of the trajectory. This also indicates the dominance of the first vibration mode in the overall dynamic response.

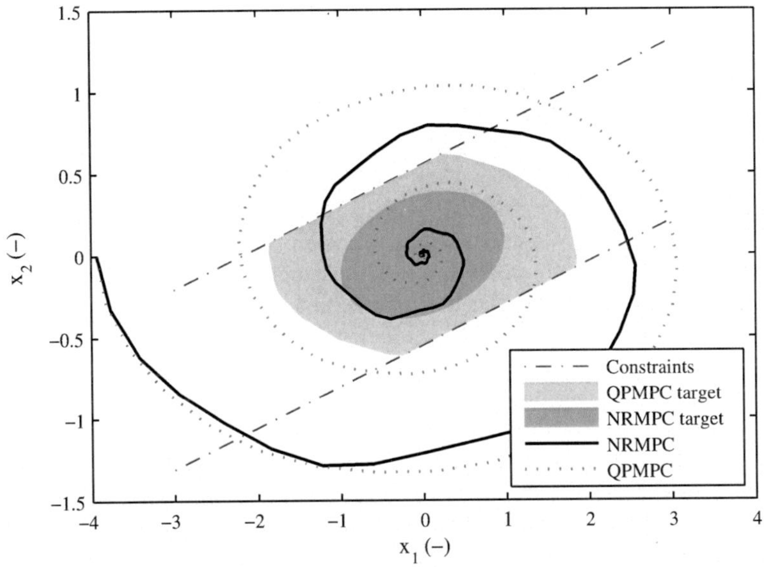

Fig. 11.13 Evolution of states projected onto the cut of target sets for NRMPC and QPMPC at x_1 and x_2

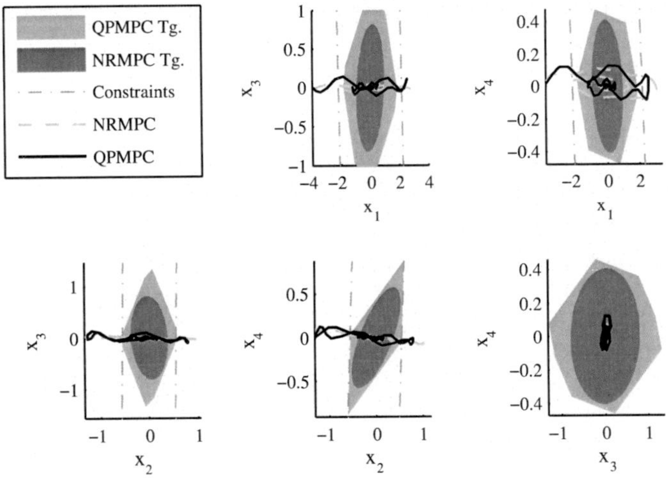

Fig. 11.14 Evolution of states projected onto the cut of target sets for NRMPC and QPMPC

11.4.1 Penalization and Optimality

The majority of simulations and experiments in this work assumes identical input penalty **R**. The choice of this tuning parameter is in most cases $\mathbf{R} = r = 1\text{E-}4$, which is based on the physical system model behavior using an LQ controller. A simple simulation[6] has been designed to determine ideal input penalization value **R**, where a fourth order model with 250 Hz sampling was utilized. State has been penalized using the identity matrix of conforming dimension; initial condition has been set to be the equivalent of a 1.5 mm deflection at the beam tip. The physical limits of the piezoelectric transducers have been kept in mind when determining a suitable **R**.

Figure 11.5 shows evolution of controller outputs using LQ controllers with different input penalization values. Fixing **R** at very low values, for example $\mathbf{R} = r = 1\text{E-}7$ produces a very aggressive simulation, where output voltages exceed 2500 V. This is not shown in the figure for clarity; only the range of ± 300 V is indicated. Setting $\mathbf{R} = r = 1\text{E-}4$ exceeds the constraints, but if one considers using a saturated controller, produces a reasonably lively output. On the other hand, with setting an input penalty of $\mathbf{R} = r = 7\text{E-}4$ one will not even reach the constraints, producing a conservative and slow controller.

Taking into account the previously introduced simulation and weighing, it is easy to see that $\mathbf{R} = r = 1\text{E-}4$ seems to be an ideal setting for the MPC control of this particular system. With an unspecified constrained MPC controller the same simulation run would hit the upper, lower and upper constraint again while avoiding constraints for the rest of the simulation run. This in fact implies an ideal setting, not sacrificing performance but maintaining a reasonable level of aggressiveness.

Determining controller output penalization has many other implications for both constrained and unconstrained MPC, although its most visible effect will still be performance. Amongst others, if constrained stable MPC control is considered penalty settings directly influence the volume of region of attraction and target set. In the case of NRMPC, these volumes are determined by the volumes of multi-dimensional augmented ellipsoid projections and intersections with the original state-space.

To better illustrate this fact, Fig. 11.6 shows the relationship between input penalty **R** and the volume of region of attraction and the target set. The most noteworthy part of the diagram is the two volumes converging to the same value, after exceeding a certain penalization level. After this, the size of the region of attraction is limited to the size of the target set and the NRMPC controller becomes a simple LQ controller.

As experienced during numerous trials with the NRMPC controller, the level of input penalty **R** has a surprisingly considerable effect on optimality and general usability of NRMPC using a fourth order model of the vibrating system. Initial conditions have been identical in all cases, emulating a tip deflection of 1.5 mm. To minimize the chance of encountering numerical problems, performance bounds were set to $\gamma = 0.5\text{E}5$ in all cases. States have been penalized by the identity matrix, actuator limits were constrained to the typical $\overline{u} = -\underline{u} = 120$ V.

---

6 See 12.1 for an experiment with different input penalty values **R**.

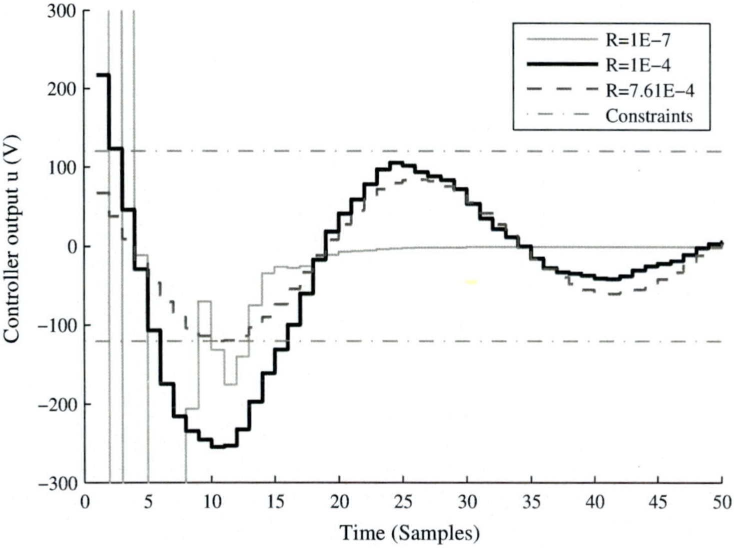

Fig. 11.15 Evolution of LQ controller outputs with different input penalization values $\mathbf{R} = r$

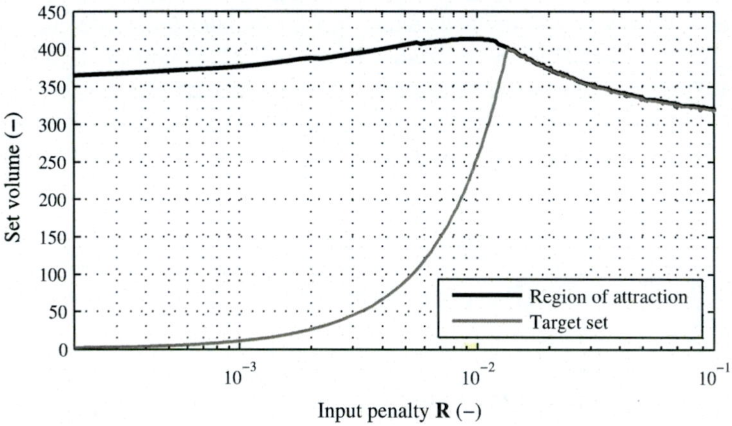

Fig. 11.16 Controller input penalty $\mathbf{R} = r$ related to the volume of the target set and region of attraction for a second order model of the physical AVC demonstrator system

Figure 11.7 shows the evolution of inputs u_k for simulations utilizing the NRMPC algorithm with different settings of $\mathbf{R}$. This simulation utilizes a fourth order prediction model to demonstrate the connection between suboptimality of the controller and penalization. For a fourth order model of the physical system the seemingly ideal $R = r = $ 1E-4 produces particularly suboptimal outputs. In this case, constraints are not even invoked and the controller does not make use of the full potential of

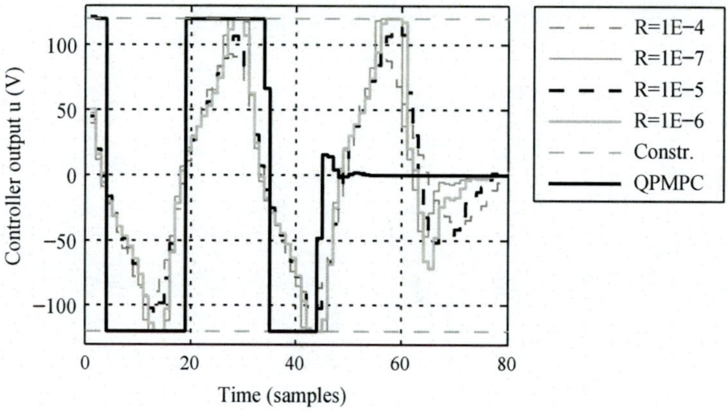

Fig. 11.17 Comparison of controller outputs produced by the extended NRMPC algorithm with various input penalty settings $\mathbf{R} = r$. A fourth order model demonstrates considerable suboptimality, although extremely low penalization seem to invoke constraints

actuators. Decreasing the level of **R** however brings an improvement, at a certain point even constraints are invoked.

Contrary to the earlier presented simulation results using LQ controllers, results demonstrated in Fig. 11.17 imply that much smaller penalization values are required for higher order examples. Penalization $\mathbf{R} = r = 1\text{E-}4$ produces a remarkably abnormal output while the seemingly too aggressive diminutive penalization at least make use of the full potential of the transducers. However, the built-in suboptimality of NRMPC is still significant, further lowering **R** does not significantly decrease closed-loop cost.

11.5 Alternate NRMPC Extension

The extension of the NRMPC algorithm introduced in Sect. 8.1.2.1 building on the optimality improvement of Kouvaritakis et al. in [29] took the concept further by using several steps ahead extrapolations the augmented state in the hope of a performance improvement. The aim of the NRMPC extension proposed by Li et al. in [41] is similar, that is to improve the optimality and thus the performance of the algorithm. Instead of iterating the augmented state $\mathbf{z}_{k+1} = \Psi \mathbf{z}_k$ several steps forward and constraining it to the invariant set E_z, Li et al. chose a one step forward iteration of the state $\mathbf{x}_k$ which was then constrained to the x-subspace projection E_{xz} of the invariant ellipsoid E_z.

Simulations have been performed to assess optimality of the NRMPC algorithm using the modified extension of applying several steps forward iterations of the augmented state $\mathbf{z}_k$ constrained into the augmented invariant ellipsoidal set E_z. A simple second order state-space model has been assumed for each simulation,

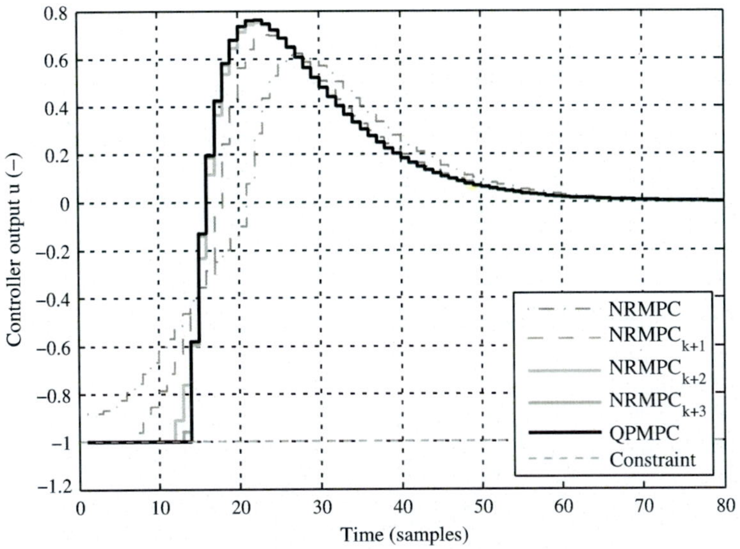

Fig. 11.18 Controller output in simulation, showing different NRMPC extensions compared to truly optimal QPMPC

having the following structure [10, 28]:

$$\mathbf{A} = \begin{bmatrix} 1 & 0.1 \\ 0 & 1 \end{bmatrix} \quad \mathbf{B} = \begin{bmatrix} 0 \\ 0.0787 \end{bmatrix}$$
$$\mathbf{C} = [1 \quad 0] \tag{11.5}$$

States have been penalized by matrix $\mathbf{Q}$, set equal to the identity matrix of conforming dimensions. Controller outputs have been penalized by $\mathbf{R} = r = 1\text{E-}4$. Prediction horizon for the QPMPC-based controller has been set to $n_c = 4$. The initial condition of $x_0 = [-0.5 \quad 0]$ was located on the boundary of the region of attraction. In the case of NRMPC with fixed prediction matrices, the same requirement calls for a horizon of $n_c = 25$ steps forward. State constraints have not been considered and controller outputs were limited to $|u| \leq 1$. In order to make sure that the invariance condition is not violated due to numerical difficulties, performance bound has been set to $\gamma = 1\text{E}5$.

Figure 11.8 shows the evolution of controller outputs for different versions of the NRMPC controller compared to truly optimal QPMPC. All simulations shown here use algorithms with optimized prediction dynamics. Understandably, QPMPC produces the best result, along with the smallest closed-loop cost. The worst evolution of outputs is acquired trough using the original NRMPC code, since constraints are not even reached. Simulation marked as $NRMPC_{k+1}$ is actually an algorithm implementing ENRMPC—the original extension introduced by Kouvaritakis et al.

Table 11.1 Comparison of closed-loop costs for the same simulation, using different NRMPC controllers. Closed-loop cost using QPMPC is $J_{QPMPC} = 97.00$, OD marks the use of optimized prediction dynamics

Extension	–[a]	1[b]	2	3	4
J_{NRMPC} (-)	105.42	105.15	103.81	99.72	-
J_{NRMPC} (OD) (-)	103.69	97.92	97.05	97.04	97.00

[a] NRMPC—no extension
[b] ENRMPC—original extension of Kouvaritakis et al.

in [29]. It provides an improvement relative to NRMPC without the extension, however there is still a possibility approaching optimal QPMPC more closely.

The remainder of simulations indicated in Fig. 11.8 implement extensions to the NRMPC algorithm, where membership of the augmented state $\mathbf{z}$ is assumed not at the next step $(k+1)$, but at steps $(k+2)$ and $(k+3)$. Improvement in the evolution of controller outputs is visually distinguishable, where $(k+3)$ produces nearly the same output as optimal MPMPC.

To quantify the level of suboptimality in comparison with QPMPC better, closed-loop costs are indicated in Table 11.1. Costs for the different adaptations of NRMPC have been calculated using the formerly introduced example and assuming the same conditions. The truly optimal closed-loop cost obtained via using QPMPC is $J_{QPMPC} = 97.00$, the NRMPC costs should ideally be as close to this as possible. NRMPC algorithms with and without optimized dynamics were evaluated where *OD* marked the use of optimized prediction dynamics at the offline stage. From this, it is implied that optimizing prediction dynamics not only enlarges the size of region of attraction, but also improves the optimality. The original formulation of NRMPC [28] produced the worst results. Using the extension and several steps ahead variations on the extension the costs gradually improve. In fact, the $(k+4)$ steps extension ensures the same cost as the QPMPC controller up to numerical precision differences.

Scalers μ resulting from different membership and feasibility conditions are plotted for the previously discussed example in Fig. 11.19. The original extension calculates scalers from the membership function for $(k+1)$ and ensures the feasibility of constraints at step (k). For each step, the higher of the two scaler values is selected. On the other hand, much less conservative scalers μ are computed for several steps ahead alternative extensions. Membership in the previous steps is not regarded, however the algorithm has to check for the feasibility of constraints for each step ahead. Thus, for an extension assuming membership at $(k+i)$ steps, scalers are compared for membership and feasibility from (k) up to $(k+i-1)$ and the highest value is used to scale the perturbation vector.

The optimal QPMPC controller output has been compared to the original and alternate NRMPC extensions for a vibration suppression example. Higher order models provide better conditions to assess optimality differences, therefore a fourth order state-space model of the vibrating beam has been considered for each trial. Figure 11.20 demonstrates the results of these simulations. A very low input penalty

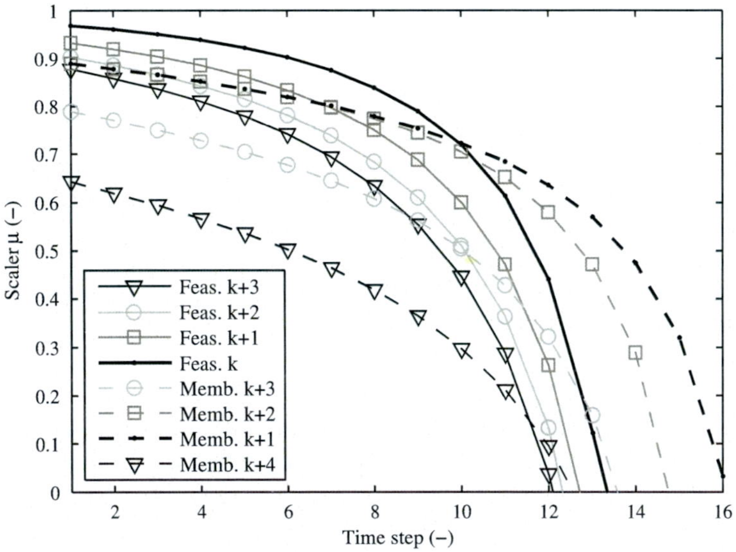

Fig. 11.19 Perturbation vector scalers μ shown for different conditions

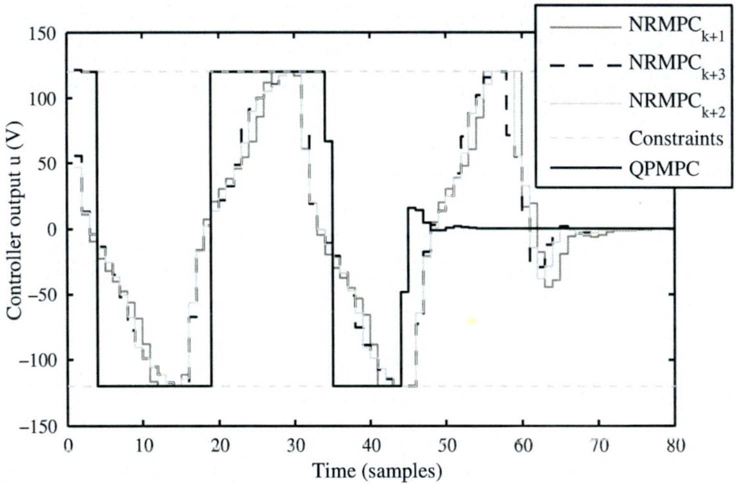

Fig. 11.20 Controller output for different alternate NRMPC extensions, using a model of the vibrating beam

has been used for the NRMPC controllers, according to the findings presented in Sect. 11.4. Controller output was constrained to ± 120 V.

As implied from the figure, alternate extensions of the NRMPC algorithm do not provide significant improvement in comparison with the original one. Two- and three-step ahead alternate extensions $(k+2)$ and $(k+3)$ produce slightly higher output voltages, theoretically being better than their original counterpart. The level of improvement is visually indistinguishable. For this reason the system output, in this case beam tip vibration is not presented here. Closed-loop costs quantify minor improvements in process optimality, where the strictly optimal QPMPC cost is $J = 133.48$. Original extended NRMPC produces a cost of $J_{(k+1)} = 195.76$, a two-step ahead modification $J_{(k+2)} = 191.09$, while three steps lowers the cost to $J_{(k+2)} = 187.66$. Two-steps ahead extension provides an optimality improvement compared to the original extension of only 2.4%, and still remains far from ideal. Even the three steps ahead alternate extension NRMPC code produces 40% worse closed-loop costs than QPMPC.[7]

Considering the formerly introduced simulation results, we may state that given the typical models used in the problem area of active vibration suppression, alternative extensions to the original NRMPC problem do not present a viable method of optimality enhancement. The gain in optimality does not justify questionable invariance properties, and the likely issues connected with model uncertainty. Although speed decrease is slight and the algorithm remains computationally efficient, several steps ahead extensions do require more computational time because of the additional feasibility conditions.

Alternate NRMPC extensions may not be suitable for improving optimality in vibration suppression, however it is possible to imagine certain models and applications where even the small optimality increase is advantageous. The simulation results presented by Li et al. in [41] suggest a similar conclusion from the viewpoint of the AVC of lightly damped systems: the optimality improvement is mainly significant with models where the effect of actuation is large, those with large input penalties **R**. Although the improvement of Li et al. does match the optimality of this modified NRMPC algorithm to QP-based MPC under a percent for the majority of randomly generated models, the results are valid for large **R**. Neither the iterated augmented state Ψz_k nor the iterated state-based algorithm of Li et al. has been later considered for the active vibration attenuation trial on the demonstrator hardware.

11.6 Comparison of QPMPC, MPMPC and NRMPC in Simulation

Beam tip vibration suppression performance through various MPC strategies has been compared in simulation and contrasted to the free response. The strategies were QPMPC, optimal MPMPC and NRMPC all with a priori stability guarantees. For this simulation study, an initial deflection of 5 mm has been considered to allow a tractable computation of the MPMPC controller structure, and shorter QPMPC simulation times. This simulation pointed out latent issues with the implementation of QPMPC

[7] This is a quite significant suboptimality especially that [29] states that for randomly generated examples, the error never rose above 2% and remained under 1% for 97% of the examples.

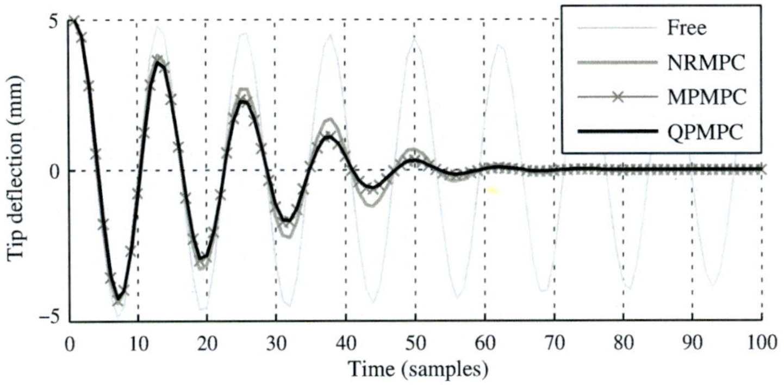

(a) Tip deflection response in mm

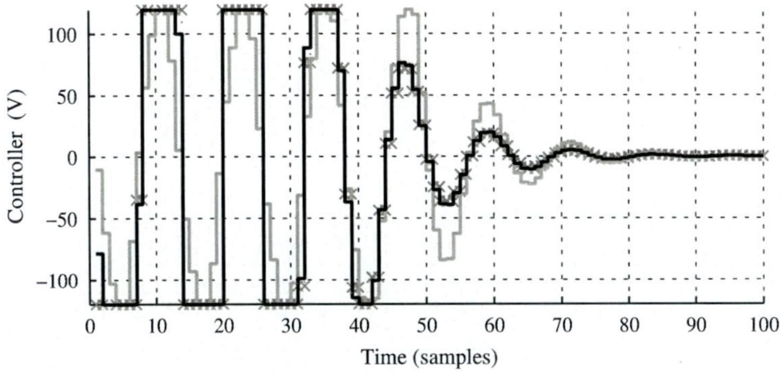

(b) Control response of QP, MP and NR based stable MPC

Fig. 11.21 Comparison of the vibration response in millimeters, measured at the beam tip (**a**) and the controller output (**b**) generated by the QPMPC, MPMPC and NRMPC algorithms in Volts. Simulated disturbance is identical initial deflection of the beam tip

and MPMPC on the real system, while at the same time validated functionality of the NRMPC algorithm. All necessary steps were taken to create as identical conditions to all three controllers as possible.

A second order model of the vibrating system has been assumed to generate predictions, sampled by 100 Hz which sufficiently includes the first vibration mode and exceeds the requirements of Shannon's theorem [23]. All simulations started with the same initial condition of $x_0 = [-7 - 1.6073]^T$, emulating a 5 mm deflection at the beam tip. The system response was simulated by the same state-space model, thus not considering model uncertainty. To avoid numerical difficulties, performance of the NRMPC controller has been bounded to $\gamma = 1\text{E}5$—affecting the size of the region of attraction. Inputs have been penalized by $\mathbf{R} = r = 1\text{E-}4$, while states used the identity matrix for $\mathbf{Q}$. System constraints were set to the typical physical limit of transducers, a maximal ± 120 V. No other process constraints were engaged. The minimal QP-based stable MPC horizon to ensure the inclusion of the given initial

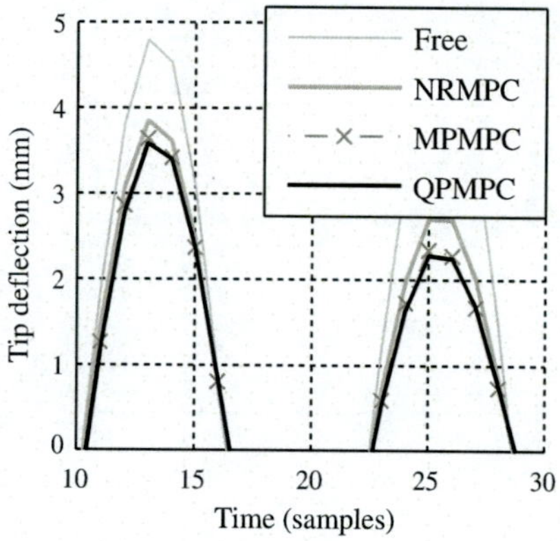

(a) Tip deflection detail

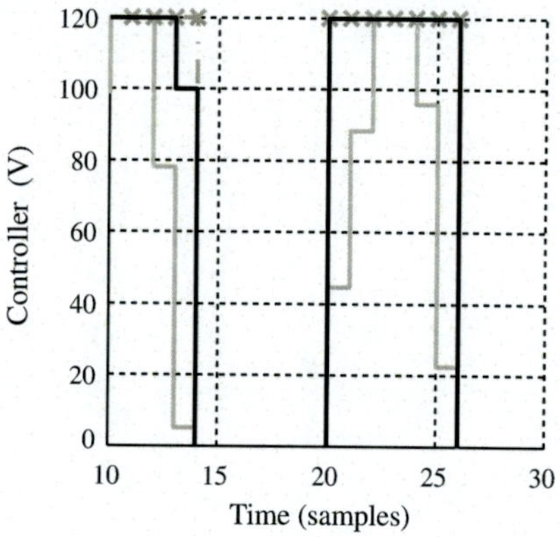

(b) Control response detail

Fig. 11.22 Detail of the input and output of the MPC controllers contrasted to the free response between samples 10 and 30. The detailed beam tip deflections are shown in (**a**), while (**b**) shows the simulation response of the control inputs

condition into the region of attraction required a prediction horizon of $n_c = 40$ steps. This same horizon was also required by the MPMPC controller, implemented as described in Chap. 10.2.

Illustrated in the example shown in Fig. 11.21, the best response is ensured by QPMPC control as it is logically expected. The controller drives actuators hard into saturation, as assumed from the physical properties of the system. QPMPC provides a strictly optimal control run with a closed-loop cost serving as an ideal lowest in comparison with suboptimal NRMPC. The QPMPC controller assumes the same quadratic cost function as the rest of the controllers, and safeguards for constraint feasibility using a constraint checking horizon. Despite the best possible response, the lengthy simulation times forecast issues at the real-time implementation.

In the case of MPMPC control, the tip vibration and the controller output is nearly indistinguishable. Essentially both controllers produce the same output, since they only differ with QPMPC only in the method of implementation. The response is very favorable, driving actuators to saturation as expected. Online simulation times are surprisingly short, however offline precomputation of the controller has been somewhat lengthy even for the limited region of attraction. Details of this example are shown in Fig. 11.22, illustrating the input and output responses between samples 10 and 30 (0.1–0.3 s).

Due to the built-in suboptimality of the formulation NRMPC performs slightly worse than the former two. This is an expected behavior, and it is visible on both figures. Increase in cost function value in MPMPC is insignificant and only due to numerical effects, while the drawback of NRMPC is a quite significant cost increase of ~15% when compared to both QPMPC and MPMPC. As discussed in Sect. 11.4, higher order models perform worse, the optimality difference can reach 40% even with the NRMPC extension[8] implemented and enabled.

Although there are some minor differences in performance, all the investigated MPC methods decrease the settling time of the beam into equilibrium significantly, ultimately improving the natural damping properties of the structure. The simulation results also suggest that the piezoceramic actuators, which contribute only very modest deflections in the static mode, considerably increase the natural damping near resonance.

References

1. Amer Y, Bauomy H (2009) Vibration reduction in a 2DOF twin-tail system to parametric excitations. Commun Nonlinear Sci Numer Simul 14(2):560–573. doi:10.1016/j.cnsns.2007.10.005, http://www.sciencedirect.com/science/article/B6X3D-4PYP723-2/2/b9d5375168fadb0b4e67857e92948bfc
2. Bandopadhya D, Bhogadi D, Bhattacharya B, Dutta A (2006) Active vibration suppression of a flexible link using ionic polymer metal composite. In: 2006 IEEE conference on robotics, automation and mechatronics, pp 1–6. doi: 10.1109/RAMECH.2006.252638
3. Baocang D (2009) Modern predictive control, 1st edn. Chapman and Hall / CRC, Boca Raton
4. Bartlett P, Eaton S, Gore J, Metheringham W, Jenner A (2001) High-power, low frequency magnetostrictive actuation for anti-vibration applications. Sens Actuators A 91(1–2):133–136. (Third European conference on magnetic sensors & actuators). doi: 10.1016/S0924-4247(01)00475-7, http://www.sciencedirect.com/science/article/B6THG-4313YT1-14/2/851ac15043a568a17313eef3042d685f
5. Bemporad A, Morari M, Dua V, Pistikopoulos EN (2002) The explicit linear quadratic regulator for constrained systems. Automatica 38(1):3–20. doi: 10.1016/S0005-1098(01)00174-1, http://www.sciencedirect.com/science/article/B6V21-44B8B5J-2/2/2a3176155886f92d43afdf1dccd128a6
6. Benaroya H, Nagurka ML (2010) Mechanical vibration: analysis, uncertainities and control, 3rd edn. CRC Press, Taylor & Francis Group, Boca Raton
7. Bittanti S, Cuzzola FA (2002) Periodic active control of vibrations in helicopters: a gain-scheduled multi-objective approach. Control Eng Pract 10(10):1043–1057. doi:10.1016/S0967-0661(02)00052-7, http://www.sciencedirect.com/science/article/B6V2H-45KSPJJ-3/2/9647861ce849d131c7d4b90cdb964751
8. Boscariol P, Gasparetto A, Zanotto V (2010) Model predictive control of a flexible links mechanism. J Intell Rob Syst 58:125–147. doi:10.1007/s10846-009-9347-5
9. Braghin F, Cinquemani S, Resta F (2010) A model of magnetostrictive actuators for active vibration control. Sens Actuators A (in press). doi:10.1016/j.sna.2010.10.019, http://www.sciencedirect.com/science/article/B6THG-51F25N5-4/2/f5cf46980d38877c74a3c4d34fbd894d
10. Cannon M (2005) Model predictive control, lecture notes. Michaelmas Term 2005 (4 Lectures), Course code 4ME44. University of Oxford, Oxford
11. Cannon M, Kouvaritakis B (2005) Optimizing prediction dynamics for robust MPC. IEEE Trans Autom Control 50(11):1892–1897. doi:10.1109/TAC.2005.858679
12. Chen H, Allgöver F (1998) A quasi-infinite horizon nonlinear model predictive control scheme with guaranteed stability. Automatica 34(10):1205–1217

[8] See Sect. 8.1.2 for more details.

13. Cychowski M, Szabat K (2010) Efficient real-time model predictive control of the drive system with elastic transmission. IET Control Theory Appl 4(1):37–49. doi: 10.1049/iet-cta.2008.0358
14. Eissa M, Bauomy H, Amer Y (2007) Active control of an aircraft tail subject to harmonic excitation. Acta Mech Sin 23:451–462. doi:10.1007/s10409-007-0077-2
15. El-Badawy AA, Nayfeh AH (2001) Control of a directly excited structural dynamic model of an F-15 tail section. J Franklin Inst 338(2–3):133–147. doi:10.1016/S0016-0032(00)00075-2, http://www.sciencedirect.com/science/article/B6V04-42HNMDV-3/2/e3bf6f797834c8e8638324be88fb78f7
16. Eure KW (1998) Adaptive predictive feedback techniques for vibration control. Doctoral dissertation, Virginia Polytechnic Institute and State University, Blacksburg
17. Fuller CR, Elliott SJ, Nelson PA (1996) Active control of vibration, 1st edn. Academic Press, San Francisco
18. Geyer T, Torrisi FD, Morari M (2008) Optimal complexity reduction of polyhedral piecewise affine systems. Automatica 44(7):1728–1740. doi: 10.1016/j.automatica.2007.11.027, http://www.sciencedirect.com/science/article/pii/S0005109807004906
19. Hassan M, Dubay R, Li C, Wang R (2007) Active vibration control of a flexible one-link manipulator using a multivariable predictive controller. Mechatronics 17(1):311–323
20. Hu Q (2009) A composite control scheme for attitude maneuvering and elastic mode stabilization of flexible spacecraft with measurable output feedback. Aerosp Sci Technol 13(2–3):81–91. doi: 10.1016/j.ast.2007.06.007, http://www.sciencedirect.com/science/article/B6VK2-4P96269-2/2/5fbc47249fdd3f1963c5ba856f071c55
21. Huang K, Yu F, Zhang Y (2010) Model predictive controller design for a developed electromagnetic suspension actuator based on experimental data. In: 2010 WASE international conference on information engineering (ICIE), vol 4, pp 152–156. doi:10.1109/ICIE.2010.327
22. Imsland L, Bar N, Foss BA (2005) More efficient predictive control. Automatica 41(8):1395–1403. doi:10.1016/j.automatica.2005.03.010, http://www.sciencedirect.com/science/article/B6V21-4G7NT35-1/2/52a9590bfe1ccc2a9561165c3fbdf872
23. Inman DJ (2007) Engineering vibrations, 3rd edn. Pearson International Education (Prentice Hall), Upper Saddle River
24. Johansen T, Grancharova A (2003) Approximate explicit constrained linear model predictive control via orthogonal search tree. IEEE Trans Autom Control 48(5):810–815. doi: 10.1016/j.cnsns.2007.10.005
25. Johansen TA, Jackson W, Schreiber R, Tøndel P (2007) Hardware synthesis of explicit model predictive controllers. IEEE Trans Control Syst Technol 15(1):191–197
26. Kang B, Mills JK (2005) Vibration control of a planar parallel manipulator using piezoelectric actuators. J Intell Rob Syst 42:51–70. doi:10.1007/s10846-004-3028-1
27. Kok J, van Heck J, Huisman R, Muijderman J, Veldpaus F (1997) Active and semi-active control of suspension systems for commercial vehicles based on preview. In: Proceedings of the 1997 American control conference, vol 5, pp 2992–2996. doi: 10.1109/ACC.1997.612006
28. Kouvaritakis B, Rossiter J, Schuurmans J (2000) Efficient robust predictive control. IEEE Trans Autom Control 45(8):1545–1549. doi:10.1109/9.871769
29. Kouvaritakis B, Cannon M, Rossiter J (2002) Who needs QP for linear MPC anyway? Automatica 38:879–884. doi:10.1016/S0005-1098(01)00263-1, http://www.sciencedirect.com/science/article/pii/S0005109801002631
30. Krishen K (2009) Space applications for ionic polymer-metal composite sensors, actuators, and artificial muscles. Acta Astronaut 64(11–12):1160–1166. doi:10.1016/j.actaastro.2009.01.008, http://www.sciencedirect.com/science/article/B6V1N-4VM2K65-3/2/f8b0b2d64f274154a5eb59da52fbf524
31. Kvasnica M (2009) Real-time model predictive control via multi-parametric programming: theory and tools, 1st edn. VDM Verlag, Saarbrücken
32. Kvasnica M (2011) Multi-parametric toolbox (MPT). In: Selected topics on constrained and nonlinear control, STU Bratislava–NTNU Trondheim, pp 101–170

33. Kvasnica M, Grieder P, Baotić M (2004) Multi-parametric toolbox (MPT). http://control.ee.ethz.ch/
34. Kvasnica M, Grieder P, Baotic M, Morari M (2004) Multi-parametric toolbox (mpt). In: Alur R, Pappas GJ (eds) Hybrid systems: computation and control, Lecture notes in computer science, vol 2993, Springer, Berlin / Heidelberg, pp 121–124
35. Kvasnica M, Grieder P, Baotic M, Christophersen FJ (2006) Multi-parametric toolbox (MPT). Extended documentation, Zürich. http://control.ee.ethz.ch/mpt/docs/
36. Kvasnica M, Christophersen FJ, Herceg M, Fikar M (2008) Polynomial approximation of closed-form MPC for piecewise affine systems. In: Proceedings of the 17th world congress of the international federation of automatic control, Seoul, pp 3877–3882
37. Kvasnica M, Fikar M, Čirka L', Herceg M (2011) Complexity reduction in explicit model predictive control. In: Selected topics on constrained and nonlinear control, STU Bratislava–NTNU Trondheim, Bratislava pp 241–288
38. Kvasnica M, Rauová I, Fikar M (2011) Real-time implementation of model predictive control using automatic code generation. In: Selected topics on constrained and nonlinear control. Preprints, STU Bratislava–NTNU Trondheim, pp 311–316
39. Kwak MK, Heo S (2007) Active vibration control of smart grid structure by multiinput and multioutput positive position feedback controller. J Sound Vib 304(1–2):230–245. doi: 10.1016/j.jsv.2007.02.021, http://www.sciencedirect.com/science/article/B6WM3-4NH6N96-2/2/ca7b43602b9d052e388f4b2a28f1ebae
40. Lau K, Zhou L, Tao X (2002) Control of natural frequencies of a clamped–clamped composite beam with embedded shape memory alloy wires. Compos Struct 58(1):39–47. doi:10.1016/S0263-8223(02)00042-9, http://www.sciencedirect.com/science/article/B6TWP-45XTP9W-N/2/07b9a065ac866d8869a4240deb918851
41. Li S, Kouvaritakis B, Cannon M (2010) Improvements in the efficiency of linear MPC. Automatica 46(1):226–229. doi:10.1016/j.automatica.2009.10.010, http://www.sciencedirect.com/science/article/B6V21-4XGCHXB-3/2/20a93fa6dd4fb88469638ac3bc2fe729
42. Lin LC, Lee TE (1997) Integrated PID-type learning and fuzzy control for flexible-joint manipulators. J Intell Rob Sys 18:47–66. doi:10.1023/A:1007942528058
43. Lu H, Meng G (2006) An experimental and analytical investigation of the dynamic characteristics of a flexible sandwich plate filled with electrorheological fluid. Int J Adv Manuf Technol 28:1049–1055. doi:10.1007/s00170-004-2433-8
44. Luo T, Hu Y (2002) Vibration suppression techniques for optical inter-satellite communications. In: IEEE 2002 international conference on communications, circuits and systems and west sino expositions, vol. 1, pp 585–589. doi:10.1109/ICCCAS.2002.1180687
45. Maciejowski JM (2000) Predictive control with constraints, 1st edn. Prentice Hall, Upper Saddle River
46. Malinauskas A (2001) Chemical deposition of conducting polymers. Polymer 42(9):3957–3972. doi: 10.1016/S0032-3861(00)00800-4, http://www.sciencedirect.com/science/article/B6TXW-42C0RR9-1/2/e8084cbb0f228b86a5cc9d061a340e22
47. Mayne DQ, Rawlings JB, Rao CV, Scokaert POM (2000) Constrained model predictive control: stability and optimality. Automatica 36(6):789–814
48. Mehra R, Amin J, Hedrick K, Osorio C, Gopalasamy S (1997) Active suspension using preview information and model predictive control. In: Proceedings of the 1997 IEEE international conference on control applications, pp 860–865. doi: 10.1109/CCA.1997.627769
49. Moon SJ, Lim CW, Kim BH, Park Y (2007) Structural vibration control using linear magnetostrictive actuators. J Sound Vib 302(4–5):875–891. doi: 10.1016/j.jsv.2006.12.023, http://www.sciencedirect.com/science/article/B6WM3-4N2M6HH-5/2/417522adfca8640acfa76e890ae0533c
50. Moon SM, Cole DG, Clark RL (2006) Real-time implementation of adaptive feedback and feedforward generalized predictive control algorithm. J Sound Vib 294(1–2):82–96. doi:10.1016/j.jsv.2005.10.017, http://www.sciencedirect.com/science/article/B6WM3-4HYMY76-1/2/50d98047187533ebe9d3ea8310446e77

51. Neat G, Melody J, Lurie B (1998) Vibration attenuation approach for spaceborne optical interferometers. IEEE Trans Control Syst Technol 6(6):689–700. doi: 10.1109/87.726529
52. Pistikopoulos EN, Georgiadis MC, Dua V (eds) (2007) Multi-parametric model-based control, vol. 2, 1st edn. Wiley-VCH Verlag GmbH & Co., Weinheim
53. Pistikopoulos EN, Georgiadis MC, Dua V (eds) (2007) Multi-parametric programming, vol 1, 1st edn. Wiley-VCH Verlag GmbH & Co., Weinheim
54. Pólik I (2005) Addendum to the SeDuMi user guide version 1.1. Technical report, Advanced Optimization Lab, McMaster University, Hamilton. http://sedumi.ie.lehigh.edu/
55. Polóni T, Takács G, Kvasnica M, Rohal'-Ilkiv B (2009) System identification and explicit control of cantilever lateral vibrations. In: Proceedings of the 17th international conference on process control
56. Pradhan S (2005) Vibration suppression of FGM shells using embedded magnetostrictive layers. Int J Solids Struct 42(9–10):2465–2488. doi:10.1016/j.ijsolstr.2004.09.049, http://www.sciencedirect.com/science/article/B6VJS-4F6SSGN-1/2/b6f9e2e6ffc65bfc0c4af5083e37df0b
57. Preumont A (2002) Vibration control of active structures, 2nd edn. Kluwer Academic Publishers, Dordrecht
58. Rajoria H, Jalili N (2005) Passive vibration damping enhancement using carbon nanotube-epoxy reinforced composites. Compos Sci Technol 65(14):2079–2093. doi:10.1016/j.compscitech.2005.05.015, http://www.sciencedirect.com/science/article/B6TWT-4GHBPN0-5/2/a67d954050aac7829a56e3e4302c8ef6
59. Richelot J, Bordeneuve-Guibe J, Pommier-Budinger V (2004) Active control of a clamped beam equipped with piezoelectric actuator and sensor using generalized predictive control. In: 2004 IEEE international symposium on industrial electronics, vol 1, pp 583–588. doi: 10.1109/ISIE.2004.1571872
60. Rossiter JA (2003) Model-based predictive control: a practical approach, 1st edn. CRC Press, Boca Raton
61. Shan J, Liu HT, Sun D (2005) Slewing and vibration control of a single-link flexible manipulator by positive position feedback (PPF). Mechatronics 15(4):487–503. doi:10.1016/j.mechatronics.2004.10.003, http://www.sciencedirect.com/science/article/B6V43-4DR87K7-4/2/2dd311fdd61308e1415cd45c1edc3076
62. Sturm JF (2001) SeDuMi 1.05 R5 user's guide. Technical report, Department of Economics, Tilburg University, Tilburg. http://sedumi.ie.lehigh.edu/
63. Sun D, Mills JK, Shan J, Tso SK (2004) A PZT actuator control of a single-link flexible manipulator based on linear velocity feedback and actuator placement. Mechatronics 14(4):381–401. doi:10.1016/S0957-4158(03)00066-7, http://www.sciencedirect.com/science/article/B6V43-49DN5K4-1/2/fa21df547f182ad568cefb2ddf3a6352
64. Takács G, Rohal'-Ilkiv B (2009) Implementation of the Newton–Raphson MPC algorithm in active vibration control applications. In: Mace BR, Ferguson NS, Rustighi E (eds) Proceedings of the 3rd international conference on noise and vibration: emerging methods, Oxford
65. Takács G, Rohal'-Ilkiv B (2009) MPC with guaranteed stability and constraint feasibility on flexible vibrating active structures: a comparative study. In: Hu H (ed) Proceedings of the eleventh IASTED international conference on control and applications, Cambridge
66. Takács G, Rohal'-Ilkiv B (2009) Newton–Raphson based efficient model predictive control applied on active vibrating structures. In: Proceedings of the European control conference, Budapest
67. Takács G, Rohal'-Ilkiv B (2009) Newton–Raphson MPC controlled active vibration attenuation. In: Hangos KM (ed) Proceedings of the 28th IASTED international conference on modeling, identification and control, Innsbruck
68. Tøndel P, Johansen TA, Bemporad A (2003) Evaluation of piecewise affine control via binary search tree. Automatica 39(5):945–950. doi: 10.1016/S0005-1098(02)00308-4, http://www.sciencedirect.com/science/article/pii/S0005109802003084

69. Tzou H, Chai W (2007) Design and testing of a hybrid polymeric electrostrictive/piezoelectric beam with bang-bang control. Mech Syst Signal Process 21(1):417–429. doi: 10.1016/j.ymssp.2005.10.008, http://www.sciencedirect.com/science/article/B6WN1-4HR75KY-1/2/73701e5908a2ea598fa7bec1ce093563
70. Wang L (2009) Model predictive control system design and implementation using MATLAB, 1st edn. Springer, London
71. Wills A, Bates D, Fleming A, Ninness B, Moheimani R (2005) Application of MPC to an active structure using sampling rates up to 25 kHz. In: 44th IEEE conference on decision and control, 2005 and 2005 European control conference. CDC-ECC '05, pp 3176–3181. doi: 10.1109/CDC.2005.1582650
72. Wills AG, Bates D, Fleming AJ, Ninness B, Moheimani SOR (2008) Model predictive control applied to constraint handling in active noise and vibration control. IEEE Trans Control Syst Technol 16(1):3–12
73. Wilson DG, Robinett RD, Parker GG, Starr GP (2002) Augmented sliding mode control for flexible link manipulators. J Intell Rob Syst 34:415–430. doi:10.1023/A:1019635709331
74. Yim W (1996) Modified nonlinear predictive control of elastic manipulators. In: Proceedings of the 1996 IEEE international conference on robotics and automation, vol 3, pp 2097–2102. doi:10.1109/ROBOT.1996.506180
75. Zhang QZ, Gan WS (2004) A model predictive algorithm for active noise control with online secondary path modelling. J Sound Vib 270(4–5):1056–1066. doi: 10.1016/S0022-460X(03)00516-9, http://www.sciencedirect.com/science/article/B6WM3-49D6XFX-4/2/805d0549ca0e60339bb5e2c798de7264
76. Zmeu K, Shipitko E (2005) Predictive controller design with offline model learning for flexible beam control. In: Proceedings of the 2005 international conference on physics and control, pp 345–350. doi: 10.1109/PHYCON.2005.1514005

Chapter 12
Experimental Model Predictive Vibration Control

This chapter presents the results of experiments comparing different model predictive control methods applied to the active vibration control laboratory device. The main aim of the experimental trials presented here is to compare damping performance and online timing properties of the different model predictive controllers. Traditional infinite horizon dual-mode quadratic programming-based MPC (QPMPC) [14, 64], optimal multi-parametric programming-based MPC (MPMPC) [4, 5], minimum-time suboptimal MPMPC [8, 43, 49] and the suboptimal Newton–Raphson MPC (NRMPC) [10, 45, 46] are included in the trials assessing the practical damping effect and task execution timing of these algorithms.

The laboratory demonstration device featured in the experiments has been introduced earlier in Chap. 5 and it conforms to this description (Fig. 12.1). As it has been earlier noted, the use of a clamped cantilever beam to demonstrate and compare damping capability of both actuators and algorithms is very common in academic literature [9, 30, 42, 56, 76, 79, 92–94]. Moreover, a simple laboratory hardware like the AVC demonstrator considered here may represent the dynamic behavior of a class of lightly damped mechanical systems[1] such as rotary or fixed wings in aircraft [3, 6, 18, 19, 53, 58], solar panels and antenna masts [1, 36, 51, 59] and others [69, 72].

Established control strategies like position, velocity or acceleration feedback [3, 7, 18, 19, 40, 54, 76], PID [2, 11, 20, 27, 29, 41, 42, 88] or other transfer function or filtering-based methods are commonly and successfully applied to vibration control systems. However, these traditional controllers have not been implemented here for the interest of preserving a fair basis of comparison with MPC. Matching PID or position feedback control to reasonably more sophisticated algorithms based on constrained online optimization would provide biased results. For example, trials involving saturated PID would not take into account the undeniable advantages of stable constrained predictive control, like preserving optimality and guaranteed stability at all times. Furthermore, a simple saturated PID controller does not require complex computational operations like the more advanced MPC, therefore

---

[1] See Sect. 5.1.1 for more details.

G. Takács and B. Rohal'-Ilkiv, *Model Predictive Vibration Control*,
DOI: 10.1007/978-1-4471-2333-0_12, © Springer-Verlag London Limited 2012

Fig. 12.1 The aluminum blade of the AVC laboratory demonstrator is vibrating violently as a test signal close to the first resonant frequency is supplied to the piezoceramic actuators

the direct comparison of the implementation particulars of these methods is simply meaningless.

The most straightforward controller, which is comparable to the underlying idea behind optimization-based MPC algorithms, is saturated linear quadratic (LQ) control. An LQ controller has been implemented on the demonstration device to serve as a basis of comparison for the damping effect and timing properties of MPC strategies. The saturated LQ control strategy is often preferred by researchers [13, 17, 26, 33, 34, 44, 48, 63, 67] for active vibration control applications, although often ignoring the ultimate effects of saturation on optimality and stability. LQ provides an undeniably good vibration suppression performance, since constraints are respected only by enforcing a software saturation limit and possibly a hardware limit setting at the amplifiers. Nevertheless, simple implementation and low computational costs shall not steer the attention away from the advantages gained by stability and constraint handling features of MPC methods. Trials with the LQ controller are introduced in the first section of this chapter.

The majority of the experiments have been motivated by evaluating the damping performance and functionality of the less known NRMPC, also in contrast with other methods [80–83]. Stable, constrained infinite horizon dual-mode QPMPC is an excellent choice in this comparison, since it is a well-known and accepted method providing optimal results [61, 74]. The computationally efficient methods are represented in this analysis by the optimal and suboptimal MPMPC controllers [49, 50, 71] and NRMPC [10, 37, 45, 46, 55]. Note, that although QPMPC represents the weakest link of all four MPC algorithms in terms of computational requirements, here it can be still regarded as a computationally efficient MPC implementation.

This is because all experiments assume the use of the qpOASES solver [22], which has been designed specifically for the use with MPC algorithms [21, 23, 25].

The most straightforward way to evaluate damping properties is to create an initial deflection at the tip and measure decay of vibrations without further interaction, which is presented in Sect. 12.2. Exciting the beam using an additional piezoelectric transducer in actuator mode offers the possibility to visualize system behavior in the frequency domain. Such a frequency domain test is described in Sect. 12.3, also with a modal shaker acting as a source of disturbance. This is followed by a trial introduced in Sect. 12.4, where damping effects were studied using a pseudo-random disturbance created by a modal shaker and turbulent airflow around the vibrating beam. Finally, a computational performance analysis is presented in Sect. 12.5, including a short reflection on possible speed improvements.

12.1 Linear Quadratic Control

The control loop and actuators of the AVC demonstrator have been tested using a linear quadratic (LQ) controller, in order to assess the improvement on the inherent natural damping properties of the structure. The LQ feedback evaluation process assumed a high order model of the structure, the result of the preliminary identification process introduced in Sect. 5.2.

A 12th order linear time-invariant state-space model was utilized first, in order to assess whether a high order model provides significantly better damping than a much simpler second order one, covering only the first resonant frequency. The high order model was covering the first five transversal bending modes. Sampling used during the experiment was 5 kHz (0.0002 s)—much higher than later NRMPC and MPMPC controllers assumed. Input was penalized with $\mathbf{R} = r = 1\text{E-}4$, and state penalty was set to $\mathbf{Q} = \mathbf{C}^T \mathbf{C}$.

The optimal steady-state feedback gain matrix $\mathbf{K}$, which minimizes a quadratic cost function has been calculated using the *dlqr* routine. This is a part of Control System Toolbox [86]. The quadratic cost function presumed in the *dlqr* command is calculated according to the theoretical basics presented earlier in Sect. 4.3. Due to the safe operation limits of the piezoelectric transducers, amplifier voltage limits were engaged [65, 66, 77]. Although no saturation limits were directly implemented in the controller algorithm and software block scheme, this can be regarded as saturated LQ control. The exact upper and lower symmetric saturation limits are unknown, and possibly higher than those used in the constrained NRMPC and MPMPC control experiments. This implies higher damping performance than it is possible with exact constrained control.

Beam tip was deflected to an initial position of –5 mm away from its equilibrium position, then without further outside force interaction was allowed to vibrate either freely or with LQ control. Figure 12.2 shows the free natural and high order LQ controlled beam tip vibration response. Uncontrolled beam vibrations are visible and measurable for much longer times than LQ controlled. With the controller engaged, the tip settles just over a second from this particular initial condition.

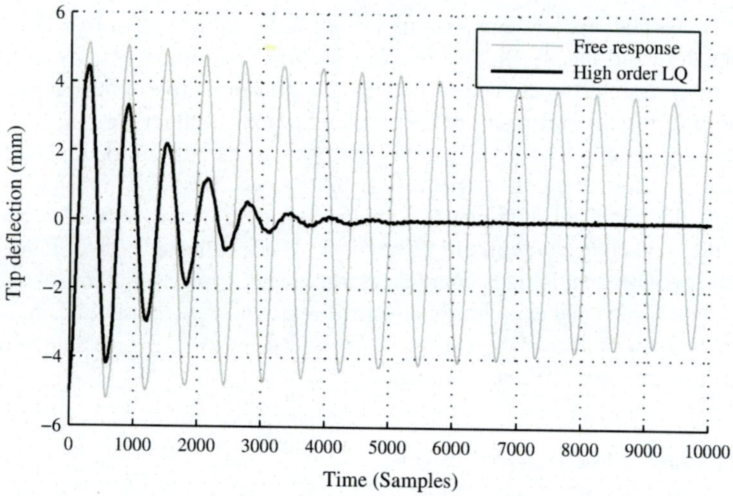

Fig. 12.2 Controlled and free beam tip vibration response. High order LQ controller, saturated by the amplifier voltage limit switch

Simulations presented in Sect. 11.4 suggested optimality issues with higher order models. The choice of input penalization $\mathbf{R} = r = 1\text{E-}4$ has been justified by a balance between aggressiveness of the control moves[2] and damping performance. Counter-intuitively, as described in Sect. 11.4.1, NRMPC control may require much lower penalization than expected.

In order to assess the effect of different input penalization values on the saturated LQ control strategy, three different controllers have been implemented on the physical system with different input penalty values $\mathbf{R} = r$. The LQ gain has been obtained using the *dlqr* routine in Matlab, utilizing a second order model of the physical system with 0.01 Hz sampling. State penalty has been set to $\mathbf{Q} = \mathbf{C}^T\mathbf{C}$ and unlike in the previous example, saturation limits were implemented both at hardware and software levels. The highest allowable voltage implemented in the controller is corresponding to the manufacturer provided specifications of ± 120 V. States have been observed using a block default Kalman filter implemented in the Matlab Signal Processing Toolbox [84].

Figure 12.3 shows different saturated LQ controllers applied to the AVC demonstrator system in comparison with the free response. As it is evident from the figure, the LQ controlled vibrations settle significantly faster than the beam using only its natural damping properties. While the controlled vibration response settles completely in approximately 4 s, the free response of the beam returns to equilibrium only after 60 s. This is more one order of magnitude faster settling time. The improved damping properties not only signalize the efficiency of the LQ controllers, but also

2 Materializing as disproportionately high control moves, severely exceeding constraints.

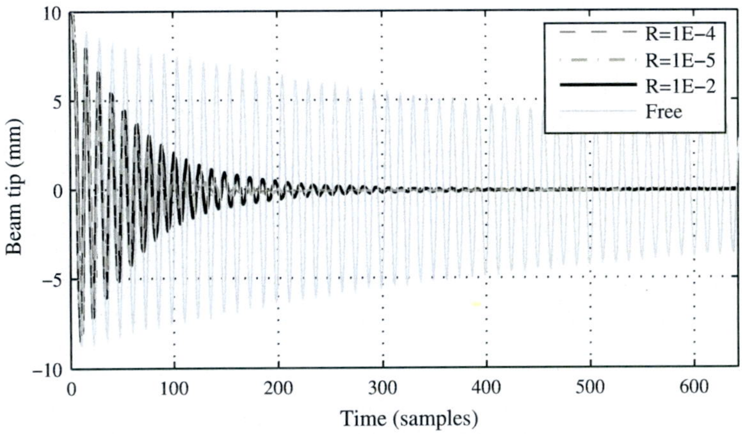

Fig. 12.3 The damping performance of low order LQ controllers with different input penalty values $\mathbf{R} = r$ is compared to the free response of the beam. The improvement is significant, exceeding more than one order of magnitude shorter settling times to equilibrium under control

validate the use of the relatively weak piezoelectric patches and demonstrate their significant damping effect in resonant modes.

Figure 12.4 indicates the details of the first 200 samples (2 s) of the same experiment, showing the measured beam tip vibration and controller voltage output as well. An input penalty set at $\mathbf{R} = r =$ 1E-4 seems to be the ideal compromise. The experiment with the higher penalty value $\mathbf{R} = r =$ 1E-3 does not make use of the full damping potential of the actuators; the decay of vibration could be improved. On the other hand, the evolution of beam tip deflections controlled by an LQ feedback using $\mathbf{R} = r =$ 1E-5 gives the impression of being overly aggressive. This is especially visible at lower vibration levels, where the controller amplifies background noise and can possibly render the system unstable. This experiment confirms the results of the simulation presented in Sect. 11.4.1, where $\mathbf{R} = r =$ 1E-4 has been established as the ideal input penalty for the given problem.

12.2 Initial Deflection Test

Initial displacement tests have been performed on the experimental device by deflecting the beam tip away from its equilibrium position, and releasing it to settle without further force interaction. This type of initial deflection or initial displacement is a common way of testing vibrating mechanical systems such as cantilevers [16, 28, 89] and plates [15, 32, 70] both in theory and experiment. In case the demonstration system is uncontrolled, the vibration resulting an initial displacement is visible and quite significant even minutes after the disturbance occurs. This is attributable to the low damping ratio of the structure. The free, uncontrolled response of the structure is

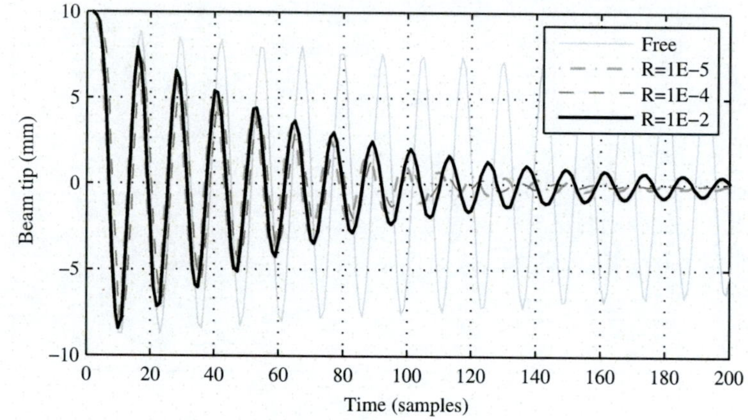

(a) Tip deflection response detail

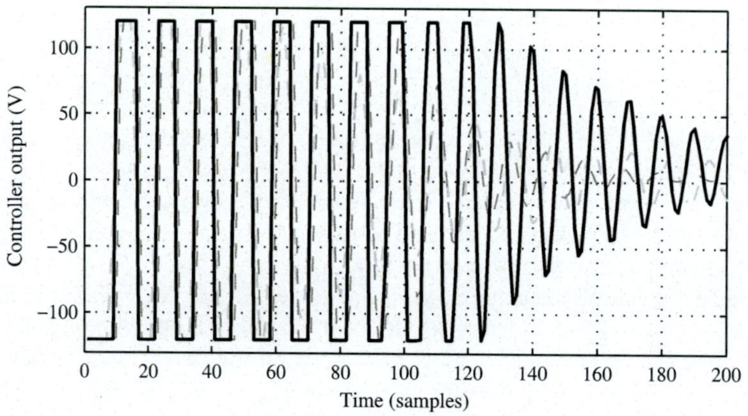

(b) LQ controller output detail

Fig. 12.4 Comparison of a detail of the vibration response at the beam tip (**a**) and the detail of the controller output (**b**) generated by low order LQ controllers with varying controller output penalization $\mathbf{R} = r$

shown for reference in Fig. 12.5, where the active structure settles back to its equilibrium more than ten times faster than without control. The natural damping of the system is improved significantly by the effect of the actuators and the control scheme.

In the interest of a meaningful damping performance and execution time comparison analysis, all experiments featured in these tests have common features. Investigated controllers utilize the same state-space model for prediction and (if applicable) use a $n_c = 70$ steps long prediction horizon. The common $n_c = 70$ steps horizon has been determined according to the maximal task execution time of the QPMPC algorithm. As QPMPC is the least computationally efficient of the four considered methods, this horizon acts as a common basis of comparison. A longer prediction

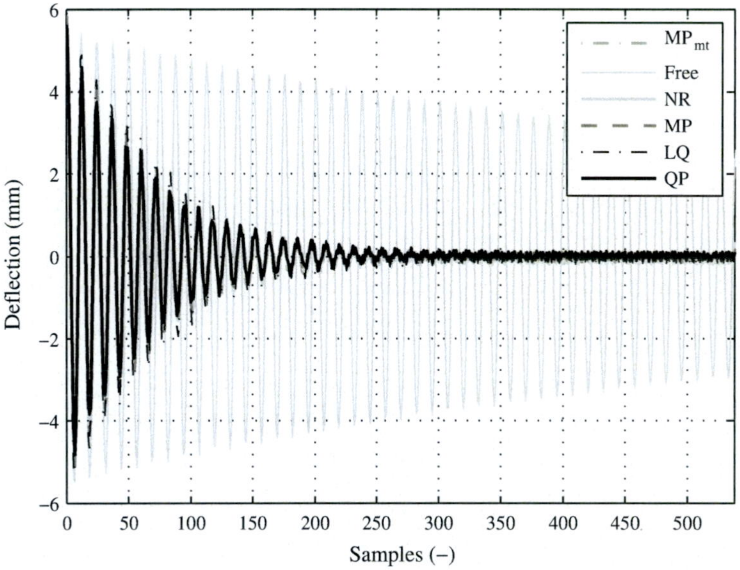

Fig. 12.5 Comparison of the free beam tip deflection without control with the MPC controlled responses. The settling time under control is shortened by more than an order of magnitude

horizon causes the QPMPC algorithm to exceed real-time sampling times and crash the xPC Target platform on the given computer.[3]

State penalty matrices in all MPC methods and in the LQ controller computation have been set to $\mathbf{Q} = \mathbf{C}^T\mathbf{C}$, which in fact just penalizes beam deflections. Input penalty has been determined to be $\mathbf{R} = r = 1\text{E-4}$ through simulation tests with different LQ controllers. This value seems to be the ideal compromise between proper performance and stability.

In all cases $\underline{u} = -\overline{u} = 120$ V constraints are enforced on the inputs and sampling time is set to $T_s = 0.01$ s. The controllers have been implemented in Matlab/Simulink and the resulting Simulink block scheme has been transferred onto the same target computer running the xPC Target environment. The block schemes have been identical in every case, except the controller algorithm itself. In addition to the controller, these block schemes contain means for A/D and D/A data conversion a Kalman filter routine utilizing the same system model, and means for data logging.

Figure 12.6 shows the closer detail of the beam response to this type of excitation. The beam tip has been deflected 5 mm away from its equilibrium, then left to vibrate freely. The four MPC methods in question are contrasted to saturated LQ control to give a basis of comparison both in damping performance and online computation time requirements.

[3] See Sect. 5.5.4.2 for exact target computer specifications and Sect. 12.5 for timing properties.

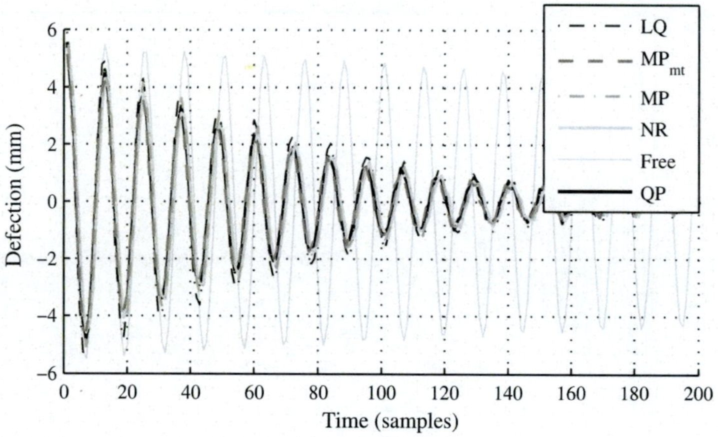

(a) Beam deflection

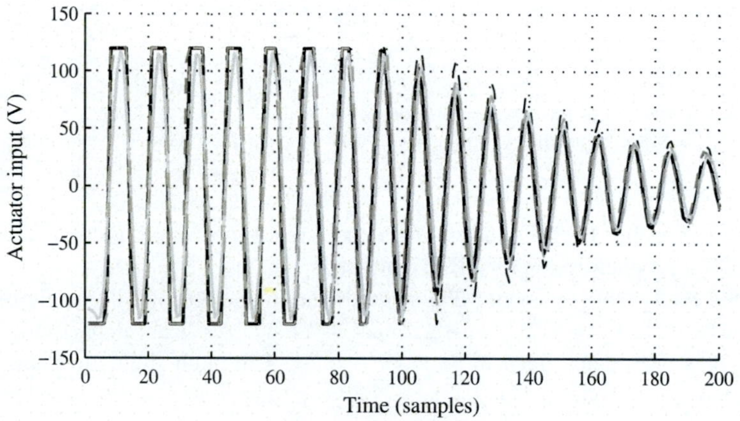

(b) Actuating voltage

Fig. 12.6 Response of the beam tip after an initial deflection of 5 mm is shown in (**a**), while the corresponding actuating signal is featured in (**b**)

Vibration of the beam tip is featured in Fig. 12.6a, where one may see that there is no substantial difference between any of the methods. The worst performance is associated with saturated LQ, which settles slower than the rest of the group. The controlled vibration response is not distinguishable, all stable MPC methods perform very similar indeed. All oscillating control courses decay to a steady-state output value at a similar exponential rate. Although the conclusion of this experiment is that the damping performance of the different MPC strategies is almost the same, Fig. 12.7a indicates a detail of the beam tip vibrations. In this figure, it is clear that QPMPC, MPMPC and mt-MPMPC give the same damping performance

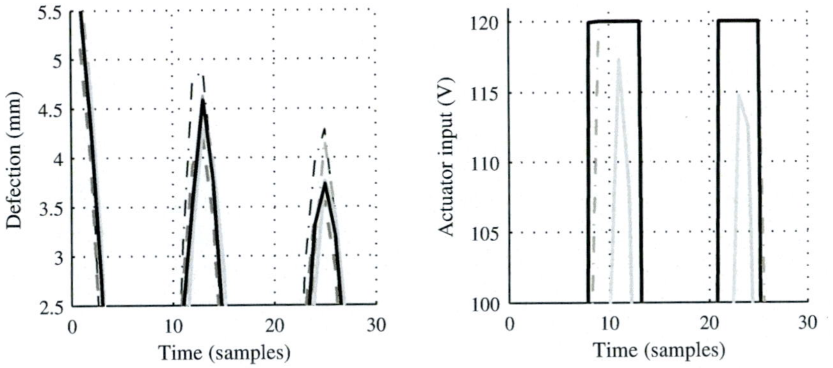

(a) Beam deflection **(b)** Actuating voltage

Fig. 12.7 Detail of the beam tip vibration output is shown (**a**), while the corresponding actuating signal is featured in (**b**). See Fig. 12.6a for a legend of line designations

while NRMPC is slightly worse. The highest displacement peaks, therefore the worst damping appears with the LQ controller.

A more dominant difference is observable in Fig. 12.6b, where the voltage passed onto the actuators by the different controllers is featured. Saturated LQ produces slightly more aggressive control moves than the MPC methods, which is especially visible after the initial saturated stage has passed. The QP, MP and minimum-time MP-based MPC methods perform very similarly, the practical differences in this test are negligible. The artifacts caused by the suboptimality of Newton–Raphson-based MPC are dominant on the figure: instead of the very saturated switching behavior resembling a square signal visible at the beginning of the test, NRMPC produces a less optimal output approximating but never fully reaching the constraints. This response resembles a sinusoidal signal on the image. Note that by implementing the one step ahead extension of the NRMPC algorithm [46, 55], the two controlled responses could possibly match more closely. However, simulation tests suggested that the extension did not provide the expected performance increase, this is believed to be caused by the given system model of the lightly damped underactuated system. Similarly to the previous case, a detail of the first 30 samples (0.3 s) of the control course focusing near the upper voltage constraint is presented in Fig. 12.7b.

A slight deviation is measurable in the control signal, drifting away from the equilibrium value of zero volts. This is due to the fact that the reference value of the laser triangulation system is arbitrary, and also tends to change by the fractions of millimeter due to deformation during experiments. To compensate for this static deflection difference, the controller attempts to supply a mean voltage shifted from zero constantly.

From this test it is evident that there is no substantial difference in the damping performance for the investigated MPC methods. All MPC methods respect the actuator constraints as expected and the natural damping properties of the beam are significantly improved. It is important to see that while the QPMPC controller was already at the verge of implementability on the given hardware and the MPMPC strategies were requiring long computation times, the NRMPC could be applied with orders of magnitude shorter sampling periods. A similar system with higher sampling rates could only be controlled by NRMPC of the four investigated MPC vibration control strategies. Taking into account the damping response of the system, the suboptimality of NRMPC is not significant in comparison with optimal QPMPC.

Although no significant performance difference has been demonstrated between the LQ controller and the MPC methods here, this is simply because this laboratory AVC model uses only a single input and a single output. The advantages of constrained MPC methods over saturated LQ control will mainly manifest with more complex MIMO vibration control systems. The constrained MPC strategy optimizes its inputs while taking into account the different limits imposed on the system. The saturated LQ method only clips the outputs to the allowable levels, therefore its performance will suffer significantly with increasing problem dimensionality in comparison to MPC.

12.3 Frequency Domain Tests

This section focuses on the behavior of the damped beam in the frequency domain. Free response is compared to the different investigated MPC methods and contrasted to LQ control as well. First, the response to a harmonic excitation by an electrodynamic modal shaker is investigated. This is followed by a brief analysis of the NRMPC and MPMPC controlled beam, which is excited by a piezoelectric transducer.

12.3.1 Disturbance by Modal Shaker

An electrodynamic modal shaker[4] has been connected directly to the beam. This modal shaker has been utilized as a source of harmonic swept disturbance through the bandwidth of interest. Swept sinusoidal disturbances are standardly used to perform frequency domain tests on mechanical structures with AVC using electrodynamic shakers [39, 52, 57, 60, 62, 78]. The linear nature of the beam dynamics is also an important assumption in the damping comparison tests performed in the frequency domain [38].

The modal shaker has been mechanically coupled to the vibrating beam via a thin copper wire. To prevent mechanical interaction in modal tests, a piano wire is usually used as the stinger mechanism, which has to be connected perpendicularly to the plane of movement [12]. The added mass shifts resonant frequencies down, while constraining the structure (thus increasing stiffness) may increase resonant frequencies [38].

4 See Sect. 5.5.5 for the description of the shaker and shaker amplifier.

As this experiment is not designed as a modal test, it is not essential to prevent all interactions. The shaker only acts as a source of disturbance. The imperfection of the connection reveals itself mainly through the shift of the first uncontrolled modal frequency to slightly higher values: from 8.1 Hz to approximately 8.8 Hz in forced vibration mode without control.

The simple second order prediction model used in the MPC controllers contains information about the first resonant frequency. Based on this fact and the sampling frequency of the controller, a chirp signal with the bandwidth of 0–20 Hz has been supplied onto the shaker to excite the beam in and around the first resonant frequency. The amplitude of the chirp signal was ± 10 V and was passed to the shaker amplifier.

A manufacturer specified 1.8 A (RMS) current limit was enforced, this however proved to be redundant. Repeated tests revealed that the amplification gain had to be significantly lower than the possible maximum. This is because the shaker in its full power drive conditions excites the beam tip over ± 15 mm, which is over the operation range of the laser triangulation system. The measurement lasted 240 s, which allowed the first mode to be excited with sufficient resolution. The sampling rate of control and measurement has been set to 0.01 s.

The same second order state-space model of the vibrating system was used for generating predictions in both controllers. Just as in the case of other trials, measures have been taken to ensure the same conditions for all algorithms. Input has been penalized by $\mathbf{R} = r = 1\text{E-}4$ and states by $\mathbf{Q} = \mathbf{C}^T\mathbf{C}$. All controllers covered a comparable region of attraction. Reference has been set to a zero deflection.

Figure 12.8 shows the result of this experiment. The periodogram of the beam tip deflection signal is featured in Fig. 12.8a, where both the controlled and free response is observable. Similarly to the experiment in Sect. 12.2 there is no observable damping performance difference for the various types of MPC control. Only the undamped response is distinguishable from the rest. We may state that the four investigated stable MPC controllers provide a very similar vibration damping performance in the bandwidth of interest.

Maximal power-frequency signal amplitude with the corresponding resonant frequency and the absolute deflection is given by Table 12.1. As one may judge by the level of peak beam tip deflections, all four types of investigated controllers reduced maximal deformations approximately to $y_{\max} = 3.3$ mm. Resonant frequency of the controlled response is shifted to higher values than free, since the active beam is essentially stiffened by the controllers.

The periodogram of the actuator voltage supplied by the controllers into the actuators is shown in Fig. 12.8a. There are three things worth noticing in the aforementioned figure: first is that there is no substantial difference in the system input for the QP, optimal MP and saturated LQ controlled experiments. The power of the NRMPC system input signal is however less powerful, this is due to the fact that NR provides suboptimal control in comparison to QP or optimal MP-based MPC. Finally it is worth noting that the minimum-time MP-based MPC system input tends to be noisy and generally more aggressive for the frequencies above and below the first resonant mode. Nevertheless, this is anticipated and the reason for it is that the minimum-time controller is behaving as an aggressive "bang–bang" controller. We may state,

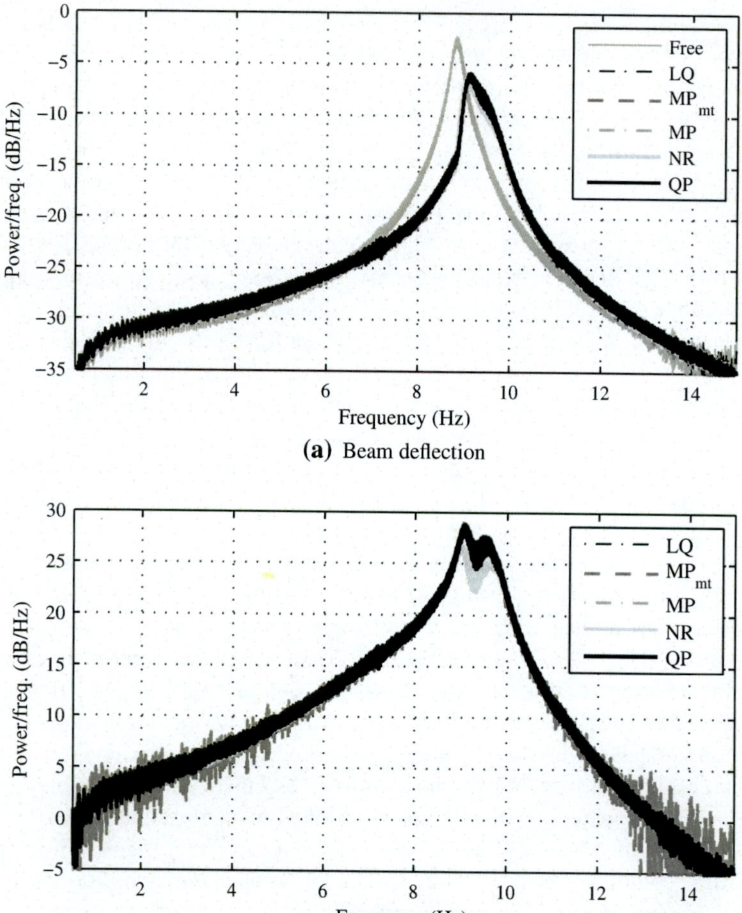

(a) Beam deflection

(b) Actuating voltage

Fig. 12.8 Narrow band periodogram of the beam tip deflection signal is shown in (**a**), while the periodogram of the corresponding actuating signal is featured in (**b**)

that the minimum-time multi-parametric controller tends to overreact small-scale disturbances outside the vicinity of the resonant frequency, but this is not significant in the overall damping response.

As it is clearly visible in Fig. 12.8a, the damping response is somewhat less efficient than in the case of vibrations forced by an additional piezoelectric transducer, as featured in the next section. The deteriorated performance of both controllers can be attributed to the fact that the effect of piezoelectric actuators is very limited in comparison with the level of the forced excitation. Harmonic excitation provided through piezoelectrics can be compensated significantly better.

Table 12.1 Summary of the damping performance analysis of the tests performed using both the chirp and PRB signal excitation

	$A(\text{dB/Hz})$	$f(\text{Hz})$	$y(\text{mm})$	$A(\text{dB/Hz})$	$f(\text{Hz})$	$y(\text{mm})$
	Chirp test			PRBS test		
Free	– 2.29	8.82	4.3	8.60	8.67	5.6
QP	– 5.82	9.13	3.4	4.76	9.11	4.7
MP	– 5.94	9.09	3.3	4.41	9.12	4.4
MP_{mt}	– 5.84	9.09	3.4	4.36	9.13	4.4
NR	– 6.25	9.08	3.3	4.70	9.14	4.7
LQ	– 5.87	9.09	3.4	4.41	9.15	4.3

Amplitude is represented as signal power/frequency and is denoted by $A(\text{dB/Hz})$, first mode resonance frequency is denoted by f(Hz) and the absolute maximal beam tip deflection is marked by y (mm) in the table

The damping performance of controllers has not been evaluated at higher frequencies, since the computer used for measurement logging is also utilized as the controller. Although, higher modes are not explicitly included in the controllers, they are expected to provide a certain level of attenuation.

12.3.2 Disturbance by PZT Actuation

In this experiment, a third piezoelectric actuator PZT3[5] has been enabled, exciting the beam through the bandwidth of interest. In addition to introducing external disturbance into vibrating structures through electrodynamic modal shakers, swept sine excitation can also be transferred to the system via piezoelectric patches [35]. The efficient suboptimal NRMPC algorithm and the optimal MPMPC algorithm have been compared to the forced free response without control.

Chirp signal with the maximal allowable amplitude has been used to drive the actuator with a frequency ranging from 0.1 to 20 Hz. The upper frequency of the chirp signal passes the first eigenfrequency, while not reaching the second. An additional test evaluated NRMPC damping performance at the second eigenfrequency around 50 Hz.

The suboptimality of the NRMPC algorithm seems to be more significant and obvious in this type of frequency domain test, as with the shaker actuated experiment or the initial deflection experiment in Sect. 12.2. This is clearly illustrated in Fig. 12.9, where the single sided logarithmic amplitude is plotted for different frequencies. As expected, the optimal MPMPC controller performs the best, suppressing the first mode by a factor of $\xi = 15.1$.[6] NRMPC provides a still significant although lower performance, damping first mode resonance by a factor of $\xi = 11.4$. Both controllers

[5] See Sect. 5.1.4 for the designation and placement of transducers.

[6] Note that ξ (xi) is used as a relative factor comparing the peak amplitudes in the periodogram and is not to be confused with the damping factor commonly denoted by ζ (zeta).

Table 12.2 First mode damping performance summary

	Freq. (Hz)	Defl. (mm)	Damp. ξ (−)
Free	8.18	12.57	–
NRMPC	8.86	1.11	11.4
MPMPC	9.3	0.83	15.1

Table listing shifted eigenfrequencies, maximal deflections and damping ratio ξ

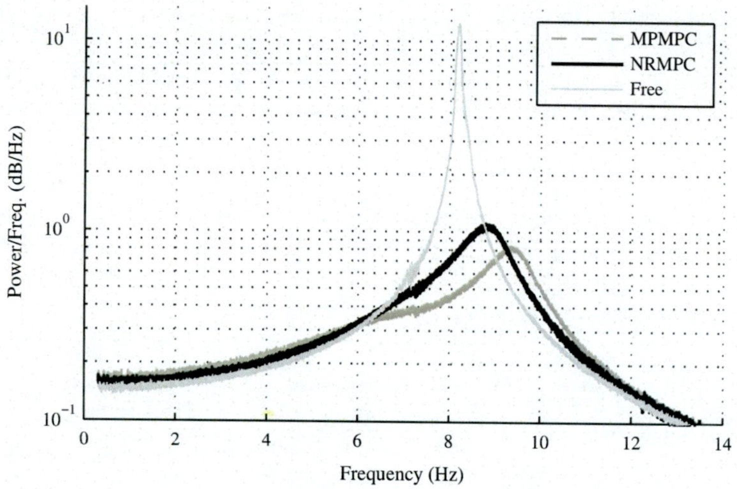

Fig. 12.9 Uncontrolled, NRMPC and MPMPC controlled beam tip response to a harmonic excitation by bonded piezoceramics in the frequency domain

shift the first mode eigenfrequency to slightly higher values, MPMPC being the furthest from the original value, followed by the NRMPC damped response. Damping performance for the first vibration mode can be summarized in Table 12.2.

It is also interesting to note that below a certain vibration displacement level [7] both controllers increase vibrations slightly. NRMPC generated more controller noise in this nearly undisturbed situation than MPMPC. At a 3 Hz excitation the mean approximate free vibration level is 0.17 mm, MPMPC is 0.18 mm and for NRMPC is 0.19 mm. Although this effect is quite insignificant, it seems that for very low-level vibrations the natural damping of the structure is more appropriate than external control. It would be advised to modify the controller to disengage when the level of vibrations fall under a given threshold. External effects present at measurement time, such as road traffic, heat convection and movement of air within the laboratory may also cause the controllers to excite the system above normal levels of vibration. Such a dead zone would possibly have an effect on controller stability, therefore its implementation is not recommended if stability guarantees are necessary.

[7] Approximately 0.2 mm, also depending on the frequency.

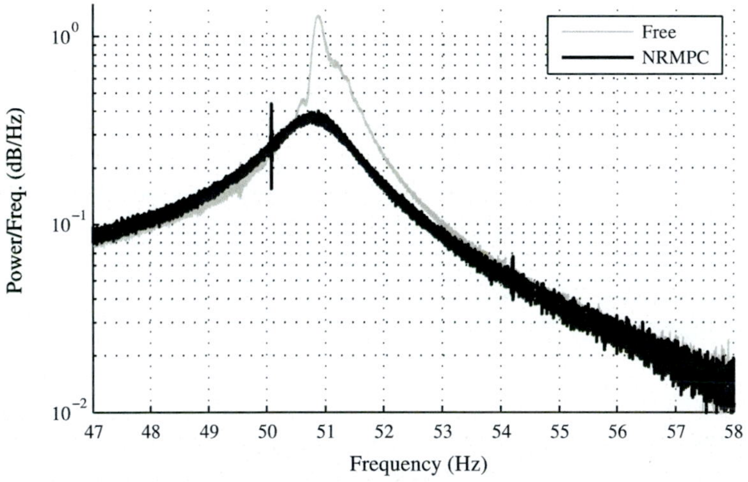

Fig. 12.10 Uncontrolled and NRMPC controlled beam tip response to a harmonic excitation in the vicinity of the second transversal modal frequency

It is worth noting that for a given example and problem parameters MPMPC differs from QPMPC only in the way of implementation and not the end result. Therefore, it is evident that this experiment performed with MPMPC has produced the same response and damping properties as QPMPC. As previously, results apply for QPMPC as well.

Although the second order model used for response prediction does not allow for the explicit handling of higher order dynamics, the NRMPC controlled tip resonance in the second mode is damped by a relative factor of more than $\xi = 3$. This is illustrated in the logarithmic amplitude spectrum of beam tip vibrations for a portion of the bandwidth of interest in Fig. 12.10. The second free vibration eigenfrequency occurs at 50.9 Hz, producing a top deflection of 1.3 mm. This is not shifted to another frequency by the NRMPC controller, resulting in a second vibration mode causing a deflection of 0.4 mm. The self-excitation at low deflections is insignificant.[8]

The extraordinary offline computational burden imposed by higher sampling rates and fourth order dynamics prevent direct comparison with MPMPC at and over second mode resonance. Given the hardware available at the time of measurement, the sampling rate of the excitation signal and measurement had to match the sampling rate of control. Therefore, this experiment used a model with higher sampling rates—which could be regarded as an oversampled model. This could lead to unexpected behavior or decremented vibration suppression performance, and has been considered only for testing purposes.

[8] At 40 Hz excitation, the difference between mean uncontrolled vibration and NRMPC controlled is approximately 0.01 mm. This is below the noise level normally encountered at the given laboratory environment.

12.4 Random Excitation

This section deals with the effect of pseudo-random excitation on the model predictive controlled active structure. First, a test is presented, where random excitation is passed onto the blade through an electrodynamic shaker. This is followed by a comparison of MPC methods on the lightly damped structure, where a fan is used to create turbulent airflow around the blade.

12.4.1 Random Excitation by a Modal Shaker

An electrodynamic modal shaker has been used to generate random vibrations on the structure. The physical setup of the shaker and amplifier settings for this test were identical to the arrangement presented for the experiment in Sect. 12.3.1. Modal shakers are commonly used to introduce random force effects to vibrating mechanical structures with control [12, 52, 68, 73, 75].

The source of the disturbance was a pseudo-random binary signal with a 0.01 Hz discrete sampling. This signal assumed two voltage levels, which were adjusted in order for the shaker to drive system states to levels, where the quadratic programming MPC algorithm requires execution times closely matching but not exceeding the sampling period of the controller. The signal has been generated utilizing the System Identification Toolbox [87].

Care has been taken to ensure the same conditions and settings for all MPC controllers. Penalization and other settings for this test were in accordance with those described in Sect. 12.3.1.

Figure 12.11 presents the results of this trial. The uncontrolled, LQ and different MPC controlled beam tip vibration power/frequency periodogram is plotted in Fig. 12.11a. The same measurement is also illustrated in the time domain for all three cases in Fig. 12.11b.

Although the resolution of the frequency domain plot in Fig. 12.11a is somewhat limited, it is evident that all controllers performed very well in these situations. As one may assess, there is no significant difference between the relative damping performance of the featured predictive control methods. The analysis given for the chirp signal test in Sect. 12.5.2 is essentially valid here as well. Maximal signal amplitudes and resonant frequencies, including maximal deflections for the individual controller schemes are featured in Table 12.1.

Controller performance is also illustrated on the time domain signal, where both the peak and average values of tip deflection were reduced in comparison with the uncontrolled situation.

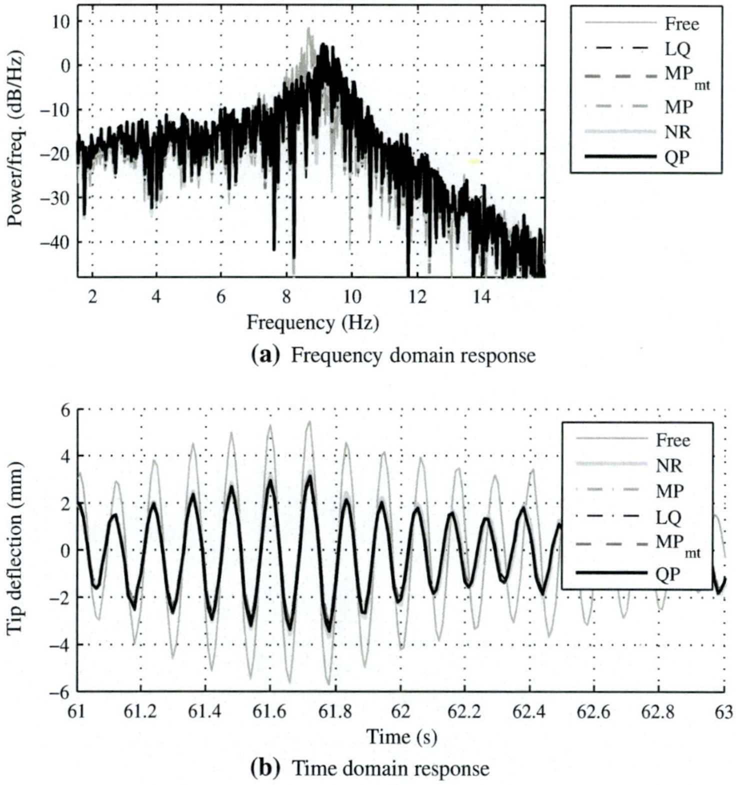

Fig. 12.11 Vibration response of the beam tip in the frequency (**a**) and time domain (**b**) to a pseudo-random excitation generated by an electrodynamic modal shaker

12.4.2 Pseudo-Random Excitation by a Medium Sized Fan

Piezoelectric actuators driven with a high amplitude random signal only trigger small deflections in the sub millimeter range. Evaluating control performance with this experimental setting is impractical. In addition to issues with noise and possible actuator overload, small deflections would prohibit the states governed by the MPC controllers to leave their terminal set and engage constraints.

Instead of using additional piezoelectric actuators, the beam has been excited by turbulent airflow produced by a medium sized fan. The fan has been placed in a stationary, perpendicular position—about 50 mm close to the free beam end. Although this excitation cannot be regarded as purely random, its effects may show additional insight in the performance of the NRMPC controller.

To engage constraints actively, the actuator limits have been modified to ± 40 V. The first portion of the graph illustrated in Fig. 12.12 shows the beam tip movement subject to the airflow with no control. Later NRMPC control was engaged, resulting

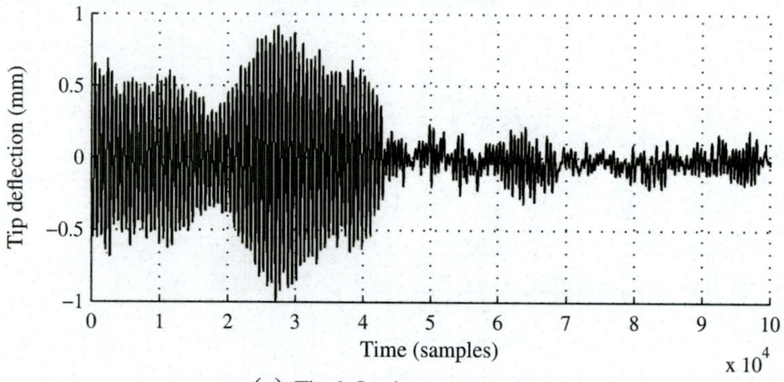

(a) Tip deflection response in mm

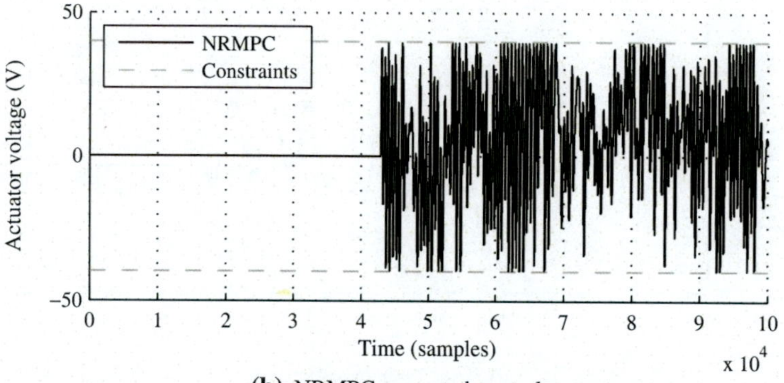

(b) NRMPC generated control response

Fig. 12.12 Vibration response of the beam tip (**a**) and the controller output (**b**) to an excitation generated by turbulent airflow around the beam

in an overall improvement in the range of tip movement. The mean absolute displacement of the uncontrolled vibration is 0.37 mm, the controlled 0.07 mm, which is a more than fivefold decrease, as visible in Fig. 12.12a. The peak uncontrolled deflection is 1.0 mm, the controlled 0.3 mm, that is a threefold decrease. The control signal indicated in Fig. 12.12b respects the given constraints, approaches but never exceeds them. The actual damping performance is significantly better, if the full voltage range of ±120 V is engaged on the actuators.

12.5 Algorithm Speed

Algorithm speeds for the different MPC implementations have been tested experimentally. All controllers featured here use common settings and are implemented on the same computer hardware, running the xPC Target rapid software prototyping platform. The prediction horizon n_c has been set according to the traditional dual-mode stable QPMPC implementation, since this is the slowest of the tested methods.[9] A $n_c = 70$ steps horizon ensures that the QPMPC controller is usable, however its top computational times are approaching the sampling instant.

12.5.1 Initial Deflection

Similarly to the experiment assessing the damping performance of the algorithms featured in Sect. 12.2, the end of the beam has been deflected 5 mm away from its initial position then left to vibrate without any further outside force interaction.

Computational times for the initial deflection test are featured in Fig. 12.13, where the horizontal axis denotes time samples and the vertical axis shows task execution times (TET) [85]. Minimal, maximal and average computational time for both inside and outside the target set is summarized in Table 12.3 as well.

As expected, this implementation of the traditional quadratic programming-based MPC [21, 22, 25] utilizes almost all the sampling period to complete its calculations. The first section of the graph is computationally more intensive, following this system states move into the target set, where a stable level of computational time is needed to evaluate the problem. With the given setup and requirements QPMPC is on the limit of practical implementability.

The rest of the controllers require significantly shorter task execution times. Optimal multi-parametric MPC achieves more than two orders of magnitude better computational times than QP even when constraints are active, indicating a substantial reserve in implementability. Minimum-time suboptimal MPMPC provides even shorter execution times, however, the less than a factor of two difference between its optimal equivalent is not very significant.

Saturated LQ is also shown in Fig. 12.13, and serves as a basis of comparison. Since all the featured MPC methods require algorithms for state observation, accessing input and output ports and data logging; the TET response of the LQ controller may be regarded as an absolute minimal computational time floor for the given hardware configuration. The laboratory measurement card manufacturer also lists a 12 μs long time for data transfer, which is included both in Fig. 12.13 and Table 12.3. In this light, it is interesting to observe the execution time graph for the NR-based controller, where NR shows no significant increase of computational time when compared to

[9] Note that the timing data for QPMPC featured in this section is given for the qpOASES sequential solver module. The sequential *qpOASES_SQProblem* module is computationally more efficient than the constant *qpOASES_QProblem*, given that the MPC formulation uses variable parameters [23]. The timing experiments featured here assume fixed prediction $\mathbf{H}, \mathbf{G}$ and fixed constraint matrices $\mathbf{A}_c, \mathbf{b}_0, \mathbf{B}_x$. The task execution times shown in the following tables and figures include only the execution time of the sequential QP solver and do not enclose any additional operation, thus QPMPC is directly comparable to the other efficient MPC methods. By using the non-sequential solver (*qpOASES_QProblem*), in theory the QPMPC method could perform slightly better than it is suggested by the timing data featured here. Nevertheless, this does not influence any of the conclusions implied by the analysis featured in the following pages.

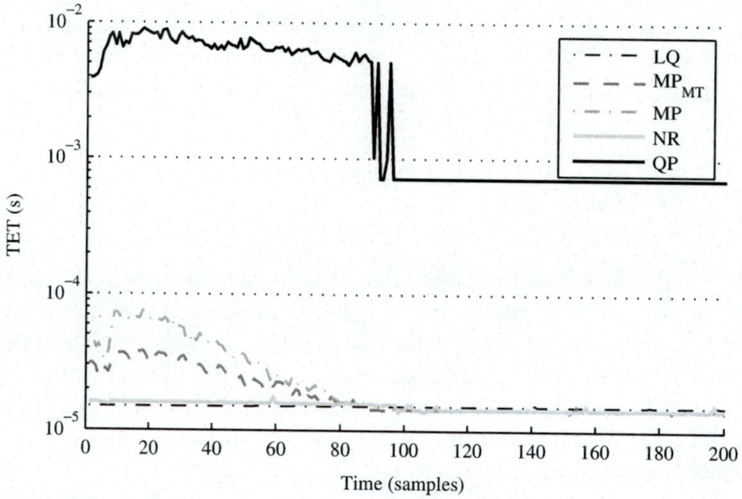

Fig. 12.13 Task execution time (TET) required to compute the previous step in seconds, for an initial deflection test

Table 12.3 Task execution time summary for the initial deflection test in microseconds. Abbreviation "t.s." denotes target set

	t_{min} (μs)	t_{max} (μs)	t_{avg} (μs)	
			Outside t.s.	Inside t.s.
QP	715	8975	6815	716
MP	14	77	42	14
MP_{mt}	14	38	26	14
NR	14	17	16	14
LQ	15	16	15	15

saturated LQ. After the observed system state enters the target set, all of the investigated controllers need shorter execution periods than it is required during control with active constraints. With inactive constraints, every investigated controller except QPMPC requires comparable computation time to the LQ controller.

A detail of the TET experiment with the efficient controllers in focus is featured in Figs. 12.6a, b, where the computational times are shown where the state is outside the target set and inside the target set.

12.5.2 Chirp Signal

A chirp signal with the bandwidth of 0–20 Hz has been supplied into a laboratory shaker to excite the beam in and around the first resonant frequency. Computational

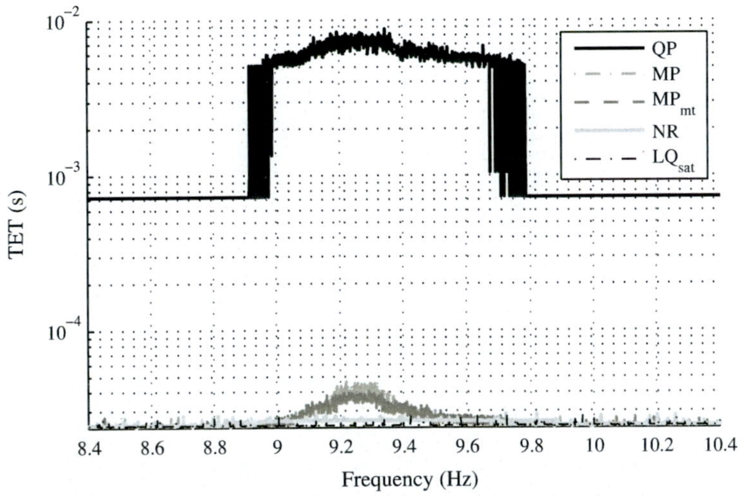

Fig. 12.14 Time required to compute one control step in the frequency domain

times were acquired for all four investigated controllers inside and outside the resonant area.

Task execution times for this experiment are featured in Fig. 12.14, while the minimal, maximal and average computational times are summarized in Table 12.4 for within and outside the resonant area. The TET values are shown in the area surrounding the first resonant frequency, as this is the region, where the disturbance introduced by the shaker may cause the system states to leave the target set. The conclusions drawn from the former experiment in Sect. 12.2 are valid for this case too. Quadratic programming-based MPC is running with execution times very close to the sampling periods of the controller. The rest of the controllers need two orders of magnitude shorter execution times. In this case, optimal MPMPC requires almost the same average execution time than its minimum-time formulation counterpart, while NR-based MPC remains the fastest of them all closely approaching the minimal possible task execution times set by the LQ controller.[10]

The task execution time for the computationally efficient algorithms is featured in detail in Fig. 12.16c, where the frequency of the excitation signal is close to that of the controlled resonant frequency of the beam. Due to the increased displacement, the system states are outside the target set and thus a longer computational time is required by all algorithms. The baseline to the computations is the LQ algorithm. Figure 12.16d shows the TET analysis for a bandwidth exceeding the first resonant frequency of the beam, with smaller displacement therefore placing the state inside the target set with the corresponding faster computational times.

---

[10] Compared to the experiment in Sect. 12.2, minimal execution time given by the LQ controller is increased by shaker controls and additional data logging.

Table 12.4 Task execution time summary for the resonant and non-resonant controlled beam response to the chirp test in microseconds

(μs)→	t_{min}	t_{max}	t_{avg}	t_{min}	t_{max}	t_{avg}
	Resonance (9.0–9.6 Hz)			Outside res.(0–8 Hz)		
QP	4925	8739	6405	725	732	728
MP	25	48	32	23	28	24
MP_{mt}	24	41	31	24	28	24
NR	25	30	26	24	28	24
LQ	24	28	24	23	28	24

Constraints are active inside the resonant area, while outside resonance the system state is located within the target set

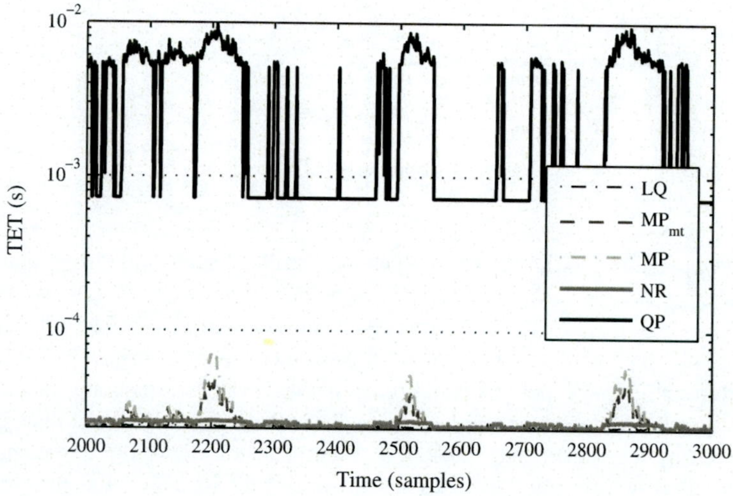

Fig. 12.15 Sample portion of a typical online computational time requirement for a disturbance initiated by a pseudo-random binary signal

12.5.3 Pseudo-Random Binary Signal

The online timing properties of the MPC methods of interest have been evaluated using a pseudo-random binary signal supplied to a modal shaker. This signal assumes two voltage levels, adjusted to drive system states to levels, where the quadratic programming MPC algorithm requires execution times closely matching but not exceeding the sampling period of the controller.

Figure 12.15 shows a sample portion of the task execution times for the individually investigated controllers, running in response to a pseudo-random binary disturbance signal. A computational time summary for a 100 s long pseudo-random test is indicated in Table 12.5 as well. The execution times are very similar to the experiments featured in Sects. 12.2 and 12.5.2, therefore the analysis will not be repeated here. Moreover, similarly to the previous cases, a detail of the task execution times for

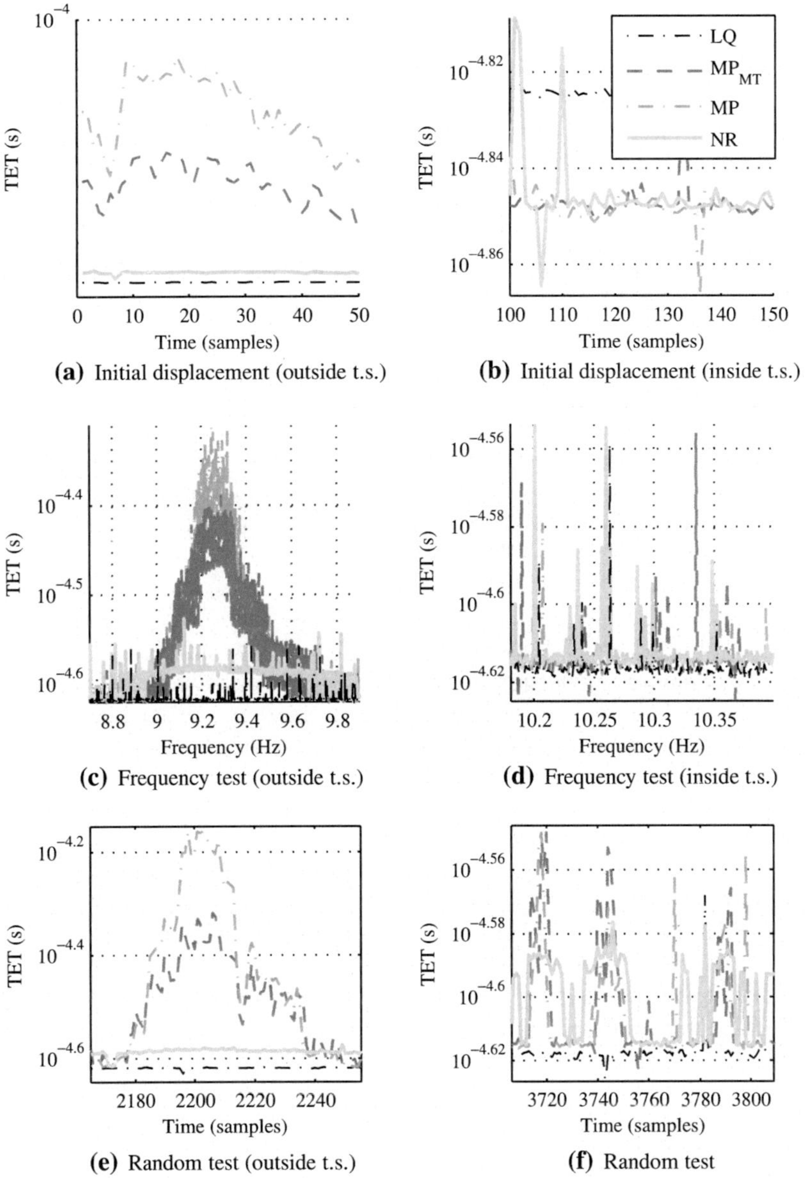

Fig. 12.16 Details of the task execution time (TET) experiments excluding the slower QPMPC algorithm and concentrating only on the computationally efficient strategies. The computational time tests are (from *top* to *bottom*): initial displacement (**a**, **b**); frequency test (**c**, **d**) and random excitation (**e**, **f**). The *left column* shows the algorithms working with a state outside the target set, while the *right column* shows the computational times if the state is inside the target set

Table 12.5 Task execution time summary for a 100 s long pseudo-random test in microseconds

	t_{min} (μs)	t_{max} (μs)	t_{avg} (μs)	$\pm t_{std}$ (μs)
QP	722	9789	2784	2603
MP	24	76	26	5
MP_{mt}	24	52	26	2
NR	24	29	25	1
LQ	23	28	24	< 1

the computationally efficient algorithms in response to the pseudo-random excitation are shown in Figs. 12.16e, f.

12.5.4 Possible Improvements on NRMPC Speed

Although the use of higher order models would increase control bandwidth, this also necessitates more time-consuming online operations involving matrices that are more complex. The computational speed achieved in the trials presented in the previous sections suggests that the sampling rate may be increased.

To achieve this, further optimization and review of online code is necessary. For the interest of rapid code deployment, the xPC Target platform has been used in this work with default code portions and blocks. On the other hand, a truly optimized code shall minimize computational overhead and use custom written functions for input–output functions, state observation and others. Process monitoring and data logging facilities also limit algorithm speed. Transcribing the C code directly into machine language, or as often referred to Assembler, is a time intensive process, however may yield very desirable results as it has been shown by a QP-based MPC approach by Wills et al. in [90, 91]. Similarly, Ferrau et al. utilized the active set QP approach also featured in this work and achieved 50 ms long sampling times[11] in [24] also without stability guarantees. Another possibility to speed up quadratic programming-based MPC is to use a multi-core parallel computing architecture [31].

Along with a more efficient algorithm implementation, the NRMPC code speed is highly dependent on the type of hardware used. A machine code optimized NRMPC algorithm deployed on a high-speed digital signal processing board presents the full potential this approach can yield.

12.6 Conclusions

Based on the experiments presented in this chapter, one might conclude that all MPC algorithms perform very similarly. Even saturated LQ provides a level of attenuation comparable to MPC, while all of the strategies significantly improve the natural

[11] This is five times longer sampling than the one utilized in this work. Moreover, Ferrau et al. utilized an 5 step prediction model instead of the 75 step assumed in this work. Note that the model in [24] has been more complex than the one used here.

damping properties of the beam. Are we suggesting that if one can afford to use an MPC control strategy with high computational demand, it does not matter which one to choose? Are we suggesting that MPC is no better than saturated LQ or any other control method with simple saturation? Absolutely not.

Let us review the question of comparing different MPC controllers in the application scenario discussed here. It is true that the experiments suggest that all algorithms perform similarly in terms of attenuation performance. However, the main issue is whether we can implement these at all. Our approach was to use the highest possible sampling speed and longest horizon, which still allows the most computationally burdensome QPMPC algorithm to be implemented. In the interest of a meaningful timing performance analysis the rest of the controllers followed this trend. Even though the damping performance was comparable, the timing data paint an entirely different picture:

- The QPMPC algorithm was already on the verge of implementability, even with the relatively slow sampling period and small model order.
- All efficient algorithms including optimal MPMPC, minimum-time MPMPC and NRMPC showed very low real-time computational needs.

That would suggest that QPMPC is not ready to be implemented in vibration control, especially with short sampling period applications or long horizons. This may change in the future by the evolution of industrial computing platforms. However, for now, that leaves us with the computationally efficient MPC algorithms. Are these all suitable for vibration control—especially the vibration control of lightly damped structures? Again, the answer is no, and this time we may turn to the offline controller computation time analysis for an explanation. The experiments were based on the longest horizon, which QPMPC could tackle on a given hardware platform. This horizon was still suitable for pre-computed explicit MPMPC controllers. Nevertheless, by looking at the offline timing data presented in the previous chapter, one may easily conclude that by increasing the horizon even further, or increasing the model order, the computation times of the MPMPC controllers become intractable. That would leave us with the NRMPC controller.

But is the NRMPC controller a clear winner of our tests? The answer is yet again, no. While NRMPC is indeed the fastest of all algorithms in the online run, moreover, its offline computation time is absolutely tractable and offers the largest possible region of attraction, it is sadly not a universal solution for vibration control. The reason for this is its suboptimality, which is detectable in the experiments comparing performance, albeit still not prohibitive for the SISO case. For a MIMO system with comparable properties but with larger model orders, the QPMPC controller along with the MPMPC controller would not be implementable. Although based on the timing data presented here NRMPC is expected to be tractable in terms of both offline and online run, its performance may degrade beyond the level of usability for increasing model orders.

We could translate the above limitations and properties of different MPC algorithms to other terms. Based on the implications of the simulations and experiments performed in the previous chapters, NRMPC provides the largest possible region of attraction. The question which algorithm may provide the largest prediction horizon with a fixed sampling period (while still being tractable for implementation) is not relevant, as this problem is in direct connection with the issue of the volume of the region of attraction. Moreover, due to the optimization of prediction dynamics, horizon length in NRMPC shall be equal to the order of the prediction model. NRMPC is also the fastest algorithm of the ones considered here, with execution times comparable to simple saturated LQ. This indicates a large reserve in terms of computational potential. What NRMPC unfortunately lacks is optimal performance, which may be a serious issue with high order SISO or more complicated MIMO systems. In case online or offline computational time and effort is not an issue but a truly optimal performance is required, the best choice is MPMPC. QPMPC is computationally the most burdensome method, thus if shorter sampling periods are required, as it is often the case with AVC, its use is not recommended. Likewise, it is not recommended to use the minimum-time variant of MPMPC in vibration control, as our simulation and experimental results did not provide a convincing improvement in off and online computational load. Thus, in summary we may state that:

- If high sampling speeds are required, NRMPC is recommended.
- If a large region of attraction is required, NRMPC is recommended.
- If the highest level of optimality is required, MPMPC is recommended.
- The use of minimum-time MPMPC or QPMPC is not recommended in AVC.

Since the performance of MPC algorithms contrasted with saturated LQ (or any other saturated method) is comparable, one might wonder, what is the reason to use the more complicated and computationally intensive predictive method, if traditional control strategies perform just as well. This question has been answered by the thousands of successful industrial users of MPC, who have preferred it to the saturated methods because of its stability guarantees and performance increase for more complicated plants. Even though the performance of the saturated LQ method is similar to MPC in the tests presented here, this property clearly diminishes with increasing model orders. The ability to handle constraints in MPC is not the same as simply cutting off the inputs of classical strategies by saturation limits. This is true both for the possible stability issues created by the non-linear saturated law and the performance, especially with higher order models with multiple inputs and outputs. To put it another way: *constraint handling* is not the same as *saturation*.

Let us attempt to provide recommendations to those who are seeking the ideal controller candidate for a vibration control system, which is constrained and medium to lightly damped:

- If constraints are not present or can be practically ignored, choose a traditional controller, such as LQ, PPF, etc.
- If constraints are present but their effect is minimal, stability issues are unlikely and the performance of the controller is not critical, choose traditional controllers like LQ, PPF and others but with saturated inputs.
- If constraints are present and they are likely to be often engaged, stability issues are likely to occur and performance is critical, choose an MPC method.

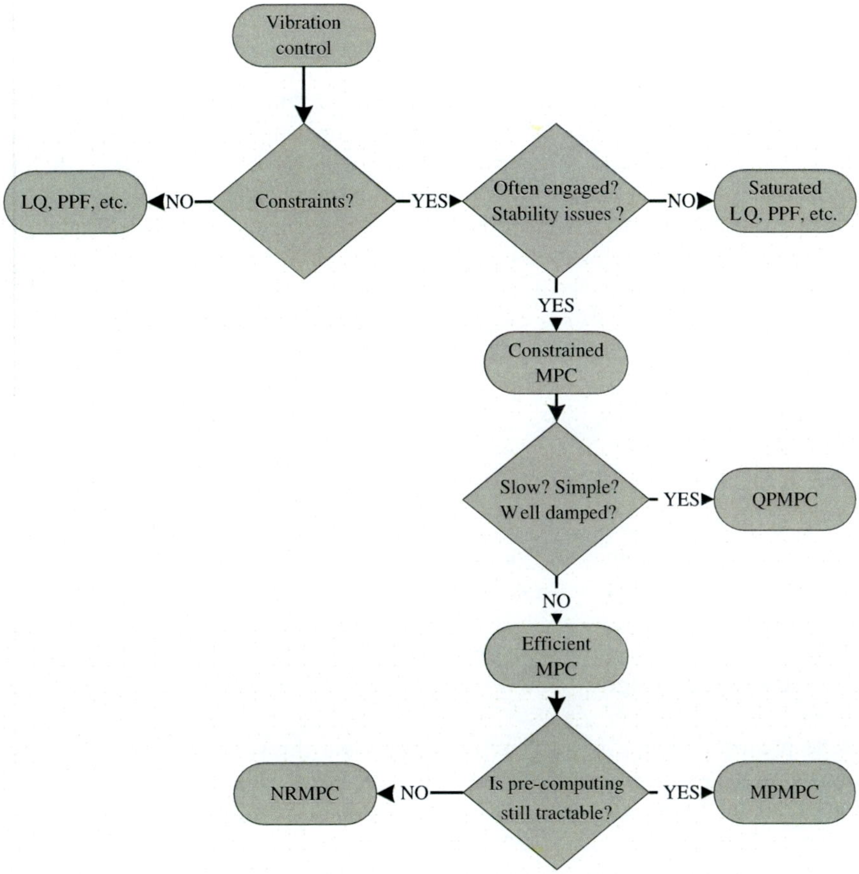

Fig. 12.17 The flowchart illustrates the line of thought one may follow in order to select the ideal controller for a vibration control system, which may have to face issues with constraints, stability and computational load

- If the vibrating system is not very lightly damped and computation speed is not an issue, choose dual-mode infinite horizon QPMPC.
- If online computational speed is an issue but the offline computation of the pre-computed explicit methods is still tractable, choose MPMPC.
- If a very large region of attraction is needed (implying a lightly damped system), very small online computational times choose NRMPC, but make sure its performance drop is not too prohibitive.

The line of thought described above is illustrated in the flowchart in Fig. 12.17.

12.7 Closing Remarks

A multi-disciplinary work has been presented here; involving the advanced topic of modern control algorithm design in vibration attenuation applications. Our ultimate aim was to successfully deploy, verify and compare computationally efficient and stabilized model predictive controllers on an experimental AVC demonstrator to actively cancel mechanical disturbances. As it is often the case with practical engineering tasks, while achieving the objectives new problems have become known. Numerous examples of the aforementioned unexpected difficulties are investigated throughout the span of this monograph. Some of these partial problems are solved or a workaround is proposed, while others were simply pointed out for giving directions for the future of this type of research.

12.7.1 Summary of Main Points

Let us now give a summary of some of the important aspects of model predictive vibration control covered in this book:

12.7.1.1 Horizon Length

Model predictive control-based vibration suppression of very flexible structures actuated by piezoceramics requires extremely long horizons, if stability is to be ensured a priori through terminal constraints. Although this fact is somewhat self-evident, the actual length of necessary prediction horizons can be surprising, as it has been presented here. These large horizons prohibit the use of QPMPC due to the heavy online computations involved. On the other hand, as the simulation data presented here suggests, the use of MPMPC in such and similar control situations is impractical especially with increased model orders. This is due to the long offline computational times required to obtain the controller, high memory needs and possibly excessive search times.[12]

12.7.1.2 NRMPC Implementation

The theory on Newton–Raphson-based suboptimal model predictive control has been established already [10, 45, 46], although the implementation in a certain physical application brings yet unknown practical aspects to light. Which semidefinite programming solver yields the best results, and what are the alternatives? How can one make the most out of the online controller code by sparing every possible elementary

[12] See Sect. 11.1 for prediction horizon lengths and Sect. 11.2 for a simulation results and discussion involving MPMPC properties on the experimental system.

operation? How is it possible to link and deploy well established matrix algebra packages to the rapid prototyping system used?[13] Relevant sections of this book propose answers to these questions and more.

12.7.1.3 NRMPC Simulation Results

It is easy to realize that if a controller fails in simulation, it cannot be counted on in a real-life application. Simulations performed using the NRMPC controller and mathematical models of the vibrating system pointed out issues involving violation of the invariance condition, causing loss of stability. This has been attributed to numerical difficulties and a possible workaround involving solver precision has been suggested. Contrary to the results of the simulations using randomly generated systems presented by Kouvaritakis et al. in [46]; the built-in suboptimality of the NRMPC algorithm may be quite significant. Unfortunately, this is true especially with state-space models of very flexible systems, like the experimental demonstrator considered here. Significant suboptimality in controller performance has been demonstrated in simulation and solution possibilities involving several steps ahead extensions discussed.[14]

12.7.1.4 Feedback Estimators

A good match has been observed between model output estimates and measurements in the feedback verification experiments. Free and LQ controlled vibrations of the beam tip are predominantly governed by the first vibration mode, therefore the considered second order models provide satisfactory tip position estimates. However, limitations of low order measurement models were evident from the frequency domain test.

Capacitive and piezoelectric sensor-based feedback has been utilized in experiments investigating the overall damping effect change due to different control schemes. Initial deflection tests demonstrated that under equivalent controllers both capacitive and piezoelectric sensor-based feedback control provide identical performance to the case of direct measurements. Moreover, the damping performance of both estimate-based controllers is on par with direct laser feedback in the frequency domain even if higher order structural vibration modes are being excited.

It may be concluded that according to the analysis presented here, industrial grade capacitive proximity sensors are suitable as position feedback sources for structural vibration control of lightly damped mechanical structures. These devices

---

[13] See Sects. 10.3.1, B.3.2.1 and B.2.3.2 for answers.

[14] See Sect. 11.3 for the numerical issues causing invariance condition violation and Sect. 11.4 for optimality issues.

present an edge over laser triangulation and laser Doppler vibrometry equipment-based feedback because of their price, availability and compactness. Since precision is not necessarily an issue in practical applications, capacitive sensors provide an adequate signal to the controller. Furthermore, capacitive proximity sensor feedback possesses the advantage of providing contact-free deformation estimates, in cases where structural integration of piezoelectric strips is not possible.

12.7.1.5 NRMPC in Experiments

The computationally efficient Newton–Raphson control strategy has been validated and evaluated in several practical tests, proving its viability in high-speed and large deformation scale vibration suppression applications. The NRMPC formulation allows recovering the largest possible region of attraction under an ellipsoidal terminal set, while having a fixed prediction horizon equal to the length of the model order. Despite of its suboptimality, NRMPC is a viable and attractive choice for the problem class in question.

Experimental measurements demonstrate excellent vibration suppression capability in wide deflection ranges, while guaranteeing stability and feasibility of constraints. NRMPC has been compared to other methods in both simulation and experiment, indicating the possibility to reach speeds up to 10 kHz without specially optimized code or hardware. Trials involving large initial deflections, frequency domain disturbances and pseudo-random excitations also proved the worth of NRMPC in such and similar applications.

12.7.1.6 MPC in Vibration Damping

Results of experiments performed on the laboratory device demonstrated no substantial difference between the vibration damping performance of the four considered stable MPC algorithms. This not only true for the theoretically identical dual-mode QPMPC and optimal explicit MPMPC, but also in the case where optimality has been traded for simplicity and in turn computational efficiency. The damping performance comparison suggests that in practice, the computationally efficient but suboptimal methods like minimum-time explicit MPC or Newton–Raphson's MPC may be implemented without a considerable loss of performance.

As it has been mentioned previously in the conclusion, the main practical distinction between the four algorithms is the computational time required to complete one cycle. From the performed tests it is evident that QP-based MPC has been on the verge of its feasible implementation, even for the fairly simple case presented in this book. Despite the fact that the solver utilized in this algorithm realization is specifically designed for the needs of MPC, it is unlikely to be useful for vibration attenuation problems with increased horizons or shortened sampling periods. The QPMPC formulation presented here may be regarded as the basis of stabilized MPC, while all others are alternatives or even approximates of the ideal method. This is why it is important to watch the development of both hardware platforms and efficient algorithms very closely. The rather pessimistic timing results presented earlier in this chapter may be somewhat alleviated by the fact that the qpOASES solver utilized in the tests is based on dense linear algebra routines. The computational effort of this class of QP routines grows quadratically to cubically with increasing horizon length. However, sparse QP solvers exist as well, allowing the computational effort to grow linearly with increasing horizon length. Sparse QP solvers may possibly be more suitable for MPC problems for extensive prediction horizons—such as the model predictive vibration control of lightly damped mechanical structures.

The optimal and pre-computed explicit MPMPC, which in theory produces equivalent outputs to QPMPC, requires extensive calculations in the offline regime. Problems of higher dimensionality (both state and horizon) are unlikely to be successfully implemented due to the likely failure of the offline computations. In short, we may conclude that given their current algorithmic formulation and hardware support, neither QP nor MP-based optimal MPC with stability and feasibility guarantees may be recommended for the active vibration damping of lightly damped structures.

Alternatively, the small loss of theoretical performance does not present problems in practice, therefore the suboptimal stable MPC methods considered in this article may be recommended for vibration attenuation purposes. Both minimum-time MPMPC and NRMPC showed damping capabilities comparable to their truly optimal counterparts. The online execution times featured in the experimental results suggest that there is a reserve for either increasing problem dimensionality or shortening sampling time.

Nevertheless, let us not forget about the possible drawbacks of these two suboptimal methods: namely, optimality decreases steeply in Newton–Raphson's MPC if the order of the prediction model is increased. This is because in higher dimensions, the true polytopic region of attraction respectively the target set cannot be effectively approximated by a hyper-ellipsoidal shape. Then again, the offline computation of the minimum-time MP-based MPC controller may fail for larger problem dimensions, and although times are significantly less than for optimal MPMPC, they may be still prohibitive for some applications.

12.7.2 Future Work

Several new ideas emerged during the individual development phases of this work. Some of them have been realized and are described in this book, while others were set aside as proposals on new directions this line of research might take, or as possible minor improvements. Hopefully, some of these initiatives can steer related work to new directions and eventually yield some interesting results. The following subsections discuss opportunities to carry on work related to this topic and suggest future applications and research directions:

12.7.2.1 Practical Applications and Uses for NRMPC

NRMPC has already proved to be a useful tool in the vibration suppression of disturbances in very flexible structures. There is a plethora of possibilities to test and match different control algorithms in other vibration attenuation applications.

Active sound insulation is a type of vibration attenuation, where the expected bandwidth of disturbances is much higher than in structural applications. Naturally, high frequency noise canceling would require much shorter sampling periods than those considered here. The question remains whether the combination of smaller disturbance amplitudes and significantly shortened sampling times would still render NRMPC as a good choice, or other methods could do better. Within the means of laboratory trials and testing, the simplest possibility is to equip a plate with several piezoelectric transducers, and use it as a sound insulation barrier. The excitations could come from a speaker or several speakers covering a wide excitation bandwidth.

12.7.2.2 Sensor Models

Future works shall investigate qualitative increase of beam deflection position precision estimates using higher order models, and whether the inclusion of higher resonant modes in the feedback model has any practical advantages in the model-based optimal control of lightly damped structures.

Increased sensing range possible with other types of capacitive proximity sensors could be beneficial for this application, but the extent of improvements needs further analysis. In addition, the exact useful bandwidth of capacitive sensors in vibration attenuation remains questionable and shall be the subject of future inquiry.

12.7.2.3 Algorithm Speed

The very low sampling rates on par with simple LQ, even with the un-optimized code and the relatively low-performance hardware suggest more room for improvement. It is desirable to investigate the performance limits of the NRMPC algorithm with different model orders. A computationally more efficient implementation may be achieved by code optimization, direct implementation into machine code, or the implementation of AAMPC [47]. Further improvement is to be gained from the use of specialized hardware—for example digital signal processing boards.

Another important issue is the performance of the NRMPC algorithm both in practical implementations and theoretically. Future works shall investigate the performance gain obtained by implementing the extension presented in [46, 47, 55] and experimentally comparing to truly optimal approaches. This extension may provide room for algorithmic improvement with a minimal increase in necessary computational time, although the level of actual performance enhancement is questionable and to be verified. This modification would involve scaling the perturbation value by stepping two or more steps outside the intersection instead of one, while verifying the preservation of invariance and constraint feasibility.

12.7.2.4 NRMPC Optimality

As the performance deterioration is much more obvious with higher order models, experimental validation has concentrated only on second order models so far. Experiments with higher order models shall investigate problems with performance, especially in comparison to other stable MPC methods. It would be interesting to distinguish whether the inclusion of higher order dynamics in the prediction model provides a significant improvement in the damping of modes above the first.

12.7.2.5 Multi-Model NRMPC

An application unrelated to vibration control is the NRMPC controlled air-fuel ratio in an internal combustion engine. Controlling the ideal Stoichiometric mixture in an automotive engine would require a nonlinear or linearized model of the process. NRMPC could be therefore involved in a multi-model control process. Conceptually several NRMPC controllers could be generated for different models, each having its own parameters. Switching between the controllers depending on engine working condition shall be realized. In addition, the effect of using NRMPC in a multi-model environment on stability and feasibility guarantees is questionable and shall be thoroughly investigated.

12.7.2.6 Adaptive NRMPC

The working version of NRMPC offline code presented in this report has been implemented in the Matlab scripting language. A software and hardware platform independent C language version could be also realized. This shall not only make system reconfiguration easier, but also offers the possibility to create an adaptive version of NRMPC. The offline optimization process does not have a chance to compete with online NRMPC sampling times, the adaptive feature could be only engaged every n_b samples. An adaptive readjustment could be performed only in the magnitude of one or several seconds. Obviously, the effects of such a feature on stability and feasibility shall be well explored.

12.7.2.7 Asymmetric Process Constraints

The NRMPC formulation introduced by Kouvaritakis et al. assumes symmetric process constraints. Practical engineering applications often require non-symmetric constraints on the output, input or states. An interesting development to the NRMPC algorithm would be therefore the inclusion of asymmetric constraints. This development shall influence numerous portions of the original algorithm, starting from the formulation of the invariance and feasibility constraints to the realization of the online optimization task. Inclusion of non-symmetric process constraints naturally must not influence guaranteed stability or constraint feasibility.

References

1. Agrawal BN, Bang H (1996) Adaptive structures for large precision antennas. Acta Astronaut 38(3):175–183. doi:10.1016/0094-5765(96)00062-8, http://www.sciencedirect.com/science/article/B6V1N-3VTW8Y7-3/2/a53f7c4acb3ee1541568e0db4062d985
2. Allaire PE, Lewis DW, Knight JD (1983) Active vibration control of a single mass rotor on flexible supports. J Franklin Inst 315(3):211–222. doi: 10.1016/0016-0032(83)90025-X, http://www.sciencedirect.com/science/article/B6V04-45D9SMR-M/2/62024de7918cc7b0b23d9703691ab67a
3. Amer Y, Bauomy H (2009) Vibration reduction in a 2DOF twin-tail system to parametric excitations. Commun Nonlinear Sci Numer Simul 14(2):560–573. doi: 10.1016/j.cnsns.2007.10.005, http://www.sciencedirect.com/science/article/B6X3D-4PYP723-2/2/b9d5375168fadb0b4e67857e92948bfc
4. Bemporad A, Bozinis NA, Dua V, Morari M, Pistikopoulos EN (2000) Model predictive control: a multi-parametric programming approach. In: Pierucci S (ed) European symposium on computer aided process engineering-10, Computer aided chemical engineering, vol 8. Elsevier, Amsterdam, pp 301–306. doi: 10.1016/S1570-7946(00)80052-8, http://www.sciencedirect.com/science/article/B8G5G-4NK5JX8-1V/2/76240158054cdb0fb1454f6e0eaa5dfe
5. Bemporad A, Morari M, Dua V, Pistikopoulos EN (2002) The explicit linear quadratic regulator for constrained systems. Automatica 38(1):3–20. doi:10.1016/S0005-1098(01)00174-1, http://www.sciencedirect.com/science/article/B6V21-44B8B5J-2/2/2a3176155886f92d43afdf1dccd128a6
6. Bittanti S, Cuzzola FA (2002) Periodic active control of vibrations in helicopters: a gain-scheduled multi-objective approach. Control Eng Pract 10(10):1043–1057. doi:10.1016/S0967-0661(02)00052-7, http://www.sciencedirect.com/science/article/B6V2H-45KSPJJ-3/2/9647861ce849d131c7d4b90cdb964751
7. Bohn C, Cortabarria A, Härtel V, Kowalczyk K (2004) Active control of engine-induced vibrations in automotive vehicles using disturbance observer gain scheduling. Control Eng Pract 12(8):1029–1039. doi:10.1016/j.conengprac.2003.09.008, http://www.sciencedirect.com/science/article/B6V2H-49Y3VWS-1/2/dd7bcefd1618f3820896ddbd6dce7430, in Special Section on Emerging Technologies for Active Noise and Vibration Control Systems
8. Borelli F, Baotic M, Bemporad A, Morari M (2003) An efficient algorithm for computing the state feedback optimal control law for discrete time hybrid systems. In: Proceeding of the American control conference, Denver, Colorado
9. Boscariol P, Gasparetto A, Zanotto V (2010) Model predictive control of a flexible links mechanism. J Intell Rob Syst 58:125–147. doi:10.1007/s10846-009-9347-5
10. Cannon M, Kouvaritakis B (2005) Optimizing prediction dynamics for robust MPC. IEEE Trans Autom Control 50(11):1892–1897. doi:10.1109/TAC.2005.858679
11. Carotti A, Lio G (1991) Experimental active control: bench tests on controller units. Eng Struct 13(3):242–252. doi:10.1016/0141-0296(91)90036-C, http://www.sciencedirect.com/science/article/B6V2Y-4829VWB-CG/2/4414a8cb4321f4e346ca04468e610264

12. Carra S, Amabili M, Ohayon R, Hutin P (2008) Active vibration control of a thin rectangular plate in air or in contact with water in presence of tonal primary disturbance. Aerosp Sci Technol 12(1):54–61. doi:10.1016/j.ast.2007.10.001, http://www.sciencedirect.com/science/article/B6VK2-4PXDM8C-1/2/db87a30acd2bfaefa3f97e3896bc9232, Aircraft Noise Reduction
13. Chang CS, Liu TS (2007) LQG controller for active vibration absorber in optical disk drive. IEEE Trans Magn 43(2):799–801. doi:10.1109/TMAG.2006.888417
14. Chen H, Allgöver F (1998) A quasi-infinite horizon nonlinear model predictive control scheme with guaranteed stability. Automatica 34(10):1205–1217
15. Chen L, Pan J, Cai G (2008) Active control of a flexible cantilever plate with multiple time delays. Acta Mech Solida Sinica 21(3):257–266. doi:10.1007/s10338-008-0829-y, http://www.sciencedirect.com/science/article/B984H-4W93GVV-8/2/ece55a791fb3bb182fa01fde13bced38
16. Choi S, Park Y, Fukuda T (1998) A proof-of-concept investigation on active vibration control of hybrid smart structures. Mechatronics 8(6):673–689. doi:10.1016/S0957-4158(98)00029-4, http://www.sciencedirect.com/science/article/B6V43-3W18XD5-3/2/7d21a042ed91fd289d4f11eb0b5dc52c
17. Choi SB, Hong SR, Sung KG, Sohn JW (2008) Optimal control of structural vibrations using a mixed-mode magnetorheological fluid mount. Int J Mech Sci 50(3):559–568. doi:10.1016/j.ijmecsci.2007.08.001, http://www.sciencedirect.com/science/article/B6V49-4PD4XHC-1/2/c491dc4a4a881e38b0e20ceef7206dec
18. Eissa M, Bauomy H, Amer Y (2007) Active control of an aircraft tail subject to harmonic excitation. Acta Mech Sin 23:451–462. doi:10.1007/s10409-007-0077-2
19. El-Badawy AA, Nayfeh AH (2001) Control of a directly excited structural dynamic model of an F-15 tail section. J Franklin Inst 338(2–3):133–147. doi:10.1016/S0016-0032(00)00075-2, http://www.sciencedirect.com/science/article/B6V04-42HNMDV-3/2/e3bf6f797834c8e8638324be88fb78f7
20. Eski I, Yıldırım S (2009) Vibration control of vehicle active suspension system using a new robust neural network control system. Simul Modell Pract Theory 17(5):778–793. doi:10.1016/j.simpat.2009.01.004, http://www.sciencedirect.com/science/article/B6X3C-4VHSDJ4-1/2/d2fe946695b369279d2e1229f15a61bd
21. Ferreau HJ (2006) An online active set strategy for fast solution of parametric quadratic programs with applications to predictive engine control. Master's thesis, University of Heidelberg
22. Ferreau HJ (2011) qpOASES—Online Active Set Strategy. Leuven, http://www.qpoases.org
23. Ferreau HJ (2011) qpOASES user's manual. Optimization in Engineering Center (OPTEC) and Department of Electrical Engineering, K. U. Leuven, Leuven
24. Ferreau HJ, Ortner P, Langthaler P, del Re L, Diehl M (2007) Predictive control of a real-world Diesel engine using an extended online active set strategy. Annu Rev Control 31(2):293–301. doi:10.1016/j.arcontrol.2007.09.001, http://www.sciencedirect.com/science/article/B6V0H-4R05C2B-2/2/23db757b09f804365ba12dc1844992d9
25. Ferreau HJ, Bock HG, Diehl M (2008) An online active set strategy to overcome the limitations of explicit MPC. Int J Robust Nonlinear Control 18(8):816–830
26. Fischer D, Isermann R (2004) Mechatronic semi-active and active vehicle suspensions. Control Eng Pract 12(11):1353–1367. doi:10.1016/j.conengprac.2003.08.003, http://www.sciencedirect.com/science/article/B6V2H-49V1CR4-2/2/0dd89d1b7760e7303a32b5bdd2cbbf9b , Mechatronic Systems
27. Fung RF, Liu YT, Wang CC (2005) Dynamic model of an electromagnetic actuator for vibration control of a cantilever beam with a tip mass. J Sound Vib 288(4-5):957–980. doi:10.1016/j.jsv.2005.01.046, http://www.sciencedirect.com/science/article/B6WM3-4G4N5VD-1/2/fc3710f0625ef69f19d16c8778a63e58
28. Grewal A, Modi VJ (1996) Robust attitude and vibration control of the space station. Acta Astronaut 38(3):139–160. doi:10.1016/0094-5765(96)00073-2, http://www.sciencedirect.com/science/article/B6V1N-3VTW8Y7-1/2/f021cc3321a4dd86a85907f0e1bc3e4c

29. Guclu R (2006) Sliding mode and PID control of a structural system against earthquake. Math Comput Modell 44(1–2):210–217. doi:10.1016/j.mcm.2006.01.014, http://www.sciencedirect.com/science/article/B6V0V-4JP9FV5-1/2/0900f85ba6e764d746c054ac040aff77 (Advances in business modeling and decision technologies, pp 1–95)
30. Hassan M, Dubay R, Li C, Wang R (2007) Active vibration control of a flexible one-link manipulator using a multivariable predictive controller. Mechatronics 17(1):311–323
31. Hassapis G (2003) Implementation of model predictive control using real-time multiprocessing computing. Microprocess Microsyst 27(7):327–340. doi:10.1016/S0141-9331(03)00063-2, http://www.sciencedirect.com/science/article/B6V0X-48FK01C-1/2/e1b3f9ea4bddf2d564bd5bcfe4a6506a
32. He XQ, Ng TY, Sivashanker S, Liew KM (2001) Active control of FGM plates with integrated piezoelectric sensors and actuators. Int J Solids Struct 38(9):1641–1655. doi:10.1016/S0020-7683(00)00050-0, http://www.sciencedirect.com/science/article/B6VJS-41WB6JY-11/2/131df87366732db91b65cd1418c2d1d9
33. Ho CC, Ma CK (2007) Active vibration control of structural systems by a combination of the linear quadratic Gaussian and input estimation approaches. J Sound Vib 301(3-5):429–449. doi:10.1016/j.jsv.2005.12.061, http://www.sciencedirect.com/science/article/B6WM3-4MV19X0-1/2/39db74e66a9494e834cdab9f0da4b886
34. Hong SR, Choi SB, Han MS (2002) Vibration control of a frame structure using electro-rheological fluid mounts. Int J Mech Sci 44(10):2027–2045. doi:10.1016/S0020-7403(02)00172-8, http://www.sciencedirect.com/science/article/B6V49-47BX3RX-4/2/53a10ce8cbf8dfa679c34e04beb688e4
35. Hsu YC, Wu CC, Lee CC, Cao GZ, Shen IY (2004) Demonstration and characterization of PZT thin-film sensors and actuators for meso- and micro-structures. Sens Actuators A 116(3):369–377. doi:10.1016/j.sna.2004.05.024, http://www.sciencedirect.com/science/article/B6THG-4CS4GDG-1/2/87f204ac4c1f3ffc13634c2fdd68a2f6
36. Hu Q (2009) A composite control scheme for attitude maneuvering and elastic mode stabilization of flexible spacecraft with measurable output feedback. Aerosp Sci Technol 13(2–3):81–91. doi:10.1016/j.ast.2007.06.007, http://www.sciencedirect.com/science/article/B6VK2-4P96269-2/2/5fbc47249fdd3f1963c5ba856f071c55
37. Imsland L, Bar N, Foss BA (2005) More efficient predictive control. Automatica 41(8):1395–1403. doi:10.1016/j.automatica.2005.03.010, http://www.sciencedirect.com/science/article/B6V21-4G7NT35-1/2/52a9590bfe1ccc2a9561165c3fbdf872
38. Inman DJ (2007) Engineering vibrations, 3rd edn. Pearson International Education (Prentice Hall), Upper Saddle River
39. Ji H, Qiu J, Zhu K, Badel A (2010) Two-mode vibration control of a beam using nonlinear synchronized switching damping based on the maximization of converted energy. J Sound Vib 329(14):2751–2767. doi:10.1016/j.jsv.2010.01.012, http://www.sciencedirect.com/science/article/B6WM3-4YG1R6R-2/2/88406f934e48ccfe56a6188409cc989e
40. Jnifene A (2007) Active vibration control of flexible structures using delayed position feedback. Syst Control Lett 56(3):215–222. doi:10.1016/j.sysconle.2006.10.005, http://www.sciencedirect.com/science/article/B6V4X-4MJC1V9-1/2/5fe33b4788d9ca97d9a9938bc7742194
41. Jung WJ, Jeong WB, Hong SR, Choi SB (2004) Vibration control of a flexible beam structure using squeeze-mode ER mount. J Sound Vib 273(1–2):185–199, doi:10.1016/S0022-460X(03)00478-4, http://www.sciencedirect.com/science/article/B6WM3-49DFFMM-1/2/1255ad59eca53b0c021632de61aef0b8
42. Kang B, Mills JK (2005) Vibration control of a planar parallel manipulator using piezoelectric actuators. J Intell Rob Syst 42:51–70. doi:10.1007/s10846-004-3028-1, http://dx.doi.org/10.1007/s10846-004-3028-1
43. Kerrigan E, Mayne D (2002) Optimal control of constrained, piecewise affine systems with bounded disturbances. In: Proceedings of the 41st IEEE conference on decision and control, 2002, vol 2, pp 1552–1557. doi:10.1109/CDC.2002.1184740

44. Kim I, Kim YS (2009) Active vibration control of trim panel using a hybrid controller to regulate sound transmission. Int J Precis Eng Manuf 10:41–47. doi:10.1007/s12541-009-0007-2, http://dx.doi.org/10.1007/s12541-009-0007-2
45. Kouvaritakis B, Rossiter J, Schuurmans J (2000) Efficient robust predictive control. IEEE Trans Autom Control 45(8):1545–1549. doi:10.1109/9.871769
46. Kouvaritakis B, Cannon M, Rossiter J (2002) Who needs QP for linear MPC anyway? Automatica 38:879–884. doi:10.1016/S0005-1098(01)00263-1, http://www.sciencedirect.com/science/article/pii/S0005109801002631
47. Kouvaritakis B, Li S, Cannon M (2010) A line search improvement of efficient MPC. Automatica 46(11):1920–1924. doi:10.1016/j.automatica.2010.07.003, http://www.sciencedirect.com/science/article/B6V21-50NH0BX-3/2/0b4491d922a7d04d1b0315edae0e8944
48. Krishnaswamy K, Rajamani R, Woo J, Cho Y (2005) Structural vibration control for broadband noise attenuation in enclosures. J Mech Sci Technol 19:1414–1423. doi:10.1007/BF03023900
49. Kvasnica M (2009) Real-time model predictive control via multi-parametric programming: theory and tools, 1st edn. VDM Verlag, Saarbrücken
50. Kvasnica M, Grieder P, Baotic M, Christophersen FJ (2006) Multi-Parametric Toolbox (MPT). Extended documentation
51. Kwak MK, Heo S (2007) Active vibration control of smart grid structure by multiinput and multioutput positive position feedback controller. J Sound Vib 304(1–2):230–245. doi:10.1016/j.jsv.2007.02.021, http://www.sciencedirect.com/science/article/B6WM3-4NH6N96-2/2/ca7b43602b9d052e388f4b2a28f1ebae
52. Landau ID, Constantinescu A, Rey D (2005) Adaptive narrow band disturbance rejection applied to an active suspension–an internal model principle approach. Automatica 41(4):563–574. doi:10.1016/j.automatica.2004.08.022, http://www.sciencedirect.com/science/article/B6V21-4FB3X55-3/2/28887440b73dcde4fdbaefe4d507e857
53. Lau K, Zhou L, Tao X (2002) Control of natural frequencies of a clamped-clamped composite beam with embedded shape memory alloy wires. Compos Struct 58(1):39–47. doi:10.1016/S0263-8223(02)00042-9, http://www.sciencedirect.com/science/article/B6TWP-45XTP9W-N/2/07b9a065ac866d8869a4240deb918851
54. Lee J, Kim J, Cheong C (1999) Piezoelectric smart structures for noise reduction in a cabin. J Mech Sci Technol 13:451–458. doi:10.1007/BF02947714
55. Li S, Kouvaritakis B, Cannon M (2010) Improvements in the efficiency of linear MPC. Automatica 46(1):226–229. doi:10.1016/j.automatica.2009.10.010, http://www.sciencedirect.com/science/article/B6V21-4XGCHXB-3/2/20a93fa6dd4fb88469638ac3bc2fe729
56. Lin LC, Lee TE (1997) Integrated PID-type learning and fuzzy control for flexible-joint manipulators. J Intell Rob Syst 18:47–66. doi:10.1023/A:1007942528058
57. Liu J, Liu K (2006) A tunable electromagnetic vibration absorber: Characterization and application. J Sound Vib 295(3–5):708–724. doi:10.1016/j.jsv.2006.01.033, http://www.sciencedirect.com/science/article/B6WM3-4JP9FXN-6/2/0b961839d0b922bbd94dcc5ce5c5f9e4
58. Lu H, Meng G (2006) An experimental and analytical investigation of the dynamic characteristics of a flexible sandwich plate filled with electrorheological fluid. Int J Adv Manuf Technol 28:1049–1055. doi:10.1007/s00170-004-2433-8
59. Luo T, Hu Y (2002) Vibration suppression techniques for optical inter-satellite communications. In: IEEE 2002 international conference on communications, circuits and systems and west sino expositions, vol 1, pp 585–589. doi:10.1109/ICCCAS.2002.1180687
60. Luo Y, Xie S, Zhang X (2008) The actuated performance of multi-layer piezoelectric actuator in active vibration control of honeycomb sandwich panel. J Sound Vib 317(3–5):496–513. doi:10.1016/j.jsv.2008.03.047, http://www.sciencedirect.com/science/article/B6WM3-4SJR2GN-1/2/04c4aad317afe74e20e6f5810f689674
61. Maciejowski JM (2000) Predictive control with constraints, 1st edn. Prentice Hall, Upper Saddle River
62. Mahmoodi SN, Jalili N, Khadem SE (2008) An experimental investigation of nonlinear vibration and frequency response analysis of cantilever viscoelastic beams. J Sound Vib

311(3–5):1409–1419. doi:10.1016/j.jsv.2007.09.027, http://www.sciencedirect.com/science/article/B6WM3-4R113SP-1/2/4baf1df12aa15dbfcbdd0e4f13149b17
63. Marzbanrad J, Ahmadi G, Jha R (2004) Optimal preview active control of structures during earthquakes. Eng Struct 26(10):1463–1471. doi: 10.1016/j.engstruct.2004.05.010, http://www.sciencedirect.com/science/article/B6V2Y-4CYNR00-1/2/271b4c49fa053fb1a95d5df632c701c8
64. Mayne DQ, Rawlings JB, Rao CV, Scokaert POM (2000) Constrained model predictive control: stability and optimality. Automatica 36(6):789–814
65. MIDÉ Technology Corporation (2007) QuickPack Actuator Catalog. MIDÉ Technology Corporation, Medford
66. MIDÉ Technology Corporation (2007) QuickPack Power Amplifier. Operator's manual, MIDÉ Technology Corporation, Medford
67. Moon SJ, Lim CW, Kim BH, Park Y (2007) Structural vibration control using linear magnetostrictive actuators. J Sound Vib 302(4–5):875–891. doi: 10.1016/j.jsv.2006.12.023, http://www.sciencedirect.com/science/article/B6WM3-4N2M6HH-5/2/417522adfca8640acfa76e890ae0533c
68. Moshrefi-Torbati M, Keane A, Elliott S, Brennan M, Anthony D, Rogers E (2006) Active vibration control (AVC) of a satellite boom structure using optimally positioned stacked piezoelectric actuators. J Sound Vib 292(1–2):203–220. doi:10.1016/j.jsv.2005.07.040, http://www.sciencedirect.com/science/article/pii/S0022460X05005171
69. Neat G, Melody J, Lurie B (1998) Vibration attenuation approach for spaceborne optical interferometers. IEEE Trans Control Syst Technol 6(6):689–700. doi:10.1109/87.726529
70. Park JS, Kim JH, Moon SH (2005) Thermal post-buckling and flutter characteristics of composite plates embedded with shape memory alloy fibers. Composites Part B 36(8):627–636. doi:10.1016/j.compositesb.2004.11.007, http://www.sciencedirect.com/science/article/B6TWK-4GDKB5G-1/2/b308d31dc4dc49f9e53da0c459ff746a
71. Pistikopoulos EN, Georgiadis MC, Dua V (eds) (2007) Multi-parametric model-based control, vol 2, 1st edn. Wiley-VCH Verlag GmbH & Co., Weinheim
72. Preumont A (2002) Vibration control of active structures, 2nd edn. Kluwer Academic Publishers, Dordrecht
73. Rashid A, Nicolescu CM (2006) Active vibration control in palletised workholding system for milling. Int J Mach Tools Manuf 46(12–13):1626–1636. doi:10.1016/j.ijmachtools.2005.08.020, http://www.sciencedirect.com/science/article/B6V4B-4HGD76C-2/2/273540b1466f54bf47cc11a241007364
74. Rossiter JA (2003) Model-based predictive control: a practical approach, 1st edn. CRC Press, Boca Raton
75. Seba B, Nedeljkovic N, Paschedag J, Lohmann B (2005) $\mathscr{H}_\infty$ feedback control and Fx-LMS feedforward control for car engine vibration attenuation. Appl Acoust 66(3):277–296. doi:10.1016/j.apacoust.2004.07.015, http://www.sciencedirect.com/science/article/B6V1S-4DTKD2W-1/2/d413b004e2a2e14e9df7fcf75f2df02f
76. Shan J, Liu HT, Sun D (2005) Slewing and vibration control of a single-link flexible manipulator by positive position feedback (PPF). Mechatronics 15(4):487–503. doi:10.1016/j.mechatronics.2004.10.003, http://www.sciencedirect.com/science/article/B6V43-4DR87K7-4/2/2dd311fdd61308e1415cd45c1edc3076
77. Spangler R (2007) Piezo sensor techical note, 2nd edn. MIDÉ Technology Corporation, Medford
78. Sumali H, Meissner K, Cudney HH (2001) A piezoelectric array for sensing vibration modal coordinates. Sens Actuators A 93(2):123–131. doi: 10.1016/S0924-4247(01)00644-6, http://www.sciencedirect.com/science/article/B6THG-43N2J67-5/2/b7e5396a71e7ee7fae9ef93cb254b7b7
79. Sun D, Mills JK, Shan J, Tso SK (2004) A PZT actuator control of a single-link flexible manipulator based on linear velocity feedback and actuator placement. Mechatronics

14(4):381–401. doi:10.1016/S0957-4158(03)00066-7, http://www.sciencedirect.com/science/article/B6V43-49DN5K4-1/2/fa21df547f182ad568cefb2ddf3a6352
80. Takács G, Rohaľ-Ilkiv B (2009) Implementation of the Newton–Raphson MPC algorithm in active vibration control applications. In: Mace BR, Ferguson NS, Rustighi E (eds) Proceedings of the 3rd international conference on noise and vibration: emerging methods, Oxford
81. Takács G, Rohaľ-Ilkiv B (2009) MPC with guaranteed stability and constraint feasibility on flexible vibrating active structures: a comparative study. In: Hu H (ed) Proceedings of the eleventh IASTED international conference on control and applications, Cambridge
82. Takács G, Rohaľ-Ilkiv B (2009) Newton–Raphson based efficient model predictive control applied on active vibrating structures. In: Proceedings of the European control conference, Budapest
83. Takács G, Rohaľ-Ilkiv B (2009) Newton–Raphson MPC controlled active vibration attenuation. In: Hangos KM (ed) Proceedings of the 28th IASTED international conference on modeling, identification and control, Innsbruck
84. The Mathworks (2007) Matlab signal processing blockset v6.6 (R2007b). Software. The Mathworks Inc., Natick
85. The MathWorks (2008) xPC target for use with real-time workshop, 6th edn. The Mathworks Inc., Natick
86. The Mathworks (2011) Matlab control system toolbox v9.1 (R2011a). Software. The Mathworks Inc., Natick. http://www.mathworks.com/products/control/
87. The Mathworks (2011) Matlab system identification toolbox v7.4.2 (R2011a). Software. The Mathworks Inc., Natick http://www.mathworks.com/help/toolbox/ident/
88. Tso SK, Yang TW, Xu WL, Sun ZQ (2003) Vibration control for a flexible-link robot arm with deflection feedback. Int J Nonlinear Mech 38(1):51–62. doi:10.1016/S0020-7462(01)00040-3, http://www.sciencedirect.com/science/article/B6TJ2-46BSCBF-5/2/db9a6ea06f0106fae187a067a96b1888
89. Vasques C, Rodrigues JD (2006) Active vibration control of smart piezoelectric beams: Comparison of classical and optimal feedback control strategies. Comput Struct 84(22–23):1402–1414. doi:10.1016/j.compstruc.2006.01.026, http://www.sciencedirect.com/science/article/B6V28-4K4219V-1/2/fc83fdc87b19e200d95c2b596f8f0201, Composite Adaptive Structures: Modelling and Simulation
90. Wills A, Bates D, Fleming A, Ninness B, Moheimani R (2005) Application of MPC to an active structure using sampling rates up to 25 kHz. In: 44th IEEE conference on decision and control, 2005 and 2005 European control conference, CDC-ECC '05, pp 3176–3181. doi:10.1109/CDC.2005.1582650
91. Wills AG, Bates D, Fleming AJ, Ninness B, Moheimani SOR (2008) Model predictive control applied to constraint handling in active noise and vibration control. IEEE Trans Control Syst Technol 16(1):3–12
92. Wilson DG, Robinett RD, Parker GG, Starr GP (2002) Augmented sliding mode control for flexible link manipulators. J Intell Rob Syst 34:415–430. doi:10.1023/A:1019635709331
93. Yim W (1996) Modified nonlinear predictive control of elastic manipulators. In: Proceedings of the 1996 IEEE international conference on robotics and automation, vol 3, pp 2097–2102. doi:10.1109/ROBOT.1996.506180
94. Zmeu K, Shipitko E (2005) Predictive controller design with offline model learning for flexible beam control. In: Proceedings of the 2005 international conference on physics and control, pp 345–350. doi:10.1109/PHYCON.2005.1514005

Appendix A
FE Modeling of the Active Structure

This appendix complements the discussion of FEM simulation results in Sect. 5.4 with a detailed account of the code used to generate those results. Thus, it is recommended to readers who are aiming to create a FEM model of a smart system equipped with piezoelectric actuators in ANSYS. ANSYS is one of the well-known commercial finite element analysis software packages available today that is capable of performing simulations including the electro-mechanical interaction from piezoelectric sensors and actuators. ANSYS has been used in numerous works aimed at investigating the vibration and damping behavior of smart structures equipped with piezoelectric actuators [3, 4, 5, 15].

The basis for the code featured here are the demonstration examples found in the ANSYS manual [1] and using the material data for the PZT actuator and the supporting aluminum beam from various sources [6, 9, 10, 12, 13, 14]. For more information about the simulation of vibrating systems in ANSYS the reader may be interested in the book by Hatch [5] or other similar publications on the topic of finite element analysis [2, 7, 8, 11, 16]. This appendix is divided into several sections discussing analysis setup, parameter declaration, creating a solid model, meshing the model, creating boundary conditions and setting up various simulation types.

The cantilever beam of the active vibration control (AVC) demonstrator—featured throughout this work as a comparison benchmark—has been modeled using the 3D structural solid element SOLID5. The higher order tetrahedral version of this element is SOLID98, although the usage of this is not justified in this application. The settings of the element properties allow for a simple mechanical use, where only the displacement of the nodes is considered.

SOLID5 is defined by eight nodes, each having three degrees of freedom. These DOF are translations in the nodal x, y and z directions [1]. Material properties required for this element in this case is Young's modulus, Poisson's ratio and density. The material properties for aluminum were obtained from engineering tables and by direct experimentation. All units used were SI default, since ANSYS utilizes the constant value for the permittivity of free space in F/m.

G. Takács and B. Rohal'-Ilkiv, *Model Predictive Vibration Control*,
DOI: 10.1007/978-1-4471-2333-0,

The piezoelectric material was modeled using a 3D coupled field solid element SOLID 5. This element was chosen because of its piezo capability. A similar element SOLID45 has also been featured in studies aimed at vibration control, for example the work by Dong et al. in [4]. This element has eight nodes with up to six degrees of freedom. This application utilizes three axial displacements and one voltage DOF. Polarization direction is assumed to be along the z axis of the element coordinate system. Required material properties to sufficiently describe the material properties of the crystal are: density, permittivity at constant stress, elastic compliance matrix and piezoelectric strain matrix. Material properties are anisotropic, meaning that it has different stress/strain behavior in x, y and z directions. The axes of anisotropy coincide with the axes of the element.

The assembly of the material property matrices proved to be difficult because of the misleading indexing conventions used in practice. Material properties for the piezoelectric material PZT5A were obtained from the manufacturer and diverse other sources [6, 12, 13]. Some elements of the matrices were numerically calculated from other known parameters, as often not all constants are readily available.

Physical dimensions and the placement of PZT patches respect the actual configuration of the experimental setup introduced in Sect. 5.1. Because the geometric design of the beam and actuator placement was initially subject to a frequent redesigning procedure due to the iterative workflow, the geometric model is working with a millimeter precision. Piezoelectric patches and the aluminum cantilever beam were meshed separately. Meshing had a uniform 2 mm density. This ensured sufficient resolution and precision without sacrificing too much computing time.

The overlapping surfaces provide coincident nodes. These nodes were merged with the relevant command, to ensure mesh connectivity. Epoxy resin used on the physical model was assumed to have an effect on the overall dynamic response and it has been explicitly modeled at first, however this assumption was later removed. Instead an ideally thin and rigid adhesive layer was assumed in more recent simulations.

A.1 Analysis Setup

Although ANSYS presents a possibility to use a graphical user interface to create FEM models and perform simulations; writing command macros has several advantages. These include the possibility to correct mistakes without tediously repeating all preceding tasks and easy reconfiguration. This work utilized the built-in ANSYS macro language ANSYS Parametric Design Language (APDL) to create models and initiate simulations.

The following sections describe the process of creating a finite element model of the considered active structure. These steps include simulation environment initiation, parameter definitions, geometric solid modeling, meshing, establishing boundary and initial conditions and of course setting up simulations.

A.1.1 Initializing

Initializing of the modeling and simulation process starts by exiting from the processor and clearing ANSYS database. The title of the job is stated, while successive commands suppress extended interpreted data input and place comments in the output. To begin the modeling phase, the pre-processor is engaged. To prevent ANSYS from displaying irrelevant warning messages, shape checking is carried out in silent mode. Shape checking limit is also increased from the default value:

```
FINISH
/CLEAR
/title
/nopr
/com
/PREP7
SHPP,SILENT,ON
SHPP,MODIFY,1,40
```

A.2 Defining Problem Variables and Parameters

Problem parameters and other variables are defined in the following code segments. These variables mainly describe the geometrical properties of transducers, support beam, adhesive layer and transducer placement. Material properties are also stated:

A.2.1 Dimensions of the Piezoelectric Transducer and its Material

Dimensions of the piezoelectric wafer are stated, along with the electric potential if needed in the actual simulation. Material properties of the piezoelectric patch are also declared. This is including density of PZT5A, piezoelectric strain coefficients[1] and relative permittivity at constant stress:

```
L=44e-3
H=0.25e-3
W=20e-3
V=120
```

[1] According to manufacturer data and available PZT5A material properties.

```
s11=16.4e-12
s22=s11
s12=-5.74e-12
s13=-7.22e-12
s23=s13
s33=18.8e-12
s44=47.5e-12
s55=47.5e-12
s66=44.3e-12
pidns=7700
d31=-179e-12
d32=-179e-12
d33=350e-12
d24=584e-12
d16=584e-12
ept11=1730
ept22=1730
ept33=1700
```

A.2.2 Support Beam Dimensions and Material

Length, thickness and width of the support beam are declared with millimeter precision. The material properties for aluminum are stated, including Young's modulus,[2] Poisson's ratio[3] and density:

```
Lb=550e-3
Hb=3e-3
Wb=40e-3
EXY=66.70e + 09
POIS=0.35
aldns=2834
```

A.2.3 Adhesive Geometry and Material Properties

Geometrical properties of the adhesive layer are stated, including length, width and bonding thickness. The offset from the piezoelectric patch sides is also declared, this is the dimension the adhesive exceeds wafer geometry.

[2] For generic aluminum this is 70 GPa, Dural has a 73 GPa modulus, while the material in question has been approximated at a softer 66 GPa modulus.

[3] From various sources: Poisson's ratio is 0.35 for aluminum.

Mechanical properties of the adhesive layer are stated. Its Young's modulus,[4] Poisson's ratio and density are conforming to generic epoxy resin properties:

```
Ha=0.05e-3
La=L
Wa=W
tmp=0e-3
Lda1=tmp
Wda1=tmp
Lda2=tmp
Wda2=tmp
Lda3=tmp
Wda3=tmp
Lda4=tmp
Wda4=tmp
EXYa=4e09
POISa=0.38
dnsa=1160
```

A.2.4 Placement of the Piezoelectric Transducers on the Beam

The following code segment states parameters defining the geometric placement of piezoelectric transducers on the surface of the aluminum support beam. All units are in meters and referenced from the default coordinate system origin[5]:

```
L1=14e-3
W1=10e-3
L2=90e-3
W2=10e-3
L3=322e-3
W3=10e-3
L4=14e-3
W4=10e-3
```

A.3 Solid Model

The following code segments create a solid volumetric model of the active structure. Due to the simple shape of the device, practically only block shapes are needed with the relevant dimensions. Base beam has to be divided into sections for

[4] See [14] for details on epoxy resin tensile strength and other mechanical properties.

[5] This is the clamped end of the blade.

practical meshing considerations, while the modeling of transducers and adhesives is straightforward.

A.3.1 Beam Model

The first command, *BLC4* creates a block with the dimensions of the support blade. Later the beam is divided into sections according to the distribution of the piezoelectric patches. This is needed for the meshing phase of the work.

Division of the original block is carried out by rotating the work plane with the *WPROTA* command and offsetting it to the desired place with the *WPOFFS* command. Finally, *VSBW* subtracts intersection of the working plane from volumes, effectively dividing volumes.

The very last *WPCSYS* command in this code segment resets the working plane into its original location by defining its position based on the working coordinate system:

```
BLC4,0,0,Lb,Wb,Hb
WPROTA,,-90,
WPOFFS,,,W1
VSBW,ALL
WPOFFS,,,W
VSBW,ALL
WPAVE,0,0,0,
WPROTA,,,90
WPOFFS,,,L1
VSBW,ALL
WPOFFS,,,L
VSBW,ALL
WPOFFS,,,L2-L1-L
VSBW,ALL
WPOFFS,,,L
VSBW,ALL
WPOFFS,,,L3-L2-L
VSBW,ALL
WPOFFS,,,L
VSBW,ALL
WPCSYS,,0
```

A.3.2 Solid Model of the Transducers

Piezoelectric transducer geometry is modeled by defining blocks at each respective location. The three upper and finally the bottom transducers are created using the previously defined dimension parameters. All dimensions relate to the particular command instruction and coordinate reference system:

```
BLOCK,L1,L1 + L,w1,w1 + W,Hb + Ha,Hb + H + Ha
BLOCK,L2,L2 + L,w2,w2 + W,Hb + Ha,Hb + H + Ha
BLOCK,L3,L3 + L,w3,w3 + W,Hb + Ha,Hb + H + Ha
BLOCK,L4,L4 + L,w4,w4 + W,0 - (Ha), - (H + Ha)
```

A.3.3 Solid Model of the Adhesive Layer

The geometry of the adhesive epoxy resin layer is modeled similar to the piezoelectric transducers. Adhesive layer is modeled by defining blocks at each respective location. All dimensions relate to the particular command instruction and coordinate reference system:

```
BLOCK,L1 - Lda1,L1 + Lda1 + La,W1-Wda1,W1 + Wda1 + Wa,Hb,Hb + Ha
BLOCK,L2 - Lda2,L2 + Lda2 + La,W2 - Wda2,w2 + Wda2 + Wa,Hb,Hb + Ha
BLOCK,L3 - Lda3,L3 + Lda3 + La,W3 - Wda3,w3 + Wda3 + Wa,Hb,Hb + Ha
BLOCK,L4 - Lda4,L4 + Lda4 + La,W4 - Wda4,w4 + Wda4 + Wa,0, - (Ha)
```

A.4 Assigning Material and Element Types

Materials for the piezoelectric transducer, aluminum beam and epoxy resin adhesive layer are created by defining a finite element type for each and assigning material properties. All three materials use the same element *SOLID5* with piezoelectric capability, although only the transducers have this feature enabled. The *ET* command inputs element type with the proper switch controlling the presence of an additional electric potential degree of freedom.

The command *MP* inputs material properties like density, Poisson's ratio and permittivity at constant stress or Young's modulus if applicable. Command *TB* activates a data table for nonlinear material properties or special element input. Fields in the command control the particular type of data input: *ANEL* assembles an elastic compliance matrix, *PIEZ* a piezoelectric strain matrix:

```
et,1,SOLID5,3
MP,DENS,1,pidns
MP,PERX,1,ept11
MP,PERY,1,ept22
MP,PERZ,1,ept33
tb,ANEL,1,,,1
tbda,1,s11,s12,s13
```

```
tbda,7,s22,s23
tbda,12,s33
tbda,16,s66
tbda,19,s55
tbda,21,s44
tb,PIEZ,1,,,1
tbda,3,d31
tbda,6,d33
tbda,9,d32
tbda,14,d24
tbda,16,d16 ET,2,SOLID5,2
MP,EX,2,EXY
MP,PRXY,2,POIS
MP,DENS,2,aldns
ET,3,SOLID5,2
MP,EX,3,EXYa
MP,PRXY,3,POISa
MP,DENS,3,dnsa
```

A.5 Meshing of the Solid Model

The following code segment assigns materials and elements to geometric shapes, and creates a finite element meshed model. First, a uniform finite element size is declared by the command *ESIZE*. Although this is not the only option, successive volumes defining beam, transducers and adhesives are selected based on their geometric location using the *VSEL* command with the *LOC* option. Material properties and elements are assigned to geometry using the *VATT* command:

```
ESIZE,4e-3
VSEL,S,LOC,Z,0,Hb
VATT,2,1,2,0
VSEL,S,LOC,Z,0,-Ha
VSEL,A,LOC,Z,Hb,Hb + Ha
VATT,3,1,3,0
VSEL,S,LOC,Z,-Ha,-Ha-H
VSEL,A,LOC,Z,Hb + Ha,Hb + Ha + H
VATT,1,1,1,0
```

All shapes are reselected with the *ALLSEL* command, and replotted to the screen using *VPLOT*. For the SOLID5 element that supports multiple shapes, *MSHAPE* specifies the 3D quadrilateral volume element shape to be used for meshing. Command *VMESH* initiates the meshing of all volumes:

```
ALLSEL
VPLOT
MSHAPE,0,3D
MSHKEY,1
VMESH,ALL
```

Finally, all nodes are selected after meshing. This is followed by the very important merging command, using *NUMMRG* with the node option and given precision. Neighboring nodes are merged into one. This actually ensures that the beam, adhesive and transducers are glued together in the simulation and act as one physical entity, thus mimicking a perfectly rigid bond:

```
nsel,all
nummrg,node,1e-5
```

A.6 Defining Boundary Conditions

The following code segments define the boundary conditions of the problem. Gravitational field or the clamp at the fixed end of the beam are present and identical to the case of each simulation type. Depending on whether one performs a modal analysis or a harmonic one, boundary conditions on the piezoelectric patches may be different.

The vector of gravitational pull is defined perpendicular to the direction of vibrations, just as in the case of the real system. The gravitational field is defined with the *ACEL* command. The left end of the blade identical with the origin of the coordinate reference system is clamped. First, the nodes at that location are selected by the *NSEL* command, then zero displacement degree of freedom constraints are engaged using the *D* command. Finally, all the nodes are reselected:

```
acel,0,-9.81,0
nsel,s,loc,x,0
d,all,ux,0,,,,uy,uz
nsel,all
```

Electrodes are defined on the piezoelectric patches by a similar methodology. First, nodes in the uppermost layer of transducer mesh are selected. Areas are picked by the *ASEL* command and then nodes in these areas selected by *NSLA*. The lowest index number node is retrieved and stored in a variable. Command *CP* defines the voltage coupled degree of freedom at these nodes, and the index node is supplied with a zero potential. This process is then repeated for all transducers:

```
asel,s,loc,z,Hb+Ha+H
asel,r,loc,x,L1,L1+L
nsla,s,1
```

```
*get,p1,node,0,num,min
cp,1,volt,all
d,p1,VOLT,V,0
allsel
asel,s,loc,z,Hb+Ha+H
asel,r,loc,x,L2,L2+L
nsla,s,1
*get,p2,node,0,num,min
!cp,2,volt,all
d,p2,VOLT,0,0
allsel
asel,s,loc,z,Hb+Ha+H
asel,r,loc,x,L3,L3+L
nsla,s,1
*get,p3,node,0,num,min
!cp,3,volt,all
d,p3,VOLT,0,0
allsel
asel,s,loc,z,-Ha-H
asel,r,loc,x,L1,L1+L
nsla,s,1
*get,p4,node,0,num,min
!cp,4,volt,all
d,p4,VOLT,0,0
allsel
```

Electrodes at the lower, bonded side are defined in a similar manner. In this case, the voltage potential is set to zero, preparing the model for a closed-circuit modal analysis. A closed-circuit modal analysis presumes that the electrode terminals are shorted, as opposed to the open state:

```
asel,s,loc,z,Hb+Ha
asel,r,loc,x,L1,L1+L
nsla,s,1
d,all,volt,0,0
nsel,all
asel,s,loc,z,Hb+Ha
asel,r,loc,x,L2,L2+L
nsla,s,1
d,all,volt,0,0
nsel,all
asel,s,loc,z,Hb+Ha
asel,r,loc,x,L3,L3+L
nsla,s,1
d,all,volt,0,0
nsel,all
```

```
asel,s,loc,z,-Ha
asel,r,loc,x,L1,L1+L
nsla,s,1 d,all,volt,0,0
nsel,all
```

Finally, the preprocessor stage is left by the *FINISH* command and the solution stage entered afterward:

```
FINISH
/ SOLU
```

A.7 Setting Up Simulations

Once a properly parameterized finite element model is created by establishing geometry, meshing and defining boundary conditions, one is presented with a plethora of analysis possibilities.

Here the setup and initiation of a modal and harmonic analysis will be presented. In addition to that, the model may be quickly altered to perform transient analyses with various initial conditions or for example a static deflection test.

A.7.1 Modal Analysis Setup

Setup and initiation of a modal analysis is very simple, given a well-configured model. First, the analysis type is determined to be modal, by the *ANTYP* command. Next, options of the modal analysis are set by the *MODOPT* command. Here the block Lanczos mode extraction method is used to retrieve the first ten modes, in the 0 to 500 Hz bandwidth. All ten modes are expanded by specifying the *MXPAND* command. Solution is started with the *SOLVE* command. Finally, the pre-solution program stage is left, one may proceed to post-processing:

```
ANTYP,MODAL
MODOPT,LANB,10,0,500
MXPAND,10
SOLVE
FINISH
```

A.7.2 Harmonic Analysis Setup

Setting up a harmonic analysis is not a particularly challenging task, given a properly configured FEM model. One has to consider the desired resolution of the outputs, since an over-meshed model and very exact required resolution may necessitate lengthy computational times. A harmonic analysis with about a thousand frequency points on this relatively simple example takes several hours to complete on average hardware.

Solution setup begins with expanding the default number of result sets to a higher number. Analysis type is specified using the *ANTYPE* command, and invoking the harmonic option. Command *HROUT* determines whether ANSYS produces real and imaginary or phase angles and degrees in the output. Solution data written to the database is controlled by *OUTRES*, which in this case is set to store everything. Rayleigh damping factors are input using *ALPHAD* and *BETAD* commands:

```
/ CONFIG,NRES,2500
ANTYPE,HARMIC
HROUT,OFF
OUTRES,ALL,ALL
ALPHAD,0
BETAD,0.001
```

The bandwidth of interest may be divided into subdomains with differing resolution of detail. In this particular example, only two portions are considered. Frequency range is input by the *HARFRQ* command, while the number of frequency substeps to be evaluated is determined by the *NSUBST* command. *KBC* specifies whether the load should be ramped or stepped, in this case it is set to be stepped. The given partial solution is defined to an indexed solution step by *LSWRITE*:

```
HARFRQ,0,70
NSUBST,500,
KBC,1
LSWRITE,1
HARFRQ,70,500
NSUBST,500,
KBC,1
LSWRITE,2
```

The solution is finally started by the *LSSOLVE* command, listing the indices of starting solution step and finishing solution step. Finally, the pre-solution program stage is left and one may proceed to post-processing:

```
LSSOLVE,1,2,1
FINISH
```

Appendix B
MPC Code Implementation Details

Appendix B of this book is an important extension of Chap. 10 and it is intended for the reader who is interested in the implementation details of MPC algorithms for vibration control or other applications. This appendix mainly contains code segments from the laboratory implementation of various MPC strategies for the experimental AVC demonstrator considered in this work. After generalization, the code featured here can be utilized for the feedback control of other dynamic systems.

This appendix may be divided into three different sections, each devoted to a certain type of MPC strategy. The first section concentrates on the offline computation of prediction matrices and formulation of constraints for the QPMPC controller. In addition to this, it also features a short discussion on simulation. As the online controller is solved using qpOASES, its parsing and internal structure is not discussed here. Each quadratic programming solver has a different input syntax, one may apply the controller to the solver according to the general rules of MPC or use the brief discussion presented in Sect. 10.1. Section B.2 features the code segments necessary both to compute an MPMPC controller offline through the MPT-Toolbox and the real-time application of the MPMPC strategy. Finally, Sect. B.3 gives a detailed account of the off and online NRMPC code with an analysis of the possible SDP solvers.

B.1 QPMPC

This section is devoted to the practical implementation of a quadratic programming-based MPC algorithm, providing constraint handling and a priori stability guarantees. The material complements the discussion in Sect. 10.1 and is based on the theoretical knowledge introduced by Chaps. 6 and 7. The theoretical basis for the formulation of the dual-mode infinite horizon quadratic programming-based MPC algorithm stabilized by a constraint checking horizon is common

G. Takács and B. Rohal'-Ilkiv, *Model Predictive Vibration Control*,
DOI: 10.1007/978-1-4471-2333-0, © Springer-Verlag London Limited 2012

and can be found in many popular works written on MPC, such as [17, 25, 34, 35, 51, 54].

The online portion of the code is parsed to the qpOASES off-the-shelf quadratic programming solver [29, 32] as discussed in Sect. 10.1. The particular way of parsing the problem to the solver always depends on the solver choice, therefore this chapter concentrates only on the problem setup such as the prediction matrices and constraint formulation. In addition to this, pointers are given to those who wish to use the algorithm in an offline simulation.

B.1.1 Setup

Let us begin with the setup of the predictive control problem. For this, the code shall be implemented in the Matlab m-script language [58, 60]. First, it is required to specify certain parameters for the QPMPC controller. Some of these parameters are: the sampling period $T_s =$ Ts, if it is an offline simulation[6] a stop time is required as well, and it is also essential to state the prediction horizon $n_c =$ nc. In case symmetric bounds on the input are assumed, the constraints are set using $\overline{u} = \underline{u} =$ umax. In the case of PZT actuation, the system constraint is the polarization voltage of the piezoelectric transducers

```
Ts=0.01;
T=0.5
nc=70;
run=T/Ts;
umax = 120;
```

where run is the runtime in case an offline simulation is needed.

In the next step, it is required to load and specify a prediction model and possibly an initial state for Kalman filtering or simulation purposes. In this example, the model is loaded from a saved system identification file:

```
load n4s2A;
A=n4s2.A;
B=n4s2.B;
C=n4s2.C;
nx=length(A);
X1(:,1)=zeros(1,nx);
```

Penalization for the inputs and states needs to be stated as well. The input penalization can be determined by direct experimentation with the algorithm, or simply evaluating different linear-quadratic controllers in simulation and determining a good balance between controller performance and aggressiveness. In this case, the input penalty has been found using an LQ controller with the

[6] Like those featured in Chap. 11.

settings $\mathbf{R} = r$ =R=1E−4 and **Q**=Q=C′*C, while balancing the input somewhat above the constraints.

```
R=1E-4;
Q=C'*C;
```

B.1.2 Prediction Matrices

To simplify the program structure it is possible to pass the information to a stand-alone function. This custom function uses the system model, penalty matrices, the constraints, a horizon and possibly system order information. Its outputs are the properly formulated prediction matrices and possibly the re-formulated constraints, which have been defined in Sects. 6.2, 6.3 and 6.7. The function creating the MPC structure including can be formulated as follows:

```
[H,F,G,Ac,b0,Bx,Ki,Nc] = predmodelqp(A,B,C,R,Q,umax,nc,
  nx);
```

Now let us begin with examining what such a function may do, in order to generate the prediction matrices and cost function prediction matrices for the online run. As in most vibration damping applications, this implementation assumes a symmetric constraint on the input:

```
ul=-uh;
```

This is followed by calculating the unconstrained linear-quadratic (LQ) optimal gain, along with the terminal weighting matrix:

```
[K,S,e]=dlqr(A,B,Q,R);
K=-K;
Qe = dlyap((A+B*K)',(Q+K'*R*K));
```

The forced and free state prediction matrices from 6.2 are calculated through a set of nested loops according to the following script:

```
N=zeros(nc*nx);
for n = 1:nc;

    M(n*nx - nx + 1:n*nx,:) = [(A^n)];
    for na = 0:nx:(nc*nx);

        N(nx*(n - 1) + (na + 1):nx*(n - 1) + (na + nx),n)=
    [(A(na/nx))*B];
        end;

end;
N=N(1:nx*nc,:);
```

where several other possible solutions may exist. These solutions can be equivalently good, and while their runtime may differ this should not be an issue in an offline problem setup process. The last n_x block row of the matrices $\mathbf{M} = \text{M}$ and $\mathbf{N} = \text{N}$ is also selected:

```
M1 = M(nx*nc-(nx-1):nx*nc,:);
N1 = N(nx*nc-(nx-1):nx*nc,:);
```

The next step is to create the cost prediction matrices as defined in Sect. 6.3. One has to begin with an initialization procedure:

```
H1=0;
F1=0;
G1=A^0*Q;
```

This is then followed by creating the cost prediction matrices $\mathbf{H}, \mathbf{F}$ and $\mathbf{G}$ first by running the following loop to get partial results:

```
for i=0:nc-2

    H1t=N(1+i*nx:i*nx+nx,:)'*Q*N(1+i*nx:i*nx+nx,:);
    H1=H1+H1t;
    F1t=N(1+i*nx:i*nx+nx,:)'*Q*M(1+i*nx:i*nx+nx,:);
    F1=F1+F1t;
    G1t=M(1+i*nx:i*nx+nx,:)'*Q*M(1+i*nx:i*nx+nx,:);
    G1=G1+G1t;

end
```

and then finally assembling cost prediction matrices $\mathbf{H} = \text{G}$, $\mathbf{F} = \text{F}$ and $\mathbf{G} = \text{G}$:

```
H=H1+N(1+(nc-1)*nx:(nc-1)*nx+nx,:)'*Qe* N(1+(nc-1)*nx:
   (nc-1)*nx+nx,:)+R*eye(nc);
F=F1+N(1+(nc-1)*nx:(nc-1)*nx+nx,:)'*Qe*M(1+(nc-1)
   *nx:(nc-1)*nx+nx,:);
G=G1+M(1+(nc-1)*nx:(nc-1)*nx+nx,:)'*Qe*M(1+(nc-1)
   *nx:(nc-1)*nx+nx,:);
```

To ensure feasibility and stability beyond the prediction horizon, the constraint checking horizon is calculated as well. This process has been described in Sect. 7.4 and in practice is started by an initialization procedure:

```
Ki=K;
Ki(2,:)=K*(A+B*K);
i=1;
Nc=0;
u=uh+1;
```

The length of the constraint checking horizon is then computed in the following loop[7]:

[7] See 7.4 for more details on the calculation of the constraint checking horizon.

```
while (u > uh);
 Ki(i+2,:)=K*(A+B*K)^(i+1);
 f=Ki(i+2,:);
 Am=[Ki(1:(i+1),:);-Ki(1:(i+1),:)];
 b=[uh*ones((i+1),1); -ul*ones((i+1),1)];
 x0=linprog(-f,Am,b);
 u=Ki(i+2,:)*x0;
 Nc=Nc+1;
 i=i+1;
end
```

B.1.3 Constraint Formulation

The routine which generates the prediction matrices also contains a simple algorithm, which is designed to define the constraints and reformulate them to be useful for direct quadratic programming solution. This formulation assumes symmetric input constraints[8]:

```
Ac1=[eye(nc)];
b0=[uh*ones(nc+Nc,1); -ul*ones(nc+Nc,1)];
Bx1=zeros((nc-1),nx);
for i=0:Nc
Ac1(nc+i,:)=[Ki(i+1,:)*M21];
Bx1(nc+i,:)=[-Ki(i+1,:)*M11];
end
Ac=[Ac1;-Ac1];
Bx=[Bx1;-Bx1];
```

The prediction matrices need to be reformulated in a way that they can be used directly with the Simulink interface of qpOASES [31, 30]. First, the cost prediction matrix **H** is reformulated, so it is suitable to pass on directly to the qpOASES problem:

```
Hqp=[];
for i=1:nc
 Hqpt=H(i,:);
 Hqp=[Hqp Hqpt];
end
```

8 See 6.7 for more details on constant formulation.

Passing on **F** is possible with the original formulation. The cost matrix $\mathbf{A}_c$ = Ac needs to be transformed likewise:

```
AcQPh=Ac(1:nc+Nc,:);
AcQP=[];
for i=1:(nc+Nc)
 AcQPt=AcQPh(i,:);
 AcQP=[AcQP AcQPt];
end
```

where the matrices need to be divided in the usual C programming style, along with constraint matrices $\mathbf{B_x}$ = Bx and $\mathbf{b}_0$ = b0:

```
BxQP=Bx(1:nc+Nc,:);
b0QP=b0(1:nc+Nc,:);
```

B.1.4 Offline Simulation

An offline simulation is often necessary the verification of results or design purposes. In this case we may use the Matlab default quadratic programming solver, named *quadprog* [59]. To do this, one needs to launch a cycle with one iteration for each sampling instant and supply the optimization problem to the solver as follows:

```
for k=1:run;
[U1(:,k),f,status1(1,k),output]=quadprog(H,F*(X1(:,k)),
 Ac,b0+Bx*X1(:,k),[],[],[],[],[],options);
 X1(:,k+1)=A*X1(:,k)+B*(U1(1,k))';
 Y1(k)=C*X1(:,k);
end
```

where X1 is the matrix containing the states response results and Y1 is the vector containing the deflection response data. The last two lines assume that there is no need for state observation; this is to make the simulation simpler. In case the cost is needed as well, one may also include:

```
J(k,:)=U1(:,k)'*H*U1(:,k)+2*(X1(:,k))'*F'*U1(:,k)+
   X1(:,k)'*G*X1(:,k);
```

where the cost can be calculated using the original prediction matrices **H**, **F** and **G** or possibly by using the definition of the cost through the penalty matrices **Q**, **R** and $\mathbf{P}_f$.

The optimization procedure can be fine-tuned by the *optimset* command:

```
options = optimset('LargeScale','off','Display','off',
  'TolFun',1e-12);
```

It is also possible to substitute the Matlab built-in *quadprog* function with the qpOASES Matlab interface [29, 32]. After compilation, the sequential qpOASES solver can be simply called by using the following code and the *qpOASES* command:

```
for k=1:run; [objOA,U2(:,k),yd,status2(1,k),nWSRout2
 (1,k)] = qpOASES('i',H,F*X2 (:,k),Ac,'','','',
   b0+Bx*X2 (:,k),10);
 X2(:,k+1)=A*X2(:,k)+B*U2(1,k);
 Y2(k)=C*X2(:,k);
end
```

where X2 and Y2 is a new or alternate set of state and deflection matrices. In case the formulation is correct, the two simulation data sets shall be equal to each other up to small and negligible numerical errors.

In case one desires to solve a sequential quadratic programming problem, where the MPC problem is formulated for an adaptive system, or it is necessary to reconfigure the constraints during the online computation process, it is advised to use the sequential module *qpOASES_sequence* of qpOASES for maximum computational efficiency [31].

B.2 MPMPC

This section of Appendix B lists the steps necessary to create an MPMPC controller via the MPT-Toolbox [40, 42, 43, 44]. The theoretical basis for multi-parametric programming has been briefly introduced in Sect. 8.2.1 while Sect. 10.2 showed the properties of the controllers as applied to the AVC demonstrator. One may find additional details on the formulation of multi-parametric controllers and the exact meaning of commands in the help of the MPT Toolbox or in the book by Kvasnica [40]. This section lists code segments that apply to both the offline and real-time code formulation. The discussion is augmented with the option to create a suboptimal but computationally efficient controller.

B.2.1 Offline Controller Computation

We begin with loading the system model to the workspace [41]:

```
load n4s2A.mat
sysStruct.A = n4s2.A;
sysStruct.B = n4s2.B;
sysStruct.C = n4s2.C;
sysStruct.D = 0;
```

This initialization process is followed by naming the state variables and setting system constraints on inputs. Output constraints $\overline{y} = -\underline{y}$ are set to infinity, therefore practically they are neglected:

```
sysStruct.StateName{1} = 'x1';
sysStruct.StateName{2}= 'x2';
sysStruct.umin =-120;
sysStruct.umax = 120;
sysStruct.ymin =-inf;
sysStruct.ymax = inf;
```

Suboptimality level is zero; this means that the software generates an optimal controller with a quadratic cost function—the norm is set to 2. Input and state penalties are defined as well, along with the controller horizon. This in the case of the controller used in experiments is $n_c = 70$ steps.

```
probStruct.norm=2;
probStruct.subopt_lev=0;
probStruct.Q=sysStruct.C'*sysStruct.C;
probStruct.R=1e-4;
probStruct.N=70;
```

The next step is the calculation of the controller regions. The main calling function takes the system and the problem structure as an argument and outputs the multi-parametric MPC problem [40, 41]:

```
ctrlex=mpt_control(sysStruct,probStruct);
```

After the computation, the controller can be saved in its original structure, so it can be later loaded into the workspace or used in offline simulations via the standard Matlab interface:

```
save ctrlex.mat ctrlex;
```

One may need data such as the volume of the region of attraction. This for example may be used to compare different controller designs or models. The total volume of the region of attraction is the sum of the individual volumes, and can be simply calculated by:

```
result.areareach=sum(volume(ctrlex.Pn))
```

Similarly, the number of regions of the controller can be determined by:

```
regions=length(ctrlex);
```

The maximal absolute deflection of the beam can be calculated by creating a convex hull around the region of attraction, transforming this into a vertex representation and by multiplying the individual edges of the region of attraction with the output matrix **C** = C we may get the direct output equivalents. The maximum of this is the maximal possible deflection at the beam tip:

```
[P,Pn]=hull(ctrlex.Pn);
result.V=extreme(P);
result.maxdef=max(abs(sysStruct.C*result.V')');
```

Non-essential tasks aimed at reducing the memory requirements and online search times of the MPMPC controller [46] such as optimal region merging [34] and binary search tree generation [62] are not considered here.

B.2.2 Real-Time Deployment

Exporting the controller into a C code via the MPT Toolbox is a very simple and straightforward process and it can be carried out using the following command [45, 47]:

```
mpt_exportc(ctrlex);
```

The function code built through the S-Function Builder takes the state vector x_k as its input and has a single output, the direct controller voltage. The core code is very simple and only involves calling the routine supplied by the MPT Toolbox in the form [41]:

```
double region;
region = mpt_getInput(x0,u);
```

where x_0 is the state vector obtained through an observer block, u is the controller output. An additional variable *region* returns the index number of the polyhedral region corresponding to the acquired current state measurement. For the correct functionality of the S-Function block the sampling rate is given, along with the MPMPC external function declaration:

```
extern double mpt_getInput(double *, double *);
```

and a declaration of the C source file as well:

```
mpt_getInput.c
```

B.2.3 Minimum-Time MPMPC

The minimum-time option is set in the MPT Toolbox by declaring:

```
[P,Pn]=hull(ctrlex.Pn);
result.V=extreme(P);
result.maxdef=max(abs(sysStruct.C*result.V')');
```

An additional output constraint is needed so the offline minimum-time MPC algorithm can terminate. This is defined through

```
sysStruct.ymin =-ym;
sysStruct.ymin = ym;
```

where $|y_{max}|$=ym is the maximal output deflection which is implied through the region of attraction of the controller.

B.3 NRMPC

The NRMPC code implementation will be introduced in the following sections, concentrating on the essential functions and steps themselves. Individual code portions taken from an actual version of the NRMPC application will be explained in detail. However, some non-essential functions will be omitted. These are mainly for diagnostic purposes, testing closed-loop stability or plotting the region of attraction and other graphical aids. This is a working version of the code, therefore some aspects certainly could be made better or more elegant, nevertheless of the several development versions this one too provides the desired parameters for the online run.

This section augments the theoretical discussion presented earlier in Sect. 8.1 and some implementation aspects in Sect. 10.3 by listing actual code segments for both the off and the online NRMPC strategy. This code has been developed for the AVC demonstrator and considers a second order prediction model. This code has been based on the publications by Kouvaritakis and Cannon et al. in [26, 37, 38].[9] One may further develop this code by introducing the theoretical findings suggested by the more recent works of Kouvaritakis et al. and Lee et al. in [39, 48] to further enhance the optimality of the NRMPC approach.

---

[9] The active help of Prof. Basil Kouvaritakis with the theoretical foundations and Dr. Mark Cannon with the code implementation is acknowledged and very much appreciated.

B.3.1 *Offline Code Implementation*

B.3.1.1 Initialization: Parameters and Penalties

The beginning portion of the offline code initializes the algorithm. Amongst others, simulation stop time is defined along with the desired deflection, which in the case of the vibration attenuation example is always zero. A linear time-invariant state-space model is loaded from a file[10] and its individual matrices are assigned. Sampling time is also defined as T_s =Ts:

```
Tstop = 60;
yd = 0;
load m2ss.mat;
A = m2ss.A; B = m2ss.B; C = m2ss.C; D = m2ss.D;
Ts = 0.01;
```

Other types of settings and tuning parameters are also declared at the beginning of the script file. Symmetric input constraints are stated as *uh*,[11] along with the state constraints if there are any.[12] Simulations and experiments performed on the active vibration attenuation system did not make use of this option. State penalties are set as Q = $\mathbf{Q}$ leaving $\mathbf{Q} = \mathbf{C}^T\mathbf{C}$ which includes the output deflection in the computed cost. Input weighting is declared as the variable R = $\mathbf{R} = r = 1E - 4$:

```
uh=120;
VX=[ ];
Q=C'*C;
R=1e-4;
```

Prediction cost performance bound γ is stated, which is necessary to be limited in order to preserve numerical stability of the process (See 11.3). A tolerance limit is also set, which is used to modify the behavior of YALMIP regarding the handling of strict inequalities [49, 50]. For constraints defined with strict constraints, a small perturbation controlled by the tolerance value is added:

```
gamma=1e6;
tolerance=1e-8;
```

Dimensionality of the problem is determined to allow the use of different state-space models for generating predictions. Number of states, inputs and other properties of the problem are acquired and stored for later use:

---

[10] See 5.2 for a discussion on the mathematical models. Most simulations and experiments in this work utilize second order models, identified from diverse data sets and identification settings. Practical experiments show no quality difference between these.

[11] *uh* meaning $\overline{u}$, as the upper and symmetric bound on the inputs.

[12] The option [] is used for no state constraints.

```
dim = size (An); dim = dim(1,1);
[w.dim] =size(B); w.dim(1,3) = w.dim(1,1);
w.dim(1,4)=w.dim(1,2)*w.dim(1,3);
w.dim(1,5)=size(Q,1);
```

Matrix square-roots of the penalization variables are taken, which are required in the construction of the invariance condition. Closed-loop linear quadratic gain is calculated, utilizing the prediction model and penalties defined previously. The closed-loop matrix of the system is also calculated:

```
sqrtR = sqrtm(R); sqrtQ = sqrtm(Q);
K=-dlqr(An,B,Q,sqrtR*sqrtR);
Phi0 = (An + B*K);
```

B.3.1.2 Variables and Constraints

Four optimization variables are declared, respecting the dimensionality of the problem. Matrix N[13] is fully parameterized and square, requiring an additional setup option. The rest of the optimization variables are real valued and symmetric:

```
Xq = sdpvar(w.dim(1,1),w.dim(1,1));
Yq = sdpvar(w.dim(1,1),w.dim(1,1));
N = sdpvar(w.dim(1,1),w.dim(1,1),'full');
M = sdpvar(w.dim(1,2),w.dim(1,1));
```

The invariance condition described by Eqs. (8.20) and (8.49) for the optimized dynamics formulation is translated into linear matrix inequalities. These LMI constrain the semidefinite programming (SDP) problem [53]. The *if–else* conditional statement constructs the problem according to whether there is a performance bound defined. Optimized prediction dynamics is required to enlarge the region of attraction to ensure sufficient deflection at the beam tip. In practice, the vibration attenuation example requires the performance bound to be always engaged.

The only command worth mentioning in this code portion is *set*, which instructs the parser YALMIP to construct a constraint in an LMI form. This high-level prototyping language formulation makes implementing changes in constraints fast and easy, letting the programmer concentrate on the high-level problem.

```
Inv1 = [Yq,Xq;Xq,Xq];
Inv3 = [Phi0*Yq+B*M Phi0*Xq;N+Phi0*Yq+B*M Phi0*Xq];
if (gamma<1/tolerance)

    gInv=gamma*eye(w.dim(1,5)+w.dim(1,2));
    zInv=zeros(w.dim(1,5)+w.dim(1,2),2*w.dim(1,1));
```

13 Refer to 8.1.3 for details and exact notation of these matrices.

```
    Inv2=blkdiag(sqrtQ,sqrtR)*[Yq,Xq; K*Yq+M,K*Xq];
    F = set([gInv,zInv,Inv2; zInv',
                        Inv1,Inv3; Inv2',Inv3',Inv1] > 0);
else
 F = set([Inv1 Inv3; Inv3' Inv1] > 0);
end
```

The *if* construct checks whether there is an input constraint defined or not. If yes, the feasibility condition defined by (8.50) is translated into the proper LMI and added to the set of constraints defining the SDP problem. Input constraints were engaged in every simulation and experiment considered in this work:

```
if ~isempty(uh)
F = F + set([uh^2 [K*Yq+M K*Xq];
                        [K*Yq + M K*Xq]' Inv1] > 0);
end
```

The simple *if* construct starting this code portion, determines whether there is a state constraint in the problem. None of the experiments or simulations using the vibration cancelling example utilized state constraints, although its inclusion is straightforward according to:

```
if ~ isempty(VX)
    for i = 1:size(VX,1)
      F = F + set(1-VX(i,:)*Yq*VX(i,:)');
    end
end
```

B.3.1.3 Solver Setup and Solution Initiation

Options are passed to the LMI parser and also to the solver, in this case SeDuMi [56].[14] Verbose diagnostic and progress output is suppressed during the optimization phase. Strict inequality constraints are relaxed and perturbed by the *shift* setting. To increase numerical precision and prevent violation of the invariance constraint during the online run, solver precision is set to the maximal level. The particular zero setting instructs the solver to search for the solution until progress is made (See 11.3.2 for the reasons of this setting.).

[14] See B.3.3.6 for possible solvers and SeDuMi.

```
ops = sdpsettings('verbose',0);
ops = sdpsettings(ops,'shift',10*tolerance);
ops = sdpsettings(ops,'solver','sedumi','sedumi.eps',0)
```

Solution of the above-defined SDP problem is initiated by the *solvesdp* YALMIP command. The LMI defining constraints are passed onto the solver as the variable F, options are contained in the *ops* parameter. The aim of this optimization problem is to maximize the volume of the ellipsoids defining the region of attraction and target set. This can be carried out by utilizing the fact that the volume of an ellipsoid is proportional to its determinant as stated in Eq. (7.77). Instead of directly minimizing (or in this case by the help of minus sign maximizing) the determinant, YALMIP utilizes the command *geomean* for such problems to minimize the geometric mean of eigenvalues:

$$\max(\mathrm{vol}E_z) = -(\det \mathbf{w})^{1/m} \tag{B.1}$$

where $\mathbf{w}$ is the optimization parameter in general and m is the dimension of $\mathbf{w}$. In this case, optimization objectives and parameters are Yq and Xq, defining the projection and intersection of the augmented ellipsoid into x space. Not only is the largest possible region of attraction desired, but also a large target set enables the system to switch to LQ as soon as possible. It is desirable to maximize the volumes of ellipsoids defined by Yq and Xq at the same time, by including them in a block diagonal construct.

In practice, the optimization task can be performed in the order of seconds. After it is completed, YALMIP passes problem diagnostics into the parameter *info*. Optimization parameters Yq, Xq, N and M are converted into the standard double precision matrix format, from the YALMIP optimization variable notation:

```
info = solvesdp(F, -geomean(blkdiag(Yq,Xq)),ops);
Yq = double(Yq); Xq = double(Xq);
N = double(N); M = double(M);
```

B.3.1.4 Factoring, Storing and Preparing Parameters for Online NRMPC

After the optimization variables are available, the parameters used in the online NRMPC run have to be factored out and stored. These tasks are largely to be attributed to the inclusion of prediction dynamics in the optimization problem, and the resulting mathematical operations to preserve convexity.

Since the augmented ellipsoid[15] is defined by matrix Γ_z and it is also obvious that $\Gamma_z^{-1}\Gamma_z = \mathbf{I}$, a resulting identity expressed in (8.45) helps to obtain variables[16] $\mathbf{V}$ and $\mathbf{X}^{-1}\mathbf{U}$ by the use of LU factorization.[17] After this, (8.41) is used to reconstruct the original matrices defining the augmented hyperellipsoid.

This is actually a code segment directly resulting from Eqs. (8.45) and (8.41), yielding matrix Γ_z from the optimization variables in several steps:

```
[V,XiU] = lu(eye(w.dim(1,1))- Yq/Xq);
XiU = XiU';
Qzi = [inv(Xq),XiU;XiU',-V\Yq*XiU)];
Qz = [Yq,V;V',-(Xq*XiU)\V];
```

Code segments *-V \(Yq*XiU)* and *-(Xq*XiU)\V* actually implement mathematical operations $-\mathbf{V}^{-1}\mathbf{Y}\mathbf{X}^{-1}\mathbf{U}$ and $-(\mathbf{X}\mathbf{X}^{-1}\mathbf{U})^{-1}\mathbf{V}$.

The full, optimized shift matrices $\mathbf{T}$ (A0) and $\mathbf{E}$ (C0) are calculated according to the relations defined in (8.43). Here the code segment *(Xq*XiU)*\(N/V') is equivalent to the operation $(\mathbf{X}\mathbf{X}^{-1}\mathbf{U})^{-1}\mathbf{K}\mathbf{V}^{T-1}$. A matrix right division is used in the segment *M/V'* to implement the operation $\Theta/(\mathbf{V}^T)^{-1}$.[18] After this, respective partitions of Γ_z are stored in variables for the needs of the online NRMPC code. It is true that partitions Γ_{xf}, Γ_{fx} are related in symmetry as $\Gamma_{xf} = \Gamma_{fx}^T$, which is true up to a numerical precision of $1E-12$ in this particular optimization task:

```
A0 = (Xq*XiU)\(N/V');
C0 = M/V';
Q11=inv(Xq)
Q12=XiU;
Q21=XiU';
Q22=-V\Yq*XiU);
Kt = [K C0];
```

B.3.1.5 Cost Transformation

The following code segment is related to cost transformation, and the resulting conversion of augmented states. The Lyapunov equations in (10.4) and (10.5) are solved successively in the offline code to obtain the cost transformation matrix

[15] This code implementation uses Q to denote the ellipsoid definition matrices Γ

[16] $\mathbf{X}^{-1}\mathbf{U}$ is noted as XiU in the code.

[17] Implemented by the command *lu* in Matlab, expressing a matrix as the product of two essentially triangular matrices, one of them a permutation of a lower triangular matrix and the other an upper triangular matrix.

[18] *Theta* in the text is denoted as M in the implemented code.

partitions. The Matlab command *dlyap* is used to evaluate the equations as follows[19]:

```
Mx = dlyap(Phi0',Q+K'*R*K);
Mc = dlyap(A0',C0'*(R+B'*Mx*B)*C0);
```

The transformation matrix *invT* is constructed using the decomposition and scaler, finally assembling the actual transformation matrix *invTT*:

```
[V,D]=eig(Mc);
d=sqrt(max(diag(D),tolerance));
invT=V*diag(1./d)/V;
invTT=blkdiag(eye(w.dim(1,1)),invT);
```

Select parameters are passed onto the online formulation, while some minor practical operations are performed in the final code segment. Computation of Δ and $\mathbf{S}$[20] as introduced in Sect. 10.3.3 is carried out, along with preparing a transformed augmented ellipsoid matrix (*Pt*) and its partition (*Q21*):

```
Pt = invTT'*Qzi*invTT;
[R,S]=eig(Pt(dim+1:2*dim,dim+1:2*dim));
Sm=diag(S);
Q21 = Pt(dim+1:2*dim,1:dim);
```

B.3.2 Online Code Implementation

B.3.2.1 BLAS Under the RTW Toolbox

The NRMPC code has been developed in Matlab, running under the Microsoft Windows operating system. Therefore, this version of the Matlab software contains an operating system and architecture optimized version of BLAS, compiled into a dynamically linked library with the file name *"libmwblas.dll"*. It is possible to utilize operations from the BLAS library if one creates a custom S-Function in Simulink using the C programming language. It is essential to refer to the BLAS documentation and call the routine within the code according to the correct syntax [18, 27]. In addition to that, the external function has to be declared: for example, a function declaration for a double precision general matrix-matrix multiplication is in the form:

[19] The text used notation **L** for the cost transformation matrix and its partitions, here M is used to denote the same concept.

[20] **S** is denoted in Sect. 10.3.3 as S_v in the code segment while Δ is denoted as R. In in this context R is not the input penalization matrix rather the result of the eigenvalue decomposition.

```
extern void dgemm_(char *,char *,int *,int *,int *,
                   double *,double *,int *,double *,
                   int *,double *,double *,int *);
```

The following BLAS functions have to be declared at compilation times:

- dgemm—double precision multiplication of two general structure matrices
- dgemv—double precision multiplication of a general structure matrix and vector
- dsymv—double precision multiplication of a symmetric matrix with a vector
- ddot—dot product of two vectors

In addition to the function declaration, one has to include the library files used by the custom C S-Functions—in this case the precompiled Matlab optimized BLAS library in the form:

```
libmwblas.dll
```

Conforming to these prerequisites, it is possible to use the BLAS routines in Simulink, if one runs the simulation on the computer on which the Matlab optimized BLAS libraries are originally included. However, if an S-Function containing BLAS routines is to be used in the Real-Time Workshop running under the xPC-Target kernel, the precompiled BLAS libraries are no longer suitable.

In case a custom Simulink S-Function making use of BLAS has been compiled for a Matlab distribution for the Microsoft Windows operating system, it will not be able to run under other platforms—unless the BLAS library files are compiled for that specific environment. This is also true for xPC target. An S-Function is perfectly suitable to be used in simulations on the computer running Microsoft Windows, but it will fail to function correctly under the xPC kernel. This is because the xPC kernel is a DOS-like environment, requiring a different BLAS library.

The problem is that neither a Windows nor a DOS optimized precompiled BLAS library will be suitable for this purpose. At the S-Function compilation stage, a Windows compatible library is required, but as soon as the controller is loaded onto the xPC machine, a DOS compiled library is needed with no Windows-specific memory calls.

To solve this dilemma, the NRMPC online implementation uses a custom compiled BLAS library [57]. A viable way to do this is to use *cygwin* a Linux style API, running under the Microsoft Windows operating system [28]. To prepare a suitable environment for the custom built BLAS packages, in addition to the core *cygwin* installation some packages have to be included:

- Download and install *cygwin*.
- Install *gcc compiler* for C and F77. (To be found in "dev".)
- Install *LAPACK native*. (To be found in "math".)

The software environment is now ready to make a custom BLAS library. It is needed to unzip the package and copy the appropriate makefile [55] by typing the following lines into the terminal:

```
tar -xvzf lapack.tar.gz
cd LAPACK
cp INSTALL/make.inc.LINUX ./make.inc
```

The next step is to modify a function called *xerbla*, which is only used for error reporting. This may cause problems throughout the build process, therefore it is advised to delete all functionality and leave only the core function. In case it is called, it would return void to the parent function. Now it is needed to modify the file *make.inc* by changing the following lines:

```
FORTRAN = g77
LOADER = g77
```

This is to specifically instruct the compiler to make a library, which is capable of running outside *cygwin*. Therefore, the previously introduced lines have to be found in *make.inc* and changed to:

```
FORTRAN = g77 -mno -cygwin
LOADER = g77 -mno -cygwin
```

The make file has to be setup to compile the default BLAS library. To do this, one needs to change the line starting with *lib* in the ***makefile*** as follows:

```
lib: blaslib lapacklib tmglib
```

The package is compiled, tested and timed by typing *make* into the *cygwin* terminal. In addition to the xPC compatible BLAS library, a LAPACK library is generated. LAPACK is used for more complex operations, currently not required for the NRMPC implementation.[21]

It is worth noting that by using a BLAS library prepared in the previous method, they are capable of running under the xPC Target kernel, even though the S-Functions are compiled using Microsoft Visual C v. 6 [52].

B.3.2.2 Custom C Functions

In addition to the main C code and BLAS functions for matrix algebra operations, there are two additional custom functions. One of them performs element-wise division of two vectors, similar to the ./ operator in Matlab. The other one is an algorithm-specific operation, scaling the elements of a vector by a scalar value and subtracting it from one.

Both these functions have to be declared similar to the BLAS routines as externals. No library declarations are needed, since instead of using a dynamically linked library, the C source codes are necessary at compilation. Function declarations are as follows:

[21] However, the one step-ahead extension not currently used would require solving sets of linear equations, which indicates the need for LAPACK functions.

```
extern double ddot_(int *,double *,int *,double *,int *);
extern void dediv_(int *, double *, double *, double *);
```

The following function performs element-wise division of two vectors. Vectors x and y have a common dimension of n. Their elements are indexed with the locally declared variable i. The function takes the dimension n and vectors x and y as its input, and places the result in vector z. Strictly speaking, this C function has no output—the results are placed in the proper memory location by pointers. The algorithm simply loops through the elements, dividing the i-th element of x with y and places it into z. The trailing underscore in the function name mimics the calling sequence of BLAS functions.

```
void dediv_(int *n, double *x, double *y, double *z)
{
  int i = 0;
  for (i; i < *n; i++)
    {
    z[i]=x[i]/y[i];
    }
}
```

The next custom is specific for the NRMPC code, its usage is hardly universal. After the unknown λ is calculated by the algorithm, each element of vector x is scaled by it. This simple scaling operation is extended by an additional step for computational efficiency. Each scaled element of x is subtracted from 1, and the result is placed in the vector y.

The function takes scalar dimension n, vector x and scalar λ as its input. The output is placed in y, where the dimensions of both vectors are n. An inside loop performs the formerly described simple operation, where the elements are indexed with the locally defined variables i. Strictly speaking, this C function produces a void output; the results are placed into the output vector by using pointers. The trailing underscore in the function name mimics the calling sequence of BLAS functions.

```
void descal_(int *n, double *lamN, double *x,double *y)
{
  int i = 0;
  for (i; i < *n; i++)
    {
    y[i]=1-*lamN*x[i];
    }
}
```

B.3.2.3 The NRMPC Code in C

This section introduces the online NRMPC code implementation in C and explains the general operations performed within. The algorithm will be divided into functional sections and commented as needed.

A prerequisite for the correct functionality of this code is the correct calling of external BLAS functions and the xPC optimized library as described in Sect. B.3.2.1. The two external custom functions described in Sect. B.3.2.2 are also needed to be present and properly declared at compilation time.

At the online control process, the following real-time NRMPC algorithm is called on and evaluated at each sampling interval:

Local variables are declared at the beginning of the code. The BLAS functions require character variables. The transposition of matrices is controlled by N and T- as in not to transpose and transpose. Some of these functions also require to mark, whether the upper or lower triangular portion of a symmetric matrix is read in.

The order of the system is declared, just as some common values as zero, one or minus one. The value of λ is set to zero at starting time, tolerance and error thresholds are stated. Finally, local matrix and vector variables are declared:

```
char *chn = ''N'',*cht = ''T'', *chu = ''U'', *chl = ''L'';
int onei = 1, order = 2;
double one = 1.0, mone = -1.0, zero = 0.0, lamN = 0,... ...tol
  = 1e-5, err = 2e-5;
double W0, W2, fval, fderval;
double tempv[2], tempm[4], vec[2], tempv2[2],... ...
  W1[2], W1d[2], W1dd[2], m[2], f[2];
```

After the local variable declarations are expressed, the following mathematical operation is performed in two steps:

$$\mathbf{W}_0 = x_0^T \Gamma_x x_0 - 1 \tag{B.2}$$

where x_0 marks the current observed state, and Γ_x is a partition of the matrix defining the invariant ellipsoid, as calculated in the offline process. $\mathbf{W}_0$ is a by-product, resulting the logical simplification of matrix operations. The first code line creates a temporary vector, a result of the matrix-vector operation $\Gamma_x x_0$ while the second finishes the task by evaluating the rest:

```
dsymv_(chu,&order,&one,Q11,&order,...
  ...x0,&onei,&zero,tempv2,&onei);
W0 = ddot_(&order,x0,&onei,tempv2,&onei)-1;
```

In case the resulting vector will be $\mathbf{W}_0 \leq 0$, the following code portion calculates the next by-product, a vector reused in later code portions:

$$\mathbf{v} = -(\Delta^T \Gamma_{xf} x_0)./\mathbf{S} \tag{B.3}$$

where v denotes the vector result of this operation, and ./ is an element-wise division. This operation is carried out in two steps. First, a general matrix-matrix multiplication saves the result of $\Delta^T \Gamma_{xf}$ into a temporary matrix. Then the expression $\mathbf{v}$ is calculated by multiplying the result with the negative of the current state measurement, and its elements divided by vector $\mathbf{S}$:

```
if(W0>=0){
 dgemm_(cht,chn,&order,&order,&order,&one,...
 ...R,&order,Q21,&order,&zero,tempm,&order);
 dgemv_(chn,&order,&order,&mone,tempm,&order,
 ...x0,&onei,&zero,W1,&onei);
 dediv_(&order,W1,Sm,vec);
```

Another partial result is calculated, by evaluating a dot product of two vectors and adding $\mathbf{W}_0$ to the result:

```
W2 = -ddot_(&order,vec,&onei,W1,&onei) + W0;
```

where in case $\mathbf{W}_2 \geq 0$, the perturbation vector can be directly calculated by evaluating:

$$\mathbf{f} = -\Delta\mathbf{v} \tag{B.4}$$

where **f** is the perturbation vector, is an input from the offline optimization, and **v** is a vector product recalculated at each sampling interval.

```
if(W2 >= tol)
 {dgemv_(chn,&order,&order,&mone,R,&order,...
 ...vec,&onei,&zero,f,&onei);}
```

The other option in the *else* construct is to evaluate for the unknown λ using the Newton–Raphson procedure. This conditional statement launches a *while* loop, which cycles through the NR procedure, until the floating point absolute value of error is larger than the preset tolerance. In practice, loop termination occurs in no more than twenty iterations.

The first part of this code segment serves only to evaluate the matrices used in the NR loop. These simplifications increase computational speed and are based on the assumptions about function $\chi(\lambda)$ and its i-th derivatives presented in Sect. 10.3.3.

The second part of the following code segment is the Newton–Raphson algorithm itself, which searches for the root of (8.26). Here the first step is to evaluate the value of $\chi(\lambda)$ and its derivative according to (10.11). The ratio of the function value and its derivative is the error, which is subtracted from the result for λ from the previous step:

```
else{
  while(fabs(err)>=tol){
    descal_(&order,&lamN,Sm,m);
    dediv_(&order,W1,m,W1d);
    dediv_(&order,W1d,m,W1dd);
    fval=ddot_(&order,vec,&onei,W1dd,&onei) + W2;
    fderval=2*ddot_(&order,W1d,&onei,W1dd,&onei);
    err=fval/fderval;
    lamN=lamN-err;}
```

Since the value of λ has been acquired in the previous step, the only task left is to evaluate for the perturbation vector **f** according to (8.25), which in this case can be also stated as:

$$\mathbf{f} = -\lambda \mathbf{T} \Delta \mathbf{W}_{1d} \tag{B.5}$$

This single mathematical operation is divided into three parts for the C code. First, the value of λ is negated, then a temporary vector is created from the product of $\mathbf{v}_{temp} - \lambda \Delta \mathbf{W}_{1d}$. The final step is to calculate **f** by multiplying this temporary vector by **T** from the left, $\mathbf{f} = \mathbf{T}\mathbf{v}_{temp}$:

```
lamN=-lamN;
dgemv_(chn,&order,&order,&lamN,R,...
  ...&order,W1d,&onei,&zero,&tempv,&onei);
dsymv_(chu,&order,&one,T,&order,...
  ...tempv,&onei,&zero,f,&onei);
```

With the perturbation value calculated in the previous step, the final task is only to evaluate the current control move according to $u = \mathbf{K}x_0 + \mathbf{C}_0\mathbf{f}$. This is performed in the C code by summing up the results of two vector dot operations:

```
u[0] = ddot_(&order,K,&onei,x0,&onei) +
  +ddot_(&order,C0,&onei,f,&onei);
```

The other option implies that the loop is already optimal, thus the perturbation $\mathbf{f} = 0$. There is no need for optimization, this is part of an *"if-than-else"* decision. In this case, the fixed feedback matrix is used to calculate the control move from the observed state by evaluating $u = \mathbf{K}\mathbf{x}_0$. This is again a simple vector dot product:

```
else{
  u[0] = ddot_(&order,K,&onei,x0,&onei);}
```

B.3.3 SDP Solvers for NRMPC

The following passages give a brief description of SDP solver choices.

B.3.3.1 DSDP

DSDP is a freely available and open source semidefinite programming solver. It is based on an implementation of an interior point method with relatively low memory requirements, also exploiting data sparsity in the problem structure [19, 21]. This software may be used as a set of subroutines in Matlab. A parallel implementation of DSDP is also available under the name PDSDP [20].

DSDP solver completed the NRMPC offline optimization task, but with warnings informing about the unbounded nature of the primal optimization problem and infeasibility of the dual. The resulting parameters could be used in simulation. Evolution of controller outputs showed no signs of numerical irregularities. However, the outputs remained highly suboptimal and not approaching constraint levels, especially in the case of higher than second order examples.

B.3.3.2 SDPLR

SDPLR is a freely available and distributed optimization software for SDP problems [24], which is suited for solving large-scale problems. Essentially, it is a C package, however a Matlab interface is also provided. Source code and architecture optimized versions are available. It is linked with either un-optimized BLAS or the Automatically Tuned Linear Algebra Software (ATLAS)—an optimized improvement of the generic BLAS.

During the evaluation of the offline NRMPC optimization problem, the solver repeatedly crashed. Simulations with the online controller could therefore not be completed. Although the performance and precision of this solver in relation to the NRMPC problem are unknown, its use has not been considered due to the concerns with reliability.

B.3.3.3 CSDP

CSDP is a freely available C library for semidefinite programming [22]. CSDP is written as a callable C subroutine, capable of running in parallel on shared memory and multi-processor systems [23]. The routine makes use of sparsity in constraint matrices, and works on systems with an ANSI C compiler and BLAS/LAPACK libraries.

The CSDP executable has to be included in the Matlab and Windows path as well. The code communicates with Matlab via text files, which is an inefficient method causing overheads for certain problems. The CSDPA solver was not usable with the NRMPC code, therefore its precision, speed and other properties could not be evaluated in relation to the given optimization problem.

B.3.3.4 SDPT3

SDPT3 is a freely distributed solver for semidefinite programming problems [61, 63]. It is a relatively well tested, although still not bug-free software. The last version implements an infeasible path-following algorithm for solving conic optimization problems involving semidefinite, second order and cone constraints. It can handle determinant maximization problems and SDP with complex data.

The simulation run produced by using online controller parameters acquired via the SDPT3 solver produced numerically unstable results. The controller output exceeded constraints by orders of magnitude higher values. The evolution of outputs did not even remotely resemble the expected oscillating behavior. Adjusting solver gap and step tolerances, increasing the maximal allowable iteration counts did not improve the situation. After the negative experience with this solver, the use of SDPT3 has not been considered for the NRMPC algorithm.

B.3.3.5 SDPA-M

SemiDefinite Programming Algorithm or SDPA is a collection of software tools for solving SDP problems [65]. Its Matlab compatible interface is called SDPA-M. The algorithm is implemented in C++ language, and it is making use of BLAS or some variant of optimized BLAS and LAPACK packages for matrix computations. The algorithm implements an infeasible primal-dual interior point method [65].

An important feature of SDPA is that it is callable as a C function, therefore readily implementable in future possible versions of adaptive NRMPC algorithms, or implementations where the offline portion is also present on the hardware at setup time. In addition to handling block diagonal and sparse data structures, it is able to exploit the sparsity of problem data matrices by using efficient methods for finding search directions. SDPA is also available for parallel computing applications. The Matlab callable version provides no callable library and at the time of completing this work and does not support 64-bit platforms [33].

Running SDPA-M with the problem and settings used to evaluate the rest of solver candidates produced no error messages. However, the resulting optimization parameters were singular matrices. Clearly, this is an unacceptable solution; simulations cannot be performed with the results obtained via SDPA-M. The use of SDPA-M for the NRMPC offline algorithm was not considered viable.

B.3.3.6 SeDuMi

SeDuMi is a second order cone and semidefinite programming problem solver software, freely distributed and available with sources and binaries [56]. SeDuMi can be compiled on any platform running Matlab. The latest version uses BLAS for improved performance and is able to run on 64-bit systems.

SeDuMi is the most commonly used solver among YALMIP users. Installation procedure of SeDuMi is as simple as copying the directory containing solver files into the Matlab path. Simulations performed with solutions acquired through SeDuMi showed no signs of serious offline solution suboptimality and therefore has been selected as the first choice of SDP solver for the offline NRMPC routine.

B.3.3.7 LMILAB and PENSDP

LMILAB and PENSDP are commercial solvers. PENSDP is a linear semidefinite programming solver, aimed at large-scale dense and sparse problems. It is available as a standalone program, a MATLAB function callable through YALMIP and as a C routine as well. PENSDP is offered for all major computer architecture systems [36]. At the time of preparing the work in this book, a student or trial license was not available, therefore the use of PENSDP as a basis of the offline NRMPC procedure has been dismissed. The efficiency and precision of this SDP solver algorithm in relation to NRMPC has not been tested and evaluated.

LMILAB is a part of the MathWorks Robust Control Toolbox. The use of LMILAB in combination with YALMIP is generally not recommended [50]. YALMIP cannot exploit a built-in feature in LMILAB, which would formulate a control specific structure increasing computational efficiency. In addition to the often slow computational times, it has a low default tolerance setting and does not return infeasibility flags to YALMIP. YALMIP reports a successful optimization even when the problem is infeasible, hence making algorithm testing and development somewhat difficult. Due to the above stated issue, the use of LMILAB as a SDP solver for the offline NRMPC problem has not been considered.

References

1. Ansys Inc (2005) Release 10.0 Documentation for ANSYS. Ansys Inc., Canonsburg
2. Bandyopadhyay B, Manjunath T, Umapathy M (2007) Modeling, control and implementation of smart structures: a FEM-state space approach, 1st edn. Springer, Berlin
3. Chu CL, Wu BS, Lin YH (2006) Active vibration control of a flexible beam mounted on an elastic base. Finite Elem Anal Des 43(1):59–67. doi:10.1016/j.finel.2006.07.001, http://www.sciencedirect.com/science/article/pii/S0168874X06001144
4. Dong X, Meng G, Peng J (2006) Vibration control of piezoelectric smart structures based on system identification technique: numerical simulation and study. J Sound Vib 297:680–693
5. Hatch MR (2000) Vibration simulation using MATLAB and ANSYS, 1st edn. Chapman and Hall/CRC, Boca Raton
6. eFunda Inc (2007) Lead zirconate titanate (PZT-5A). http://www.efunda.com/materials/piezo/material_data/matdata_output.cfm?Material_ID=PZT-5A
7. Kim NH, Sankar BV (2008) Introduction to finite element analysis and design, 1st edn. Wiley, Hoboken
8. Logan DL (2011) A first course in the finite element method, 3rd edn. CL-Engineering, Stamford
9. MIDÉ Technology Corporation (2007) QuickPack Actuator Catalog. MIDÉ Technology Corporation. MIDÉ Technology Corporation, Medford
10. MIDÉ Technology Corporation (2008) Attaching the Quickpack/PowerAct Transducer to a Structure with Epoxy. MIDÉ Technology Corporation, quick Pack Technical Notes. MIDÉ Technology Corporation, Medford
11. Moaveni S (2007) Finite element analysis theory and application with ANSYS, 3rd edn. Prentice Hall, London

12. Piezosystem-Jena (2007) Piezoline theory. http://www.piezojena.com/files.php4?dl_mg_id=229&file=dl_mg_119514214..pdf&SID=125nfb5prkpt35d2k9cops3021
13. Spangler R (2007) Piezo sensor techical note, 2nd edn. MIDÉ Technology Corporation. MIDÉ Technology Corporation, Medford
14. SP Systems (2001) Resin properties for composite materials. http://www.azom.com/details.asp?ArticleID=997, Newport
15. Šolek P (2009) Numerical analyses of piezoelectric elements, 1st edn. Slovenská technická univerzita v Bratislave, Nakladatel'stvo STU, Bratislava
16. Zimmerman WBJ (2006) Multiphysics modeling with finite element methods, 1st edn. World Scientific Publishing Company, London
17. Baocang D (2009) Modern predictive control, 1st edn. Chapman and Hall/CRC, Boca Raton
18. Basic Linear Algebra Subprograms Technical (BLAST) Forum (2001) Basic Linear Algebra Subprograms Technical (BLAST) Forum Standard. http://www.netlib.org/blas/blast-forum/blas-report.pdf
19. Benson SJ, Ye Y (2005) DSDP5: Software for semidefinite programming. Tech. Rep. ANL/MCS-P1289-0905, Mathematics and Computer Science Division, Argonne National Laboratory, Argonne, IL. http://www.mcs.anl.gov/ benson/dsdp, submitted to ACM Transactions on Mathematical Software
20. Benson SJ, Ye Y (2005) DSDP5 user guide—software for semidefinite programming. Tech. Rep. ANL/MCS-TM-277, Mathematics and Computer Science Division, Argonne National Laboratory, Argonne, IL. http://www.mcs.anl.gov/benson/dsdp
21. Benson SJ, Ye Y, Zhang X (2000) Solving large-scale sparse semidefinite programs for combinatorial optimization. SIAM J Optim 10(2):443–461
22. Borchers B (1999) CSDP, a C library for semidefinite programming. Optim Method Softw 11(1–4):613–623
23. Borchers B, Young JG (2007) Implementation of a primal-dual method for SDP on a shared memory parallel architecture. Comput Optim Appl 37(3):355–369
24. Burer S, Monteiro RDC (2003) A nonlinear programming algorithm for solving semidefinite programs via low-rank factorization. Math Program Ser B 95(2):329–357
25. Camacho EF, Bordons C (2007) Model predictive control, 2nd edn. Springer, London
26. Cannon M, Kouvaritakis B (2005) Optimizing prediction dynamics for robust MPC. IEEE Trans Autom Control 50(11):1892–1897, doi:10.1109/TAC.2005.858679
27. Dongarra J (2002) Basic linear algebra subprograms technical forum standard. Int J High Perform Appl Supercomput 16:1–111
28. Faylor C, Vinschen C, Chamberlain S, Okhapkin S, Giacinti P, Taylor IL (2009) Cygwin. http://www.cygwin.com/
29. Ferreau HJ (2006) An online active set strategy for fast solution of parametric quadratic programs with applications to predictive engine control. Master's thesis, University of Heidelberg
30. Ferreau HJ (2011) qpOASES-Online Active Set Strategy. Leuven, http://www.qpoases.org
31. Ferreau HJ (2011) qpOASES User's Manual. Optimization in Engineering Center (OPTEC) and Department of Electrical Engineering, K. U. Leuven, Leuven
32. Ferreau HJ, Bock HG, Diehl M (2008) An online active set strategy to overcome the limitations of explicit MPC. Int J Robust Nonlinear Control 18(8):816–830
33. Fujisawa K, Futakata Y, Kojima M, Matsuyama S, Nakamura S, Nakata K, Yamashita M (2005) SDPA-M (SemiDefinite Programming Algorithm in MATLAB). Tech. Rep. B-359, Dept. Math. & Comp. Sciences, Tokyo Institute of Technology, Tokyo. http://sdpa.indsys.chuo-u.ac.jp/sdpa/files/sdpamManual.pdf, user's manual version 6.2.0
34. Geyer T, Torrisi FD, Morari M (2008) Optimal complexity reduction of polyhedral piecewise affine systems. Automatica 44(7):1728–1740, doi:10.1016/j.automatica.2007.11.027, http://www.sciencedirect.com/science/article/pii/S0005109807004906
35. Hyun-Kwon W, Hee-Han S (2005) Receding horizon control: model predictive control for state models, 1st edn. Springer, London

36. Kocvara M, Stingl M (2006) PENSDP User's Guide. Igensdorf OT Stöckach. http://www.penopt.com/doc/pensdp2_2.pdf
37. Kouvaritakis B, Rossiter J, Schuurmans J (2000) Efficient robust predictive control. IEEE Trans Autom Control 45(8):1545–1549, doi:10.1109/9.871769
38. Kouvaritakis B, Cannon M, Rossiter J (2002) Who needs QP for linear MPC anyway? Automatica 38:879–884, doi:10.1016/S0005-1098(01)00263-1, http://www.sciencedirect.com/science/article/pii/S0005109801002631
39. Kouvaritakis B, Li S, Cannon M (2010) A line search improvement of efficient MPC. Automatica 46(11):1920–1924, doi:10.1016/j.automatica.2010.07.003, http://www.sciencedirect.com/science/article/B6V21-50NH0BX-3/2/0b4491d922a7d04d1b0315edae0e8944
40. Kvasnica M (2009) Real-time model predictive control via multi-parametric programming: theory and tools, 1st edn. VDM Verlag, Saarbrücken
41. Kvasnica M (2011) Selected topics on constrained and nonlinear control. Workbook, STU Bratislava-NTNU Trondheim, chap Multi-Parametric Toolbox, pp 101–170
42. Kvasnica M, Grieder P, Baotić M (2004) Multi-parametric toolbox (MPT). http://control.ee.ethz.ch/
43. Kvasnica M, Grieder P, Baotic M, Morari M (2004) Multi-parametric toolbox (mpt). In: Alur R, Pappas GJ (eds) Hybrid systems: computation and control, Lecture Notes in Computer Science, vol 2993. Springer, Berlin, pp 121–124
44. Kvasnica M, Grieder P, Baotic M, Christophersen FJ (2006) Multi-parametric toolbox (MPT). Extended documentation
45. Kvasnica M, Rauová I, Fikar M (2010) Automatic code generation for real-time implementation of model predictive control. In: Proceedings of the 2010 IEEE international symposium on computer-aided control system design, Yokohama, pp 993–998
46. Kvasnica M, Fikar M, Čirka L', Herceg M (2011) Selected topics on constrained and nonlinear control. Textbook, STU Bratislava-NTNU Trondheim, chap Complexity Reduction in Explicit Model Predictive Control, pp 241–288
47. Kvasnica M, Rauová I, Fikar M (2011) Real-time implementation of model predictive control using automatic code generation. In: Selected Topics on Constrained and Nonlinear Control. Preprints, STU Bratislava-NTNU Trondheim, pp 311–316
48. Li S, Kouvaritakis B, Cannon M (2010) Improvements in the efficiency of linear MPC. Automatica 46(1):226–229. doi:10.1016/j.automatica.2009.10.010, http://www.sciencedirect.com/science/article/B6V21-4XGCHXB-3/2/20a93fa6dd4fb88469638ac3bc2fe729
49. Lofberg J (2004) YALMIP: a toolbox for modeling and optimization in MATLAB. In: Proceedings of the CACSD Conference, Taipei
50. Lofberg J (2009) YALMIP wiki. http://control.ee.ethz.ch/~joloef/wiki/pmwiki.php?n=Main.HomePage
51. Maciejowski JM (2000) Predictive Control with Constraints, 1st edn. Prentice Hall, Upper Saddle River
52. Microsoft Corporation (1998) Visual C++ 6.0. Software
53. Nesterov Y, Nemirovskii A (1994) Interior-point polynomial methods in convex programming, Studies in Applied Mathematics, vol 13. SIAM, Philadelphia
54. Rossiter JA (2003) Model-based predictive control: a practical approach, 1st edn. CRC Press, Boca Raton
55. Souder JS (2003) An investigation into BLAS and LAPACK for Real-Time Engine control. http://vehicle.me.berkeley.edu/~jsouder/enginectlATLAS.p.
56. Sturm JF (1999) Using SeDuMi 1.02, a MATLAB toolbox for optimization over symmetric cones. Optimization Methods and Software-Special issue on Interior Point Methods 11–12:625–653
57. Takács G, Rohal'-Ilkiv B (2009) Newton–Raphson MPC controlled active vibration attenuation. In: Hangos KM (ed) Proceedings of the 28. IASTED International Conference on Modeling, Identification and Control, Innsbruck

58. The Mathworks (2007) Matlab 7.5 R2007b. Software, Natick
59. The Mathworks (2011) Matlab optimization toolbox v6.0 (R2011a). Software, Natick, http://www.mathworks.com/products/optimization
60. The Mathworks (2012) Matlab 7.12 R2011a. Software, Natick
61. Toh K, Todd M, Tutuncu R (1999) SDPT3–a Matlab software package for semidefinite programming. Optim Method Softw 11:545–581
62. Tøndel P, Johansen TA, Bemporad A (2003) Evaluation of piecewise affine control via binary search tree. Automatica 39(5):945–950. doi: 10.1016/S0005-1098(02)00308-4, http://www.sciencedirect.com/science/article/pii/S0005109802003084
63. Tutuncu R, Toh K, Todd M (2003) Solving semidefinite-quadratic-linear programs using SDPT3. Math Program Ser B 95:189–217
64. Wang L (2009) Model predictive control system design and implementation using MATLAB, 1st edn. Springer, London
65. Yamashita M, Fujisawa K, Kojima M (2003) Implementation and evaluation of SDPA 6.0. Optim Method Softw 18:491–505

Legal Information

Disclaimer

This book has been written by members of the faculty of the Slovak University of Technology in Bratislava as a part of ongoing active research projects. It has been altered and corrected within the means of the usual scientific peer-review process, however, may still contain errors or mistakes.

The book is "as-is" without any guarantees to accuracy, reliability or validity of the expressed views and published data. The book is provided without warranty of any kind, and no other warranties, either expressed or implied are made with respect to this book. The authors of this book and the publisher bear no responsibility and cannot be held liable, including but not limited to: health damage, potential loss of life, loss of property or finances or any other inconvenience as a result of using information published in this monograph.

Endorsement

This work has not been endorsed nor sponsored by any company, corporation, domestic or foreign government entity except the *Slovak Research and Development Agency*. Company names and government entities are only used in this book for the purpose of identification and reference.

Trademarks

This work makes no claims to trademarks, company or product names. Protected and registered trademarks are only used in the text to refer to products, applications or companies without any intent to infringe. Registered trademarks, company and product names are the properties of their respective owners.

Software Licensing

All work has been carried out on software licensed to the Slovak Technical University in Bratislava. Inquiries on licensing shall be made to the University or its institutes. The authors take no responsibility on the content of the agreements

DOI: 10.1007/978-1-4471-2333-0,

and contracts between the Slovak Technical University in Bratislava and its respective institutes and software vendors.

Graphic Images

Images, photos, diagrams, illustrations, typesetting, the graphic layout and structure of this document are the work of the authors, unless otherwise indicated. The typesetting of the document makes use of the layout materials provided by the publisher.

This book contains reprinted photographs and illustrators as well. The authors have made every reasonable effort to trace and contact the copyright owners of these materials in order to ask for permission. Despite our numerous efforts we have not received responses in all cases. The authors apologize if permission has not been obtained or copyrighted material is used without due acknowledgment. Please notify us so we can correct these unintentional mistakes in future editions.

Photos released by the United States National Aeronautics and Space Administration are used according to the general terms and conditions of the owner. Images released by NASA to the public domain are not copyrighted and may be used for educational or informational purposes[1]. Written permissions have been obtained to use images from the United State National Aeronautics and Space Administration (NASA), from the European Space Agency (ESA), the Smithsonian Institution, Liebherr-International Deutschland GmbH, the CEDRAT Group, the Japan Society of Maintenology, the Noise and Vibration Control Ltd, Dr. Thomas Huber, Dr. Bishakh Bhattacharya[2]. The image reproduced from Popular Science Monthly is in the public domain because its copyright has expired[3]. Photographs and illustrations reproduced from the Wikimedia Commons database are cleared to use in publications, limited copyright notices are indicated in the references–if applicable[4]. The image obtained from the morgueFile database is free to use for commercial purposes without attributing the original author[5]. The use of photographs and images published under the Creative Commons Attribution 3.0 Unported license permits copy, distribution, adaptation for commercial purposes with attribution[6]. The image released by the author to the public domain is free for use for any purpose without restriction under copyright[7].

NASA[1,2]	4,74, 76,81,87, 91,120,376,379,380,381
European Space Agency (ESA)[2]	4, 30
The Boeing Company	9,163
Smithsonian Institution[2]	3
Liebherr International[2]	373
CEDRAT Group[2]	84
Noise & Vibration Control Ltd[2]	377
Japan Society of Maintenology[2]	367
Popular Science Monthly[4,3]	2
Martin Helfer / FKFS Stuttgart[4,7]	365
Thomas Huber[2]	208

Armand du Plessis[4,6]	358
Bishakh Bhattacharya[2]	102
Wikimedia Commons[4,6]	368, 369
morgueFile[5]	232

Credit, Citations and Referencing

Every effort has been made to properly reference and give credit to authors whose work has been used here. Published books, articles, papers, theses, manuals and other printed and electronic works are listed in the References sections of individual chapters. In certain select cases, credit may be given in footnotes for personal consultations, ideas and others. Special attention has been devoted to keep the list of references accurate, however if you feel something is missing, its absence is entirely unintentional. Please contact us so we can correct our mistakes in the upcoming editions.

Index

DOI: 10.1007/978-1-4471-2333-0, © Springer-Verlag London Limited 2012